Algebraic Codes for Data Transmission

The need to transmit and store massive amount of data reliably and without error is a vital part of modern communications systems. Error-correcting codes play a fundamental role in minimizing data corruption caused by defects such as noise, interference, cross talk, and packet loss. This book provides an accessible introduction to the basic elements of algebraic codes, and discusses their use in a variety of applications. The author describes a range of important coding techniques, including Reed–Solomon codes, BCH codes, trellis codes, and turbocodes. Throughout the book, mathematical theory is illustrated by reference to many practical examples. The book is aimed at graduate students of electrical and computer engineering, and at practising engineers whose work involves communications or signal processing.

Professor Richard E. Blahut is Head of the Department of Electrical and Computer Engineering at the University of Illinois, Urbana Champaign. He is a Fellow of the Institute of Electrical and Electronics Engineers and the recipient of many awards including the IEEE Alexander Graham Bell Medal (1998), the Tau Beta Pi Daniel C. Drucker Eminent Faculty Award, and the IEEE Millennium Medal. He was named a Fellow of the IBM Corporation in 1980 (where he worked for over 30 years) and was elected to the National Academy of Engineering in 1990.

Algebraic Codes for Data Transmission

Richard E. Blahut

Henry Magnuski Professor in Electrical and Computer Engineering,
University of Illinois at Urbana – Champaign

CAMBRIDGE
UNIVERSITY PRESS

PUBLISHED BY THE PRESS SYNDICATE OF THE UNIVERSITY OF CAMBRIDGE
The Pitt Building, Trumpington Street, Cambridge, United Kingdom

CAMBRIDGE UNIVERSITY PRESS
The Edinburgh Building, Cambridge CB2 2RU, UK
40 West 20th Street, New York, NY 10011-4211, USA
477 Williamstown Road, Port Melbourne, VIC 3207, Australia
Ruiz de Alarcón 13, 28014 Madrid, Spain
Dock House, The Waterfront, Cape Town 8001, South Africa

http://www.cambridge.org

First published 2003

Printed in the United Kingdom at the University Press, Cambridge

Typefaces Times 10.5/14 pt and Helvetica Neue *System* LaTeX 2ε [TB]

A catalogue record for this book is available from the British Library

ISBN 0 521 55374 1 hardback

Contents

Preface

This book is a second edition of my 1983 book *Theory and Practice of Error Control Codes*. Some chapters from that earlier book reappear here with minor changes. Most chapters, however, have been completely rewritten. Some old topics have been removed, and some new topics have been inserted.

During the two decades since the publication of that first edition, error-control codes have become commonplace in communications and storage equipment. Many such communication and storage devices, including the compact disk, that are now in general use could not exist, or would be much more primitive, if it were not for the subject matter covered by that first edition and repeated in this edition.

The second edition retains the original purpose of the first edition. It is a rigorous, introductory book to the subject of algebraic codes for data transmission. In fact, this phrase, "algebraic codes for data transmission," has been chosen as the title of the second edition because it reflects a more modern perspective on the subject.

Standing alongside the class of algebraic codes that is the subject of this book is another important class of codes, the class of nonalgebraic codes for data transmission. That rapidly developing branch of the subject, which is briefly treated in Chapter 11 of this edition, deserves a book of its own; this may soon appear now that the topic is reaching a more mature form.

This book is a companion to my more advanced book *Algebraic Codes on Lines, Planes, and Curves*. Although both books deal with algebraic codes, they are written to be read independently, and by different audiences. Consequently, there is some overlap in the material, which is necessary so that each book stands alone. I regard the two books as belonging to the general field of *informatics*, an emerging collection of topics that is not quite mathematics and not quite engineering, but topics that form the intellectual bedrock for the information technology that is now under rapid development.

The preparation of the second edition has benefited greatly from the comments, advice, and criticism of many people over many years. These comments came from the reviewers of the first edition, from some readers of the first edition, and from Dr.

Irina Grusko who kindly and capably translated that book into Russian. Critical remarks that helped this second edition came from Professor Dilip Sarwate, Dr. G. David Forney, Professor Joseph A. O'Sullivan, Dr. Weishi Feng, Professor William Weeks IV, Dr. Gottfried Ungerboeck, Professor Ralf Koetter, Professor Steve McLaughlin, Dr. Dakshi Agrawal, and Professor Alon Orlitsky. The quality of the presentation has much to do with the editing and composition skills of Mrs Helen Metzinger and Mrs Francie Bridges. And, as always, Barbara made it possible.

Urbana, Illinois

"Words alone are nothing."

– MOTTO OF THE ROYAL SOCIETY

1 Introduction

A profusion and variety of communication systems, which carry massive amounts of digital data between terminals and data users of many kinds, exist today. Alongside these communication systems are many different magnetic tape storage systems, and magnetic and optical disk storage systems. The received signal in any communication or recording system is always contaminated by thermal noise and, in practice, may also be contaminated by various kinds of defects, nongaussian noise, burst noise, interference, fading, dispersion, cross talk, and packet loss. The communication system or storage system must transmit its data with very high reliability in the presence of these channel impairments. Bit error rates as small as one bit error in 10^{12} bits (or even smaller) are routinely specified.

Primitive communication and storage systems may seek to keep bit error rates small by the simple expedient of transmitting high signal power or by repeating the message. These simplistic techniques may be adequate if the required bit error rate is not too stringent, or if the data rate is low, and if errors are caused by noise rather than by defects or interference. Such systems, however, buy performance with the least expendable resources: Power and bandwidth.

In contrast, modern communication and storage systems obtain high performance via the use of elaborate message structures with complex cross-checks built into the waveform. The advantage of these modern communication waveforms is that high data rates can be reliably transmitted while keeping the transmitted power and spectral bandwidth small. This advantage is offset by the need for sophisticated computations in the receiver (and in the transmitter) to recover the message. Such computations, however, are now regarded as affordable by using modern electronic technology. For example, current telephone-line data modems use microprocessors in the demodulator with well over 500 machine cycles of computation per received data bit. Clearly, with this amount of computation in the modem, the waveforms may have a very sophisticated structure, allowing each individual bit to be deeply buried in the waveform. In some systems it may be impossible to specify where a particular user bit resides in the channel waveform; the entire message is modulated into the channel waveform as a package, and an individual bit appears in a diffuse but recoverable way.

The data-transmission codes described in this book are codes used for the prevention of error. The phrase "prevention of error" has a positive tone that conveys the true role such codes have in modern systems. The more neutral term, "error-control code," is also suitable. The older and widespread term, "error-correcting code," is used as well, but suffers from the fact that it has a negative connotation. It implies that the code is used only to correct an unforeseen deficiency in the communication system whereas, in modern practice, the code is an integral part of any high-performance communication or storage system. Furthermore, in many applications, the code is so tightly integrated with the demodulation that the point within the system where the errors occur and are corrected is really not visible to any external observer. It is a better description to say that the errors are prevented because the preliminary estimates of the data bits within the receiver are accompanied by extra information that cross-checks these data bits. In this sense, the errors never really happen because they are eliminated when the preliminary estimate of the datastream is replaced by the final estimate of the datastream that is given to the user.

1.1 The discrete communication channel

A communication system connects a data source to a data user through a channel. Microwave links, coaxial cables, telephone circuits, and even magnetic and optical disks are examples of channels. A discrete communication channel may transmit binary symbols, or symbols in an alphabet of size 2^m, or even symbols in an alphabet of size q where q is not a power of 2. Indeed, digital communication theory teaches that discrete channels using a larger symbol alphabet are usually more energy efficient than channels that use a binary alphabet.

The designer of the communication system develops devices that prepare the codestream for the input to the discrete channel and process the output of the discrete channel to recover the user's datastream. Although user data may originate as a sequence of bits, within the communication system it is often treated as a sequence of symbols. A symbol may consist of eight bits; then it is called a *byte*. In other cases, a communication system may be designed around a symbol of r bits for some value of r other than eight; the symbol then is called an r-bit symbol. The choice of symbol structure within the communication system is transparent to the user because the datastream is reformatted at the input and output of the communication system.

A *datastream* is a sequence of data symbols, which could be bits, bytes, or other symbols at the input of an encoder. A *codestream* is a sequence of channel symbols, which could be bits, bytes, or other symbols at the output of an encoder. The user perceives that the datastream is being sent through the channel, but what is actually sent is the codestream.

The encoder maps the datastream into the codestream. Codes are of two types: block codes and tree codes. The distinction between them is based on the way that data

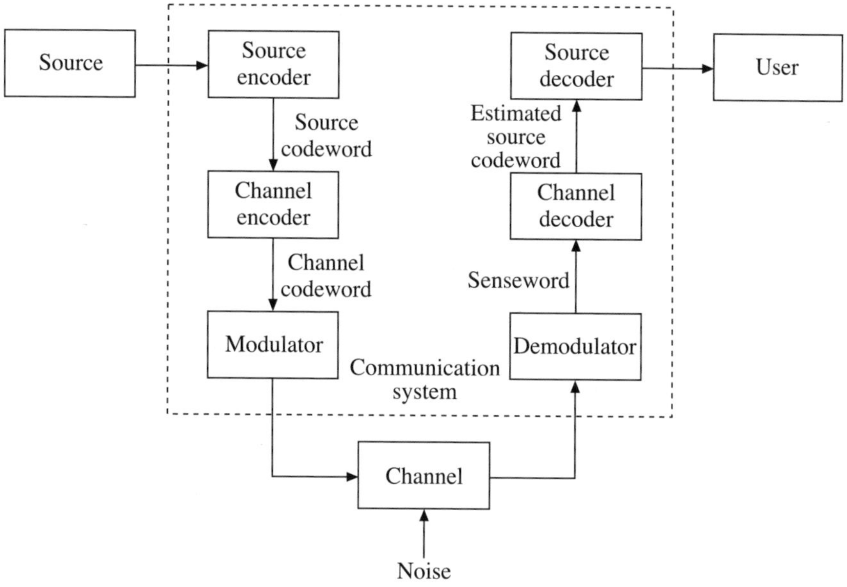

Figure 1.1. Block diagram of a digital communication system

memory is used in the encoder. For constructing the codestream, additional structure is defined on the datastream by segmenting it into pieces called *datawords* or *dataframes*. Likewise, the codestream is segmented into pieces called *codewords* or *codeframes*. The codewords or codeframes are serially concatenated to form the codestream.

It is traditional to partition the major functions of the digital communication system as in the block diagram of Figure 1.1. Data, which enters the communication system from the data source, is first processed by a source encoder designed to represent the source data more compactly. This interim representation is a sequence of symbols called the *source codestream*. The source codestream becomes the input datastream to the channel encoder, which transforms the sequence of symbols into another sequence called the *channel codestream*. The channel codestream is a new, longer sequence that has more redundancy than the source codestream. Each symbol in the channel codestream might be represented by a bit, or perhaps by a group of bits. Next, the modulator converts each symbol of the channel codestream into a corresponding symbol from a finite set of symbols known as the channel alphabet. This sequence of analog symbols from the channel alphabet is transmitted through the channel.

Because the channel is subject to various types of noise, distortion, and interference, the channel output differs from the channel input. The demodulator may convert the received channel output signal into a sequence of the symbols of the channel codestream. Then each demodulated symbol is a best estimate of that code symbol, though the demodulator may make some errors because of channel noise. The demodulated sequence of symbols is called the *senseword* or the *received word*. Because of errors, the symbols of the senseword do not always match those of the channel codestream. The channel decoder uses the redundancy in the channel codestream to correct the errors in the received word and then produces an estimate of the user datastream. If all errors are corrected, the estimated user datastream matches the original user datastream. The source decoder performs the inverse operation of the source encoder and delivers its output datastream to the user.

Alternatively, some functions of the demodulator may be moved into the channel decoder in order to improve performance. Then the demodulator need not make hard decisions on individual code symbols but may give the channel decoder something closer to the raw channel data.

This book deals only with the design of the channel encoder and decoder, a subject known as the subject of *error-control codes*, or *data-transmission codes*, or perhaps, *error-prevention codes*. The emphasis is on the algebraic aspects of the subject; the interplay between algebraic codes and modulation is treated only lightly. The data compression or data compaction functions performed by the source encoder and source decoder are not discussed within this book, nor are the modulator and the demodulator. The channel encoder and the channel decoder will be referred to herein simply as the encoder and the decoder, respectively.

1.2 The history of data-transmission codes

The history of data-transmission codes began in 1948 with the publication of a famous paper by Claude Shannon. Shannon showed that associated with any communication channel or storage channel is a number C (measured in bits per second), called the *capacity* of the channel, which has the following significance. Whenever the information transmission rate R (in bits per second) required of a communication or storage system is less than C then, by using a data-transmission code, it is possible to design a communication system for the channel whose probability of output error is as small as desired. In fact, an important conclusion from Shannon's theory of information is that it is wasteful to make the raw error rate from an uncoded modulator–demodulator too good; it is cheaper and ultimately more effective to use a powerful data-transmission code.

Shannon, however, did not tell us how to find suitable codes; his contribution was to prove that they exist and to define their role. Throughout the 1950s, much effort was devoted to finding explicit constructions for classes of codes that would produce the promised arbitrarily small probability of error, but progress was meager. In the 1960s, for the most part, there was less obsession with this ambitious goal; rather, coding research began to settle down to a prolonged attack along two main avenues.

The first avenue has a strong algebraic flavor and is concerned primarily with block codes. The first block codes were introduced in 1950 when Hamming described a class of single-error-correcting block codes. Shortly thereafter Muller (1954) described a class of multiple-error-correcting codes and Reed (1954) gave a decoding algorithm for them. The Hamming codes and the Reed–Muller codes were disappointingly weak compared with the far stronger codes promised by Shannon. Despite diligent research, no better class of codes was found until the end of the decade. During this period, codes of short blocklength were found, but without any general theory. The major advances came when Bose and Ray-Chaudhuri (1960) and Hocquenghem (1959) found a large class of multiple-error-correcting codes (the BCH codes), and Reed and Solomon (1960) and, independently, Arimoto (1961) found a related class of codes for nonbinary channels. Although these remain among the most important classes of codes, the theory of the subject since that time has been greatly strengthened, and new codes continue to be discovered.

The discovery of BCH codes led to a search for practical methods of designing the hardware or software to implement the encoder and decoder. The first good algorithm was found by Peterson (1960). Later, a powerful algorithm for decoding was discovered by Berlekamp (1968) and Massey (1969), and its implementation became practical as new digital technology became available. Now many varieties of algorithms are available to fit different codes and different applications.

The second avenue of coding research has a more probabilistic flavor. Early research was concerned with estimating the error probability for the best family of block codes despite the fact that the best codes were not known. Associated with these studies were attempts to understand encoding and decoding from a probabilistic point of view, and these attempts led to the notion of sequential decoding. Sequential decoding required the introduction of a class of nonblock codes of indefinite length, which can be represented by a tree and can be decoded by algorithms for searching the tree. The most useful tree codes are highly structured codes called *convolutional codes*. These codes can be generated by a linear shift-register circuit that performs a convolution operation on the data sequence. Convolutional codes were successfully decoded by sequential decoding algorithms in the late 1950s. It is intriguing that the Viterbi algorithm, a much simpler algorithm for decoding them, was not developed until 1967. The Viterbi algorithm gained widespread popularity for convolutional codes of modest complexity, but it is impractical for stronger convolutional codes.

During the 1970s, these two avenues of research began to draw together in some ways and to diverge further in others. Development of the algebraic theory of convolutional codes was begun by Massey and Forney, who brought new insights to the subject of convolutional codes. In the theory of block codes, schemes were proposed to construct good codes of long blocklength. Concatenated codes were introduced by Forney (1966), and Justesen used the idea of a concatenated code to devise a completely constructive class of long block codes with good performance. Meanwhile, Goppa (1970) defined a class of codes that is sure to contain good codes, though without saying how to identify the good ones.

The 1980s saw encoders and decoders appear frequently in newly designed digital communication systems and digital storage systems. A visible example is the compact disk, which uses a simple Reed–Solomon code for correcting double byte errors. Reed–Solomon codes also appear frequently in many magnetic tape drives and network modems, and now in digital video disks. In other applications, such as telephone-line modems, the role of algebraic codes has been displaced by euclidean-space codes, such as the trellis-coded modulation of Ungerboeck (1982). The success of these methods led to further work on the design of nonalgebraic codes based on euclidean distance. The decade closed with widespread applications of data-transmission codes. Meanwhile, mathematicians took the search for good codes based on the Hamming distance into the subject of algebraic geometry and there started a new wave of theoretical progress that continues to grow.

The 1990s saw a further blurring of the walls between coding, signal processing, and digital communications. The development of the notion of turbo decoding and the accompanying codes of Berrou (1993) can be seen as the central event of this period. This work did as much for communications over the wideband channel as Ungerboeck's work did the previous decade for communications over the bandlimited channel. Practical iterative algorithms, such as the "two-way algorithm," for soft-decision decoding of large binary codes are now available to achieve the performance promised by Shannon. The Ungerboeck codes and the Berrou codes, together with their euclidean-space decoding algorithms, have created a body of techniques, still in rapid development, that lie midway between the subjects of modulation theory and of data transmission codes. Further advances toward the codes promised by Shannon are awaited.

This decade also saw the development of algorithms for hard-decision decoding of large nonbinary block codes defined on algebraic curves. Decoders for the codes known as hermitian codes are now available and these codes may soon appear in commercial products. At the same time, the roots of the subject are growing even deeper into the rich soil of mathematics.

1.3 Applications

Because the development of data-transmission codes was motivated primarily by problems in communications, much of the terminology of the subject has been drawn from the subject of communication theory. These codes, however, have many other applications. Codes are used to protect data in computer memories and on digital tapes and disks, and to protect against circuit malfunction or noise in digital logic circuits.

Applications to communication problems are diversified. Binary messages are commonly transmitted between computer terminals, in communication networks, between aircraft, and from spacecraft. Codes can be used to achieve reliable communication even when the received signal power is close to the thermal noise power. And, as the electromagnetic spectrum becomes ever more crowded with man-made signals, data-transmission codes will become even more important because they permit communication links to function reliably in the presence of interference. In military applications, it often is essential to employ a data-transmission code to protect against intentional enemy interference.

Many communication systems have limitations on transmitted power. For example, power may be very expensive in communication relay satellites. Data-transmission codes provide an excellent tool with which to reduce power needs because, with the aid of the code, the messages received weakly at their destinations can be recovered correctly.

Transmissions within computer systems usually are intolerant of even very low error rates because a single error can destroy the validity of a computer program. Error-control coding is important in these applications. Bits can be packed more tightly into some kinds of computer memories (magnetic or optical disks, for example) by using a data-transmission code.

Another kind of communication system structure is a multiaccess system, in which each of a number of users is preassigned an access slot for the channel. This access may be a time slot or frequency slot, consisting of a time interval or frequency interval during which transmission is permitted, or it may be a predetermined coded sequence representing a particular symbol that the user is permitted to transmit. A long binary message may be divided into packets with one packet transmitted within an assigned access slot. Occasionally packets become lost because of collisions, synchronization failure, or routing problems. A suitable data-transmission code protects against these losses because missing packets can be deduced from known packets.

Communication is also important within a large system. In complex digital systems, a large data flow may exist between subsystems. Digital autopilots, digital process-control systems, digital switching systems, and digital radar signal processing all are systems that involve large amounts of digital data which must be shared by multiple interconnected subsystems. This data transfer might be either by dedicated lines or by a more sophisticated, time-shared data-bus system. In either case, error-control techniques are important to ensure proper performance.

Eventually, data-transmission codes and the circuits for encoding and decoding will reach the point where they can handle massive amounts of data. One may anticipate that such techniques will play a central role in all communication systems of the future. Phonograph records, tapes, and television waveforms of the near future will employ digital messages protected by error-control codes. Scratches in a record, or interference in a received signal, will be completely suppressed by the coding as long as the errors are less serious than the capability designed into the error-control code. (Even as these words were written for the first edition in 1981, the as yet unannounced compact disk was nearing the end of its development.)

1.4 Elementary concepts

The subject of data-transmission codes is both simple and difficult at the same time. It is simple in the sense that the fundamental problem is easily explained to any technically trained person. It is difficult in the sense that the development of a solution – and only a partial solution at that – occupies the length of this book. The development of the standard block codes requires a digression into topics of modern algebra before it can be studied.

Suppose that all data of interest can be represented as binary (coded) data, that is, as a sequence of zeros and ones. This binary data is to be transmitted through a binary channel that causes occasional errors. The purpose of a code is to add extra check symbols to the data symbols so that errors may be found and corrected at the receiver. That is, a sequence of data symbols is represented by some longer sequence of symbols with enough redundancy to protect the data.

A binary code of size M and *blocklength* n is a set of M binary words of length n called *codewords*. Usually, $M = 2^k$ for an integer k, and the code is referred to as an (n, k) binary code.

For example, we can make up the following code

$$\mathcal{C} = \left\{ \begin{array}{ccccc} 1 & 0 & 1 & 0 & 1 \\ 1 & 0 & 0 & 1 & 0 \\ 0 & 1 & 1 & 1 & 0 \\ 1 & 1 & 1 & 1 & 1 \end{array} \right\}.$$

This is a very poor (and very small) code with $M = 4$ and $n = 5$, but it satisfies the requirements of the definition, so it is a code. We can use this code to represent two-bit binary numbers by using the following (arbitrary) correspondence:

$$\begin{array}{cccccccc} 0 & 0 & \leftrightarrow & 1 & 0 & 1 & 0 & 1 \\ 0 & 1 & \leftrightarrow & 1 & 0 & 0 & 1 & 0 \\ 1 & 0 & \leftrightarrow & 0 & 1 & 1 & 1 & 0 \\ 1 & 1 & \leftrightarrow & 1 & 1 & 1 & 1 & 1. \end{array}$$

If one of the four five-bit codewords is received, we may then suppose that the corresponding two data bits are the original two data bits. If an error is made, we receive a different five-bit senseword. We then attempt to find the most likely transmitted codeword to obtain our estimate of the original two data bits.

For example, if we receive the senseword $(0, 1, 1, 0, 0)$, then we may presume that $(0, 1, 1, 1, 0)$ was the transmitted codeword, and hence 10 is the two-bit dataword. If we receive the "soft" senseword consisting of the real numbers $(0.1, 1.1, 0.9, 0.4, 0.2)$ then we may presume that $(0, 1, 1, 1, 0)$ was the transmitted codeword because it is closest in euclidean distance, and hence 10 is the two-bit dataword. The decoding of soft sensewords is treated in Chapter 11.

The code of the example is not a good code because it is not able to correct many patterns of errors. We want to design a code so that every codeword is as different as possible from every other codeword, and we want to do this especially when the blocklength is long.

The first purpose of this book is to find good codes. Although, superficially, this may seem like a simple task, it is, in fact, exceedingly difficult, and many good codes are as yet undiscovered.

To the inexperienced, it may seem that it should suffice to define the requirements of a good code and then let a computer search through the set of all possible codes. But how many binary codes are there for a given (n, k)? Each codeword is a sequence of n binary symbols, and there are 2^k such codewords in an (n, k) binary code. Therefore a code is described by $n \cdot 2^k$ binary symbols. Altogether there are $2^{n \cdot 2^k}$ ways of picking these binary symbols. Hence the number of different (n, k) codes is $2^{n \cdot 2^k}$. Of course, a great many of these codes are of little value (as when two codewords are identical), but either the computer search must include these codes or some theory must be developed for excluding them.

For example, take $(n, k) = (40, 20)$, which is a very modest code by today's standards. The number of such codes is much larger than $10^{10,000,000}$ – an inconceivably large number. Hence undisciplined search procedures are worthless.

In general, we define block codes over an arbitrary finite alphabet, say the alphabet with q symbols $\{0, 1, 2, \ldots, q-1\}$. At first sight, it might seem to be an unnecessary generalization to introduce alphabets other than the binary alphabet. For reasons such as energy efficiency, however, many channels today are nonbinary, and codes for these channels must be nonbinary. In fact, data-transmission codes for nonbinary channels are often quite good, and this can reinforce the reasons for using a nonbinary channel. It is a trivial matter to represent binary source data in terms of a q-ary alphabet, especially if q is a power of 2, as usually it is in practice.

Definition 1.4.1. *A block code of size M over an alphabet with q symbols is a set of M q-ary sequences of length n called* codewords.

If $q = 2$, the symbols are called bits. Usually, $M = q^k$ for some integer k, and we shall be interested only in this case, calling the code an (n, k) code. Each sequence of k q-ary data symbols can be associated with a sequence of n q-ary symbols comprising a codeword.

There are two basic classes of codes: *block codes* and *trellis codes*. These are illustrated in Figure 1.2. A block code represents a block of k data symbols by an n-symbol codeword. The rate R of a block code[1] is defined as $R = k/n$. Initially, we shall restrict our attention to block codes.

A trellis code is more complicated. It takes a nonending sequence of data symbols arranged in k-symbol segments called *dataframes*, and puts out a continuous sequence of code symbols arranged in n-symbol segments called *codeframes*. The distinction with

[1] This rate is dimensionless, or perhaps measured in units of bits/bit or symbols/symbol. It should be distinguished from another use of the term *rate* measured in bits/second through a channel. Yet another definition, $R = (k/n)\log_e q$, which has the units of nats/symbol, with a nat equaling $\log_2 e$ bits, is in use. The definition $R = (k/n)\log_2 q$, which has the units of bits/symbol, is also popular.

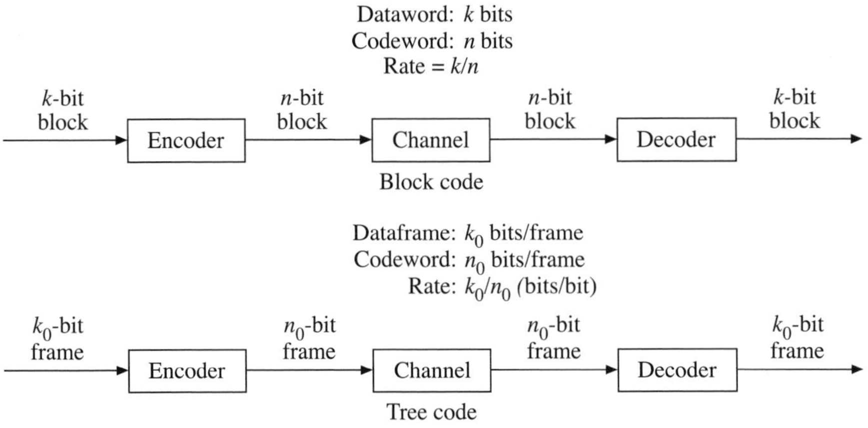

Figure 1.2. Basic classes of codes

block codes is that in a trellis code, a k-symbol dataframe can affect all succeeding codeword frames, whereas in a block code, a k-symbol datablock determines only the next n-symbol codeblock, but no others. We shall defer the study of trellis codes, specifically convolutional codes, until Chapter 9.

Whenever a message consists of a large number of bits, it is better, in principle, to use a single block code of large blocklength than to use a succession of codewords from a shorter block code. The nature of statistical fluctuations is such that a random pattern of errors usually exhibits some clustering of errors. Some segments of the random pattern contain more than the average number of errors, and some segments contain less. Long codewords are considerably less sensitive to random errors than are short codewords of the same rate, because a segment with many errors can be offset by a segment with few errors, but of course, the encoder and decoder may be more complex.

As an example, suppose that 1000 data bits are transmitted with a (fictitious) 2000-bit binary codeword that can correct 100 bit errors. Compare this with a scheme for transmitting 100 data bits at a time with a 200-bit binary codeword that can correct 10 bit errors per block. Ten such blocks are needed to transmit 1000 bits. This latter scheme can also correct a total of 100 errors, but only if they are properly distributed – ten errors to a 200-bit block. The first scheme can correct 100 errors no matter how they are distributed within the 2000-bit codeword. It is far more powerful.

This heuristic argument can be given a sound theoretical footing, but that is not our purpose here. We only wish to make plausible the fact that good codes are of long blocklength, and that very good codes are of very long blocklength. Such codes can be very hard to find and, when found, may require complex devices to implement the encoding and decoding operations.

Given two sequences of the same length of symbols from some fixed symbol alphabet, perhaps the binary alphabet $\{0, 1\}$, we shall want to measure how different those two sequences are from each other. The most suggestive way to measure the difference between the two sequences is to count the number of places in which they differ. This is called the *Hamming distance* between the sequences.

Definition 1.4.2. *The Hamming distance $d(\boldsymbol{x}, \boldsymbol{y})$ between two q-ary sequences $\boldsymbol{x}$ and $\boldsymbol{y}$ of length n is the number of places in which $\boldsymbol{x}$ and $\boldsymbol{y}$ differ.*

For example, take $\boldsymbol{x} = 10101$, $\boldsymbol{y} = 01100$, then $d(10101, 01100) = 3$. For another example, take $\boldsymbol{x} = 30102$, $\boldsymbol{y} = 21103$, then $d(30102, 21103) = 3$.

The reason for choosing the term "distance" is to appeal to geometric intuition when constructing codes. It is obvious that the Hamming distance is nonnegative and symmetric. It is easy to verify that the Hamming distance also satisfies the triangle inequality $d(\boldsymbol{x}, \boldsymbol{y}) \leq d(\boldsymbol{x}, \boldsymbol{z}) + d(\boldsymbol{y}, \boldsymbol{z})$. This means that geometric reasoning and intuition based on these properties are valid.

Definition 1.4.3. *Let $\mathcal{C} = \{\boldsymbol{c}_\ell \mid \ell = 0, \ldots, M-1\}$ be a code. Then the minimum Hamming distance $d_{\min}$ (or d) of $\mathcal{C}$ is the Hamming distance between the pair of codewords with smallest Hamming distance. That is,*

$$d_{\min} = \min_{\substack{\boldsymbol{c}_i, \boldsymbol{c}_j \in \mathcal{C} \\ i \neq j}} d(\boldsymbol{c}_i, \boldsymbol{c}_j).$$

Block codes are judged by three parameters: the blocklength n, the datalength k, and the minimum distance $d_{\min}$. An (n, k) block code with minimum distance $d_{\min}$ is also described as an $(n, k, d_{\min})$ block code.

In the block code $\mathcal{C}$, given in the first example of this section,

$$\begin{aligned}
&d(10101, 10010) = 3\\
&d(10101, 01110) = 4\\
&d(10101, 11111) = 2\\
&d(10010, 01110) = 3\\
&d(10010, 11111) = 3\\
&d(01110, 11111) = 2.
\end{aligned}$$

Hence $d_{\min} = 2$ for this code.

We may also have two infinitely long sequences over some symbol alphabet. Again, the Hamming distance is defined as the number of places in which the two sequences are different. The Hamming distance between two infinite sequences will be infinite unless the sequences are different only on a finite segment.

Suppose that a block codeword is transmitted and a single error is made by the channel in that block. Then the Hamming distance from the senseword to the transmitted codeword is equal to 1. If the distance to every other codeword is larger than 1, then the decoder will properly correct the error if it presumes that the closest codeword to the senseword is the codeword that was actually transmitted.

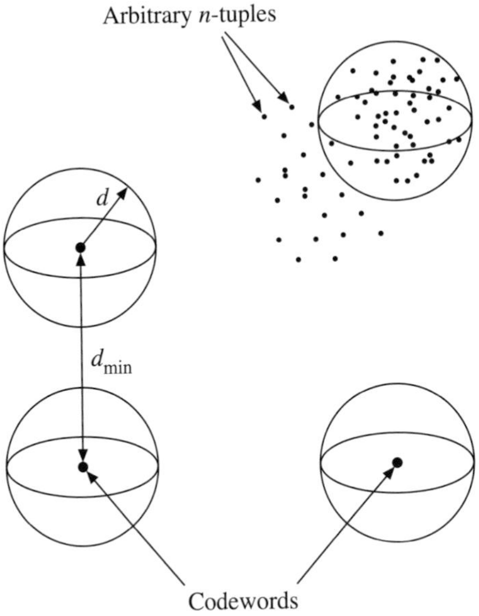

Figure 1.3. Decoding spheres

More generally, if t errors occur, and if the distance from the senseword to every other codeword is larger than t, then the decoder will properly correct the t errors if it presumes that the closest codeword to the senseword was actually transmitted. This always occurs if

$$d_{\min} \geq 2t + 1.$$

It may be possible, sometimes, to correct certain error patterns with t errors even when this inequality is not satisfied. However, correction of t errors cannot be guaranteed if $d_{\min} < 2t + 1$ because then it depends on which codeword is transmitted and on the actual pattern of the t errors within the block.

We shall often describe coding and decoding by using the language of geometry because geometric models are intuitive and powerful aids to reasoning. Figure 1.3 illustrates the geometric situation. Within the space of all q-ary n-tuples, the *Hamming sphere* of radius t (a nonnegative integer), with the center at the sequence $\boldsymbol{v}$, is the set of all sequences $\boldsymbol{v}'$ such that $d(\boldsymbol{v}, \boldsymbol{v}') \leq t$. To define a code within the space of q-ary n-tuples, a set of n-tuples is selected, and these n-tuples are designated as codewords of code $\mathcal{C}$. If $d_{\min}$ is the minimum distance of this code and t is the largest integer satisfying

$$d_{\min} \geq 2t + 1,$$

then nonintersecting Hamming spheres of radius t can be drawn about each of the codewords. A senseword contained in a sphere is decoded as the codeword at the center of that sphere. If t or fewer errors occur, then the senseword is always in the proper sphere, and the decoding is correct.

Some sensewords that have more than t errors will be in a decoding sphere about another codeword and, hence, will be decoded incorrectly. Other sensewords that have more than t errors will lie in the interstitial space between decoding spheres. Depending on the requirements of the application, these can be treated in either of two ways.

A *bounded-distance decoder* decodes only those sensewords lying in one of the decoding spheres about one of the codewords. Other sensewords have more errors than a bounded-distance decoder can correct and are so declared by the decoder. Such error patterns in a bounded-distance decoder are called *uncorrectable error patterns*. When a decoder encounters an uncorrectable error pattern, it declares a *decoding failure*. A bounded-distance decoder is an example of an *incomplete decoder*, which means that it has uncorrectable error patterns. Most error-correcting decoders in use are bounded-distance decoders.

A *complete decoder* decodes every received word into a closest codeword. In geometrical terms, the complete decoder carves up the interstices between spheres and attaches portions to each of the spheres so that each point in an interstice is attached to a closest sphere located nearby. (Some points are equidistant from several spheres and are arbitrarily assigned to one of the closest spheres.) When more than t (but not *too* many) errors occur in a codeword of large blocklength, the complete decoder will usually decode correctly, but occasionally will produce an incorrect codeword. A complete decoder may be preferred for its performance, but for a large code the issue of complexity leads to the use of an incomplete decoder. An incomplete decoder may also be preferred as a way to reduce the probability of decoding error in exchange for a larger probability of decoding failure.

We shall also deal with channels that make *erasures* – or both errors and erasures – as well as channels, called *soft-output channels*, whose output for each symbol is a real number, such as a likelihood measure. A soft-output channel has an input alphabet of size q and an output alphabet consisting of real numbers, or vectors of real numbers. An error-control code can be used with a soft-output channel. The output of the channel then is called a *soft senseword* and the decoder is called a *soft decoder* or a *soft-input decoder*. A soft-input decoder is more tightly interconnected with the modulator and, for this reason, often has very good performance.

For an *erasure channel*, the receiver is designed to declare a symbol erased when that symbol is received ambiguously, as when the receiver recognizes the presence of interference or a transient malfunction. An erasure channel has an input alphabet of size q and an output alphabet of size $q + 1$; the extra symbol is called an *erasure*. For example, an erasure of the third symbol from the message 12345 gives $12 - 45$. This should not be confused with another notion known as a *deletion*, which would give 1245.

An error-control code can be used with an erasure channel. If the code has a minimum distance $d_{\min}$, then any pattern of ρ erasures can be filled if $d_{\min} \geq \rho + 1$. Furthermore, any pattern of ν errors and ρ erasures can be decoded, provided

$$d_{\min} \geq 2\nu + 1 + \rho$$

is satisfied. To prove this statement, delete the ρ components that contain erasures in the senseword from all codewords of the code. This process gives a new code, called a punctured code, whose minimum distance is not smaller than $d_{\min} - \rho$; hence ν errors can be corrected, provided $d_{\min} - \rho \geq 2\nu + 1$ is satisfied. In this way we can recover the punctured codeword, which is equivalent to the original codeword with ρ components erased. Finally, because $d_{\min} \geq \rho + 1$, there is only one codeword that agrees with the unerased components; thus the entire codeword can be recovered.

1.5 Elementary codes

Some codes are simple enough to be described at the outset.

Parity-check codes

These are high-rate codes with poor error performance on a binary output channel. Given k data bits, add a $(k + 1)$th bit so that the total number of ones in each codeword is even. Thus for example, with $k = 4$,

$$\begin{array}{cccccccccc} 0 & 0 & 0 & 0 & \leftrightarrow & 0 & 0 & 0 & 0 & 0 \\ 0 & 0 & 0 & 1 & \leftrightarrow & 0 & 0 & 0 & 1 & 1 \\ 0 & 0 & 1 & 0 & \leftrightarrow & 0 & 0 & 1 & 0 & 1 \\ 0 & 0 & 1 & 1 & \leftrightarrow & 0 & 0 & 1 & 1 & 0, \end{array}$$

and so forth. This is a $(k + 1, k)$ or an $(n, n - 1)$ code. The minimum distance is 2, and hence no errors can be corrected. A simple parity-check code is used to detect (but not correct) a single error.

Repetition codes

These are low-rate codes with good error performance on a binary output channel. Given a single data bit, repeat it n times. Usually, n is odd

$$\begin{array}{ccccccc} 0 & \leftrightarrow & 0 & 0 & 0 & 0 & 0 \\ 1 & \leftrightarrow & 1 & 1 & 1 & 1 & 1. \end{array}$$

This is an $(n, 1)$ code. The minimum distance is n, and $\frac{1}{2}(n - 1)$ errors can be corrected by assuming that the majority of the received bits agrees with the correct data bit.

Hamming codes

These are codes that can correct a single error. For each m, there is a $(2^m - 1, 2^m - 1 - m)$ binary Hamming code. When m is large, the code rate is close to 1, but the fraction of the total number of bits that can be in error is very small. In this section, we will introduce the (7, 4) Hamming codes via a direct descriptive approach. The (7, 4) Hamming code can be described by the implementation in Figure 1.4(a). Given four data bits (a_0, a_1, a_2, a_3), let the first four bits of the codeword equal the four

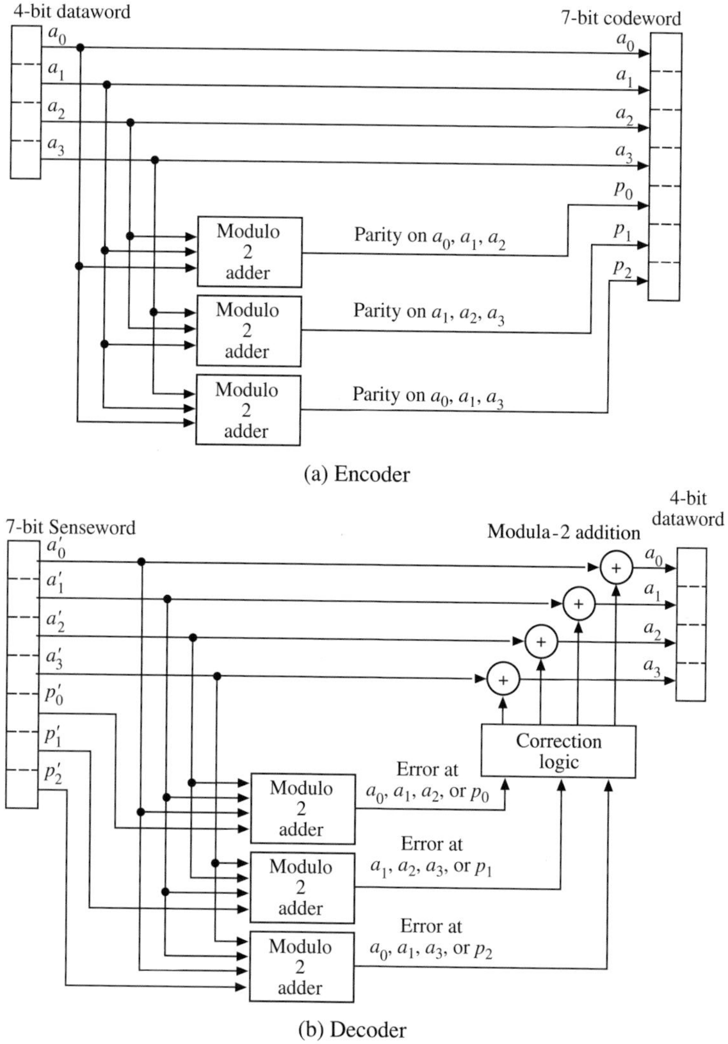

Figure 1.4. A simple encoder/decoder for a (7, 4) Hamming code

Table 1.1. *The* (7, 4) *Hamming code*

0	0	0	0	0	0	0
0	0	0	1	0	1	1
0	0	1	0	1	1	0
0	0	1	1	1	0	1
0	1	0	0	1	1	1
0	1	0	1	1	0	0
0	1	1	0	0	0	1
0	1	1	1	0	1	0
1	0	0	0	1	0	1
1	0	0	1	1	1	0
1	0	1	0	0	1	1
1	0	1	1	0	0	0
1	1	0	0	0	1	0
1	1	0	1	0	0	1
1	1	1	0	1	0	0
1	1	1	1	1	1	1

data bits. Append three check bits (p_0, p_1, p_2), defined by

$$p_0 = a_0 + a_1 + a_2$$
$$p_1 = a_1 + a_2 + a_3$$
$$p_2 = a_0 + a_1 + a_3.$$

Here + denotes modulo-2 addition $(0 + 0 = 0, 0 + 1 = 1, 1 + 0 = 1, 1 + 1 = 0)$. The sixteen codewords of the (7, 4) Hamming code are shown in Table 1.1. Of course, the idea of the code is not changed if the bit positions are permuted. All of these variations are equivalent, and all are called the (7, 4) Hamming code.

The decoder receives a seven-bit senseword $\boldsymbol{v} = (a_0', a_1', a_2', a_3', p_0', p_1', p_2')$. This corresponds to a transmitted codeword with at most one error. The decoder, shown in Figure 1.4(b), computes

$$s_0 = p_0' + a_0' + a_1' + a_2'$$
$$s_1 = p_1' + a_1' + a_2' + a_3'$$
$$s_2 = p_2' + a_0' + a_1' + a_3'.$$

The three-bit pattern (s_0, s_1, s_2) is called the *syndrome*. It does not depend on the actual data bits, but only on the error pattern. There are eight possible syndromes: one that corresponds to no error, and one for each of the seven possible patterns with a single error. Inspection shows that each of these error patterns has a unique syndrome, as shown in Table 1.2.

It is a simple matter to design binary logic that will complement the bit location indicated by the syndrome. After correction is complete, the check bits can be discarded.

Table 1.2. *Syndrome table*

Syndrome			Error						
0	0	0	0	0	0	0	0	0	0
0	0	1	0	0	0	0	0	0	1
0	1	0	0	0	0	0	0	1	0
0	1	1	0	0	0	1	0	0	0
1	0	0	0	0	0	0	1	0	0
1	0	1	1	0	0	0	0	0	0
1	1	0	0	0	1	0	0	0	0
1	1	1	0	1	0	0	0	0	0

If two or more errors occur, then the design specification of the code is exceeded and the code will miscorrect. That is, it will make a wrong correction and put out incorrect data bits.

Because the (7, 4) Hamming code is a very simple code, it is possible to describe it in this elementary way. A more compact description, which we will eventually prefer, is to use vector space methods, writing the codeword as a vector–matrix product

$$\begin{bmatrix} a_0 \\ a_1 \\ a_2 \\ a_3 \\ p_0 \\ p_1 \\ p_2 \end{bmatrix} = \begin{bmatrix} 1 & 0 & 0 & 0 \\ 0 & 1 & 0 & 0 \\ 0 & 0 & 1 & 0 \\ 0 & 0 & 0 & 1 \\ 1 & 1 & 1 & 0 \\ 0 & 1 & 1 & 1 \\ 1 & 1 & 0 & 1 \end{bmatrix} \begin{bmatrix} a_0 \\ a_1 \\ a_2 \\ a_3 \end{bmatrix},$$

and the syndrome as another matrix–vector product

$$\begin{bmatrix} s_0 \\ s_1 \\ s_2 \end{bmatrix} = \begin{bmatrix} 1 & 1 & 1 & 0 & 1 & 0 & 0 \\ 0 & 1 & 1 & 1 & 0 & 1 & 0 \\ 1 & 1 & 0 & 1 & 0 & 0 & 0 \end{bmatrix} \begin{bmatrix} a'_0 \\ a'_1 \\ a'_2 \\ a'_3 \\ p'_0 \\ p'_1 \\ p'_2 \end{bmatrix}.$$

Problems

1.1 a. By trial and error, find a set of four binary words of length 3 such that each word is at least a distance of 2 from every other word.

b. Find a set of sixteen binary words of length 7 such that each word is at least a distance of 3 from every other word.

1.2 a. Describe how to cut 88 circles of 1-inch diameter out of a sheet of paper of width 8.5 inches and length 11 inches. Prove that it is not possible to cut out more than 119 circles of 1-inch diameter.

b. Prove that it is not possible to find 32 binary words, each of length 8 bits, such that every word differs from every other word in at least three places.

1.3 A single-error-correcting Hamming code has $2^m - 1$ bits of which m bits are check bits.

a. Write (n, k) for the first five nontrivial Hamming codes (starting at $m = 3$).

b. Calculate their rates.

c. Write an expression for the probability of decoding error, p_e, when the code is used with a binary channel that makes errors with probability q. How does the probability of error behave with n?

1.4 Design an encoder/decoder for a (15, 11) Hamming code by reasoning as in Figure 1.4. There is no need to show repetitive details (that is, show the principle).

1.5 For any (n, k) block code with minimum distance $2t + 1$ or greater, the number of data symbols satisfies

$$n - k \geq \log_q \left[1 + \binom{n}{1}(q-1) + \binom{n}{2}(q-1)^2 + \cdots + \binom{n}{t}(q-1)^t \right].$$

Prove this statement, which is known as the *Hamming bound*.

1.6 The simplest example of a kind of code known as a *product code* is of the form:

$$\begin{array}{cccc|c}
a_{00} & a_{01} & \ldots & a_{0,k_1-1} & p_{0,k_1} \\
a_{10} & a_{11} & & & p_{1,k_1} \\
\vdots & & & \vdots & \vdots \\
a_{k_2-1,0} & \ldots & & a_{k_2-1,k_1-1} & p_{k_2-1,k_1} \\
\hline
p_{k_2,0} & & \ldots & p_{k_2,k_1-1} & p_{k_2,k_1}
\end{array}$$

where the k_1k_2 symbols in the upper left block are binary data symbols, and each row (and column) is a simple parity-check code. This gives a $((k_1 + 1)(k_2 + 1), k_1k_2)$ binary product code.

a. Show that p_{k_2,k_1} is a check on both its column and its row.

b. Show that this is a single-error-correcting code.

c. Show that this code is also a double-error-detecting code. Give two double-error patterns that cannot be distinguished from one another when using this code and so cannot be corrected.

d. What is the minimum distance of the code?

1.7 Show that Hamming distance has the following three properties:

(i) $d(\boldsymbol{x}, \boldsymbol{y}) \geq 0$ with equality if and only if $\boldsymbol{x} = \boldsymbol{y}$;

(ii) $d(\boldsymbol{x}, \boldsymbol{y}) = d(\boldsymbol{y}, \boldsymbol{x})$.

(iii) Triangle inequality

$$d(\boldsymbol{x}, \boldsymbol{y}) \leq d(\boldsymbol{x}, \boldsymbol{z}) + d(\boldsymbol{y}, \boldsymbol{z}).$$

A distance function with these three properties is called a *metric*.

1.8 a. Show that a code $\mathcal{C}$ is capable of detecting any pattern of d or fewer errors if and only if the minimum distance of the code $\mathcal{C}$ is greater than d.

b. Show that a code is capable of correcting any pattern of t or fewer errors if and only if the minimum distance of the code is at least $2t + 1$.

c. Show that a code can be used to correct all patterns of t or fewer errors and, simultaneously, detect all patterns of d or fewer errors ($d \geq t$) if the minimum distance of the code is at least $t + d + 1$.

d. Show that a code can be used to fill ρ erasures if the minimum distance of the code is at least $\rho + 1$.

1.9 A *soft senseword* $\boldsymbol{v}$ is a vector of real numbers, one corresponding to each bit. To decode a soft senseword, one may choose the codeword $\boldsymbol{c}$ that lies closest to the senseword in euclidean distance

$$d(\boldsymbol{v}, \boldsymbol{c}) = \sum_{i=0}^{n-1} (v_i - c_i)^2.$$

Let $v_i = c_i + e_i$ where the noise components e_i are independent, white, and identical gaussian random variables of variance σ^2 and zero mean. Let $E_m = \sum_{i=0}^{n-1} c_i^2$.

Prove that a binary repetition code using the real numbers ± 1 to represent the code bits has the same energy and the same probability of bit error as an uncoded bit that uses the two real numbers $\pm n$ to represent the value of the single bit.

1.10 a. Show that if the binary (15, 11) Hamming code is used to correct single errors on a channel that makes two errors, the decoder output is always wrong.

b. Show that if the two errors are in check bits, the decoder will always miscorrect a data bit.

c. By appending an overall check bit, show how to extend the (15, 11) Hamming code to a (16, 11) code that corrects all single errors and detects all double errors. What is the minimum distance of this code?

1.11 Show that the list of codewords in Table 1.1 is unchanged by the permutation

$$(c_0, c_1, c_2, c_3, c_4, c_5, c_6) \rightarrow (c_0, c_4, c_1, c_5, c_2, c_6, c_3).$$

Such a permutation is called an automorphism of the code.

2 Introduction to Algebra

The search for good data-transmission codes has relied, to a large extent, on the powerful and beautiful structures of modern algebra. Many important codes, based on the mathematical structures known as Galois fields, have been discovered. Further, this algebraic framework provides the necessary tools with which to design encoders and decoders. This chapter and Chapter 4 are devoted to developing those topics in algebra that are significant to the theory of data-transmission codes. The treatment is rigorous, but it is limited to material that will be used in later chapters.

2.1 Fields of characteristic two

Real numbers form a familiar set of mathematical objects that can be added, subtracted, multiplied, and divided. Similarly, complex numbers form a set of objects that can be added, subtracted, multiplied, and divided. Both of these arithmetic systems are of fundamental importance in engineering disciplines. We will need to develop other, less familiar, arithmetic systems that are useful in the study of data-transmission codes. These new arithmetic systems consist of sets together with operations on the elements of the sets. We shall call the operations "addition," "subtraction," "multiplication," and "division," although they need not be the same operations as those of elementary arithmetic.

Modern algebraic theory classifies the many arithmetic systems it studies according to their mathematical strength. Later in this chapter, these classifications will be defined formally. For now, we have the following loose definitions.

1. **Abelian group.**[1] A set of mathematical objects that can be "added" and "subtracted."
2. **Ring.** A set of mathematical objects that can be "added," "subtracted," and "multiplied."
3. **Field.** A set of mathematical objects that can be "added," "subtracted," "multiplied," and "divided."

[1] An abelian group is a special case of a group. The arithmetic operation in a general group is too weak to be called "addition."

The names of these operations are enclosed in quotation marks because, in general, they are not the conventional operations of elementary arithmetic; these names are used because the operations resemble the conventional operations.

Before we study these concepts formally, let us do some sample calculations in the simplest of all possible fields, namely, the field with only two elements. (The real field has an infinite number of elements.) Let the symbols 0 and 1 denote the two elements in the field. Define the operations of addition and multiplication by

$$\begin{array}{ll} 0+0=0 & 0\cdot 0=0 \\ 0+1=1 & 0\cdot 1=0 \\ 1+0=1 & 1\cdot 0=0 \\ 1+1=0 & 1\cdot 1=1. \end{array}$$

The addition and multiplication defined here are called modulo-2 addition and modulo-2 multiplication. Note that $1+1=0$ implies that $-1=1$, and $1\cdot 1=1$ implies that $1^{-1}=1$. With these observations, it is easy to verify that subtraction and division are always defined, except for division by zero. The alphabet of two symbols, 0 and 1, together with modulo-2 addition and modulo-2 multiplication is called the field of two elements, and is denoted by the label $GF(2)$.

The familiar ideas of algebra can be used with the above arithmetic. The following set of equations in $GF(2)$ provides an example

$$\begin{aligned} x+y+z&=1 \\ x+y&=0 \\ x+z&=1. \end{aligned}$$

This set can be solved by subtracting the third equation from the first equation to get $y=0$. Then from the second equation, $x=0$, and from the first equation, $z=1$. Substitution of this solution back into the original set of equations verifies that it does, indeed, satisfy the set of equations.

For an alternative method of solution, we assume that all the usual techniques of linear algebra also hold in $GF(2)$. The determinant is computed as follows:

$$\begin{aligned} D=\det\begin{bmatrix} 1 & 1 & 1 \\ 1 & 1 & 0 \\ 1 & 0 & 1 \end{bmatrix} &= 1\cdot\begin{vmatrix} 1 & 0 \\ 0 & 1 \end{vmatrix} - 1\cdot\begin{vmatrix} 1 & 0 \\ 1 & 1 \end{vmatrix} + 1\cdot\begin{vmatrix} 1 & 1 \\ 1 & 0 \end{vmatrix} \\ &= 1\cdot 1 - 1\cdot 1 - 1\cdot 1 = -1 = 1. \end{aligned}$$

We then solve the set of equations by Cramer's rule:

$$x=D^{-1}\begin{vmatrix} 1 & 1 & 1 \\ 0 & 1 & 0 \\ 1 & 0 & 1 \end{vmatrix}=0, \quad y=D^{-1}\begin{vmatrix} 1 & 1 & 1 \\ 1 & 0 & 0 \\ 1 & 1 & 1 \end{vmatrix}=0, \quad z=D^{-1}\begin{vmatrix} 1 & 1 & 1 \\ 1 & 1 & 0 \\ 1 & 0 & 1 \end{vmatrix}=1.$$

This is the same answer as before.

+	0	1	2	3	4	5	6	7	6	9	A	B	C	D	E	F
0	0	1	2	3	4	5	6	7	6	9	A	B	C	D	E	F
1	1	0	3	2	5	4	7	6	9	8	B	A	D	C	F	E
2	2	3	0	1	6	7	4	5	A	B	8	9	E	F	C	D
3	3	2	1	0	7	6	5	4	B	A	9	8	F	E	D	C
4	4	5	6	7	0	1	2	3	C	D	E	F	8	9	A	B
5	5	4	7	6	1	0	3	2	D	C	F	E	9	8	B	A
6	6	7	4	5	2	3	0	1	E	F	C	D	A	B	8	9
7	7	6	5	4	3	2	1	0	F	E	D	C	B	A	9	8
8	8	9	A	B	C	D	E	F	0	1	2	3	4	5	6	7
9	9	8	B	A	D	C	F	E	1	0	3	2	5	4	7	6
A	A	B	8	9	E	F	C	D	2	3	0	1	6	7	4	5
B	B	A	9	8	F	E	D	C	3	2	1	0	7	6	4	5
C	C	D	E	F	8	9	A	B	4	5	6	7	0	1	2	3
D	D	C	F	E	9	8	B	A	5	4	7	6	1	0	3	2
E	E	F	C	D	A	B	8	9	6	7	4	5	2	3	0	1
F	F	E	D	C	B	A	9	8	7	6	5	4	3	2	1	0

(a) Addition Table

×	0	1	2	3	4	5	6	7	6	9	A	B	C	D	E	F
0	0	0	0	0	0	0	0	0	0	0	0	0	0	0	0	0
1	0	1	2	3	4	5	6	7	6	9	A	B	C	D	E	F
2	0	2	4	6	8	A	C	E	3	1	7	5	B	9	F	D
3	0	3	6	5	C	F	A	9	B	8	D	E	7	4	1	2
4	0	4	8	C	3	7	B	F	6	2	E	A	5	1	D	9
5	0	5	A	F	7	2	D	8	E	B	4	1	9	C	3	6
6	0	6	C	A	B	D	7	1	5	3	9	F	E	8	2	4
7	0	7	E	9	F	8	1	6	D	A	3	4	2	5	C	B
8	0	8	3	B	6	E	5	D	C	4	F	7	A	2	9	1
9	0	9	1	8	2	B	3	A	4	D	5	C	6	F	7	E
A	0	A	7	D	E	4	9	3	F	5	8	2	1	B	6	C
B	0	B	5	3	A	1	F	4	7	C	2	9	D	6	8	3
C	0	C	B	7	5	9	E	2	A	6	1	D	F	3	4	8
D	0	D	9	4	1	C	8	5	2	F	B	6	3	E	A	7
E	0	E	F	1	D	3	2	C	9	7	6	8	4	A	B	5
F	0	F	D	2	9	6	4	B	1	E	C	3	8	7	5	A

(b) Multiplication Table

Figure 2.1. The field $GF(16)$ in hexadecimal notation

A second example of a field is known as $GF(16)$. This field has exactly sixteen elements, which can be denoted by the sixteen four-bit numbers, or can be denoted by the sixteen symbols of the hexadecimal alphabet $\{0, 1, 2, 3, 4, 5, 6, 7, 8, 9, A, B, C, D, E, F\}$. The addition and multiplication tables for $GF(16)$ are shown in Figure 2.1. Notice that addition and multiplication are quite different from the familiar operations on the integers. The tables are internally consistent, however, and allow subtraction and division. For subtraction, $x - y = x + (-y)$ where $-y$ is that field element satisfying $y + (-y) = 0$. For division, $x \div y = x \cdot (y^{-1})$ where y^{-1} is that element of the field satisfying $y \cdot y^{-1} = 1$. Inspection of the multiplication table shows that every nonzero element has an inverse, and hence division is always defined except for division by zero.

In general, the finite field with q elements is called $GF(q)$. If q is a power of two, then $GF(2^m)$ is called a binary field, or a field of *characteristic* two. Most of the techniques of linear algebra, such as matrix operations, can be justified for an arbitrary field. Because of this, fields with a finite number of elements will prove very useful. We shall study these fields and find a method of constructing the addition and multiplication tables that will produce a field even when the number of elements is large. In Chapter 4, we shall see that finite fields with q elements can be constructed when, and only when, q is equal to p^m, where p is a prime number and m is an arbitrary positive integer. But first, we must develop the concepts of groups and rings.

2.2 Groups

A group is a mathematical abstraction of an algebraic structure that may occur frequently in many forms. There are many concrete examples of interesting groups. The abstract idea is introduced into mathematics because it is easier to study all mathematical systems with a common structure at the same time rather than to study them one by one.

Definition 2.2.1. *A group G is a set, together with an operation on pairs of elements of the set (denoted by $*$), satisfying the following four properties.*

1. *Closure: For every a, b in the set, $c = a * b$ is in the set.*
2. *Associativity: For every a, b, c in the set,*

   $$a * (b * c) = (a * b) * c.$$

3. *Identity: There is an element e, called the* identity element, *that satisfies*

   $$a * e = e * a = a$$

   for every a in the set.
4. *Inverse: If a is in the set, then there is some element b in the set, called an* inverse *of a, such that*

   $$a * b = b * a = e.$$

If G has a finite number of elements, then it is called a finite group, *and the number of elements in G is called the* order *of G.*

Some groups satisfy the additional property that for all a, b in the group,

$$a * b = b * a.$$

This property is called the *commutativity property*. Groups with this additional property are called *commutative groups*, or *abelian groups*. With the exception of some material in this section, we shall always deal with abelian groups.

In the case of an abelian group, the symbol for the group operation is written $+$ and is called "addition" (even though it might not be the usual arithmetic addition). In this case, the identity element e is called "zero" and written 0, and the inverse element of a is written $-a$, so that

$$a + (-a) = (-a) + a = 0.$$

Sometimes the symbol for the group operation is written $\cdot$ and called "multiplication" (even though it might not be the usual arithmetic multiplication). In this case, the identity element e is called "one" and written 1, and the inverse element of a is written a^{-1}, so that

$$a \cdot a^{-1} = a^{-1} \cdot a = 1.$$

Theorem 2.2.2. *In every group, the identity element is unique. Also, the inverse of each group element is unique, and* $(a^{-1})^{-1} = a$.

Proof: Suppose that e and e' are identity elements. Then $e = e * e' = e'$. Next, suppose that b and b' are inverses for element a. Then

$$b = b * (a * b') = (b * a) * b' = b'.$$

Finally, $a^{-1}a = aa^{-1} = 1$, so a is an inverse for a^{-1}. But because inverses are unique, $(a^{-1})^{-1} = a$. □

There is a limitless supply of examples of groups. Many groups have an infinite number of elements. Examples are: the integers under addition; the positive rationals under multiplication;[2] and the set of two-by-two, real-valued matrices under matrix addition. Many other groups have only a finite number of elements. Examples are: The two-element set $\{0, 1\}$ under the exclusive-or operation (modulo-2 addition); the set $\{0, 1, \ldots, 8, 9\}$ under modulo-10 addition; and so forth.

For a less familiar example, we shall construct a finite nonabelian group that is not a familiar structure. One way of constructing groups with interesting algebraic structures is to study transformations of simple geometrical shapes and mimic these with the algebra. For example, an equilateral triangle with vertices A, B, and C (labeled clockwise) can be rotated or reflected into itself in exactly six different ways, and each of these has a rotation or reflection inverse. By making use of some obvious facts in this geometrical situation, we can quickly construct an algebraic group. Let the six transformations be denoted by the labels $1, a, b, c, d$, and e as follows:

$1 = (ABC \rightarrow ABC)$ (no change)
$a = (ABC \rightarrow BCA)$ (counterclockwise rotation)
$b = (ABC \rightarrow CAB)$ (clockwise rotation)

[2] In general the arithmetic operation in a group need not be commutative, and in that case is regarded as multiplication rather than addition. In a concrete case, even a commutative group operation might be called multiplication.

$$c = (ABC \to ACB) \quad \text{(reflection about bisector of angle } A)$$
$$d = (ABC \to CBA) \quad \text{(reflection about bisector of angle } B)$$
$$e = (ABC \to BAC) \quad \text{(reflection about bisector of angle } C),$$

where the transformation $(ABC \to BCA)$ means that vertex A goes into vertex B, vertex B goes into vertex C, and vertex C goes into vertex A. That is, the triangle is rotated by 120°. Let the group $(G, *)$ be defined by

$$G = \{1, a, b, c, d, e\},$$

and $y * x$ is that group element that denotes the transformation one obtains by first performing sequentially the transformation denoted by x and then the transformation denoted by y. Thus, for example,

$$\begin{aligned} a * d &= (ABC \to BCA) * (ABC \to CBA) \\ &= (ABC \to BAC) = e. \end{aligned}$$

In this way, one can construct a table for $y * x$:

$y \backslash x$	1	a	b	c	d	e
1	1	a	b	c	d	e
a	a	b	1	d	e	c
b	b	1	a	e	c	d
c	c	e	d	1	b	a
d	d	c	e	a	1	b
e	e	d	c	b	a	1

Once the table is constructed, we can discard the geometrical scaffolding. The table alone defines the group. It is a nonabelian group because $a * c \neq c * a$. Notice, as in any finite group, that every element appears once in each column and once in each row.

Our final example is the group of permutations on n letters. Let X be the set $\{1, 2, 3, \ldots, n\}$. A one-to-one map of this set onto itself is called a *permutation*. There are $n!$ such permutations. We can define a group called the *symmetric group*, denoted by the label S_n, whose elements are the permutations of X. (This may be a little confusing at first because the elements of the group are operators – the permutation operators on X. In fact, the previous example obtained from the transformations of an equilateral triangle is the permutation group S_3.) If we take a permutation of the integers and permute it, we end up with just another permutation of the integers. The group operation $*$ is taken to be this composition of permutations. For example, take $n = 4$. There are $4! = 24$ permutations in S_4. A typical element of S_4 is

$$a = [(1234) \to (3142)],$$

which is the permutation that replaces 1 by 3, 2 by 1, 3 by 4, and 4 by 2. Another such permutation is

$$b = [(1234) \to (4132)].$$

Then in S_4, the product $b * a$ is the permutation obtained by applying first a, then b. That is,

$$b * a = [(1234) \to (2341)],$$

which is an element of S_4. With this definition of multiplication, the permutation group S_4 is a nonabelian group with twenty-four elements.

Let G be a group and let H be a subset of G. Then H is called a *subgroup* of G if H is a group with respect to the restriction of $*$ to H. To prove that a nonempty set H is a subgroup of G, it is only necessary to check that $a * b$ is in H whenever a and b are in H (closure), and that the inverse of each a in H is also in H. The other properties required of a group then will be inherited from the group G. If the group is finite, then even the inverse property is satisfied automatically if the closure property is satisfied, as we shall see shortly in the discussion of cyclic subgroups.

As an example, in the set of integers (positive, negative, and zero) under addition, the set of even integers is a subgroup, as is the set of multiples of three.

One way to obtain a subgroup H of a finite group G is to take any element h from G, and let H be the set of elements obtained by multiplying h by itself an arbitrary number of times. That is, form the sequence of elements

$$h, h * h, h * h * h, h * h * h * h, \ldots,$$

denoting these elements more simply by $h, h^2, h^3, h^4, \ldots$. Because G is a finite group, only a finite number of these elements can be distinct, so the sequence must eventually repeat. The first element repeated must be h itself, because if two other elements h^i and h^j are equal, they can be multiplied by the inverse of h, and thus h^{i-1} and h^{j-1} are also equal. Next, notice that if $h^j = h$, then $h^{j-1} = 1$, the group identity element. The set H is called the subgroup generated by h. The number c of elements in H is called the *order* of the element h. The set of elements $h, h^2, h^3, \ldots, h^c = 1$ is called a *cycle*. A cycle is a subgroup because a product of two such elements is another of the same form, and the inverse of h^i is h^{c-i}, and hence is one of the elements of the cycle. A group that consists of all the powers of one of its elements is called a *cyclic group*.

Given a finite group G and a subgroup H, there is an important construction known as the *coset decomposition* of G, which illustrates certain relationships between H and G. Let the elements of H be denoted by $h_1, h_2, h_3, \ldots$, and choose h_1 to be the identity element. Construct the array as follows: The first row consists of the elements of H, with the identity element at the left and every other element of H appearing once and only once. Choose any element of G not appearing in the first row. Call it g_2 and use it as the first element of the second row. The rest of the elements of the second row are obtained by multiplying each subgroup element by this first element on the left. Continue in this way to construct a third, fourth, and subsequent rows, if possible, each time choosing a previously unused group element for the element in the first column. Stop the process when there is no previously unused group element. The process must stop because G is finite. The final array is:

$$\begin{array}{cccccc}
h_1 = 1 & h_2 & h_3 & h_4 & \dots & h_n \\
g_2 * h_1 = g_2 & g_2 * h_2 & g_2 * h_3 & g_2 * h_4 & \dots & g_2 * h_n \\
g_3 * h_1 = g_3 & g_3 * h_2 & g_3 * h_3 & g_3 * h_4 & \dots & g_3 * h_n \\
\vdots & \vdots & \vdots & \vdots & \vdots & \vdots \\
g_m * h_1 = g_m & g_m * h_2 & g_m * h_3 & g_m * h_4 & \dots & g_m * h_n.
\end{array}$$

The first element on the left of each row is known as a *coset leader*. Each row in the array is known as a *left coset*, or simply as a coset when the group is abelian. Alternatively, if the coset decomposition is defined with the elements of G multiplied on the right, the rows of the array are known as *right cosets*. The coset decomposition is always rectangular, with all rows completed, because it is constructed that way. We shall prove that we always obtain an array in which every element of G appears exactly once.

Theorem 2.2.3. *Every element of G appears once and only once in a coset decomposition of G.*

Proof: Every element appears at least once because, otherwise, the construction is not halted. We will prove that an element cannot appear twice in the same row, and then prove that an element cannot appear in two different rows.

Suppose that two entries in the same row, $g_i * h_j$ and $g_i * h_k$, are equal. Then multiplying each by g_i^{-1} gives $h_j = h_k$. This is a contradiction because each subgroup element appears only once in the first row.

Suppose that two entries in different rows, $g_i * h_j$ and $g_k * h_\ell$, are equal and that $k < i$. Multiplying on the right by h_j^{-1} gives $g_i = g_k * h_\ell * h_j^{-1}$. Then g_i is in the kth coset because $h_\ell * h_j^{-1}$ is in the subgroup. This contradicts the rule of construction that coset leaders must be previously unused. □

Corollary 2.2.4. *If H is a subgroup of G, then the number of elements in H divides the number of elements in G. That is,*

(Order of H)(Number of cosets of G with respect to H) = (Order of G).

Proof: This follows immediately from the rectangular structure of the coset decomposition. □

Theorem 2.2.5. *The order of a finite group is divisible by the order of any of its elements.*

Proof: The group contains the cyclic subgroup generated by any element, and thus Corollary 2.2.4 proves the theorem. □

2.3 Rings

The next algebraic structure we will need is that of a ring. A ring is an abstract set that is an abelian group and also has an additional structure.

Definition 2.3.1. *A ring R is a set with two operations defined: The first is called addition (denoted by $+$); the second is called* multiplication *(denoted by juxtaposition); and the following axioms are satisfied.*

1. *R is an abelian group under addition ($+$).*
2. *Closure: For any a, b in R, the product ab is in R.*
3. *Associativity:*

$$a(bc) = (ab)c.$$

4. *Distributivity:*

$$a(b + c) = ab + ac,$$
$$(b + c)a = ba + ca.$$

The addition operation is always commutative in a ring, but the multiplication operation need not be commutative. A *commutative ring* is one in which multiplication is commutative, that is, $ab = ba$ for all a, b in R.

The distributivity axiom in the definition of a ring links the addition and multiplication operations. This axiom has several immediate consequences, as follows.

Theorem 2.3.2. *For any elements a, b in a ring R,*

(i) $a0 = 0a = 0$,
(ii) $a(-b) = (-a)b = -(ab)$.

Proof:

(i) $a0 = a(0 + 0) = a0 + a0$.
Hence subtracting $a0$ from both sides gives $0 = a0$. The second half of (i) is proved in the same way.

(ii) $0 = a0 = a(b - b) = ab + a(-b)$.
Hence

$$a(-b) = -(ab).$$

The second half of (ii) is proved in the same way. □

The addition operation in a ring has an identity element called "zero" and written 0. The multiplication operation need not have an identity element, but if there is an identity element, it is unique. A ring that has an identity element under multiplication is called a *ring with identity*. The identity element is called "one" and written 1. Then

$$1a = a1 = a$$

for all a in R.

Every element in a ring has an inverse under the addition operation. Under the multiplication operation an inverse is defined only in a ring with identity. In such a ring inverses may exist but need not. That is, given an element a, there may exist an element b with $ab = 1$. If so, b is called a *right inverse* for a. Similarly, if there is an element c such that $ca = 1$, then c is called a *left inverse* for a.

Theorem 2.3.3. *In a ring with identity,*

(i) The identity is unique.
(ii) If an element a has both a right inverse b and a left inverse c, then $b = c$. In this case, the element a is said to have an inverse (denoted by a^{-1}). The inverse is unique.
(iii) $(a^{-1})^{-1} = a$.

Proof: The argument is similar to that used in Theorem 2.2.2. □

An element that has an inverse under multiplication is called a *unit*. The set of all units is closed under multiplication because if a and b are units, then $c = ab$ has inverse $c^{-1} = b^{-1}a^{-1}$.

Theorem 2.3.4.

(i) Under ring multiplication, the set of units of a ring forms a group.
(ii) If $c = ab$ and c is a unit, then a has a right inverse and b has a left inverse.

Proof: Straightforward. □

There are many familiar examples of rings, some of which follow. It is instructive to review Theorems 2.3.3 and 2.3.4 in terms of these examples.

1. The set of all real numbers under the usual addition and multiplication is a commutative ring with identity. Every nonzero element is a unit.
2. The set of all integers (positive, negative, and zero) under the usual addition and multiplication is a commutative ring with identity. This ring is conventionally denoted by the label $\mathbf{Z}$. The only units in this ring are ± 1.
3. The set of all n by n matrices with real-valued elements under matrix addition and matrix multiplication is a noncommutative ring with identity. The identity element is the n by n identity matrix. The units are the nonsingular matrices.

4. The set of all n by n matrices with integer-valued elements under matrix addition and matrix multiplication is a noncommutative ring with identity. The units are those matrices with determinant ± 1.
5. The set of all polynomials in x with real-valued coefficients under polynomial addition and polynomial multiplication is a commutative ring with identity. The identity element is the zero-degree polynomial $p(x) = 1$. The units are the real-valued polynomials of degree zero.

2.4 Fields

Loosely speaking, an abelian group is a set in which one can add and subtract, and a ring is a set in which one can add, subtract, and multiply. A more powerful algebraic structure, known as a field, is a set in which one can add, subtract, multiply, and divide.

Definition 2.4.1. *A field F is a set that has two operations defined on it: Addition and multiplication, such that the following axioms are satisfied.*

1. *The set is an abelian group under addition.*
2. *The set is closed under multiplication, and the set of nonzero elements is an abelian group under multiplication.*
3. *The distributive law*

   $$(a+b)c = ac + bc$$

   holds for all a, b, c in the set.

It is conventional to denote by 0 the identity element under addition and to call it "zero;" to denote by $-a$ the additive inverse of a; to denote by 1 the identity element under multiplication and to call it "one;" and to denote by a^{-1} the multiplicative inverse of a. By subtraction $a - b$, we mean $a + (-b)$; by division a/b, we mean $b^{-1}a$.

The following examples of fields are well known.

1. $\boldsymbol{R}$, the set of real numbers.
2. $\boldsymbol{C}$, the set of complex numbers.
3. $\boldsymbol{Q}$, the set of rational numbers.

All of these fields have an infinite number of elements. We are interested in fields with a finite number of elements. A field with q elements, if it exists, is called a *finite field*, or a *Galois field*, and is denoted by the label $GF(q)$.

What is the smallest field? It must have an element zero and an element one. In fact, these suffice with the addition and multiplication tables

+	0	1
0	0	1
1	1	0

·	0	1
0	0	0
1	0	1

This is the field $GF(2)$, which we have already seen in Section 2.1. No other field with two elements exists.

In Chapter 4, we shall study finite fields in great detail. For now, we will be content with two additional simple examples. These are described by their addition and multiplication tables. Subtraction and division are defined implicitly by the addition and multiplication tables.

The field $GF(3)$ is the set $\{0, 1, 2\}$ together with the operations:

+	0	1	2
0	0	1	2
1	1	2	0
2	2	0	1

·	0	1	2
0	0	0	0
1	0	1	2
2	0	2	1

The field $GF(4)$ is the set $\{0, 1, 2, 3\}$ together with the operations:

+	0	1	2	3
0	0	1	2	3
1	1	0	3	2
2	2	3	0	1
3	3	2	1	0

·	0	1	2	3
0	0	0	0	0
1	0	1	2	3
2	0	2	3	1
3	0	3	1	2

Notice in $GF(4)$ that multiplication is *not* modulo-4 multiplication, and addition is *not* modulo-4 addition. Notice also that $GF(2)$ is contained in $GF(4)$ because in $GF(4)$ the two elements zero and one add and multiply just as they do in $GF(2)$. However, $GF(2)$ is not contained in $GF(3)$.

Even though these examples are very small fields, it is not easy by inspection to see what is promised by the structure. Many other Galois fields exist. An understanding of the structure of these and larger fields will be developed in Chapter 4.

Definition 2.4.2. *Let F be a field. A subset of F is called a subfield if it is a field under the inherited addition and multiplication. The original field F is then called an extension field of the subfield.*

To prove that a subset of a finite field is a subfield, it is only necessary to prove that it contains a nonzero element and that it is closed under addition and multiplication. All other necessary properties are inherited from F. Inverses under addition or multiplication of an element β are contained in the cyclic group generated by β under the operation of addition or multiplication.

A field has all the properties of a ring. It also has an additional important property – it is always possible to cancel.

Theorem 2.4.3. *In any field, if $ab = ac$ and $a \neq 0$, then $b = c$.*

Proof: Multiply by a^{-1}. □

Some rings may satisfy this cancellation law and yet not be fields. The ring of integers is a simple example. Cancellation is possible in the ring of integers, but cannot be proved as in Theorem 2.4.3 because a^{-1} does not exist in this ring. There is a special name for rings in which cancellation is always possible – a domain.

Definition 2.4.4. A domain *(or integral domain) is a commutative ring in which $b = c$ whenever $ab = ac$ and a is nonzero.*

2.5 Vector spaces

A familiar example of a vector space is the three-dimensional euclidean space that arises in problems of physics. This can be extended mathematically to an n-dimensional vector space over the real numbers. The concept of an n-dimensional vector space is closely related to the ideas of linear algebra and matrix theory, and is important in many applications. Vector spaces also can be defined abstractly with respect to any field.

Definition 2.5.1. *Let F be a field. The elements of F will be called scalars. A set V is called a vector space, and its elements are called vectors, if there is defined an operation called vector addition (denoted by $+$) on pairs of elements from V, and an operation called scalar multiplication (denoted by juxtaposition) on an element of F and an element of V to produce an element of V provided the following axioms are satisfied.*

1. *V is an abelian group under vector addition.*
2. *Distributivity: For any vectors $\boldsymbol{v}_1$, $\boldsymbol{v}_2$ and any scalar c,*

$$c(\boldsymbol{v}_1 + \boldsymbol{v}_2) = c\boldsymbol{v}_1 + c\boldsymbol{v}_2.$$

3. *Distributivity: For any vector $\boldsymbol{v}$, $1\boldsymbol{v} = \boldsymbol{v}$ and for any scalars c_1, c_2,*

$$(c_1 + c_2)\boldsymbol{v} = c_1\boldsymbol{v} + c_2\boldsymbol{v}.$$

4. *Associativity: For any vector $\boldsymbol{v}$ and any scalars c_1, c_2,*

$$(c_1c_2)\boldsymbol{v} = c_1(c_2\boldsymbol{v}).$$

The zero element of V is called the origin of V and is denoted by $\mathbf{0}$. *Note that, for all* v, $0v = \mathbf{0}$.

Notice that we have two different uses for the symbol $+$: Vector addition and addition within the field. Also notice that the symbol $\mathbf{0}$ is used for the origin of the vector space, and the symbol 0 is used for the zero of the field. In practice, these ambiguities cause no confusion.

Given a field F, the quantity $(a_1, a_2, \ldots, a_n)$, composed of field elements, is called an n-tuple of elements from the field F. Under the operations of componentwise addition and componentwise scalar multiplication, the set of n-tuples of elements from a field F is a vector space and is denoted by the label F^n. A familiar example of a vector space is the space of n-tuples over the real numbers. This vector space is denoted $\boldsymbol{R}^n$. Another familiar example is the vector space of n-tuples over the complex numbers, denoted $\boldsymbol{C}^n$.

As a less familiar example of a vector space, take V to be the set of polynomials in x with coefficients in $GF(q)$, and take $F = GF(q)$. In this space, the vectors are polynomials. Vector addition is polynomial addition, and scalar multiplication is multiplication of a polynomial by a field element.

In a vector space V, a sum of the form

$$\boldsymbol{u} = a_1\boldsymbol{v}_1 + a_2\boldsymbol{v}_2 + \cdots + a_k\boldsymbol{v}_k,$$

where the a_i are scalars, is called a *linear combination* of the vectors $\boldsymbol{v}_1, \ldots, \boldsymbol{v}_k$. A set of vectors $\{\boldsymbol{v}_1, \ldots, \boldsymbol{v}_k\}$ is called *linearly dependent* if there is a set of scalars $\{a_1, \ldots, a_k\}$, not all zero, such that

$$a_1\boldsymbol{v}_1 + a_2\boldsymbol{v}_2 + \cdots + a_k\boldsymbol{v}_k = 0.$$

A set of vectors that is not linearly dependent is called *linearly independent*. No vector in a linearly independent set can be expressed as a linear combination of the other vectors. Note that the all-zero vector $\mathbf{0}$ cannot belong to a linearly independent set; every set containing $\mathbf{0}$ is linearly dependent.

A set of vectors is said to *span* a vector space if every vector in the space equals at least one linear combination of the vectors in the set. A vector space that is spanned by a finite set of vectors is called a *finite-dimensional vector space*. We are interested primarily in finite-dimensional vector spaces.

Theorem 2.5.2. *If a vector space V is spanned by a finite set of k vectors* $\mathcal{A} = \{\boldsymbol{v}_1, \ldots, \boldsymbol{v}_k\}$, *and V contains a set of m linearly independent vectors* $\mathcal{B} = \{\boldsymbol{u}_1, \ldots, \boldsymbol{u}_m\}$, *then* $k \geq m$.

Proof: We shall describe the construction of a sequence of sets, $\mathcal{A}_0, \mathcal{A}_1, \mathcal{A}_2, \ldots, \mathcal{A}_m$, in such a way that each set spans V; each set has k elements chosen from $\mathcal{A}$ and $\mathcal{B}$; and the set $\mathcal{A}_r$ contains $u_1, \ldots, u_r$. Consequently, $\mathcal{A}_m$ contains $\boldsymbol{u}_1, \ldots, \boldsymbol{u}_m$ among its k elements, and thus $k \geq m$.

Because no nonzero linear combination of vectors of $\mathcal{B}$ is equal to zero, no element of $\mathcal{B}$ can be expressed as a linear combination of the other elements of $\mathcal{B}$. If the set $\mathcal{A}_{r-1}$ does not contain vector $\boldsymbol{u}_r$ and $\mathcal{A}_{r-1}$ spans V, then it must be possible to express $\boldsymbol{u}_r$ as a linear combination of elements of $\mathcal{A}_{r-1}$, including at least one vector of $\mathcal{A}_{r-1}$ (say $\boldsymbol{v}_j$) that is not also in $\mathcal{B}$. The equation describing the linear combination can be solved to express vector $\boldsymbol{v}_j$ as a linear combination of $\boldsymbol{u}_r$ and the other elements of $\mathcal{A}_{r-1}$. This is the key to the construction.

The construction is as follows. Let $\mathcal{A}_0 = \mathcal{A}$. If $\mathcal{A}_{r-1}$ contains $\boldsymbol{u}_r$, then let $\mathcal{A}_r = \mathcal{A}_{r-1}$. Otherwise, $\boldsymbol{u}_r$ does not appear in $\mathcal{A}_{r-1}$, but it can be expressed as a linear combination of the elements of $\mathcal{A}_{r-1}$, involving some element $\boldsymbol{v}_j$ of $\mathcal{A}$ not in $\mathcal{B}$. Form $\mathcal{A}_r$ from $\mathcal{A}_{r-1}$ by replacing $\boldsymbol{v}_j$ with $\boldsymbol{u}_r$.

Any vector $\boldsymbol{v}$ is a linear combination of the elements of $\mathcal{A}_{r-1}$ and so, too, of $\mathcal{A}_r$ because $\boldsymbol{v}_j$ can be eliminated by using the linear equation that relates $\boldsymbol{v}_j$ to $\boldsymbol{u}_r$ and the other elements of $\mathcal{A}_{r-1}$. Therefore the set $\mathcal{A}_r$ spans V. From $\mathcal{A}_{r-1}$ we have constructed $\mathcal{A}_r$ with the desired properties, so we can repeat the process to eventually construct $\mathcal{A}_m$, and the proof is complete. □

Theorem 2.5.3. *Two linearly independent sets of vectors that span the same finite-dimensional vector space have the same number of vectors.*

Proof: If one set has m vectors and the other set has k vectors, then by Theorem 2.5.2, $m \geq k$ and $k \geq m$, and thus $m = k$. □

The number of linearly independent vectors in a set that spans a finite-dimensional vector space V is called the *dimension* of V. A set of k linearly independent vectors that spans a k-dimensional vector space is called a *basis* of the space. From Theorem 2.5.2, every set of more than k vectors in a k-dimensional vector space is linearly dependent.

Theorem 2.5.4. *In a k-dimensional vector space V, any set of k linearly independent vectors is a basis for V.*

Proof: Let $\{\boldsymbol{v}_1, \boldsymbol{v}_2, \ldots, \boldsymbol{v}_k\}$ be any set of k linearly independent vectors in V. If the set does not span V, then one can find a vector $\boldsymbol{v}$ in V that is not a linear combination of $\{\boldsymbol{v}_1, \boldsymbol{v}_2, \ldots, \boldsymbol{v}_k\}$. The set $\{\boldsymbol{v}, \boldsymbol{v}_1, \boldsymbol{v}_2, \ldots, \boldsymbol{v}_k\}$ is linearly independent and contains $k + 1$ vectors in V, which contradicts Theorem 2.5.3. Therefore $\{\boldsymbol{v}_1, \boldsymbol{v}_2, \ldots, \boldsymbol{v}_k\}$ spans V and is a basis. □

If a linearly independent set of vectors in a k-dimensional vector space is not a basis, then it must have fewer than k vectors. Adjoining vectors to such a set in order to make it into a basis is called a *completion* of the basis.

Theorem 2.5.5. *Given a set of linearly independent vectors in a finite-dimensional vector space, it is always possible to complete the set to form a basis.*

Proof: If the set is not a basis, then some vector in the space is not a linear combination of vectors in the set. Choose any such vector and append it to the set, making the size of the set larger by one. If it is still not a basis, repeat the process. The process must stop eventually because the number of linearly independent vectors in a set is not larger than the dimension of the space. The final set of vectors satisfies the requirements of the theorem. □

A nonempty subset of a vector space is called a *vector subspace* if it is also a vector space under the original vector addition and scalar multiplication. Under the operation of vector addition, a vector space is a group, and a vector subspace is a subgroup. In order to check whether a nonempty subset of a vector space is a subspace, it is only necessary to check for closure under vector addition and under scalar multiplication. Closure under scalar multiplication ensures that the zero vector is in the subset. All other required properties are always inherited from the original space.

Theorem 2.5.6. *In any vector space V, the set of all linear combinations of a nonempty set of vectors $\{\boldsymbol{v}_1, \ldots, \boldsymbol{v}_k\}$ is a subspace of V.*

Proof: Every linear combination of $\boldsymbol{v}_1, \ldots, \boldsymbol{v}_k$ is a vector in V, and thus W, the set of all linear combinations, is a subset. Because $\mathbf{0}$ is in W, the subset is not empty. We must show that W is a subspace. If $\boldsymbol{w} = b_1\boldsymbol{v}_1 + \cdots + b_k\boldsymbol{v}_k$ and $\boldsymbol{u} = c_1\boldsymbol{v}_1 + \cdots + c_k\boldsymbol{v}_k$ are any two elements of W, then $\boldsymbol{w} + \boldsymbol{u} = (b_1 + c_1)\boldsymbol{v}_1 + \cdots + (b_k + c_k)\boldsymbol{v}_k$ is also in W. Next, for any $\boldsymbol{w}$, any scalar multiple of $\boldsymbol{w}$, $a\boldsymbol{w} = ab_1\boldsymbol{v}_1 + \cdots + ab_k\boldsymbol{v}_k$, is in W. Because W is closed under vector addition and scalar multiplication, it is a vector subspace of V. □

Theorem 2.5.7. *If W, a vector subspace of a finite-dimensional vector space V, has the same dimension as V, then $W = V$.*

Proof: Let k be the dimension of the two spaces. Choose a basis for W. This is a set of k linearly independent vectors in V, and thus it is a basis for V. Therefore every vector in V is also in W. □

Any finite-dimensional vector space can be represented as an n-tuple space by choosing a basis $\{\boldsymbol{v}_1, \ldots, \boldsymbol{v}_n\}$, and representing a vector $\boldsymbol{v} = a_1\boldsymbol{v}_1 + \cdots + a_n\boldsymbol{v}_n$ by the n-tuple of coefficients $(a_1, \ldots, a_n)$. Hence we need consider only vector spaces of n-tuples.

The *inner product* of two n-tuples of F^n

$$\boldsymbol{u} = (a_1, \ldots, a_n)$$
$$\boldsymbol{v} = (b_1, \ldots, b_n)$$

is defined as the scalar

$$\boldsymbol{u} \cdot \boldsymbol{v} = (a_1, \ldots, a_n) \cdot (b_1, \ldots, b_n) = a_1b_1 + \cdots + a_nb_n.$$

We can verify immediately that $\boldsymbol{u} \cdot \boldsymbol{v} = \boldsymbol{v} \cdot \boldsymbol{u}$, that $(c\boldsymbol{u}) \cdot \boldsymbol{v} = c(\boldsymbol{u} \cdot \boldsymbol{v})$, and also that $\boldsymbol{w} \cdot (\boldsymbol{u} + \boldsymbol{v}) = (\boldsymbol{w} \cdot \boldsymbol{u}) + (\boldsymbol{w} \cdot \boldsymbol{v})$. If the inner product of two vectors is zero, they are said to be *orthogonal*. It should be noted that it is possible for a nonzero vector over $GF(q)$ to be orthogonal to itself. For this reason many geometric properties of $\boldsymbol{R}^n$ or $\boldsymbol{C}^n$ do not carry over to $GF(q)^n$. A vector orthogonal to every vector in a set is said to be orthogonal to the set.

Theorem 2.5.8. *Let V be the vector space of n-tuples over a field F, and let W be a subspace. The set of vectors orthogonal to W is itself a subspace.*

Proof: Let U be the set of all vectors orthogonal to W. Because $\mathbf{0}$ is in U, U is not empty. Let $\boldsymbol{w}$ be any vector in W, and let $\boldsymbol{u}_1$ and $\boldsymbol{u}_2$ be any vectors in U. Then $\boldsymbol{w} \cdot \boldsymbol{u}_1 = \boldsymbol{w} \cdot \boldsymbol{u}_2 = 0$, and $\boldsymbol{w} \cdot \boldsymbol{u}_1 + \boldsymbol{w} \cdot \boldsymbol{u}_2 = 0 = \boldsymbol{w} \cdot (\boldsymbol{u}_1 + \boldsymbol{u}_2)$; thus $\boldsymbol{u}_1 + \boldsymbol{u}_2$ is in U. Also, $\boldsymbol{w} \cdot (c\boldsymbol{u}_1) = c(\boldsymbol{w} \cdot \boldsymbol{u}_1) = 0$, and thus $c\boldsymbol{u}_1$ is in U. Therefore U is a subspace. □

The set of vectors orthogonal to W is called the *orthogonal complement* (or the *dual space*) of W and is denoted by $W^{\perp}$. In a finite-dimensional vector space over the real numbers, the intersection of W and $W^{\perp}$ contains only the all-zero vector, but in a vector space over $GF(q)$, $W^{\perp}$ may have a nontrivial intersection with W, or may even lie within W or contain W. In fact, one can construct examples of subspaces that are their own orthogonal complements. For example, in $GF(2)^2$, the subspace {00, 11} is its own orthogonal complement.

Theorem 2.5.9. *A vector that is orthogonal to every vector of a set that spans W is in the orthogonal complement of W.*

Proof: Suppose that set $\{\boldsymbol{w}_1, \ldots, \boldsymbol{w}_n\}$ spans W. A vector $\boldsymbol{w}$ in W can be written in the form $\boldsymbol{w} = c_1\boldsymbol{w}_1 + \cdots + c_n\boldsymbol{w}_n$. Then

$$\boldsymbol{w} \cdot \boldsymbol{u} = (c_1\boldsymbol{w}_1 + \cdots + c_n\boldsymbol{w}_n) \cdot \boldsymbol{u} = c_1\boldsymbol{w}_1 \cdot \boldsymbol{u} + \cdots + c_n\boldsymbol{w}_n \cdot \boldsymbol{u}.$$

If $\boldsymbol{u}$ is orthogonal to each of the $\boldsymbol{w}_i$, it is orthogonal to every $\boldsymbol{w}$ in W. □

If a vector space of n-tuples has a subspace W of dimension k, then the orthogonal complement $W^{\perp}$ has dimension $n - k$. This fact, which will be used frequently in later chapters, will be proved at the end of the next section. We refer to this fact now in proving the following theorem.

Theorem 2.5.10. *Let W be a subspace of the space of n-tuples, and let $W^{\perp}$ be the orthogonal complement of W. Then $(W^{\perp})^{\perp} = W$.*

Proof: Every vector of W is orthogonal to $W^{\perp}$, so $W \subset (W^{\perp})^{\perp}$. Let k be the dimension of W, then by Theorem 2.6.10 (in the next section), $W^{\perp}$ has dimension $n - k$ and $(W^{\perp})^{\perp}$ has dimension k. Therefore W is contained in the orthogonal complement of $W^{\perp}$ and has the same dimension. Hence they are equal. □

2.6 Linear algebra

The topic of linear algebra, particularly matrix theory, is a much-used topic in applied mathematics that is commonly studied only for the field of real numbers and the field of complex numbers. Most of the familiar operations of linear algebra are also valid in an arbitrary field. We shall outline the development of this subject partly for review and partly to prove that the techniques remain valid over an arbitrary field (sometimes even over an arbitrary ring).

Definition 2.6.1. *An n by m matrix $\boldsymbol{A}$ over a ring R consists of nm elements from R arranged in a rectangular array of n rows and m columns*

$$\boldsymbol{A} = \begin{bmatrix} a_{11} & a_{12} & \dots & a_{1m} \\ a_{21} & a_{22} & \dots & a_{2m} \\ \vdots & \vdots & & \vdots \\ a_{n1} & a_{n2} & \dots & a_{nm} \end{bmatrix} = [a_{ij}].$$

In most applications, the ring R is actually a field, and we shall usually restrict our attention to this case. We are mostly concerned with matrices over a finite field $GF(q)$.

The set of elements a_{ii}, for which the column number and row number are equal, is called the *main diagonal*. If all elements not on the main diagonal are zero, the matrix is called a *diagonal matrix*. If n equals m, the matrix is called a *square matrix* of *size n*. An n by n matrix with the field element one in every entry of the main diagonal, and the field element zero in every other matrix entry, is called an *n by n identity matrix*. An identity matrix is denoted by $\boldsymbol{I}$. An example of an identity matrix is

$$\boldsymbol{I} = \begin{bmatrix} 1 & 0 & 0 \\ 0 & 1 & 0 \\ 0 & 0 & 1 \end{bmatrix}.$$

Two n by m matrices, $\boldsymbol{A}$ and $\boldsymbol{B}$, over $GF(q)$ can be added by the rule

$$\boldsymbol{A} + \boldsymbol{B} = \begin{bmatrix} a_{11} + b_{11} & a_{12} + b_{12} & \dots & a_{1m} + b_{1m} \\ \vdots & & & \vdots \\ a_{n1} + b_{n1} & a_{n2} + b_{n2} & \dots & a_{nm} + b_{nm} \end{bmatrix}.$$

An n by m matrix $\boldsymbol{A}$ can be multiplied by a field element β by the rule

$$\beta\boldsymbol{A} = \begin{bmatrix} \beta a_{11} & \beta a_{12} & \dots & \beta a_{1m} \\ \vdots & & & \vdots \\ \beta a_{n1} & \beta a_{n2} & \dots & \beta a_{nm} \end{bmatrix}.$$

An ℓ by n matrix $\boldsymbol{A}$ and an n by m matrix $\boldsymbol{B}$ can be multiplied to produce an ℓ by m matrix $\boldsymbol{C}$ by using the following rule:

$$c_{ij} = \sum_{k=1}^{n} a_{ik} b_{kj} \qquad \begin{array}{l} i = 1, \ldots, \ell \\ j = 1, \ldots, m. \end{array}$$

This matrix product is denoted as

$$\boldsymbol{C} = \boldsymbol{A}\boldsymbol{B}.$$

With this definition of matrix multiplication, and the earlier definition of matrix addition, the set of n by n square matrices over any field F forms a ring, as can be easily verified. It is a noncommutative ring, but it does have an identity, namely the n by n identity matrix.

A matrix can be partitioned as follows:

$$\boldsymbol{A} = \left[\begin{array}{c|c} \boldsymbol{A}_{11} & \boldsymbol{A}_{12} \\ \hline \boldsymbol{A}_{21} & \boldsymbol{A}_{22} \end{array}\right]$$

where $\boldsymbol{A}_{11}$, $\boldsymbol{A}_{12}$, $\boldsymbol{A}_{21}$, and $\boldsymbol{A}_{22}$ are smaller matrices whose dimensions add up to the dimensions of $\boldsymbol{A}$ in the obvious way. That is, the number of rows of $\boldsymbol{A}_{11}$ (or $\boldsymbol{A}_{12}$), plus the number of rows of $\boldsymbol{A}_{21}$ (or $\boldsymbol{A}_{22}$), equals the number of rows of $\boldsymbol{A}$, and a similar statement holds for columns.

Matrices may be multiplied in blocks. That is, if

$$\boldsymbol{A} = \left[\begin{array}{c|c} \boldsymbol{A}_{11} & \boldsymbol{A}_{12} \\ \hline \boldsymbol{A}_{21} & \boldsymbol{A}_{22} \end{array}\right] \quad \text{and} \quad \boldsymbol{B} = \left[\begin{array}{c|c} \boldsymbol{B}_{11} & \boldsymbol{B}_{12} \\ \hline \boldsymbol{B}_{21} & \boldsymbol{B}_{22} \end{array}\right]$$

and $\boldsymbol{C} = \boldsymbol{A}\boldsymbol{B}$, then

$$\boldsymbol{C} = \left[\begin{array}{c|c} \boldsymbol{A}_{11}\boldsymbol{B}_{11} + \boldsymbol{A}_{12}\boldsymbol{B}_{21} & \boldsymbol{A}_{11}\boldsymbol{B}_{12} + \boldsymbol{A}_{12}\boldsymbol{B}_{22} \\ \hline \boldsymbol{A}_{21}\boldsymbol{B}_{11} + \boldsymbol{A}_{22}\boldsymbol{B}_{21} & \boldsymbol{A}_{21}\boldsymbol{B}_{12} + \boldsymbol{A}_{22}\boldsymbol{B}_{22} \end{array}\right],$$

provided the block sizes are compatible in the sense that all matrix products and additions are defined. This decomposition can be readily verified as a simple consequence of the associativity and distributivity axioms of the underlying field.

The *transpose* of an n by m matrix $\boldsymbol{A}$ is an m by n matrix, denoted $\boldsymbol{A}^{\mathrm{T}}$, such that $a^{\mathrm{T}}_{ij} = a_{ji}$. That is, the rows of $\boldsymbol{A}^{\mathrm{T}}$ are the columns of $\boldsymbol{A}$, and the columns of $\boldsymbol{A}^{\mathrm{T}}$ are the rows of $\boldsymbol{A}$. The *inverse* of the square matrix $\boldsymbol{A}$ is the square matrix $\boldsymbol{A}^{-1}$, if it exists, such that $\boldsymbol{A}^{-1}\boldsymbol{A} = \boldsymbol{A}\boldsymbol{A}^{-1} = \boldsymbol{I}$. The set of all square n by n matrices for which an inverse exists is a group under matrix multiplication, as can be easily checked. Therefore whenever a matrix has an inverse, it is unique because we have seen in Theorem 2.2.2 that this property holds in any group. A matrix that has an inverse is called *nonsingular*; otherwise, it is called *singular*. If $\boldsymbol{C} = \boldsymbol{A}\boldsymbol{B}$, then $\boldsymbol{C}^{-1} = \boldsymbol{B}^{-1}\boldsymbol{A}^{-1}$, provided the inverses of $\boldsymbol{A}$ and $\boldsymbol{B}$ exist because $(\boldsymbol{B}^{-1}\boldsymbol{A}^{-1})\boldsymbol{C} = \boldsymbol{I} = \boldsymbol{C}(\boldsymbol{B}^{-1}\boldsymbol{A}^{-1})$. We shall see later that if the inverse of either $\boldsymbol{A}$ or $\boldsymbol{B}$ does not exist, then neither does the inverse of $\boldsymbol{C}$.

Definition 2.6.2. *Let the field F be given. For each n, the determinant of a square n by n matrix $\mathbf{A}$ is the value assumed by* $\det(\mathbf{A})$, *a function from the set of n by n matrices over F into the field F. The function* $\det(\mathbf{A})$ *is given by*

$$\det(\mathbf{A}) = \sum \xi_{i_1 \ldots i_n} a_{1i_1} a_{2i_2} a_{3i_3} \ldots a_{ni_n}$$

where $i_1, i_2, \ldots, i_n$ is a permutation of the integers $1, 2, \ldots, n$; the sum is over all possible permutations; and $\xi_{i_1 \ldots i_n}$ is ± 1 according to whether the permutation is an even or odd permutation.

An odd permutation is one that can be obtained as a product of an odd number of pairwise transpositions. An even permutation is one that cannot be obtained as a product of an odd number of pairwise transpositions. A transposition is an interchange of two terms.

One way of visualizing the definition is to take the set of all matrices that can be obtained by permuting the rows of $\mathbf{A}$. Then for each such matrix, take the product of terms down the main diagonal. Reverse the sign if the permutation is odd, and add all such product terms together. Because this is computationally clumsy, one should not actually compute the determinant this way, but it is a good way to establish properties.

The following theorem contains some properties of the determinant that follow easily from the definition.

Theorem 2.6.3.

(i) *If all elements of any row of a square matrix are zero, the determinant of the matrix is zero.*

(ii) *The determinant of a matrix equals the determinant of its transpose.*

(iii) *If two rows of a square matrix are interchanged, the determinant is replaced by its negative.*

(iv) *If two rows of a square matrix are equal, then the determinant is zero.*

(v) *If all elements of one row of a square matrix are multiplied by a field element c, the value of the determinant is multiplied by c.*

(vi) *If two square matrices, $\mathbf{A}$ and $\mathbf{B}$, differ only in row i, the sum of their determinants equals the determinant of a matrix $\mathbf{C}$ whose ith row is the sum of the ith rows of $\mathbf{A}$ and $\mathbf{B}$, and whose other rows equal the corresponding rows of $\mathbf{A}$ or $\mathbf{B}$.*

(vii) *If k times the elements of any row of a square matrix are added to the corresponding elements of any other row, the determinant is unchanged. Proof: Combine properties (iv), (v), and (vi).*

(viii) *The determinant of a square matrix is nonzero if and only if its rows (or columns) are linearly independent.*

Proof: Exercise. (**Note:** In general, line (iv) cannot be proved by interchanging the two equal rows and using line (iii). Why?) □

If the row and column containing an element a_{ij} in a square matrix are deleted, then the determinant of the remaining $(n-1)$ by $(n-1)$ square array is called the *minor* of a_{ij} and is denoted M_{ij}. The *cofactor* of a_{ij}, denoted here by C_{ij}, is defined by

$$C_{ij} = (-1)^{i+j} M_{ij}.$$

We see from the definition of the determinant that the cofactor of a_{ij} is the coefficient of a_{ij} in the expansion of $\det(\boldsymbol{A})$. That is, for any i,

$$\det(\boldsymbol{A}) = \sum_{k=1}^{n} a_{ik} C_{ik}.$$

This is known as the *Laplace expansion formula* for determinants. It gives the determinant of an n by n matrix in terms of n determinants of $(n-1)$ by $(n-1)$ matrices. The Laplace expansion formula is used as a recursive method of computing the determinant.

If a_{ik} is replaced by a_{jk}, then $\sum_{k=1}^{n} a_{jk} C_{ik}$ is the determinant of a new matrix in which the elements of the ith row are replaced by the elements of the jth row, and hence it is zero if $j \neq i$. Thus

$$\sum_{k=1}^{n} a_{jk} C_{ik} = \begin{cases} \det(\boldsymbol{A}) & i = j \\ 0 & i \neq j. \end{cases}$$

Therefore the matrix $\boldsymbol{A} = [a_{ij}]$ has the *matrix inverse*

$$\boldsymbol{A}^{-1} = \left[\frac{C_{ij}}{\det(\boldsymbol{A})} \right],$$

provided that $\det(\boldsymbol{A}) \neq 0$. When $\det(\boldsymbol{A}) = 0$, an inverse does not exist.[3]

The rows of an n by m matrix $\boldsymbol{A}$ over $GF(q)$ may be thought of as a set of vectors of length m. The *row space* of $\boldsymbol{A}$ is the subspace of $GF(q)^m$ consisting of all linear combinations of the row vectors of $\boldsymbol{A}$. The dimension of the row space is called the *row rank*. Similarly, the columns of $\boldsymbol{A}$ may be thought of as a set of vectors of length n. The *column space* of $\boldsymbol{A}$ is the subspace of $GF(q)^n$ consisting of all linear combinations of column vectors of $\boldsymbol{A}$. The dimension of the column space is called the *column rank*.

The set of vectors $\boldsymbol{v}$ such that $\boldsymbol{A}\boldsymbol{v}^{\mathrm{T}} = \boldsymbol{0}$ is called the *null space* of the matrix $\boldsymbol{A}$. It is clear that the null space is a vector subspace of $GF(q)^m$. In particular, the null space of matrix $\boldsymbol{A}$ is the orthogonal complement of the row space of $\boldsymbol{A}$ because the null space can be described as the set of all vectors that are orthogonal to all vectors of the row space.

The following are the *elementary row operations* on a matrix.

1. Interchange of any two rows.
2. Multiplication of any row by a nonzero field element.
3. Replacement of any row by the sum of that row and a multiple of any other row.

[3] If a matrix is defined over a commutative ring with identity, the matrix inverse exists if and only if $\det(\boldsymbol{A})$ is a unit of the ring.

Each elementary row operation is inverted by an elementary row operation of the same kind. Each elementary row operation on an n by m matrix $\boldsymbol{A}$ can be effected by multiplying the matrix $\boldsymbol{A}$ from the left by an appropriate n by n matrix $\boldsymbol{F}$, called an *elementary matrix*. (Similarly, *elementary column operations* can be effected by multiplying the matrix $\boldsymbol{A}$ from the right by an elementary matrix.) Each elementary matrix has one of the following three forms, one form corresponding to each elementary row operation.

$$\begin{bmatrix} 1 & & & & & \\ & \ddots & & & & \\ & & 0 & \dots & 1 & & \\ & & \vdots & \ddots & \vdots & & \\ & & 1 & \dots & 0 & & \\ & & & & & \ddots & \\ & & & & & & 1 \end{bmatrix}, \quad \begin{bmatrix} 1 & & & & \\ & \ddots & & & \\ & & a & & \\ & & & \ddots & \\ & & & & 1 \end{bmatrix}, \quad \text{or} \quad \begin{bmatrix} 1 & & & & & \\ & \ddots & & & & \\ & & 1 & & & \\ & & \vdots & \ddots & & \\ & & a & \dots & 1 & \\ & & & & \ddots & \\ & & & & & 1 \end{bmatrix}$$

Each of these matrices has all remaining diagonal elements equal to one and all remaining nondiagonal elements equal to zero. Each matrix has an inverse matrix of a similar form.

Elementary row operations are used to put a matrix in a standard form, called the *row–echelon form*, which is defined as follows.

1. The leading nonzero term of every nonzero row is a one.
2. Every column containing such a leading term has a zero for each of its other entries.
3. The leading term of any row is to the right of the leading term in every higher row.
4. All zero rows are below all nonzero rows.

An example of a matrix in row–echelon form is the matrix

$$\boldsymbol{A} = \begin{bmatrix} 1 & 1 & 0 & 1 & 3 & 0 \\ 0 & 0 & 1 & 1 & 0 & 0 \\ 0 & 0 & 0 & 0 & 0 & 1 \\ 0 & 0 & 0 & 0 & 0 & 0 \end{bmatrix}.$$

Notice that this example has an all-zero row at the bottom. Also notice that if the last row is deleted, then all columns of a 3 by 3 identity matrix appear as columns, but are scattered about the matrix. In general, if there are k nonzero rows and at least this many columns, then all columns of a k by k identity matrix will appear within a row–echelon matrix.

A special case of a matrix in row–echelon form is a matrix of the form

$$\boldsymbol{A} = [\boldsymbol{I} \quad \boldsymbol{P}]$$

where $\boldsymbol{I}$ is an identity matrix. Every matrix with at least as many columns as rows can be put in row–echelon form by elementary row operations, but not every such matrix can be put in this special form.

Theorem 2.6.4. *If two matrices,* $\mathbf{A}$ *and* $\mathbf{A}'$, *are related by a succession of elementary row operations, both matrices have the same row space.*

Proof: Each row of $\boldsymbol{A}'$ is also a linear combination of rows of $\boldsymbol{A}$. Therefore any linear combination of rows of $\boldsymbol{A}'$ is also a linear combination of rows of $\boldsymbol{A}$, and thus the row space of $\boldsymbol{A}$ contains the row space of $\boldsymbol{A}'$. But $\boldsymbol{A}$ can be obtained from $\boldsymbol{A}'$ by the inverse operation, and thus the row space of $\boldsymbol{A}'$ contains the row space of $\boldsymbol{A}$. Therefore $\boldsymbol{A}$ and $\boldsymbol{A}'$ have equal row spaces. □

Theorem 2.6.5. *If two matrices,* $\mathbf{A}$ *and* $\mathbf{A}'$, *are related by a succession of elementary row operations, then any set of columns that is linearly independent in* $\mathbf{A}$ *is also linearly independent in* $\mathbf{A}'$.

Proof: It suffices to prove the theorem for each single elementary row operation, and the theorem is obvious if it is the first or second kind of elementary row operation. Hence suppose $\boldsymbol{A}'$ is formed from $\boldsymbol{A}$ by adding a multiple of row α to row β. In any linearly dependent combination of columns of $\boldsymbol{A}'$, the elements in row α must combine to give zero, and thus can have no effect in row β. That is, this set of columns is also linearly dependent in $\boldsymbol{A}$. □

Theorem 2.6.6. *A* k *by* n *matrix* $\mathbf{A}$ *whose* k *rows are linearly independent has* k *linearly independent columns.*

Proof: Put $\boldsymbol{A}$ in row–echelon form $\boldsymbol{A}'$. Because the rows are linearly independent, no row has all zeros. Hence each row has a column where it is equal to one and every other row is equal to zero. This set of k columns of $\boldsymbol{A}'$ is linearly independent, and thus by Theorem 2.6.5, this same set of columns of $\boldsymbol{A}$ is linearly independent. □

The following theorem states that the row rank and the column rank of a matrix are equal. Hence, this value is called, simply, the *rank* of the matrix. A k by k matrix is called a *full-rank matrix* if its rank is equal to the smaller of k and n.

Theorem 2.6.7. *The row rank of a matrix* $\mathbf{A}$ *equals its column rank, and both are equal to the size of any largest square submatrix with a determinant not equal to zero.*

Proof: It is only necessary to show that the row rank of $\boldsymbol{A}$ is equal to the size of a largest square submatrix with a nonzero determinant. The same proof applied to the transpose of $\boldsymbol{A}$ then proves the same for the column rank of $\boldsymbol{A}$, and thus proves that the row rank equals the column rank.

A submatrix of $\boldsymbol{A}$ is a matrix obtained by deleting any number of rows and columns from $\boldsymbol{A}$. Let $\boldsymbol{M}$ be a nonsingular square submatrix of $\boldsymbol{A}$ of the largest size. The rows of $\boldsymbol{M}$ are linearly independent by Theorem 2.6.3(viii), and thus the rows of $\boldsymbol{A}$ that give rise to these rows of $\boldsymbol{M}$ must be linearly independent. Therefore the row rank of $\boldsymbol{A}$ is at least as large as the size of $\boldsymbol{M}$.

On the other hand, choose any set of k linearly independent rows. A matrix of these rows, by Theorem 2.6.6, has k linearly independent columns. Hence choosing these k columns from these k rows gives a matrix with a nonzero determinant. Therefore the size of a largest nonsingular submatrix of $\boldsymbol{A}$ is at least as large as the row rank of $\boldsymbol{A}$. This completes the proof. □

Let $\boldsymbol{A}$ be a square n by n matrix with a nonzero determinant. Then by Theorems 2.6.4 and 2.6.7, the row–echelon form is an n by n matrix with no all-zero rows. Hence it is the identity matrix. Because $\boldsymbol{A}$ can be obtained from $\boldsymbol{I}$ by the inverse of the sequence of elementary row operations, $\boldsymbol{A}$ can be written as a product of elementary matrices as

$$\boldsymbol{A} = \boldsymbol{F}_1 \boldsymbol{F}_2 \cdots \boldsymbol{F}_r.$$

Theorem 2.6.8. *In the ring of n by n matrices over a field F, let $\boldsymbol{C} = \boldsymbol{AB}$. Then*

$$\det(\boldsymbol{C}) = \det(\boldsymbol{A}) \det(\boldsymbol{B}).$$

Proof:
Step 1. First, we show that $\det(\boldsymbol{C})$ equals zero if either $\det(\boldsymbol{A})$ or $\det(\boldsymbol{B})$ equals zero. Suppose $\det(\boldsymbol{B})$ equals zero. Then by Theorem 2.6.3(viii), the rows of $\boldsymbol{B}$ are linearly dependent. But the rows of $\boldsymbol{C}$ are linear combinations of the rows of $\boldsymbol{B}$. Hence the rows of $\boldsymbol{C}$ are linearly dependent and $\det(\boldsymbol{C})$ equals zero. A similar argument is used if $\det(\boldsymbol{A})$ equals zero.
Step 2. Suppose $\det(\boldsymbol{A})$ is not zero. Then it is possible to write $\boldsymbol{A}$ as a product of elementary matrices:

$$\boldsymbol{A} = \boldsymbol{F}_1 \boldsymbol{F}_2 \cdots \boldsymbol{F}_r.$$

Each of the $\boldsymbol{F}_\ell$ corresponds to an elementary row operation, and thus by Theorem 2.6.3(iii), (v), and (vii),

$$\begin{aligned}\det(\boldsymbol{AB}) = \det[(\boldsymbol{F}_1 \boldsymbol{F}_2 \cdots \boldsymbol{F}_r)\boldsymbol{B}] &= \det[\boldsymbol{F}_1(\boldsymbol{F}_2 \cdots \boldsymbol{F}_r \boldsymbol{B})] \\ &= (\det \boldsymbol{F}_1) \det(\boldsymbol{F}_2 \cdots \boldsymbol{F}_r \boldsymbol{B}) \\ &= (\det \boldsymbol{F}_1)(\det \boldsymbol{F}_2) \cdots (\det \boldsymbol{F}_r)(\det \boldsymbol{B}).\end{aligned}$$

When $\boldsymbol{B} = \boldsymbol{I}$ this gives

$$\det(\boldsymbol{A}) = (\det \boldsymbol{F}_1)(\det \boldsymbol{F}_2) \cdots (\det \boldsymbol{F}_r).$$

Substituting this into the formula for a general $\boldsymbol{B}$ gives $\det(\boldsymbol{AB}) = \det(\boldsymbol{A}) \det(\boldsymbol{B})$, as was to be proved. □

One consequence of this theorem is that if $\boldsymbol{C} = \boldsymbol{AB}$, then $\boldsymbol{C}$ has an inverse if and only if both $\boldsymbol{A}$ and $\boldsymbol{B}$ have inverses because a square matrix has an inverse if and only if its determinant is nonzero.

Any square matrix of the form

$$A = \begin{bmatrix} 1 & 1 & \dots & 1 \\ X_1 & X_2 & \dots & X_\mu \\ X_1^2 & X_2^2 & \dots & X_\mu^2 \\ \vdots & & \vdots & \\ X_1^{\mu-1} & X_2^{\mu-1} & \dots & X_\mu^{\mu-1} \end{bmatrix}$$

is called a *Vandermonde matrix.*

Theorem 2.6.9. *The Vandermonde matrix has a nonzero determinant if and only if all of the X_i for $i = 1, \dots, \mu$ are distinct.*

Proof: The converse is obvious, because if any two of the X_i are equal, then the matrix has two columns the same, and hence it has determinant zero.

To prove the direct part, use induction. The theorem is true if $\mu = 1$. We will show that if it is true for all $\mu - 1$ by $\mu - 1$ Vandermonde matrices, then it is also true for all μ by μ Vandermonde matrices. Replace X_1 by the indeterminate x and transpose the matrix. Then the determinant is a function of x, given by

$$D(x) = \det \begin{bmatrix} 1 & x & x^2 & \dots & x^{\mu-1} \\ 1 & X_2 & X_2^2 & \dots & X_2^{\mu-1} \\ \vdots & & & \vdots & \\ 1 & X_\mu & X_\mu^2 & \dots & X_\mu^{\mu-1} \end{bmatrix}.$$

The determinant can be expanded in terms of the elements of the first row multiplied by the cofactors of these elements of the first row. This gives a polynomial in x of degree $\mu - 1$, which can be written

$$D(x) = d_{\mu-1}x^{\mu-1} + \dots + d_1 x + d_0.$$

The polynomial $D(x)$ has at most $\mu - 1$ zeros. The coefficient $d_{\mu-1}$ is itself the determinant of a Vandermonde matrix, and by the induction hypothesis, is nonzero. If for any i, $2 \le i \le \mu$, we set $x = X_i$, then two rows of the matrix are equal, and $D(X_i) = 0$. Thus for each $i \ne 1$, X_i is a zero of $D(x)$, and because they are all distinct and there are $\mu - 1$ of them, the polynomial can be easily factored:

$$D(x) = d_{\mu-1} \left[\prod_{i=2}^{\mu} (x - X_i) \right].$$

Therefore the determinant of the original Vandermonde matrix is

$$D(X_1) = d_{\mu-1} \left[\prod_{i=2}^{\mu} (X_1 - X_i) \right].$$

This is nonzero because $d_{\mu-1}$ is nonzero and X_1 is different from each of the remaining X_i. Hence the determinant of the μ by μ Vandermonde matrix is nonzero, and by induction, the theorem is true for all μ. □

To end this section, we will finish a piece of work left over from the last section.

Theorem 2.6.10. *If W, a subspace of a vector space of n-tuples, has dimension k, then $W^{\perp}$, the dual space of W, has dimension $n-k$.*

Proof: Let $\{\boldsymbol{g}_1, \ldots, \boldsymbol{g}_k\}$ be a basis for the subspace W, and define the matrix $\boldsymbol{G}$ by

$$\boldsymbol{G} = \begin{bmatrix} \boldsymbol{g}_1 \\ \vdots \\ \boldsymbol{g}_k \end{bmatrix}$$

where the basis vectors appear as rows. This matrix has rank k, so the column space of $\boldsymbol{G}$ has dimension k. A vector $\boldsymbol{v}$ is in $W^{\perp}$ if

$$\boldsymbol{G}\boldsymbol{v}^{\mathrm{T}} = \boldsymbol{0}$$

because then it is orthogonal to every vector of a basis for W.

Let $\{\boldsymbol{h}_1, \ldots, \boldsymbol{h}_r\}$ be a basis for $W^{\perp}$. Extend this to a basis for the whole space, $\{\boldsymbol{h}_1, \ldots, \boldsymbol{h}_r, \boldsymbol{f}_1, \ldots, \boldsymbol{f}_{n-r}\}$. Every vector $\boldsymbol{v}$ in the column space of $\boldsymbol{G}$ is expressible as $\boldsymbol{v} = \boldsymbol{G}\boldsymbol{B}^{\mathrm{T}}$ where $\boldsymbol{b}$ is a linear combination of the basis vectors $\{\boldsymbol{h}_1, \ldots, \boldsymbol{h}_r, \boldsymbol{f}_1, \ldots, \boldsymbol{f}_{n-r}\}$. Hence every vector $\boldsymbol{v}$ in the column space of $\boldsymbol{G}$ must be expressible as a linear combination of $\{\boldsymbol{G}\boldsymbol{h}_1^{\mathrm{T}}, \boldsymbol{G}\boldsymbol{h}_2^{\mathrm{T}}, \ldots, \boldsymbol{G}\boldsymbol{h}_r^{\mathrm{T}}, \boldsymbol{G}\boldsymbol{f}_1^{\mathrm{T}}, \ldots, \boldsymbol{G}\boldsymbol{f}_{n-r}^{\mathrm{T}}\}$.

To find the row rank of $\boldsymbol{G}$, it suffices by Theorem 2.6.7 to find the column rank of $\boldsymbol{G}$. Because $\boldsymbol{G}\boldsymbol{h}_i^{\mathrm{T}} = \boldsymbol{0}$, the set $\{\boldsymbol{G}\boldsymbol{f}_1^{\mathrm{T}}, \ldots, \boldsymbol{G}\boldsymbol{f}_{n-r}^{\mathrm{T}}\}$ spans the column space of $\boldsymbol{G}$. Further, the elements are independent because if

$$a_1(\boldsymbol{G}\boldsymbol{f}_1^{\mathrm{T}}) + \cdots + a_{n-r}(\boldsymbol{G}\boldsymbol{f}_{n-r}^{\mathrm{T}}) = \boldsymbol{0},$$

then

$$\boldsymbol{G}(a_1\boldsymbol{f}_1^{\mathrm{T}} + \cdots + a_{n-r}\boldsymbol{f}_{n-r}^{\mathrm{T}}) = \boldsymbol{0}.$$

Hence $a_1 = a_2 = \cdots = a_{n-r} = 0$ since $\boldsymbol{0}$ is the only linear combination of $\boldsymbol{f}_1, \ldots, \boldsymbol{f}_{n-r}$ in the null space of $\boldsymbol{G}$. Therefore $\{\boldsymbol{G}\boldsymbol{f}_1^{\mathrm{T}}, \ldots, \boldsymbol{G}\boldsymbol{f}_{n-r}^{\mathrm{T}}\}$ is a basis for the column space of $\boldsymbol{G}$. Hence $n-r=k$, which proves the theorem. □

Problems

2.1 A group can be constructed by using the rotations and reflections of a pentagon into itself.

a. How many elements are in this group? Is it an abelian group?

b. Construct the group.
c. Find a subgroup with five elements and a subgroup with two elements.
d. Are there any subgroups with four elements? Why?

2.2 a. Show that only one group exists with three elements. Construct this group and show that it is abelian.
b. Show that only two groups exist with four elements. Construct them and show that they are abelian. Show that one of the two groups with four elements has no element of order 4. This group is called the Klein four-group. Is one of these the addition table for $GF(4)$?

2.3 a. Let the group operation in the groups in Problem 2.2 be called addition.
b. Define multiplication to make the three-element group a ring. Is it unique?
c. For each of the two four-element groups, define multiplication to make it a ring. Is each unique?

2.4 Which of the three rings in Problem 2.3 are also fields? Can multiplication be defined differently to get a field?

2.5 Show that the set of all integers (positive, negative, and zero) is not a group under the operation of subtraction.

2.6 Give an example of a ring without identity.

2.7 Consider the set $S = \{0, 1, 2, 3\}$ with the operations

+	0	1	2	3
0	0	1	2	3
1	1	2	3	0
2	2	3	0	1
3	3	0	1	2

·	0	1	2	3
0	0	0	0	0
1	0	1	2	3
2	0	2	3	1
3	0	3	1	2

Is this a field?

2.8 Let G be an arbitrary group (not necessarily finite). For convenience, call the group operation "multiplication" and the identity "one." Let g be any element of G, and suppose that ν is the smallest integer, if it exists, such that $g^\nu = 1$ where g^ν means $g \cdot g \cdot \ldots \cdot g$, ν times. Then ν is called the *order* of g. Prove that the subset $\{g, g^2, g^3, \ldots, g^{\nu-1}, g^\nu\}$ is a subgroup of G. Prove that the subgroup is abelian even when G is not.

2.9 Prove Theorems 2.3.3 and 2.3.4.

2.10 Given a ring with identity one, and given $a \cdot b = 1$, prove that the following are equivalent:

(i) b is a left inverse for a;
(ii) $ax = 0$ implies $x = 0$;
(iii) $yb = 0$ implies $y = 0$.

Note: In some rings, condition (ii) is not true. In such a ring, an element may have a right inverse or a left inverse that only works on one side.

2.11 The field with four elements, $GF(4)$, is given by the arithmetic tables

+	0	1	2	3
0	0	1	2	3
1	1	0	3	2
2	2	3	0	1
3	3	2	1	0

·	0	1	2	3
0	0	0	0	0
1	0	1	2	3
2	0	2	3	1
3	0	3	1	2

In $GF(4)$, solve:

$$2x + y = 3,$$
$$x + 2y = 3.$$

2.12 The field with three elements, $GF(3)$, is given by the arithmetic tables

+	0	1	2
0	0	1	2
1	1	2	0
2	2	0	1

·	0	1	2
0	0	0	0
1	0	1	2
2	0	2	1

Calculate the determinant of the following matrix and show that its rank is 3.

$$\begin{bmatrix} 2 & 1 & 2 \\ 1 & 1 & 2 \\ 1 & 0 & 1 \end{bmatrix}$$

2.13 Put the matrix

$$\boldsymbol{A} = \begin{bmatrix} 1 & 1 & 0 & 1 \\ 1 & 0 & 1 & 1 \\ 0 & 1 & 1 & 0 \\ 0 & 1 & 0 & 1 \end{bmatrix}$$

into row–echelon form. Can the problem be solved without specifying the field? Why?

2.14 The set of all 7 by 7 matrices over $GF(2)$ is a vector space. Give a basis for this space. How many vectors are in the space? Is the set of all such matrices with zeros on the diagonal a subspace? If so, how many vectors are in this subspace?

2.15 How many vectors are there in the vector space $GF(2)^n$?

2.16 Is it true that if $\boldsymbol{x}$, $\boldsymbol{y}$, and $\boldsymbol{z}$ are linearly independent vectors over $GF(q)$, then so are $\boldsymbol{x} + \boldsymbol{y}$, $\boldsymbol{y} + \boldsymbol{z}$, and $\boldsymbol{z} + \boldsymbol{x}$?

2.17 Does a vector space exist with twenty-four elements over some finite field $GF(q)$?

2.18 Given that S and T are distinct two-dimensional subspaces of a three-dimensional vector space, show that their intersection is a one-dimensional subspace.

2.19 Prove that the null space of a matrix is a vector space.

2.20 Let S be any finite set. Let G be the set of subsets of S. If A and B are two subsets, let $A \cup B$ denote the set of elements in either A or B, let $A \cap B$ denote the set of elements in both A and B, and let $A - B$ denote the set of elements in A but not in B.

a. Show that if the operation $*$ is the set union $\cup$, then G is not a group.

b. The set operation of symmetric difference Δ is given by

$$A \Delta B = (A - B) \cup (B - A).$$

Show that G with $*$ as the operation of symmetric difference does form a group. Is it abelian?

c. Show that G with the operations Δ and $\cap$ forms a ring. Is it a commutative ring? Is there an identity?

2.21 A field with an infinite number of elements that contains a finite field is also called a Galois field. Let F be the set of formal expressions

$$F = \left\{ \frac{\sum_{i=0}^{I-1} a_i x^i}{\sum_{j=0}^{J-1} b_j x^j} \right\}$$

where I and J are any positive integers, and a_i and b_j are elements of $GF(3)$. Provide your own definitions of addition and multiplication in this set so that F is a Galois field with an infinite number of elements.

Notes

This chapter deals with standard topics in modern algebra. Many textbooks can be found that cover the material more thoroughly. The book by Birkhoff and MacLane (1953) is intended as an introductory text and is easily understood at the level of this book. The text by Van der Waerden (1949, 1953), a more advanced work addressed primarily to mathematicians, goes more deeply into many topics. The material on linear algebra and matrix theory is standard and can be found in many textbooks written specifically for these topics, though often referring only to the real or complex fields. The book by Thrall and Tornheim (1957) is especially suitable since it does not presuppose any particular underlying field.

The Galois fields are named after Evariste Galois (1811–1832). Abelian groups are named after Niels Henrik Abel (1802–1829).

3 Linear Block Codes

Most known good codes belong to the class of codes called *linear codes*. This class of codes is defined by imposing a strong structural property on the codes. The structure provides guidance in the search for good codes and also helps to make the encoders and decoders practical.

We must emphasize that we do not study linear codes because the best possible codes are necessarily linear. Rather, we have not yet found general methods to find good nonlinear codes. There are good linear codes, however, and they are attractive for many reasons. Almost all of the best codes we know are linear. Most of the strongest theoretical techniques are useful only for linear codes. Thus the study of codes is usually restricted to the class of linear codes.

3.1 Structure of linear block codes

Recall that under componentwise vector addition and componentwise scalar multiplication the set of n-tuples of elements from $GF(q)$ is the vector space called $GF(q)^n$. A special case of major importance is $GF(2)^n$, which is the vector space of all binary words of length n with two such vectors added by modulo-2 addition in each component.

Definition 3.1.1. *A linear code $\mathcal{C}$ is a subspace of $GF(q)^n$.*

That is, a linear code $\mathcal{C}$ is a nonempty set of n-tuples over $GF(q)$, called codewords, such that the sum of two codewords is a codeword, and the product of any codeword with a field element is a codeword. In any linear code, the all-zero word, as the vector-space origin, is always a codeword. More directly, if $\boldsymbol{c}$ is a codeword, then $(-\boldsymbol{c})$ is a codeword, and hence $\boldsymbol{c} + (-\boldsymbol{c})$ is a codeword.

Let k be the dimension of $\mathcal{C}$. The code $\mathcal{C}$ is referred to as an (n, k) code where n is the *blocklength* of the code and k is the *dimension* of the code.

A linear code is invariant under translation by a codeword. That is, for any codeword $\boldsymbol{c}$, $\mathcal{C} + \boldsymbol{c} = \mathcal{C}$. This means every codeword in a linear code bears a relationship to the rest of the code that is completely equivalent to the relationship any other codeword bears to the rest of the code. The arrangement of neighboring codewords about the all-zero codeword is typical of the arrangement of the neighboring codewords about any other codeword. For example, suppose that $\boldsymbol{c}$ is any codeword, and $\boldsymbol{c}_1, \ldots, \boldsymbol{c}_r$ are codewords at distance d from $\boldsymbol{c}$; then $\boldsymbol{c} - \boldsymbol{c}$ is the all-zero codeword, and $\boldsymbol{c}_1 - \boldsymbol{c}$, $\boldsymbol{c}_2 - \boldsymbol{c}, \ldots, \boldsymbol{c}_r - \boldsymbol{c}$ are codewords at distance d from the all-zero codeword. Hence to determine the minimum distance of a linear code, it suffices to determine the distance from the all-zero codeword to the codeword closest to it.

Definition 3.1.2. *The Hamming weight, or weight, $w(\boldsymbol{c})$ of a vector $\boldsymbol{c}$ is equal to the number of nonzero components in the vector. The minimum Hamming weight $w_{\min}$ of a code $\mathcal{C}$ is the smallest Hamming weight of any nonzero codeword of $\mathcal{C}$.*

Theorem 3.1.3. *For a linear code, the minimum distance $d_{\min}$ satisfies*

$$d_{\min} = \min_{\boldsymbol{c} \neq \boldsymbol{0}} w(\boldsymbol{c}) = w_{\min}$$

where the minimum is over all codewords except the all-zero codeword.

Proof:

$$d_{\min} = \min_{\substack{\boldsymbol{c}_i, \boldsymbol{c}_j \in \mathcal{C} \\ i \neq j}} d(\boldsymbol{0}, \boldsymbol{c}_i - \boldsymbol{c}_j) = \min_{\substack{\boldsymbol{c} \in \mathcal{C} \\ \boldsymbol{c} \neq \boldsymbol{0}}} w(\boldsymbol{c}).$$

□

Hence to find a linear code that can correct t errors, one must find a linear code whose minimum weight satisfies

$$w_{\min} \geq 2t + 1.$$

The study of the distance structure of a linear code is much easier than that of a nonlinear code because the study of the distance structure is equivalent to the study of the weight structure.

3.2 Matrix description of linear block codes

A linear code $\mathcal{C}$ is a subspace of $GF(q)^n$. The theory of vector spaces can be used to study these codes. Any set of basis vectors for the subspace can be used as rows to form a k by n matrix $\boldsymbol{G}$ called the *generator matrix* of the code. The row space of $\boldsymbol{G}$ is the linear code $\mathcal{C}$; any codeword is a linear combination of the rows of $\boldsymbol{G}$. The set of q^k codewords is called an (n, k) linear code.

The rows of $\boldsymbol{G}$ are linearly independent, while the number of rows k is the dimension of the code. The matrix $\boldsymbol{G}$ has rank k. The code is a subspace of the whole space $GF(q)^n$, a vector space of dimension n. There are q^k codewords in $\mathcal{C}$, and the q^k distinct k-tuples over $GF(q)$ can be mapped onto the set of codewords.

Any one-to-one pairing of k-tuples and codewords can be used as an encoding procedure, but the most natural approach is to use the following:

$$\boldsymbol{c} = \boldsymbol{a}\boldsymbol{G}$$

where $\boldsymbol{a}$, the dataword,[1] is a k-tuple of data symbols to be encoded, and the codeword $\boldsymbol{c}$ is an n-tuple. With this expression defining the encoder, the pairing between datawords and codewords depends on the choice of basis vectors appearing as rows of $\boldsymbol{G}$, but of course, the total set of codewords is unaffected.

For a simple example of a binary linear code, take the generator matrix

$$\boldsymbol{G} = \begin{bmatrix} 1 & 0 & 0 & 1 & 0 \\ 0 & 1 & 0 & 0 & 1 \\ 0 & 0 & 1 & 1 & 1 \end{bmatrix}.$$

The data vector

$$\boldsymbol{a} = [0 \quad 1 \quad 1]$$

is encoded into the codeword

$$[0 \quad 1 \quad 1] \begin{bmatrix} 1 & 0 & 0 & 1 & 0 \\ 0 & 1 & 0 & 0 & 1 \\ 0 & 0 & 1 & 1 & 1 \end{bmatrix} = [0 \quad 1 \quad 1 \quad 1 \quad 0].$$

The generator matrix is a concise way of describing a linear code. Compare a binary (100, 50) linear code, which is described by $100 \times 50 = 5000$ bits as the 5000 elements of the matrix $\boldsymbol{G}$, with an arbitrary (100, 50) code that has 2^{50} codewords, requiring approximately 10^{17} bits to list.

Because $\mathcal{C}$ is a subspace of $GF(q)^n$, it has a dimension k. This is equal to the number of rows in $\boldsymbol{G}$. Because $\mathcal{C}$ is a subspace, it has an orthogonal complement $\mathcal{C}^\perp$, which is the set of all vectors orthogonal to $\mathcal{C}$. The orthogonal complement is also a subspace of $GF(q)^n$, and thus itself is a code. Whenever $\mathcal{C}^\perp$ itself is thought of as a code, it is called the *dual code* of $\mathcal{C}$. By Theorem 2.6.10, the dual code $\mathcal{C}^\perp$ has dimension $n - k$. Any basis of $\mathcal{C}^\perp$ contains $n - k$ vectors. Let $\boldsymbol{H}$ be a matrix with any set of basis vectors of $\mathcal{C}^\perp$ as rows. Then an n-tuple $\boldsymbol{c}$ is a codeword in $\mathcal{C}$ if and only if it is orthogonal to every row vector of $\boldsymbol{H}$. That is,

$$\boldsymbol{c}\boldsymbol{H}^{\mathrm{T}} = \boldsymbol{0}.$$

This *check equation* provides a way for testing whether a word is a codeword of $\mathcal{C}$.

[1] The symbol d is reserved to denote distance.

The matrix $\boldsymbol{H}$ is called a *check matrix* of the code $\mathcal{C}$. It is an $(n-k)$ by n matrix over $GF(q)$. Because $\boldsymbol{c}\boldsymbol{H}^{\mathrm{T}} = \boldsymbol{0}$ holds whenever $\boldsymbol{c}$ is equal to any row of $\boldsymbol{G}$, we have

$$\boldsymbol{G}\boldsymbol{H}^{\mathrm{T}} = \boldsymbol{0}.$$

Using the same example of a generator matrix $\boldsymbol{G}$ as before, we have

$$\boldsymbol{H} = \begin{bmatrix} 1 & 0 & 1 & 1 & 0 \\ 0 & 1 & 1 & 0 & 1 \end{bmatrix}$$

as one choice for $\boldsymbol{H}$. Notice that just as there is more than one choice for $\boldsymbol{G}$, there is more than one choice for $\boldsymbol{H}$.

Theorem 3.2.1. *A generator matrix for $\mathcal{C}$ is a check matrix for the dual code $\mathcal{C}^{\perp}$.*

Proof: The proof is immediate. □

The minimum weight of a code is related to the check matrix by the following theorem.[2]

Theorem 3.2.2. *The code $\mathcal{C}$ contains a nonzero codeword of Hamming weight w or less if and only if a linearly dependent set of w columns of $\boldsymbol{H}$ exists.*

Proof: For any codeword $\boldsymbol{c}$, $\boldsymbol{c}\boldsymbol{H}^{\mathrm{T}} = \boldsymbol{0}$. Let $\boldsymbol{c}$ have weight w or less. Drop $n-w$ components of $\boldsymbol{c}$ that are zero. What remains is a linear dependence relationship in w columns of $\boldsymbol{H}$. Hence $\boldsymbol{H}$ has a linearly dependent set of w columns.

Conversely, if $\boldsymbol{H}$ has a linearly dependent set of w columns, then a linear combination of w columns is equal to zero. These w coefficients, not all zero, define a nonzero vector $\boldsymbol{c}$ of weight w or less for which $\boldsymbol{c}\boldsymbol{H}^{\mathrm{T}} = \boldsymbol{0}$. □

Corollary 3.2.3. *If $\boldsymbol{H}$ is any check matrix for a linear code $\mathcal{C}$, then the minimum distance of code $\mathcal{C}$ is equal to the fewest number of columns of $\boldsymbol{H}$ that form a linearly dependent set.*

Hence to find an (n, k) code that can correct t errors, it suffices to find an $(n-k)$ by n matrix $\boldsymbol{H}$ such that every set of $2t$ columns is linearly independent.

Given an (n, k) code with minimum distance $d_{\min}$, a new code with the same parameters can be obtained by choosing any two components and transposing the symbols in these two components of every codeword. This, however, gives a code that is only trivially different from the original code; it is said to be equivalent to the original code. In general, two linear codes that are the same except for a permutation of

[2] The theorem can be used to re-express the minimum weight of a code by coining a term dual to the *rank* of a matrix. "The *rank* of $\boldsymbol{H}$ is the largest r such that some set of r columns of $\boldsymbol{H}$ is linearly independent. The *heft* of $\boldsymbol{H}$ is the largest r such that every set of r columns of $\boldsymbol{H}$ is linearly independent." The minimum distance of a code is one plus the heft of any check matrix for the code ($d_{\min} = 1 + \text{heft } \boldsymbol{H}$).

components are called *equivalent* codes. Indeed, they are usually considered to be the same code.

The generator matrices $\boldsymbol{G}$ and $\boldsymbol{G}'$ of equivalent codes, called *equivalent generator matrices*, can be simply related. The code itself is the row space of $\boldsymbol{G}$, and thus it is unchanged under elementary row operations. Permutation of the components of the code corresponds to permutation of the columns of $\boldsymbol{G}$. Hence two codes are equivalent if and only if they can be formed by generator matrices that are related by column permutations and elementary row operations.

Every generator matrix $\boldsymbol{G}$ can be converted to a generator matrix in row–echelon form, and because the rows are linearly independent, no row will contain all zeros. Then, by column permutations, every generator matrix can be converted to a generator matrix for an equivalent code with a k by k identity in the first k columns. That is,

$$\boldsymbol{G} = [\boldsymbol{I} \quad \boldsymbol{P}]$$

where $\boldsymbol{P}$ is a k by $(n-k)$ matrix. Every generator matrix can be reduced to this special form by a sequence of elementary row operations, followed by a sequence of column permutations. We call this the *systematic form* of a generator matrix.

Suppose that $\boldsymbol{G} = [\boldsymbol{I} \quad \boldsymbol{P}]$. Then clearly, an appropriate expression for a check matrix is $\boldsymbol{H} = [-\boldsymbol{P}^{\mathrm{T}} \quad \boldsymbol{I}]$ because

$$\boldsymbol{G}\boldsymbol{H}^{\mathrm{T}} = [\boldsymbol{I} \quad \boldsymbol{P}]\begin{bmatrix} -\boldsymbol{P} \\ \boldsymbol{I} \end{bmatrix} = -\boldsymbol{P} + \boldsymbol{P} = \boldsymbol{0}.$$

Definition 3.2.4. *A systematic encoder*[3] *for a block code is one that maps each dataword into a codeword with the k data symbols unmodified in the first k symbols of the codeword. The remaining symbols are called check symbols.*

One sometimes speaks loosely of a systematic code, although what is always meant is a systematic encoding of the code. The same code can be used with a systematic encoder or with a nonsystematic encoder.

Theorem 3.2.5. *Every linear block code is equivalent to a code that can be encoded by a systematic encoder.*

Proof: A systematically encoded linear code is encoded by multiplying the data vector by a generator matrix $\boldsymbol{G}$ in systematic form, and every generator matrix can be converted to a generator matrix in systematic form. □

[3] When dealing with cyclic codes in Chapter 5, it will be convenient and conventional to think of a codeword as beginning with the high-index component c_{n-1} and ending with c_0. Then we should use the equivalent matrices of the form $\boldsymbol{G} = [\boldsymbol{P}\ \boldsymbol{I}]$ and $\boldsymbol{H} = [\boldsymbol{I}\ -\boldsymbol{P}^{\mathrm{T}}]$ for some new $\boldsymbol{P}$. These are obtained by reversing all rows and reversing all columns of the earlier form. They also are called systematic matrices.

For an example, choose

$$G = \begin{bmatrix} 1 & 0 & 0 & | & 1 & 0 \\ 0 & 1 & 0 & | & 0 & 1 \\ 0 & 0 & 1 & | & 1 & 1 \end{bmatrix}, \quad H = \begin{bmatrix} 1 & 0 & 1 & | & 1 & 0 \\ 0 & 1 & 1 & | & 0 & 1 \end{bmatrix},$$

then $\boldsymbol{a} = [0\ 1\ 1]$ is systematically encoded as $\boldsymbol{c} = [0\ 1\ 1\ 1\ 0]$.

By looking at a code in a systematic form, it is possible to obtain a simple inequality relating the parameters of a code. Consider an arbitrary systematic linear (n, k) code with minimum distance $d_{\min}$. The possible values of $(n, k, d_{\min})$ are limited by the following theorem.

Theorem 3.2.6 (Singleton Bound). *The minimum distance of any linear (n, k) block code satisfies*

$$d_{\min} \leq 1 + n - k.$$

Proof: The smallest-weight, nonzero codeword has weight $d_{\min}$. Systematic codewords exist with only one nonzero data symbol and $n - k$ check symbols. Such a codeword cannot have a weight larger than $1 + (n - k)$. Hence the minimum weight of the code cannot be larger than $1 + n - k$. □

Definition 3.2.7. *Any code whose minimum distance satisfies*

$$d_{\min} = 1 + n - k$$

is called a maximum-distance code (or a maximum-distance separable code).

The Singleton bound is very loose in that most codes – even the best codes – have a minimum distance well below the bound, but it is sometimes useful. The bound tells us that to correct t errors, a code must have at least $2t$ check symbols – two check symbols per error to be corrected. Most codes, even optimum codes, have considerably more check symbols than required by the Singleton bound, but some meet it with equality. A maximum-distance code has exactly $2t$ check symbols.

3.3 Hamming codes

A code whose minimum distance is at least 3 must, by Corollary 3.2.3, have a check matrix all of whose columns are distinct and nonzero. If a check matrix for a binary code has m rows, then each column is an m-bit binary number. There are $2^m - 1$ possible columns. Hence if the check matrix of a binary code with $d_{\min}$ at least 3 has m rows, then it can have $2^m - 1$ columns, but no more. This defines a $(2^m - 1, 2^m - 1 - m)$ code. These codes are called binary *Hamming codes*.

Table 3.1. *Parameters (n, k) for some Hamming codes*

$GF(2)$	$GF(4)$	$GF(8)$	$GF(16)$	$GF(27)$
(7, 4)	(5, 3)	(9, 7)	(17, 15)	(28, 26)
(15, 11)	(21, 18)	(73,70)	(273, 270)	(757, 754)
(31, 26)	(85, 81)	(585, 581)		
(63, 57)	(341, 336)			
(127, 120)				

The simplest nontrivial example is the binary Hamming code with $m = 3$. In systematic form, such a check matrix is

$$\boldsymbol{H} = \begin{bmatrix} 1 & 1 & 0 & 1 & 1 & 0 & 0 \\ 1 & 0 & 1 & 1 & 0 & 1 & 0 \\ 0 & 1 & 1 & 1 & 0 & 0 & 1 \end{bmatrix}.$$

The corresponding generator matrix is

$$\boldsymbol{G} = \begin{bmatrix} 1 & 0 & 0 & 0 & 1 & 1 & 0 \\ 0 & 1 & 0 & 0 & 1 & 0 & 1 \\ 0 & 0 & 1 & 0 & 0 & 1 & 1 \\ 0 & 0 & 0 & 1 & 1 & 1 & 1 \end{bmatrix}.$$

Clearly for the general case of a check matrix for a binary Hamming code, every pair of columns of $\boldsymbol{H}$ is independent (because no pair of distinct binary vectors can sum to zero), and some sets of three columns are dependent. Hence by Theorem 3.2.2, the minimum weight is 3, and the code can correct a single error.

Hamming codes can be defined easily for larger alphabet sizes. One merely notes that the main idea is to define a matrix $\boldsymbol{H}$ such that every pair of columns is linearly independent. Over $GF(q)$, $q \neq 2$, one cannot use all nonzero m-tuples as columns because some columns will be multiples of each other, and so are linearly dependent. To force linear independence, choose as columns all nonzero m-tuples that have a one in the topmost nonzero component. Then two columns can never be linearly dependent. Because three columns can be linearly dependent, the code has a minimum distance equal to 3.

There are $(q^m - 1)/(q - 1)$ distinct nonzero columns satisfying the rule that the topmost nonzero element is a one. Hence the code is a $[(q^m - 1)/(q - 1), (q^m - 1)/(q - 1) - m]$ code. A single-error-correcting Hamming code with these parameters exists for each q for which a field $GF(q)$ exists, and for each m. Some sample parameters of the Hamming codes are shown in Table 3.1.

For example, a (13, 10) Hamming code over $GF(3)$ is given by the check matrix

$$\boldsymbol{H} = \begin{bmatrix} 1 & 1 & 1 & 1 & 1 & 1 & 1 & 1 & 0 & 0 & 1 & 0 & 0 \\ 0 & 0 & 1 & 1 & 1 & 2 & 2 & 2 & 1 & 1 & 0 & 1 & 0 \\ 1 & 2 & 0 & 1 & 2 & 0 & 1 & 2 & 1 & 2 & 0 & 0 & 1 \end{bmatrix}$$

and the generator matrix

$$G = \begin{bmatrix} 1 & 0 & 0 & 0 & 0 & 0 & 0 & 0 & 0 & 0 & 2 & 0 & 2 \\ 0 & 1 & 0 & 0 & 0 & 0 & 0 & 0 & 0 & 0 & 2 & 0 & 1 \\ 0 & 0 & 1 & 0 & 0 & 0 & 0 & 0 & 0 & 0 & 2 & 2 & 0 \\ 0 & 0 & 0 & 1 & 0 & 0 & 0 & 0 & 0 & 0 & 2 & 2 & 2 \\ 0 & 0 & 0 & 0 & 1 & 0 & 0 & 0 & 0 & 0 & 2 & 2 & 1 \\ 0 & 0 & 0 & 0 & 0 & 1 & 0 & 0 & 0 & 0 & 2 & 1 & 0 \\ 0 & 0 & 0 & 0 & 0 & 0 & 1 & 0 & 0 & 0 & 2 & 1 & 2 \\ 0 & 0 & 0 & 0 & 0 & 0 & 0 & 1 & 0 & 0 & 2 & 1 & 1 \\ 0 & 0 & 0 & 0 & 0 & 0 & 0 & 0 & 1 & 0 & 0 & 2 & 2 \\ 0 & 0 & 0 & 0 & 0 & 0 & 0 & 0 & 0 & 1 & 0 & 2 & 1 \end{bmatrix}.$$

The next example is a Hamming code over $GF(16)$. Taking $q = 16$ and $m = 2$, we see that one can have a (17, 15) hexadecimal Hamming code, based on the arithmetic of $GF(16)$. Using this field, we can construct the generator matrix for the Hamming (17, 15) code over $GF(16)$. From the matrix, we can read off the following equations for the check symbols:

$$p_0 = a_0 + a_1 + a_2 + \cdots + a_{13} + a_{14},$$
$$p_1 = a_0 + 2a_1 + 3a_2 + \cdots + Ea_{13} + Fa_{14}.$$

After every block of fifteen data symbols, these two check symbols are inserted. The decoder uses the check symbols to correct a single symbol error in the block of seventeen symbols.

Of course, our example over the field $GF(16)$ makes sense only if the field $GF(16)$ exists. This field was given in Figure 2.1 without much explanation. Before we can use the above construction in any field $GF(q)$, we must prove that $GF(q)$ exists and also find how to add and multiply in that field. These will be the major tasks of Chapter 4.

3.4 The standard array

The Hamming sphere of radius t about a codeword $\boldsymbol{c}$ is the set of all words at Hamming distance from $\boldsymbol{c}$ not larger than t. That is,

$$\mathcal{S}_c = \{\boldsymbol{v} \mid d(\boldsymbol{c}, \boldsymbol{v}) \le t\}.$$

If all Hamming spheres of radius t are disjoint, then we may correctly decode all sensewords with up to t errors. If $d_{\min}$ is odd, we may choose t so that $d_{\min} = 2t + 1$. We call such nonintersecting spheres *decoding spheres* and such a decoder a *bounded-distance decoder.*

Because the difference of two codewords in a linear code is a codeword, the all-zero word is always a codeword. For a linear code, if we know which sensewords lie closest to the all-zero codeword, then we know which sensewords lie closest to any other codeword simply by translating the origin.

The sphere of radius t about the all-zero word is

$$\mathcal{S}_0 = \{\boldsymbol{v} \mid d(\boldsymbol{0}, \boldsymbol{v}) \leq t\}.$$

The sphere contains all sensewords that will be decoded into the all-zero codeword. The sphere of radius t about codeword $\boldsymbol{c}$ can be written

$$\mathcal{S}_c = \{\boldsymbol{v} \mid d(\boldsymbol{0}, \boldsymbol{v} - \boldsymbol{c}) \leq t\}.$$

Then we have

$$\mathcal{S}_c = \mathcal{S}_0 + \boldsymbol{c} = \{\boldsymbol{v} + \boldsymbol{c} \mid \boldsymbol{v} \in \mathcal{S}_0\}.$$

Hence we can record either the elements of every decoding sphere, or more efficiently, the elements of the decoding sphere about the all-zero codeword, and compute the other decoding spheres, as needed, by a simple translation.

The *standard array* is a way of tabulating all decoding spheres. Let $\boldsymbol{0}, \boldsymbol{c}_2, \boldsymbol{c}_3, \ldots, \boldsymbol{c}_{q^k}$ be the q^k codewords in an (n, k) linear code. Form the table of Figure 3.1 as follows. Write all the codewords in the first row. Of the noncodewords in $GF(q)^n$ lying closest to the all-zero codeword, choose any word and call it $\boldsymbol{v}_1$. Write $\boldsymbol{0} + \boldsymbol{v}_1, \boldsymbol{c}_2 + \boldsymbol{v}_1, \boldsymbol{c}_3 + \boldsymbol{v}_1, \ldots, \boldsymbol{c}_{q^k} + \boldsymbol{v}_1$ in the second row. Continue in this way to form additional rows. At the jth step, choose $\boldsymbol{v}_j$ as close as possible to the all-zero word and previously unused, and write $\boldsymbol{0} + \boldsymbol{v}_j, \boldsymbol{c}_2 + \boldsymbol{v}_j, \boldsymbol{c}_3 + \boldsymbol{v}_j, \ldots, \boldsymbol{c}_{q^k} + \boldsymbol{v}_j$ for the jth row. Stop when no unused element of $GF(q)^n$ remains to start a new row.

If the vector space $GF(q)^n$ is regarded as a group under vector addition, then the code is a subgroup. By Theorem 2.2.3, the process of Figure 3.1 generates cosets, halting with each element of $GF(q)^n$ appearing exactly once. By Corollary 2.2.4, because there are q^k columns, there will be q^{n-k} rows. The words in the first column are called *coset leaders*.

Because the rows are constructed from the shells around codewords by starting at the all-zero word, the first column must include all words in the decoding sphere about

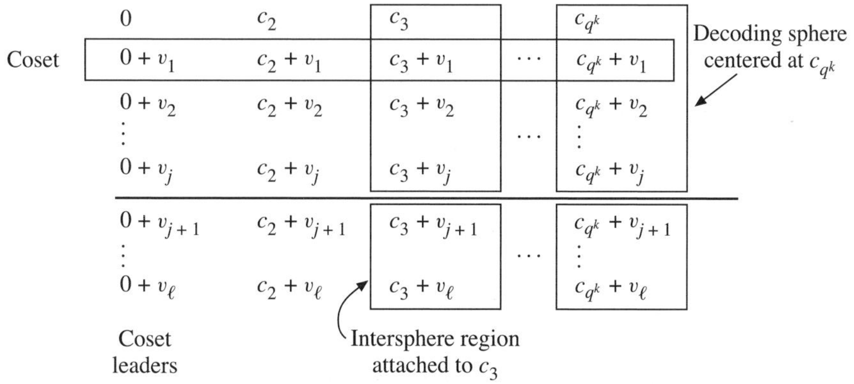

Figure 3.1. The standard array

the all-zero codeword, that is, all words within distance t of the all-zero codeword. In the same way, every column contains a decoding sphere about the codeword at the top of that column. When the spheres of radius t are complete, draw a horizontal line across the standard array. There may be points still unused at this point. The construction of the standard array then goes on to attach these points to a nearby codeword, but this is partly arbitrary.

The two basic classes of decoders can be described in terms of the standard array. A complete decoder assigns every senseword to a nearby codeword. When a senseword is received, find it in the standard array and decode it into the codeword at the top of its column. A bounded-distance decoder assigns every senseword to a codeword within distance t, if there is one. Otherwise it refuses to decode and declares a decoding failure. When a senseword is received, find it in the standard array. If it lies above the horizontal line, decode it into the codeword at the top of its column. If it is below the line, flag the senseword as uncorrectable – it has more than t errors.

For an example, consider the (5, 2) code with

$$\boldsymbol{G} = \begin{bmatrix} 1 & 0 & 1 & 1 & 1 \\ 0 & 1 & 1 & 0 & 1 \end{bmatrix}.$$

This code corrects one error. The standard array is

$$\begin{array}{ccccc} 0\;0\;0\;0\;0 & & 1\;0\;1\;1\;1 & & 0\;1\;1\;0\;1 & & 1\;1\;0\;1\;0 \\ 0\;0\;0\;0\;1 & & 1\;0\;1\;1\;0 & & 0\;1\;1\;0\;0 & & 1\;1\;0\;1\;1 \\ 0\;0\;0\;1\;0 & & 1\;0\;1\;0\;1 & & 0\;1\;1\;1\;1 & & 1\;1\;0\;0\;0 \\ 0\;0\;1\;0\;0 & & 1\;0\;0\;1\;1 & & 0\;1\;0\;0\;1 & & 1\;1\;1\;1\;0 \\ 0\;1\;0\;0\;0 & & 1\;1\;1\;1\;1 & & 0\;0\;1\;0\;1 & & 1\;0\;0\;1\;0 \\ 1\;0\;0\;0\;0 & & 0\;0\;1\;1\;1 & & 1\;1\;1\;0\;1 & & 0\;1\;0\;1\;0 \\ \hline 0\;0\;0\;1\;1 & & 1\;0\;1\;0\;0 & & 0\;1\;1\;1\;0 & & 1\;1\;0\;0\;1 \\ 0\;0\;1\;1\;0 & & 1\;0\;0\;0\;1 & & 0\;1\;0\;1\;1 & & 1\;1\;1\;0\;0 \end{array}$$

There are four disjoint spheres each of radius 1, six points per sphere, and eight points outside of any sphere. For a large n and k, such a table would be impractical to list. The standard array is of value only as a conceptual tool.

The storage of the standard array can be simplified if we store only the first column and compute the remaining columns as needed. We do this by introducing the concept of the syndrome of the error pattern.

For any senseword $\boldsymbol{v}$, define the *syndrome* of $\boldsymbol{v}$ by

$$s = \boldsymbol{v}\boldsymbol{H}^{\mathrm{T}}.$$

Theorem 3.4.1. *All vectors in the same coset have the same syndrome that is unique to that coset.*

Proof: If $\boldsymbol{v}$ and $\boldsymbol{v}'$ are in the same coset, then $\boldsymbol{v} = \boldsymbol{c} + \boldsymbol{y}$, and $\boldsymbol{v}' = \boldsymbol{c}' + \boldsymbol{y}$ for some $\boldsymbol{y}$ and for some codewords $\boldsymbol{c}$ and $\boldsymbol{c}'$. For any codeword $\boldsymbol{c}$, $\boldsymbol{c}\boldsymbol{H}^{\mathrm{T}} = \boldsymbol{0}$. Hence

$$\boldsymbol{s} = \boldsymbol{v}\boldsymbol{H}^{\mathrm{T}} = \boldsymbol{y}\boldsymbol{H}^{\mathrm{T}}$$
$$\boldsymbol{s}' = \boldsymbol{v}'\boldsymbol{H}^{\mathrm{T}} = \boldsymbol{y}\boldsymbol{H}^{\mathrm{T}}$$

and $\boldsymbol{s} = \boldsymbol{s}'$. Conversely, suppose $\boldsymbol{s} = \boldsymbol{s}'$, then $(\boldsymbol{v} - \boldsymbol{v}')\boldsymbol{H}^{\mathrm{T}} = \boldsymbol{0}$, and thus $\boldsymbol{v} - \boldsymbol{v}'$ is a codeword. Hence $\boldsymbol{v}$ and $\boldsymbol{v}'$ are in the same coset. □

Any two vectors in the same coset have the same syndrome. Hence we need only to tabulate syndromes and coset leaders. Then we can decode as follows. Given a senseword $\boldsymbol{v}$, compute the syndrome $\boldsymbol{s}$ and look up its associated coset leader. This coset leader is the difference between the senseword and the center of its decoding sphere; it is the error vector. Hence subtract the coset leader from $\boldsymbol{v}$ to correct the error.

For the example introduced above, the check matrix is

$$\boldsymbol{H} = \begin{bmatrix} 1 & 1 & 1 & 0 & 0 \\ 1 & 0 & 0 & 1 & 0 \\ 1 & 1 & 0 & 0 & 1 \end{bmatrix}.$$

The new table is

Coset Leader	Syndrome
0 0 0 0 0	0 0 0
0 0 0 0 1	0 0 1
0 0 0 1 0	0 1 0
0 0 1 0 0	1 0 0
0 1 0 0 0	1 0 1
1 0 0 0 0	1 1 1
0 0 0 1 1	0 1 1
0 0 1 1 0	1 1 0

This table is simpler than the standard array. Suppose $\boldsymbol{v} = 10010$ is the senseword. Then $\boldsymbol{s} = \boldsymbol{v}\boldsymbol{H}^{\mathrm{T}} = 101$. The coset leader is 01000. Therefore the transmitted word is $10010 - 01000 = 11010$, and the dataword is 11.

3.5 Hamming spheres and perfect codes

The Hamming sphere in $GF(q)^n$ of radius t about a point $\boldsymbol{v}$ is the set whose points are at distance t from $\boldsymbol{v}$ or closer. The *volume* of a Hamming sphere is the number of points that it contains. The volume is

$$V = \sum_{\ell=0}^{t} \binom{n}{\ell} (q-1)^{\ell}$$

because there are $\binom{n}{\ell}$ ways to choose ℓ places where a point may differ from $\boldsymbol{v}$, and $q-1$ ways in which it can be different in each of those places.

Visualize a small Hamming sphere about each of the codewords of a code, each sphere with the same radius (an integer). Allow these spheres to increase in radius by integer amounts until they cannot be made larger without causing some spheres to intersect. That value of the radius is equal to the number of errors that can be corrected by the code. It is called the *packing radius* of the code. Now allow the radii of the spheres to continue to increase by integer amounts until every point in the space is contained in at least one sphere. That radius is called the *covering radius* of the code.

The packing radius and the covering radius may be equal; if so, then the construction of the standard array stops just when the spheres of radius t are completed. All points of the space are used in these spheres; none are left over.

Definition 3.5.1. *A perfect code is one for which there are spheres of equal radius about the codewords that are disjoint and that completely fill the space.*

A perfect code satisfies the Hamming bound of Problem 1.5 with equality. A Hamming code with blocklength $n = (q^m - 1)/(q - 1)$ is perfect. This is because there are $1 + n(q-1) = q^m$ points in each sphere of radius 1, and the number of points in the space, divided by the number of spheres, is $q^n/q^k = q^m$ because $n - k = m$. Perfect codes, when they exist, have certain properties that are aesthetically pleasing, but their rarity reduces their importance in practical applications.

Other than distance 3 codes, such as the Hamming codes, the only nontrivial perfect codes of importance are those known as the *Golay codes*. There are two Golay codes: the binary (23, 12, 7) Golay code and the ternary (11, 6, 5) Golay code.

Anyone disposed to study a table of binomial coefficients might notice that

$$\left[\binom{23}{0} + \binom{23}{1} + \binom{23}{2} + \binom{23}{3}\right] 2^{12} = 2^{23}.$$

This is a necessary (but not sufficient [4]) condition for the existence of a perfect (23, 12) triple-error-correcting code over $GF(2)$ because: (1) the number of points in a decoding sphere is given by the number in brackets; (2) there are 2^{12} such spheres; and (3) there are 2^{23} points in the space. Therefore one might suspect the existence of a (23, 12, 7)

[4] Although

$$\binom{90}{0} + \binom{90}{1} + \binom{90}{2} = 2^{12},$$

there is no perfect binary (90, 78) double-error-correcting code.

code. Such a code does exist; it is the binary Golay code. We shall introduce the binary Golay code as a binary cyclic code in Chapter 5.

Definition 3.5.2. *A quasi-perfect code is a code with packing radius t and covering radius $t + 1$.*

Quasi-perfect codes are more common than perfect codes. When a quasi-perfect code exists for a given n and k (and no such perfect code exists), then for this n and k no other code can have a larger $d_{\min}$. However, quasi-perfect codes are scarce and have no special importance in practical applications.

For large codes it is convenient to have an approximation to the volume of a Hamming sphere. One useful approximation is expressed in terms of the *binary entropy function*, which is defined for $x \in [0, 1]$ as

$$H_b(x) = -x \log_2 x - (1 - x) \log_2(1 - x).$$

Theorem 3.5.3. *The volume of a Hamming sphere in $GF(2)^n$ of radius t not larger than $\frac{n}{2}$ satisfies*

$$\sum_{\ell=0}^{t} \binom{n}{\ell} \leq 2^{nH_b(t/n)}.$$

Proof: Let λ be any positive number. Because $\binom{n}{\ell} = \binom{n}{n-\ell}$, we have the following string of inequalities:

$$\begin{aligned} 2^{\lambda(n-t)} \sum_{\ell=0}^{t} \binom{n}{\ell} &= 2^{\lambda(n-t)} \sum_{\ell=(n-t)}^{n} \binom{n}{\ell} \leq \sum_{\ell=(n-t)}^{n} 2^{\lambda\ell} \binom{n}{\ell} \\ &\leq \sum_{\ell=0}^{n} 2^{\lambda\ell} \binom{n}{\ell} = (1 + 2^{\lambda})^n. \end{aligned}$$

Let $p = t/n$, and summarize this inequality as

$$\sum_{\ell=0}^{t} \binom{n}{\ell} \leq \left(2^{-\lambda(1-p)} + 2^{\lambda p}\right)^n.$$

Now choose $\lambda = \log_2[(1 - p)/p]$. Then

$$\begin{aligned} \sum_{\ell=0}^{t} \binom{n}{\ell} &\leq \left[2^{-(1-p)\log_2[(1-p)/p]} + 2^{p \log_2[(1-p)/p)]}\right]^n \\ &= \left[2^{-(1-p)\log_2(1-p) - p\log_2 p}\left(2^{\log_2 p} + 2^{\log_2(1-p)}\right)\right]^n \\ &= 2^{nH(p)}(p + 1 - p)^n, \end{aligned}$$

which proves the theorem. □

Although in this theorem we state and prove only an upper bound, Stirling's approximation to $n!$ can be used to show the bound is asymptotically tight. To see the usefulness of the theorem, suppose that $n = 1000$ and $t = 110$. Although the left side would be very difficult to calculate, the right side is quickly seen to be 2^{500}.

3.6 Simple modifications to a linear code

There are a number of simple changes one might make to a linear code to obtain a new code. If the new code is also linear, these changes correspond to simple changes that can be made to the generator matrix $\boldsymbol{G}$: Namely, one can add a column or a row; delete a column or a row; add both a column and a row; or delete both a column and a row. These changes affect the code in ways that are simply described, although it might not be a simple matter to find such modifications that improve the code.

The blocklength n can be increased by increasing the dimension, or by increasing the redundancy. We call these *lengthening* and *extending* and we subsume both notions under the term *expanding*. By expanding, we mean increasing the blocklength either by extending or by lengthening. The blocklength n can be decreased by decreasing the dimension, or by decreasing the redundancy. We call these *puncturing* and *shortening*, and we subsume both notions under the term *restricting*. By restricting, we mean decreasing the blocklength either by puncturing or by shortening. We may also increase or decrease the number of data symbols without changing the blocklength.

The six basic changes are as follows.

Extending a code. Increasing the length by appending more check symbols. This corresponds to increasing the larger dimension of the generator matrix.

Lengthening a code. Increasing the length by appending more data symbols. This corresponds to increasing both dimensions of the generator matrix by the same amount.

Puncturing a code. Decreasing the length by dropping check symbols. This corresponds to eliminating columns of the generator matrix $\boldsymbol{G}$, thereby decreasing its larger dimension.

Shortening a code. Decreasing the length by dropping data symbols. This corresponds to eliminating columns of the check matrix $\boldsymbol{H}$, thereby decreasing both dimensions of the generator matrix by the same amount.

Augmenting a code. Increasing the number of data symbols without changing the length. This corresponds to increasing the smaller dimension of the generator matrix.

Expurgating a code. Decreasing the number of data symbols without changing the length. This corresponds to eliminating rows of the generator matrix $\boldsymbol{G}$, thereby decreasing the smaller dimension of the generator matrix.

These modifications can be used to remodel a known code to fit a specific application. They might also be used to devise new classes of good codes, or to study properties of one code by relating it to another. An expurgated code is also referred to as a *subcode* of the original code, while an augmented code is also called a *supercode* of the original code.

Any binary $(n, k, d_{\min})$ code whose minimum distance is odd can be expanded to an $(n+1, k, d_{\min}+1)$ code by appending the sum of all other codeword components as an overall check symbol. When the original codeword has odd weight, the new bit will be a 1. Hence all codewords of weight $d_{\min}$ become codewords of weight $d_{\min}+1$. If $\boldsymbol{H}$ is the check matrix of the original code, then the expanded code has check matrix

$$\boldsymbol{H}' = \begin{bmatrix} 1 & 1 \ldots 1 \\ 0 & \\ \vdots & \boldsymbol{H} \\ 0 & \end{bmatrix}.$$

Specifically, every $(2^m - 1, 2^m - 1 - m)$ Hamming code can be extended into a $(2^m, 2^m - m)$ single-error-correcting, double-error-detecting code, also called a *Hamming code*, or more precisely, an *extended Hamming code*. The minimum distance of this code is four.

One can also reduce a code over a given field to a code over a smaller field. Start with a code $\mathcal{C}$ over a field $GF(q^m)$ and gather together all those codewords that have components only in the subfield $GF(q)$. This code over $GF(q)$ is called a *subfield-subcode* of the original code. The subfield-subcode is $\mathcal{C} \cap GF(q)^n$. If the original code is linear, then the subfield-subcode is also linear because over the subfield $GF(q)$ all linear combinations of codewords must be in the subfield-subcode. It is not, however, a subspace of the original code because a subspace must contain all multiples of codewords by elements of $GF(q^m)$. Any set of basis vectors for the subfield-subcode is also linearly independent – even over $GF(q^m)$. Thus the dimension of the original code is at least as large as the dimension of the subfield-subcode. In general, however, the original code has a larger dimension; a subfield-subcode usually has a smaller rate than the original code.

Problems

3.1 The generator matrix for a code over $GF(2)$ is given by

$$\boldsymbol{G} = \begin{bmatrix} 1 & 0 & 1 & 0 & 1 & 1 \\ 0 & 1 & 1 & 1 & 0 & 1 \\ 0 & 1 & 1 & 0 & 1 & 0 \end{bmatrix}.$$

a. Find the generator matrix and the check matrix for an equivalent systematic code.

b. List the vectors in the orthogonal complement of the code.

c. Form the standard array for this code.

d. How many codewords are there of weight $0, \ldots, 6$?

e. Find the codeword with 101 as data symbols. Decode the received word 111001.

3.2 Given a code with check matrix $\boldsymbol{H}$, show that the coset with syndrome $\boldsymbol{s}$ contains a vector of weight w if and only if some linear combination of w columns of $\boldsymbol{H}$ equals $\boldsymbol{s}$.

3.3 This problem will show you that a good decoder must have some nonlinear operations (even though the code itself may be linear).

a. Show that the operation of forming the syndrome is a linear function of the error pattern: If $\boldsymbol{s} = F(\boldsymbol{e})$, then

$$F(a\boldsymbol{e}_1 + b\boldsymbol{e}_2) = aF(\boldsymbol{e}_1) + bF(\boldsymbol{e}_2).$$

b. A linear decoder is a decoder for which the decoder estimate of the error pattern from the syndrome $\hat{\boldsymbol{e}} = f(\boldsymbol{s})$ satisfies

$$f(a\boldsymbol{s}_1 + b\boldsymbol{s}_2) = af(\boldsymbol{s}_1) + bf(\boldsymbol{s}_2).$$

Show that a linear decoder can correct at most $(n-k)(q-1)$ of the $n(q-1)$ possible single-error patterns.

c. Show that the decoder estimate of the error pattern $\hat{\boldsymbol{e}} = f(\boldsymbol{s})$ must be nonlinear if the decoder is to correct all possible single-error patterns.

3.4 a. For a linear block code over $GF(2)$, prove that either no codeword has odd weight or that half of the codewords have odd weight.

b. Show that if a linear binary code has a minimum distance that is odd, extending the code by appending an overall check symbol increases the minimum distance by one.

3.5 Define a linear (5, 3) code over $GF(4)$ by the generator matrix

$$\boldsymbol{G} = \begin{bmatrix} 1 & 0 & 0 & 1 & 1 \\ 0 & 1 & 0 & 1 & 2 \\ 0 & 0 & 1 & 1 & 3 \end{bmatrix}.$$

a. Find the check matrix.

b. Prove that this is a single-error-correcting code.

c. Prove that it is a double-erasure-correcting code.

d. Prove that it is a perfect code.

3.6 What is the rate of the (maximal) binary subfield-subcode of the $GF(4)$-ary code given in Problem 3.5? Give a systematic generator matrix for this subfield-subcode.

3.7 By a counting argument, show that one should conjecture the existence of a perfect (11, 6, 5) double-error-correcting code over $GF(3)$. (Golay found such a code in 1949.)

3.8 By a counting argument, show that one should conjecture the existence of a perfect $(q+1, q-1)$ single-error-correcting code over $GF(q)$. (These codes are called extended single-error-correcting Reed–Solomon codes. They are equivalent to Hamming codes.)

3.9 Find a procedure for restoring double erasures for the binary Hamming (7, 4) code. (An erasure is a position in which the transmitted symbol is lost. It differs from an error because the position where the erasure occurs is known.)

3.10 Find a check matrix and a generator matrix for the (21, 18) Hamming code over $GF(4)$. Design an encoder and decoder using a syndrome look-up technique by sketching circuits that will implement the decoder. It is not necessary to fill in all details of the decoder. Is the look-up table too large to be practical?

3.11 Prove that the sum of two binary m-vectors of even weight has even weight. Is this statement also true for nonbinary vectors?

3.12 Ball weighing. You are given a pan balance and twelve pool balls where one ball may be either too light or too heavy.

a. Use the check matrix of a shortened Hamming (12, 9) code over $GF(3)$ to devise a procedure for determining with three uses of the balance which ball, if any, is faulty and whether it is too light or too heavy.

b. Can the problem be solved in this way for thirteen pool balls? Can it be solved for thirteen pool balls if a fourteenth ball, known to be good, is made available?

c. By sphere-packing arguments, show that the problem cannot be solved if there are fourteen balls, any one of which may be too heavy or too light.

3.13 A code that is equal to its dual code is called a *self-dual code*.

a. Show that a linear code with check matrix $[(-\boldsymbol{P})\ \boldsymbol{I}]$ is a self-dual code if and only if $\boldsymbol{P}$ is a square matrix satisfying $\boldsymbol{P}\boldsymbol{P}^{\mathrm{T}} = -\boldsymbol{I}$.

b. Construct binary self-dual codes of blocklengths 4 and 8.

3.14 Can every nonbinary (n, k, d) Hamming code be expanded by a simple check symbol to obtain an $(n+1, k, d+1)$ code? Give a proof or a counterexample.

3.15 Construct the Hamming (7, 4) binary code from a "Venn diagram" as follows. Intersect three circles to form seven regions, assigning data bits to four regions and check bits to three regions. Using this picture, discuss error correction and erasure filling.

3.16 a. Two linear codes have generator matrices

$$\boldsymbol{G} = \begin{bmatrix} 1 & 0 & 0 & 1 & 1 & 1 \\ 0 & 1 & 0 & 1 & 1 & 1 \\ 0 & 0 & 1 & 1 & 1 & 1 \end{bmatrix}$$

and

$$G' = \begin{bmatrix} 1 & 0 & 0 & 1 & 0 & 0 \\ 0 & 1 & 0 & 0 & 1 & 0 \\ 0 & 0 & 1 & 0 & 0 & 1 \end{bmatrix}.$$

For each code, list the number of codewords of each weight from 1 to 6.

b. If two linear codes have the same weight distribution, does this necessarily mean that the codes are equivalent?

3.17 Show that expanding a code by appending an overall check, then puncturing that code by deleting the new check symbol, reproduces the original code. Show that this need not be so if the operations are reversed.

3.18 Under what condition is puncturing a code by b places equivalent to shortening a code by b places?

3.19 a. Prove that $\log_e(1 + x) \le x$ with equality if and only if $x = 0$.

b. Prove that for any n, $(\frac{n+1}{n})^n < e$.

Notes

The study of linear codes begins with the early papers of Hamming (1950) and Golay (1949). Most of the formal setting now used to introduce linear codes is from Slepian (1956, 1960). The structure of the first three sections of this chapter depend heavily on his work. Earlier, Kiyasu (1953) had noticed the relationship between linear codes and subspaces of vector spaces. The maximum-distance codes were first studied by Singleton (1964).

The binary Hamming codes as error-correcting codes are from Hamming, although the combinatorial structure appeared earlier in problems of statistics. Nonbinary Hamming codes were developed by Golay (1958) and Cocke (1959).

The notion of a perfect code is from Golay, although he did not use this name. Early on, there was much excitement about the search for perfect codes, but few were found. In a series of difficult papers, Tietäväinen and van Lint (concluding in 1974 and 1975, respectively) proved that no nontrivial perfect codes over a field exist other than the Hamming codes and the Golay codes, except for nonlinear single-error-correcting codes such as the codes of Vasilýev (1962) and Schönheim (1968), and no nonlinear (nontrivial) perfect codes exist.

4 The Arithmetic of Galois Fields

The most powerful and important ideas of coding theory are based on the arithmetic systems of Galois fields. Since these arithmetic systems may be unfamiliar to many, we must develop a background in this branch of mathematics before we can proceed with the study of coding theory.

In this chapter, we return to the development of the structure of Galois fields begun in Chapter 2. There we introduced the definition of a field, but did not develop procedures for actually constructing Galois fields in terms of their addition and multiplication tables. In this chapter, we shall develop such procedures. Galois fields will be studied by means of two constructions: one based on the integer ring and one based on polynomial rings. Because the integer ring and the polynomial rings have many properties in common we will find that the two constructions are very similar. Later, after the constructions of Galois fields are studied, we shall prove that all finite fields can be constructed in this way.

4.1 The integer ring

The set of integers (positive, negative, and zero) forms a ring under the usual operations of addition and multiplication. This ring is conventionally denoted by the label **Z**. We shall study the structure of the integer ring in this section.

We say that the integer s is *divisible* by the integer r, or that r *divides* s (or that r is a *factor* of s), if $ra = s$ for some integer a. Whenever r both divides s and is divisible by s, then $r = \pm s$. This is because $r = sa$ and $s = rb$ for some a and b. Therefore $r = rab$, and ab must equal one. Since a and b are integers, a and b must both be either one or minus one.

A positive integer p greater than one that is divisible only by $\pm p$ or ± 1 is called a *prime integer*. A positive integer greater than one that is not prime is called *composite*. The *greatest common divisor* of two integers, r and s, denoted by GCD(r, s), is the largest positive integer that divides both of them. The least common multiple of two integers, r and s, denoted by LCM(r, s), is the smallest positive integer that is divisible by both of them. Two integers are said to be *coprime* (or *relatively prime*) if their greatest common divisor is one.

Within the ring of integers, division is not possible in general. We do, however, have the two next-best things: *cancellation* and *division with remainder*. Because cancellation is possible, the integer ring is a domain. Division with remainder (known as the division algorithm) is usually proved by a constructive procedure. We state it as a self-evident theorem.

Theorem 4.1.1 (Division Algorithm). *For every pair of integers c and d, with d nonzero, there is a unique pair of integers Q (the quotient) and s (the remainder) such that $c = dQ + s$ and $0 \leq s < |d|$.*

Usually, we will be more interested in the remainder than in the quotient. The remainder will often be expressed as

$$s = R_d[c],$$

which is read as "s is the remainder or residue of c when divided by d." Another closely related notation is

$$s \equiv c(\text{mod } d).$$

In this form, the expression is called a *congruence* and is read as "s is congruent to c modulo d." It means that s and c have the same remainder when divided by d, but s is not necessarily smaller than d.

The computation of the remainder of a complicated expression involving addition and multiplication is facilitated by noting that the process of computing a remainder can be interchanged with addition and multiplication. That is,

Theorem 4.1.2.

(i) $R_d[a + b] = R_d[R_d[a] + R_d[b]]$.
(ii) $R_d[a \cdot b] = R_d[R_d[a] \cdot R_d[b]]$.

Proof: The proof is a consequence of the uniqueness of the remainder in the division algorithm, and is left as an exercise. □

Using the division algorithm, we can find the greatest common divisor of two integers. As an example, GCD (814, 187) is found as follows:

$$814 = 4 \times 187 + 66$$
$$187 = 2 \times 66 + 55$$
$$66 = 1 \times 55 + 11$$
$$55 = 5 \times 11 + 0.$$

In the first line, because GCD(814, 187) divides 814 and 187, it must divide the remainder 66. Because it divides 187 and 66, it divides 55. Because it divides 66 and 55, it divides 11. On the other hand, starting at the bottom line, 11 divides 55, and therefore 66, and therefore 187, and finally also 814. Therefore GCD(814, 187) must be 11.

We now can express 11 as a linear combination of 814 and 187 by starting at the bottom of the above sequence of equations and working as follows:

$$\begin{aligned}
11 &= 66 - 1 \times 55 \\
&= 66 - 1 \times (187 - 2 \times 66) = 3 \times 66 - 1 \times 187 \\
&= 3 \times (814 - 4 \times 187) - 1 \times 187 \\
&= 3 \times 814 - 13 \times 187.
\end{aligned}$$

Hence we have found that GCD(814, 187) can be expressed as a linear combination of 814 and 187 with coefficients from the integer ring, that is,

$$\text{GCD}(814, 187) = 3 \times 814 - 13 \times 187.$$

The argument can be restated in general terms for arbitrary integers r and s to prove the following theorem and corollary.

Theorem 4.1.3 (Euclidean Algorithm). *Given two distinct nonzero integers r and s, their greatest common divisor can be computed by an iterative application of the division algorithm. Suppose that $r < s$ and that both are positive; the algorithm is*

$$\begin{aligned}
s &= Q_1 r + r_1 \\
r &= Q_2 r_1 + r_2 \\
r_1 &= Q_3 r_2 + r_3 \\
&\vdots \\
r_{n-1} &= Q_{n+1} r_n
\end{aligned}$$

where the process stops when a remainder of zero is obtained. The last nonzero remainder, r_n, is the greatest common divisor.

The following corollary, sometimes called the *extended euclidean algorithm*, is an important and nonintuitive conclusion of number theory.

Corollary 4.1.4. *For any integers r and s, there exist integers a and b such that*

$$\text{GCD}(r, s) = ar + bs.$$

Proof: The last remainder r_n in Theorem 4.1.3 is $r_{n-2} = Q_n r_{n-1} + r_n$. Use the set of equations to successively eliminate all other remainders. This gives r_n as a linear combination of r and s with integer coefficients. □

The final topic of this section gives a condition under which a set of remainders uniquely specify an integer.

Theorem 4.1.5 (Chinese Remainder Theorem for Integers). *For any set of pairwise coprime positive integers* $\{m_0, m_1, \ldots, m_{K-1}\}$, *the set of congruences*

$$c \equiv c_k \pmod{m_k}, \qquad k = 0, \ldots, K-1$$

has exactly one nonnegative solution smaller than the product $M = \prod_k m_k$, *which is given by*

$$c = \sum_{k=0}^{K-1} c_k N_k M_k \pmod{M}$$

where $M_k = M/m_k$, *and* N_k *is the integer that satisfies*

$$N_k M_k + n_k m_k = 1.$$

Proof: The specified N_k must exist by Corollary 4.1.4 because $\text{GCD}(M_k, m_k) = 1$.

To show uniqueness, suppose that c and c' are solutions to the theorem, both smaller than M. Then the residues c_k and c'_k are equal for every k, so $c - c'$ is a multiple of m_k for every k. Consequently, $c - c'$ must be zero because it is a multiple of M and is between $-M$ and M.

To complete the proof, it is necessary to show that $c' (\text{mod } m_k) = c_k$ where

$$c' = \sum_{k=0}^{K-1} c_k N_k M_k \pmod{M}.$$

But $M_{k'} = 0 \pmod{m_k}$ if $k' \neq k$, so for $k = 0, \ldots, K-1$, this equation becomes

$$\begin{aligned} c' (\text{mod } m_k) &= \sum_{k'=0}^{K-1} c_{k'} N_{k'} M_{k'} \pmod{m_k} \\ &= c_k N_k M_k \pmod{m_k} \\ &= c_k (1 - n_k m_k) \pmod{m_k} \\ &= c_k \end{aligned}$$

as was to be proved. □

4.2 Finite fields based on the integer ring

There is an important method that can be used to construct a new ring, called a *quotient ring*, from a given commutative ring. For an arbitrary commutative ring, the quotient ring is defined in a somewhat technical way that involves the construction of cosets. In the ring of integers, however, the construction of the quotient ring is easy. This construction, in some cases, will also yield a field (when the ring is a domain).

Definition 4.2.1. *Let q be a positive integer. The quotient ring called the ring of integers modulo q, denoted by $\mathbf{Z}/(q)$ or $\mathbf{Z}_q$, is the set $\{0, \ldots, q-1\}$ with addition and multiplication defined by*

$$a + b = R_q[a + b],$$
$$a \cdot b = R_q[ab].$$

Any element a of $\mathbf{Z}$ can be mapped into $\mathbf{Z}/(q)$ by $a' = R_q[a]$. Two elements a and b of $\mathbf{Z}$ that map into the same element of $\mathbf{Z}/(q)$ are congruent modulo q, and $a = b + mq$ for some integer m. One may think of the elements of $\mathbf{Z}/(q)$ as the same objects as the first q elements of $\mathbf{Z}$, but it is formally correct to think of them as some other objects, namely cosets.

Theorem 4.2.2. *The quotient ring $\mathbf{Z}/(q)$ is a ring.*

Proof: Exercise. □

We can see in the examples of Section 2.4 that the arithmetic of the Galois fields $GF(2)$ and $GF(3)$ can be described as addition and multiplication modulo 2 and 3, respectively, but the arithmetic of $GF(4)$ cannot be so described. In symbols, $GF(2) = \mathbf{Z}/(2)$, $GF(3) = \mathbf{Z}/(3)$, and $GF(4) \neq \mathbf{Z}/(4)$. This general fact is given by the following theorem.

Theorem 4.2.3. *The quotient ring $\mathbf{Z}/(q)$ is a field if and only if q is a prime integer.*

Proof: Suppose that q is a prime. To prove that the ring is a field, we must show that every nonzero element has a multiplicative inverse. Let s be a nonzero element of the ring. Then

$$1 \leq s \leq q - 1.$$

Because q is prime, GCD$(s, q) = 1$. Hence by Corollary 4.1.4,

$$1 = aq + bs$$

for some integers a and b. Therefore

$$\begin{aligned} 1 &= R_q[1] = R_q[aq + bs] = R_q\{R_q[aq] + R_q[bs]\} \\ &= R_q[bs] = R_q\{R_q[b]R_q[s]\} \\ &= R_q\{R_q[b]s\}. \end{aligned}$$

Hence $R_q[b]$ is a multiplicative inverse for s under modulo-q multiplication.

Now suppose that q is composite. Then $q = rs$ where r and s are nonzero. If the ring is a field, then r has an inverse r^{-1}. Hence

$$s = R_q[s] = R_q[r^{-1}rs] = R_q[R_q[r^{-1}]R_q[rq]] = 0.$$

But $s \neq 0$, and thus we have a contradiction. Hence the ring is not a field. □

Whenever the quotient ring $\mathbf{Z}/(q)$ is a field, it is also known by the name $GF(q)$, which emphasizes that it is a field. Thus $GF(q) = \mathbf{Z}/(q)$ if and only if q is a prime integer.

4.3 Polynomial rings

A *polynomial* over any field F is a mathematical expression

$$f(x) = f_{n-1}x^{n-1} + f_{n-2}x^{n-2} + \cdots + f_1x + f_0$$

where the symbol x is an *indeterminate*, the coefficients $f_{n-1}, \ldots, f_0$ are elements of F, the *index* of f_i is the integer i, and exponents of x are integers. We are interested primarily in polynomials over the finite field $GF(q)$. The *zero polynomial* is

$$f(x) = 0.$$

A *monic polynomial* is a polynomial with leading coefficient f_{n-1} equal to one. Two polynomials $f(x)$ and $g(x)$ are equal if coefficients f_i and g_i are equal for each i.

The *degree* of a nonzero polynomial $f(x)$, denoted $\deg f(x)$, is the index of the leading coefficient f_{n-1}. The degree of a nonzero polynomial is always finite. By convention, the degree of the zero polynomial is negative infinity $(-\infty)$.

The set of all polynomials over $GF(q)$ forms a ring if addition and multiplication are defined as the usual addition and multiplication of polynomials. We define such a polynomial ring for each Galois field $GF(q)$. This ring is denoted by the label $GF(q)[x]$. In discussions about the ring $GF(q)[x]$, the elements of the field $GF(q)$ are sometimes called *scalars*.

The sum of two polynomials $f(x)$ and $g(x)$ in $GF(q)[x]$ is another polynomial in $GF(q)[x]$, defined by

$$f(x) + g(x) = \sum_{i=0}^{\infty}(f_i + g_i)x^i$$

where, of course, terms whose index is larger than the larger of the degrees of $f(x)$ and $g(x)$ are all zero. The degree of the sum is not greater than the larger of these two degrees.

As an example, over $GF(2)$,

$$(x^3 + x^2 + 1) + (x^2 + x + 1) = x^3 + (1 + 1)x^2 + x + (1 + 1) = x^3 + x.$$

The product of two polynomials in $GF(q)[x]$ is another polynomial in $GF(q)[x]$, defined by

$$f(x)g(x) = \sum_i \left(\sum_{j=0}^{i} f_j g_{i-j}\right) x^i.$$

The degree of a product is equal to the sum of the degrees of the two factors.

As an example, over $GF(2)$,

$$(x^3 + x^2 + 1)(x^2 + x + 1) = x^5 + x + 1.$$

A polynomial ring is analogous in many ways to the ring of integers. To make this evident, this section will closely follow Section 4.1. We say that the polynomial $s(x)$ is *divisible* by the polynomial $r(x)$, or that $r(x)$ is a *factor* of $s(x)$ if there is a polynomial $a(x)$ such that $r(x)a(x) = s(x)$. A polynomial $p(x)$ that is divisible only by $\alpha p(x)$ and by α, where α is any nonzero field element in $GF(q)$, is called an *irreducible polynomial.* A monic irreducible polynomial of degree at least 1 is called a *prime polynomial.*

The *greatest common divisor* of two polynomials $r(x)$ and $s(x)$, denoted $\text{GCD}[r(x), s(x)]$, is the monic polynomial of the largest degree that divides both of them. The *least common multiple* of two polynomials $r(x)$ and $s(x)$, denoted by $\text{LCM}[r(x), s(x)]$, is the monic polynomial of the smallest degree divisible by both of them. We shall see that the greatest common divisor and the least common multiple of $r(x)$ and $s(x)$ are unique, and thus our wording is appropriate. If the greatest common divisor of two polynomials is one, then they are said to be *coprime* or *relatively prime.*

Whenever $r(x)$ both divides $s(x)$ and is divisible by $s(x)$, then $r(x) = \alpha s(x)$ where α is a field element of $GF(q)$. This is proved as follows. There must exist polynomials $a(x)$ and $b(x)$ such that $r(x) = s(x)a(x)$ and $s(x) = r(x)b(x)$. Therefore $r(x) = r(x)b(x)a(x)$. But the degree of the right side is the sum of the degrees of $r(x)$, $b(x)$, and $a(x)$. Because this must equal the degree of the left side, $a(x)$ and $b(x)$ must have degree 0, that is, they are scalars.

For polynomials over the real field, the notion of differentiation is both elementary and useful. It is not possible to so define differentiation of polynomials over a finite field in the sense of a limiting operation. Nevertheless, it is convenient simply to define an operation on polynomials that behaves the way we expect derivatives to behave. This is called the *formal derivative* of a polynomial.

Definition 4.3.1. *Let* $r(x) = r_{n-1}x^{n-1} + r_{n-2}x^{n-2} + \cdots + r_1x + r_0$ *be a polynomial over* $GF(q)$. *The formal derivative of* $r(x)$ *is a polynomial* $r'(x)$, *given by*

$$r'(x) = ((n-1))r_{n-1}x^{n-2} + ((n-2))r_{n-2}x^{n-3} + \cdots + r_1.$$

The coefficients $((i))$ are called *integers* of the field $GF(q)$ and are given by

$$((i)) = 1 + 1 + \cdots + 1,$$

a sum of i terms in $GF(q)$. It is easy to verify that many of the usual properties of derivatives hold for the formal derivative, namely that

$$[r(x)s(x)]' = r'(x)s(x) + r(x)s'(x),$$

and that if $a(x)^2$ divides $r(x)$, then $a(x)$ divides $r'(x)$.

Every polynomial $p(x) = \sum_{i=0}^{r} p_i x^i$ is associated with a *reciprocal polynomial* defined as $\widetilde{p}(x) = \sum_{i=0}^{r} p_{r-i} x^i$. The reciprocal polynomial can be written $\widetilde{p}(x) = x^r p(x^{-1})$, which makes it clear that if $p(x) = a(x)b(x)$, then $\widetilde{p}(x) = \widetilde{a}(x)\widetilde{b}(x)$. If one specifies that p_r is not zero, then r is the degree of $p(x)$. The definition works, however, as long as r is chosen to be at least as large as the degree of $p(x)$.

Within a polynomial ring, as in the integer ring, division is not possible in general. For polynomials over a field, however, we again have the two next-best things: Cancellation and division with remainder. The latter is known as the division algorithm for polynomials.

Theorem 4.3.2 (Division Algorithm for Polynomials). *For every pair of polynomials $c(x)$ and $d(x)$, with $d(x)$ not equal to zero, there is a unique pair of polynomials $Q(x)$, the quotient polynomial, and $s(x)$, the remainder polynomial, such that*

$$c(x) = d(x)Q(x) + s(x)$$

and

$$\deg s(x) < \deg d(x).$$

Proof: A quotient polynomial and remainder polynomial can be found by elementary long division of polynomials. They are unique because if

$$c(x) = d(x)Q_1(x) + s_1(x) = d(x)Q_2(x) + s_2(x),$$

then

$$-d(x)[Q_1(x) - Q_2(x)] = s_1(x) - s_2(x).$$

If the right side is nonzero, it has a degree less than $\deg d(x)$, whereas if the left side is nonzero, it has degree at least as large as $\deg d(x)$. Hence both are zero and the representation is unique. □

In practice, one can compute the quotient polynomial and the remainder polynomial by simple long division of polynomials. Usually, we will be more interested in the remainder polynomial than the quotient polynomial. The remainder polynomial will also be written

$$s(x) = R_{d(x)}[c(x)].$$

The remainder polynomial $s(x)$ is also called the *residue* of $c(x)$ when divided by $d(x)$. A slightly different concept is the *congruence*

$$s(x) \equiv c(x) \pmod{d(x)},$$

which means that $s(x)$ and $c(x)$ have the same remainder under division by $d(x)$, but the degree of $s(x)$ is not necessarily smaller than that of $d(x)$.

Computation of a remainder is sometimes made more convenient if the division can be broken down into steps. We can do this with the aid of the following theorem.

Theorem 4.3.3. *Let $d(x)$ be a multiple of $g(x)$. Then for any $a(x)$,*

$$R_{g(x)}[a(x)] = R_{g(x)}\big[R_{d(x)}[a(x)]\big].$$

Proof: Let $d(x) = g(x)h(x)$ for some $h(x)$. Expanding the meaning of the right side gives

$$\begin{aligned} a(x) &= Q_1(x)d(x) + R_{d(x)}[a(x)] \\ &= Q_1(x)h(x)g(x) + Q_2(x)g(x) + R_{g(x)}\big[R_{d(x)}[a(x)]\big] \end{aligned}$$

where the remainder has a degree less than $\deg g(x)$. Expanding the meaning of the left side gives

$$a(x) = Q(x)g(x) + R_{g(x)}[a(x)],$$

and the division algorithm says there is only one such expansion where the remainder has a degree less than $\deg g(x)$. The theorem follows by identifying like terms in the two expansions. □

Theorem 4.3.4.

(i) $R_{d(x)}[a(x) + b(x)] = R_{d(x)}[a(x)] + R_{d(x)}[b(x)]$.
(ii) $R_{d(x)}[a(x) \cdot b(x)] = R_{d(x)}\big\{R_{d(x)}[a(x)] \cdot R_{d(x)}[b(x)]\big\}$.

Proof: Exercise: Use the division algorithm on both sides of the equation and equate the two remainders. □

Just as it is often useful to express positive integers as products of primes, it is often useful to express monic polynomials as products of prime polynomials. The following theorem states that the *prime factorization* is unique.

Theorem 4.3.5 (Unique Factorization Theorem). *A nonzero polynomial $p(x)$ over a field has a unique factorization (up to the order of the factors) into a field element times a product of prime polynomials over the field.*

Proof: Clearly, the field element must be the coefficient p_{n-1} where $n-1$ is the degree of the polynomial $p(x)$. We can factor this field element from the polynomial and prove the theorem for monic polynomials.

Suppose the theorem is false. Let $p(x)$ be a monic polynomial of the lowest degree for which the theorem fails. Then there are two factorizations:

$$p(x) = a_1(x)a_2(x)\cdots a_K(x) = b_1(x)b_2(x)\cdots b_J(x)$$

where the $a_k(x)$ and $b_j(x)$ are prime polynomials. All of the $a_k(x)$ must be different from all of the $b_j(x)$ because, otherwise, the common terms could be canceled to give a polynomial of lower degree that could be factored in two different ways.

Without loss of generality, suppose that $b_1(x)$ has a degree not larger than that of $a_1(x)$. Then

$$a_1(x) = b_1(x)h(x) + s(x)$$

where $\deg s(x) < \deg b_1(x) \leq \deg a_1(x)$. Substitute this for $a_1(x)$ to obtain

$$s(x)a_2(x)a_3(x)\cdots a_K(x) = b_1(x)[b_2(x)\cdots b_J(x) - h(x)a_2(x)\cdots a_K(x)].$$

Factor both $s(x)$ and the bracketed term into their prime factors and, if necessary, divide by a field element to make all factors monic. Because $b_1(x)$ does not appear on the left side, we have two different factorizations of a monic polynomial whose degree is smaller than the degree of $p(x)$. The contradiction proves the theorem. □

From the unique factorization theorem, it is clear that for any polynomials $r(x)$ and $s(x)$, GCD$[r(x), s(x)]$ and LCM$[r(x), s(x)]$ are unique because the greatest common divisor is the product of all prime factors common to both $r(x)$ and $s(x)$, with each factor raised to the smallest power with which it appears in either $r(x)$ or $s(x)$. Also, the least common multiple is the product of all prime factors that appear in either $r(x)$ or $s(x)$, with each factor raised to the largest power that appears in either $r(x)$ or $s(x)$. Further, any polynomial that divides both $r(x)$ and $s(x)$ divides GCD$[r(x), s(x)]$, and any polynomial that both $r(x)$ and $s(x)$ divide is divided by LCM$[r(x), s(x)]$.

The division algorithm for polynomials has an important consequence known as the *euclidean algorithm for polynomials.*

Theorem 4.3.6 (Euclidean Algorithm for Polynomials). *Given two polynomials $r(x)$ and $s(x)$ over $GF(q)$, their greatest common divisor can be computed by an iterative application of the division algorithm for polynomials. If* $\deg s(x) \geq \deg r(x) \geq 0$, *this computation is*

$$\begin{aligned}
s(x) &= Q_1(x)r(x) + r_1(x) \\
r(x) &= Q_2(x)r_1(x) + r_2(x) \\
r_1(x) &= Q_3(x)r_2(x) + r_3(x) \\
&\ \vdots \\
r_{n-1}(x) &= Q_{n+1}(x)r_n(x)
\end{aligned}$$

where the process stops when a remainder of zero is obtained. Then $r_n(x) =$ αGCD$[r(x), s(x)]$ where α is a scalar.

Proof: Starting with the top equation, GCD$[r(x), s(x)]$ divides both the dividend and divisor, and thus it divides the remainder. Push this observation down through the equations to see that GCD$[r(x), s(x)]$ divides $r_n(x)$. Starting with the bottom equation, $r_n(x)$ divides the divisor and remainder, and thus divides the dividend. Push this observation up through the equations to see that $r_n(x)$ divides GCD$[r(x), s(x)]$. Because $r_n(x)$ both divides and is divided by GCD$[r(x), s(x)]$, the theorem follows. □

Corollary 4.3.7.

$$\text{GCD}[r(x), s(x)] = a(x)r(x) + b(x)s(x)$$

where $a(x)$ and $b(x)$ are polynomials over $GF(q)$.

Proof: In the statement of the theorem, the last equation with a nonzero remainder expresses $r_n(x)$ in terms of $r_{n-1}(x)$ and $r_{n-2}(x)$. By working the list of equations from the bottom up, eliminate $r_{n-1}(x)$, then $r_{n-2}(x)$, and so on until only $r(x)$ and $s(x)$ remain in the expression for $r_n(x)$. □

Theorem 4.3.8 (Chinese Remainder Theorem for Polynomials). *For any set of pairwise coprime polynomials $\{m_0(x), m_1(x), \ldots, m_{K-1}(x)\}$, the set of congruences*

$$c(x) \equiv c_k(x) \quad (\mathit{mod}\ \ m_k(x)), \qquad k = 0, \ldots, K-1$$

has exactly one solution of a degree smaller than the degree of $M(x) = \prod_k m_k(x)$, which is given by

$$c(x) = \sum_{k=0}^{K-1} c_k(x)N_k(x)M_k(x) \quad (\mathit{mod}\ \ M(x))$$

where $M_k(x) = M(x)/m_k(x)$, and $N_k(x)$ is the polynomial that satisfies

$$N_k(x)M_k(x) + n_k(x)m_k(x) = 1.$$

Proof: The specified $N_k(x)$ must exist by Corollary 4.3.7 because GCD$[M_k(x), m_k(x)] = 1$.

To show uniqueness, suppose that $c(x)$ and $c'(x)$ are solutions to the theorem. Then the residues $c_k(x)$ and $c'_k(x)$ are equal, so $c_k(x) - c'_k(x)$ is a multiple of $m_k(x)$ for all k. Because $c(x) - c'(x)$ is a multiple of $M(x)$ with smaller degree, it must be zero, and it follows that $c(x) = c'(x)$.

To complete the proof, it is necessary to show that $c'(x)$ (mod $m_k(x)) = c_k(x)$ where

$$c'(x) = \sum_{k=0}^{K-1} c_k(x)N_k(x)M_k(x) \quad (\text{mod}\ \ M(x)).$$

But $M_{k'}(x) = 0 \pmod{m_k(x)}$ if $k' \neq k$, so this equation becomes

$$\begin{aligned} c'(x) \pmod{m_k(x)} &= \sum_{k'=0}^{K-1} c_{k'}(x) N_{k'}(x) M_{k'}(x) \pmod{m_k(x)} \\ &= c_k(x) N_k(x) M_k(x) \pmod{m_k(x)} \\ &= c_k(x)[1 - n_k(x) m_k(x)] \pmod{m_k(x)} \\ &= c_k(x) \end{aligned}$$

as was to be proved. □

A polynomial over $GF(q)$ can be evaluated at any element β of $GF(q)$. This is done by substituting the field element β for the indeterminate x. For example, over $GF(3)$, let

$$p(x) = 2x^5 + x^4 + x^2 + 2.$$

The field $GF(3) = \{0, 1, 2\}$ has modulo-3 arithmetic. It is easy to evaluate $p(x)$ at the three elements of $GF(3)$:

$$\begin{aligned} p(0) &= 2 \cdot 0^5 + 0^4 + 0^2 + 2 = 2 \\ p(1) &= 2 \cdot 1^5 + 1^4 + 1^2 + 2 = 0 \\ p(2) &= 2 \cdot 2^5 + 2^4 + 2^2 + 2 = 2. \end{aligned}$$

In the case of the real field, the evaluation of a polynomial in an extension field is a familiar concept because polynomials with real coefficients are commonly evaluated over the complex field. Similarly, a polynomial over $GF(q)$ can be evaluated in an extension of $GF(q)$. This is done by substituting the element of the extension field for the indeterminate x and performing the computations in the extension field. For example, over $GF(2)$, let

$$p(x) = x^3 + x + 1.$$

Then for elements in $GF(4)$, using the arithmetic in Figure 4.1, we can compute

$$\begin{aligned} p(0) &= 0^3 + 0 + 1 = 1 \\ p(1) &= 1^3 + 1 + 1 = 1 \\ p(2) &= 2^3 + 2 + 1 = 2 \\ p(3) &= 3^3 + 3 + 1 = 3. \end{aligned}$$

If $p(\beta) = 0$, the field element β is called a *zero* of a polynomial $p(x)$, or a *root* of the equation $p(x) = 0$. A polynomial does not necessarily have zeros in its own field. The polynomial $x^3 + x + 1$ has no zeros in $GF(2)$ and also has no zeros in $GF(4)$. It does have zeros in $GF(8)$.

$GF(4)$									
$+$	0	1	2	3	$\cdot$	0	1	2	3
0	0	1	2	3	0	0	0	0	0
1	1	0	3	2	1	0	1	2	3
2	2	3	0	1	2	0	2	3	1
3	3	2	1	0	3	0	3	1	2

Figure 4.1. Example of a finite field

Theorem 4.3.9 (Fundamental Theorem of Algebra). *A polynomial $p(x)$ of degree n has at most n zeros. In particular, the polynomial $p(x)$ has field element β as a zero if and only if $(x - \beta)$ is a factor of $p(x)$.*

Proof: Suppose β is a zero of $p(x)$. From the division algorithm,

$$p(x) = (x - \beta)Q(x) + s(x)$$

where the degree of $s(x)$ is less than one. That is, $s(x)$ is a field element, s_0. Hence

$$0 = p(\beta) = (\beta - \beta)Q(\beta) + s_0,$$

and thus $s(x) = s_0 = 0$. Conversely, if $(x - \beta)$ is a factor, then

$$p(x) = (x - \beta)Q(x)$$

and $p(\beta) = (\beta - \beta)Q(\beta) = 0$, and thus β is a zero of $p(x)$. Next, factor $p(x)$ into a field element times a product of prime polynomials. The degree of $p(x)$ equals the sum of the degrees of the prime factors, and one such prime factor exists for each zero. Hence there are at most n zeros. □

4.4 Finite fields based on polynomial rings

Finite fields can be obtained from polynomial rings by using constructions that mimic those used to obtain finite fields from the integer ring. Suppose that we have $F[x]$, the ring of polynomials over the field F. Just as we constructed quotient rings in the ring $\mathbf{Z}$, so, too, can we construct quotient rings in $F[x]$. Choosing any polynomial $p(x)$ from $F[x]$, we define the quotient ring by using $p(x)$ as a modulus for polynomial arithmetic. We will restrict the discussion to monic polynomials because this restriction eliminates needless ambiguity in the construction.

Definition 4.4.1. *For any monic polynomial $p(x)$ of nonzero degree over the field F, the ring of polynomials modulo $p(x)$ is the set of all polynomials with a degree smaller than that of $p(x)$, together with polynomial addition and polynomial multiplication modulo $p(x)$. This ring is conventionally denoted by $F[x]/\langle p(x)\rangle$.*

Any element $r(x)$ of $F[x]$ can be mapped into $F[x]/\langle p(x)\rangle$ by $r(x) \to R_{p(x)}[r(x)]$. Two elements $a(x)$ and $b(x)$ of $F[x]$ that map into the same element of $F[x]/\langle p(x)\rangle$ are called *congruent*:

$$a(x) \equiv b(x) \pmod{p(x)}.$$

Then $b(x) = a(x) + Q(x)p(x)$ for some polynomial $Q(x)$.

Theorem 4.4.2. *$F[x]/\langle p(x)\rangle$ is a ring.*

Proof: Exercise. □

As an example, in the ring of polynomials over $GF(2)$, choose $p(x) = x^3 + 1$. Then the ring of polynomials modulo $p(x)$ is $GF(2)[x]/\langle x^3 + 1\rangle$. It consists of the set $\{0, 1, x, x+1, x^2, x^2+1, x^2+x, x^2+x+1\}$. In this ring, an example of multiplication is as follows:

$$\begin{aligned}(x^2+1)\cdot(x^2) &= R_{x^3+1}[(x^2+1)\cdot x^2]\\ &= R_{x^3+1}[x(x^3+1)+x^2+x] = x^2+x\end{aligned}$$

where we have used the reduction $x^4 = x(x^3+1) + x$.

Theorem 4.4.3. *The ring of polynomials modulo a monic polynomial $p(x)$ is a field if and only if $p(x)$ is a prime polynomial.*[1]

Proof: Suppose that $p(x)$ is prime. To prove that the ring is a field, we must show that every nonzero element has a multiplicative inverse. Let $s(x)$ be a nonzero element of the ring. Then

$$\deg s(x) < \deg p(x).$$

Because $p(x)$ is a prime polynomial, GCD$[s(x), p(x)] = 1$. By Corollary 4.3.7,

$$1 = a(x)p(x) + b(x)s(x)$$

for some polynomials $a(x)$ and $b(x)$. Hence

$$\begin{aligned}1 = R_{p(x)}[1] &= R_{p(x)}[a(x)p(x) + b(x)s(x)]\\ &= R_{p(x)}\big\{R_{p(x)}[b(x)]\cdot R_{p(x)}[s(x)]\big\}\\ &= R_{p(x)}\big\{R_{p(x)}[b(x)]\cdot s(x)\big\}.\end{aligned}$$

Therefore $R_{p(x)}[b(x)]$ is a multiplicative inverse for $s(x)$ in the ring of polynomials modulo $p(x)$.

Now suppose that $p(x)$, whose degree is at least 2, is not prime. Then $p(x) = r(x)s(x)$ for some $r(x)$ and $s(x)$, each of degree at least 1. If the ring is a field, then $r(x)$ has an

[1] Recall that a prime polynomial is both monic and irreducible. To get a field, it is enough for $p(x)$ to be irreducible, but we insist on the convention of using a polynomial that is monic as well so that later results are less arbitrary.

Representations of *GF*(4)

Polynomial notation	Binary notation	Integer notation	Exponential notation
0	00	0	0
1	01	1	x^0
x	10	2	x^1
$x+1$	11	3	x^2

Arithmetic Tables

$+$	0	1	x	$x+1$
0	0	1	x	$x+1$
1	1	0	$x+1$	x
x	x	$x+1$	0	1
$x+1$	$x+1$	x	1	0

$\cdot$	0	1	x	$x+1$
0	0	0	0	0
1	0	1	x	$x+1$
x	0	x	$x+1$	1
$x+1$	0	$x+1$	1	x

Figure 4.2. Structure of $GF(4)$

inverse under multiplication, denoted $r^{-1}(x)$. Hence

$$s(x) = R_{p(x)}[s(x)] = R_{p(x)}[r^{-1}(x)r(x)s(x)] = R_{p(x)}[r^{-1}(x)p(x)] = 0.$$

But $s(x) \neq 0$, and thus we have a contradiction. Hence the ring is not a field. □

Using the theory of this section, whenever we can find a prime polynomial of degree m over $GF(q)$, then we can construct a finite field with q^m elements. In this construction, the elements are represented by polynomials over $GF(q)$ of a degree less than m. There are q^m such polynomials, and hence q^m elements in the field.

As an example, we will construct $GF(4)$ from $GF(2)$, using the prime polynomial $p(x) = x^2 + x + 1$. This polynomial is easily verified to be irreducible by testing all possible factorizations. The field elements are represented by the set of polynomials $\{0, 1, x, x+1\}$. The addition and multiplication tables, shown in Figure 4.2, are readily constructed. Of course, once the arithmetic tables have been constructed, one can replace the polynomial notation by an integer notation or by any other convenient notation.

Table 4.1 gives a list of prime polynomials over $GF(2)$. One way to verify that these polynomials are prime is by trial and error, testing all possible factorizations – although this will require a computer for polynomials of large degree. The particular prime polynomials selected for Table 4.1 are a special kind of prime polynomial known as a *primitive polynomial*. Using a primitive polynomial to construct an extension field gives an especially desirable representation, as will be described in the next section.

Table 4.1. *Prime polynomials over GF(2)*

Degree	Prime polynomial (Note that all entries are primitive polynomials)
2	$x^2 + x + 1$
3	$x^3 + x + 1$
4	$x^4 + x + 1$
5	$x^5 + x^2 + 1$
6	$x^6 + x + 1$
7	$x^7 + x^3 + 1$
8	$x^8 + x^4 + x^3 + x^2 + 1$
9	$x^9 + x^4 + 1$
10	$x^{10} + x^3 + 1$
11	$x^{11} + x^2 + 1$
12	$x^{12} + x^6 + x^4 + x + 1$
13	$x^{13} + x^4 + x^3 + x + 1$
14	$x^{14} + x^{10} + x^6 + x + 1$
15	$x^{15} + x + 1$
16	$x^{16} + x^{12} + x^3 + x + 1$
17	$x^{17} + x^3 + 1$
18	$x^{18} + x^7 + 1$
19	$x^{19} + x^5 + x^2 + x + 1$
20	$x^{20} + x^3 + 1$
21	$x^{21} + x^2 + 1$
22	$x^{22} + x + 1$
23	$x^{23} + x^5 + 1$
24	$x^{24} + x^7 + x^2 + x + 1$
25	$x^{25} + x^3 + 1$
26	$x^{26} + x^6 + x^2 + x + 1$
27	$x^{27} + x^5 + x^2 + x + 1$
28	$x^{28} + x^3 + 1$

To conclude this section, we will summarize where we are. We have developed the main constructions needed to obtain the finite fields that will be used later. Additional topics must be developed for a full understanding of the subject. In particular, we need to establish the following facts. (1) Prime polynomials of every degree exist over every finite field. (2) The constructions put forth are sufficient to construct all finite fields – there are no others.[2] (3) Certain preferred field elements, called *primitive elements*, exist in every field. We will spend the remainder of this chapter establishing most of these facts and introducing new terms. Figure 4.3 summarizes the principal facts about finite fields.

[2] Mathematical precision requires a more formal statement here. The technically correct phrase is that there are no others up to isomorphism. Informally, this means that any two finite fields with the same number of elements are the same field, but expressed in different notation. Possibly the notation is a permutation of the same symbols and so creates the illusion of a different structure.

1. In any finite field, the number of elements is a power of a prime.
2. If p is prime and m is a positive integer, the smallest subfield of $GF(p^m)$ is $GF(p)$. The elements of $GF(p)$ are called the *integers* of $GF(p^m)$, and p is called its *characteristic*.
3. In a finite field of characteristic 2, $-\beta = \beta$ for every β in the field.
4. If p is a prime and m is an integer, then there is a finite field with p^m elements.
5. Every finite field $GF(q)$ has at least one primitive element.
6. Every finite field $GF(q)$ has at least one primitive polynomial over it of every positive degree.
7. Every primitive element of $GF(q)$ has, over any subfield of $GF(q)$, a minimal polynomial that is a prime polynomial.
8. Two finite fields with the same number of elements are isomorphic.
9. For any prime power q and positive integer m, $GF(q)$ is a subfield of $GF(q^m)$, and $GF(q^m)$ is an extension field of $GF(q)$.
10. $GF(q^n)$ is not a subfield of $GF(q^m)$ if n does not divide m.
11. The degree of the minimal polynomial over $GF(q)$ of any element of $GF(q^m)$ is a divisor of m.
12. The additive structure of the field $GF(q^m)$ is isomorphic to the vector space $[GF(q)]^m$.

Figure 4.3. Some basic properties of finite fields

4.5 Primitive elements

In the previous section, we constructed $GF(4)$. By inspection of Figure 4.1, we see that the field element represented by the polynomial x can be used as a kind of logarithm base. All field elements, except zero, can be expressed as a power of x.

Definition 4.5.1. *A primitive element of the field $GF(q)$ is an element α such that every field element except zero can be expressed as a power of α.*

For example, in the field $GF(5)$, we have

$$2^1 = 2,\ 2^2 = 4,\ 2^3 = 3,\ 2^4 = 1,$$

and thus 2 is a primitive element of $GF(5)$. In contrast, 4 is not a primitive element in $GF(5)$ because 2 cannot be expressed as a power of 4. Primitive elements are useful for constructing fields because, if we can find a primitive element, then we can construct a multiplication table by multiplying powers of the primitive element. In this section we shall prove that every finite field contains a primitive element.

A field forms an abelian group in two ways: The set of field elements forms an abelian group under the addition operation, and the set of field elements excluding the zero element forms an abelian group under the multiplication operation. We will work with the group under multiplication. By Theorem 2.2.5, the order of this group is divisible by the order of any of its elements.

Theorem 4.5.2. *Let $\beta_1, \beta_2, \ldots, \beta_{q-1}$ denote the nonzero field elements of $GF(q)$. Then*

$$x^{q-1} - 1 = (x - \beta_1)(x - \beta_2) \cdots (x - \beta_{q-1}).$$

Proof: The set of nonzero elements of $GF(q)$ is a finite group under the operation of multiplication. Let β be any nonzero element of $GF(q)$, and let h be the order of β under the operation of multiplication. Then by Theorem 2.2.5, h divides $q-1$. Hence

$$\beta^{q-1} = (\beta^h)^{(q-1)/h} = 1^{(q-1)/h} = 1,$$

and thus β is a zero of $x^{q-1} - 1$. □

Theorem 4.5.3. *The group of nonzero elements of $GF(q)$ under multiplication is a cyclic group.*

Proof: If $q-1$ is a prime, the theorem is trivial because then every element except zero and one has order $q-1$, and thus every element is primitive. We need to prove the theorem only for composite $q-1$. Consider the prime factorization of $q-1$:

$$q - 1 = \prod_{i=1}^{s} p_i^{\nu_i}.$$

Because $GF(q)$ is a field, of the $q-1$ nonzero elements of $GF(q)$, there must be at least one that is not a zero of $x^{(q-1)/p_i} - 1$ since this polynomial has at most $(q-1)/p_i$ zeros. Hence for each i, a nonzero element a_i of $GF(q)$ can be found for which $a_i^{(q-1)/p_i} \neq 1$. Let $b_i = a_i^{(q-1)/p_i^{\nu_i}}$, and let $b = \prod_{i=1}^{s} b_i$. We shall prove that b has order $q-1$, and therefore the group is cyclic.

Step 1. The element b_i has order $p_i^{\nu_i}$. **Proof:** Clearly, $b_i^{p_i^{\nu_i}} = 1$, so the order of b_i divides $p_i^{\nu_i}$; it is of the form $p_i^{n_i}$. But if n_i were less than ν_i, then $b_i^{p_i^{\nu_i - 1}} = 1$. Because $b_i^{p_i^{\nu_i-1}} = a_i^{(q-1)/p_i} \neq 1$, we conclude that b_i has order $p_i^{\nu_i}$.

Step 2. The element b has order $q-1$. **Proof:** Suppose $b^n = 1$ for some n. We first show that this implies $n = 0 \pmod{p_i^{\nu_i}}$ for $i = 1, \ldots, s$. For each i, we can write

$$b^{\left(n \prod_{j \neq i} p_j^{\nu_j}\right)} = 1.$$

Replacing b by $\prod_{i=1}^{s} b_i$ and using $b_j^{p_j^{\nu_j}} = 1$, we find

$$b_i^{\left(n \prod_{j\neq i} p_j^{\nu_j}\right)} = 1.$$

Therefore

$$n \prod_{j\neq i} p_j^{\nu_j} = 0 \pmod{p_i^{\nu_i}}.$$

Because the p_i are distinct primes, it follows that $n = 0 \pmod{p_i^{\nu_i}}$ for each i. Hence $n = \prod_{i=1}^{s} p_i^{\nu_i}$. The proof is complete. □

This theorem provides an important key to the understanding of the structure of finite fields as follows.

Theorem 4.5.4. *Every finite field has a primitive element.*

Proof: As a cyclic group, the nonzero elements of $GF(q)$ include an element of order $q - 1$. This is a primitive element. □

The use of a primitive element for multiplication is shown by the following examples.

1. In $GF(8)$, every nonzero element has an order that divides seven. Because seven is prime, every element (except zero and one) has order 7 and is thus primitive. We can construct $GF(8)$ with the polynomial $p(z) = z^3 + z + 1$. Based on the primitive element $\alpha = z$, we have

$$\begin{aligned}
\alpha &= \quad z \\
\alpha^2 &= z^2 \\
\alpha^3 &= \quad z + 1 \\
\alpha^4 &= z^2 + z \\
\alpha^5 &= z^2 + z + 1 \\
\alpha^6 &= z^2 \quad + 1 \\
\alpha^7 &= \quad\quad 1 = \alpha^0.
\end{aligned}$$

With this representation, multiplication is easy. For example,

$$\alpha^4 \cdot \alpha^5 = \alpha^7 \cdot \alpha^2 = \alpha^2.$$

2. In $GF(16)$, every nonzero element has an order that divides fifteen. An element may have order 1, 3, 5, or 15. An element with order 15 is primitive. We can construct $GF(16)$ with the polynomial $p(z) = z^4 + z + 1$, and the element $\alpha = z$ is primitive.

We have

$$
\begin{aligned}
\alpha &= && && z \\
\alpha^2 &= && z^2 && \\
\alpha^3 &= z^3 && && \\
\alpha^4 &= && && z + 1 \\
\alpha^5 &= && z^2 && + z \\
\alpha^6 &= z^3 && + z^2 && \\
\alpha^7 &= z^3 && + && z + 1 \\
\alpha^8 &= && z^2 && + 1 \\
\alpha^9 &= z^3 && && + z \\
\alpha^{10} &= && z^2 && + z + 1 \\
\alpha^{11} &= z^3 && + z^2 && + z \\
\alpha^{12} &= z^3 && + z^2 && + z + 1 \\
\alpha^{13} &= z^3 && + z^2 && + 1 \\
\alpha^{14} &= z^3 && && + 1 \\
\alpha^{15} &= && && 1.
\end{aligned}
$$

Again, with this representation, multiplication in $GF(16)$ is easy. For example,

$$\alpha^{11} \cdot \alpha^{13} = \alpha^{24} = \alpha^{15} \cdot \alpha^9 = \alpha^9.$$

When constructing an extension field as a set of polynomials, it is usually convenient if the polynomial x corresponds to a primitive element of the field. Then one can use x as a logarithm base to construct a multiplication table. This can be done by choosing a special prime polynomial, called a *primitive polynomial*, to construct the field.

Definition 4.5.5. *A primitive polynomial $p(x)$ over $GF(q)$ is a prime polynomial over $GF(q)$ with the property that in the extension field constructed modulo $p(x)$, the field element represented by x is a primitive element.*

Primitive polynomials of every degree exist over every finite field, but we will not provide a proof of this until the end of the next section. Anticipating this result, we can say that a primitive polynomial is a prime polynomial having a primitive element as a zero.

4.6 The structure of finite fields

Earlier in this chapter, we studied how to construct a finite field with p^m elements by first finding a prime polynomial of degree m over $GF(p)$. In this section, instead of constructing our own field, we will assume that we are given a finite field $GF(q)$. We will prove that wherever this field came from, it must have the same structure as the field with q elements constructed according to the methods of the earlier sections, though it may use different notation. Two fields differing only in notation are said to be *isomorphic*.

As we work through this section, we will gain further understanding of the structure of finite fields. The structural properties will be useful in various applications. We will also prove that prime polynomials of every degree exist over every finite field.

Definition 4.6.1. *The number of elements in the smallest subfield of $GF(q)$ is called the characteristic of $GF(q)$.*

Theorem 4.6.2. *Each finite field contains a unique smallest subfield, which has a prime number of elements. Hence the characteristic of every Galois field is a prime number.*

Proof: The field contains the elements zero and one. To define the subfield, consider the subset $G = \{0, 1, 1+1, 1+1+1, \ldots\}$, denoting this by $\{0, 1, 2, 3, \ldots\}$. This subset is a cyclic subgroup under addition. It must contain a finite number, p, of elements. We will show that p is a prime and $G = GF(p)$. In G, addition is modulo p because it is a cyclic group under addition. Because of the distributive law,

$$\begin{aligned}\alpha \cdot \beta &= (1 + \cdots + 1) \cdot \beta \\ &= \beta + \cdots + \beta\end{aligned}$$

where there are α copies of β in the sum, and the addition is modulo p. Hence multiplication is also modulo p. Each nonzero element β has an inverse in G under multiplication because the sequence $\beta, \beta^2, \beta^3, \ldots$ is a cyclic subgroup of G. The sequence contains 1, so for some α in G, $\alpha\beta = 1$. Thus the subset G contains the identity element, is closed under addition and multiplication, and contains all inverses under addition and multiplication. Hence it is a subfield, and it has modulo-p arithmetic. But this is just the field described by Theorem 4.2.3, and hence p must be a prime. □

In the finite field $GF(q)$, we found the subfield $GF(p)$ with p a prime. In particular, if q is a prime p to start with, then we see that $GF(q)$ can be interpreted as the field of integers modulo p. Hence for a given prime p, there is really only one field with p elements, although of course it may be represented by many different notations. Every field $GF(q)$ must have a subfield $GF(p)$ for some prime p.

We shall now study polynomials over $GF(p)$ that have a selected element of $GF(q)$ as a zero. We start with the following definition.

Definition 4.6.3. *Let $GF(q)$ be a field, and let $GF(Q)$ be an extension field of $GF(q)$. Let β be in $GF(Q)$. The prime polynomial $f(x)$ of smallest degree over $GF(q)$ with $f(\beta) = 0$ is called the minimal polynomial of β over $GF(q)$.*

We must prove that the minimal polynomial always exists and is unique.

Theorem 4.6.4. *Every element β of $GF(Q)$ has a unique minimal polynomial over $GF(q)$. Further, if β has minimal polynomial $f(x)$ and a polynomial $g(x)$ has β as a zero, then $f(x)$ divides $g(x)$.*

Proof: First of all, each of the Q elements of $GF(Q)$ is a zero of $x^Q - x$, which is a polynomial over $GF(q)$. In particular, β is a zero of $x^Q - x$. Now use the unique factorization theorem:

$$x^Q - x = f_1(x)f_2(x)\cdots f_k(x)$$

where the factors on the right side are all prime polynomials over $GF(q)$. Because β is a zero of the left side, it must be a zero of some term on the right side. It can be a zero of only one term on the right side because the left side has Q distinct zeros and so every polynomial on the right side must have a number of zeros equal to its degree.

To prove the second part of the theorem, write

$$g(x) = f(x)h(x) + s(x)$$

where $s(x)$ has a smaller degree than $f(x)$, and hence cannot have β as a zero. But

$$0 = g(\beta) = f(\beta)h(\beta) + s(\beta) = s(\beta).$$

Hence $s(x)$ must be zero, and the theorem is proved. □

Corollary 4.6.5. *If $f_1(x), \ldots, f_k(x)$ are the distinct polynomials that are minimal polynomials over $GF(q)$ for one or more elements in $GF(Q)$, then*

$$x^Q - x = f_1(x)f_2(x)\cdots f_k(x).$$

Proof: This follows from the theorem because every β is a zero of $x^Q - x$. □

When Q is equal to q, this reduces to

$$x^q - x = x(x - \beta_1)(x - \beta_2)\cdots(x - \beta_{q-1}),$$

which we have already seen in Theorem 4.5.2. The minimal polynomial over $GF(q)$ of an element β of $GF(q)$ is the first-degree polynomial $f(x) = x - \beta$.

Theorem 4.6.6. *Let $g(x)$ be any polynomial over $GF(q)$. Then an extension field $GF(Q)$ exists in which $g(x)$ can be expressed as the product of a constant and monic polynomials of degree 1.*

Proof: Without loss of generality, suppose that $g(x)$ is a monic polynomial. Construct a sequence of extension fields

$$GF(q) \subset GF(Q_1) \subset GF(Q_2) \subset \cdots \subset GF(Q)$$

as follows. At the jth step, write $g(x)$ as a product of prime polynomials over the field $GF(Q_j)$. If some of the polynomials are not of first degree, then choose any factor $g_i(x)$ with a degree greater than one and construct an extension of $GF(Q_j)$ using $g_i(y)$ as the prime polynomial. In this extension field, $g_i(x)$ can be further factored because the new element $\beta = y$ is a zero of $g_i(y)$. Continue in this way (redefining unique notation for the polynomials as necessary) until all factors are linear. The process must halt in a finite number of steps because $g(x)$ has a finite degree. □

Definition 4.6.7. *Any extension field of $GF(q)$ in which $g(x)$, a polynomial over $GF(q)$, factors into linear and constant terms is called a splitting field of $g(x)$.*

We now have developed all the tools that are necessary to dissect the structure of an arbitrary finite field.

Theorem 4.6.8. *Let α be a primitive element in a finite field $GF(Q)$, an extension field of $GF(q)$. Let m be the degree of $f(x)$, the minimal polynomial of α over $GF(q)$. Then the number of elements in the field is $Q = q^m$, and each element β can be written*

$$\beta = a_{m-1}\alpha^{m-1} + a_{m-2}\alpha^{m-2} + \cdots + a_1\alpha + a_0$$

where $a_{m-1}, \ldots, a_0$ are elements of $GF(q)$.

Proof: Clearly, any element β written in the form

$$\beta = a_{m-1}\alpha^{m-1} + a_{m-2}\alpha^{m-2} + \cdots + a_1\alpha + a_0$$

is an element of $GF(Q)$. Each is unique because if

$$\beta = b_{m-1}\alpha^{m-1} + b_{m-2}\alpha^{m-2} + \cdots + b_1\alpha + b_0$$

is another representation of element β, then

$$0 = (a_{m-1} - b_{m-1})\alpha^{m-1} + \cdots + (a_1 - b_1)\alpha + (a_0 - b_0),$$

and thus α is a zero of a polynomial of degree $m - 1$, contrary to the definition of m. There are q^m such β, and therefore Q is at least as large as q^m.

On the other hand, every nonzero field element can be expressed as a power of α. But if $f(x)$ is the minimal polynomial of α, then $f(\alpha) = 0$. Hence

$$\alpha^m + f_{m-1}\alpha^{m-1} + \cdots + f_1\alpha + f_0 = 0.$$

This can be used to express α^m in terms of the lower powers of α:

$$\alpha^m = -f_{m-1}\alpha^{m-1} - \cdots - f_1\alpha - f_0.$$

This relationship can be used repeatedly to reduce any power of α to a linear combination of $\alpha^{m-1}, \ldots, \alpha^1$, and α^0. That is,

$$\alpha^{m+1} = -f_{m-1}(-f_{m-1}\alpha^{m-1} - \cdots - f_1\alpha - f_0) - f_{m-2}\alpha^{m-1} - \cdots - f_1\alpha^2 - f_0\alpha,$$

and so forth. Hence every element of $GF(Q)$ can be expressed as a distinct linear combination of $\alpha^{m-1}, \alpha^{m-2}, \ldots, \alpha^0$. Consequently, Q is not larger than q^m, and the theorem is proved. □

Corollary 4.6.9. *Every finite field has p^m elements for some positive integer m and prime p.*

Proof: Every finite field has a subfield with p elements to which Theorem 4.6.8 applies. □

Notice that we can use the theorem to associate with each field element a polynomial of degree at most $m - 1$ simply by replacing α with the indeterminate x. These polynomials may be regarded as the field elements. They are added and multiplied modulo $f(x)$, the minimal polynomial of α. This is just the field we would obtain from the construction of Theorem 4.4.3 by using $f(x)$ as the prime polynomial. Hence the number of elements in each finite field is a prime power, and each finite field can be constructed by polynomial arithmetic modulo a prime polynomial.

Finally, we must show the converse – that for every prime p and positive integer m, there is such a field. We will proceed through a series of preliminary theorems.

Theorem 4.6.10. *Let $GF(q)$ have characteristic p. Then for any positive integer m and for any elements α and β in $GF(q)$,*

$$(\alpha \pm \beta)^{p^m} = \alpha^{p^m} \pm \beta^{p^m}.$$

Proof: Suppose the theorem is true for $m = 1$. Then

$$(\alpha \pm \beta)^p = \alpha^p \pm \beta^p.$$

This can be raised to the pth power,

$$((\alpha \pm \beta)^p)^p = (\alpha^p \pm \beta^p)^p,$$

and again using the theorem for $m = 1$,

$$(\alpha \pm \beta)^{p^2} = \alpha^{p^2} \pm \beta^{p^2}.$$

This can be repeated $m - 1$ times to get

$$(\alpha \pm \beta)^{p^m} = \alpha^{p^m} \pm \beta^{p^m}.$$

Hence it is necessary only to prove the theorem for $m = 1$. But by the binomial theorem

$$(\alpha \pm \beta)^p = \sum_{i=0}^{p} \binom{p}{i} a^i (\pm\beta)^{p-i},$$

and so it suffices to show that, in $GF(q)$,

$$\binom{p}{i} = 0 \quad i = 1, \ldots, p - 1.$$

But for such i,

$$\binom{p}{i} = \frac{p!}{i!(p-i)!} = \frac{p(p-1)!}{i!(p-i)!}$$

is an integer, and p is a prime. Every factor in the denominator is smaller than p and so does not divide p. Hence the denominator divides $(p-1)!$, and $\binom{p}{i}$ is a multiple of p. That is, $\binom{p}{i} = 0 \pmod{p}$, and because all integer arithmetic is modulo p in $GF(q)$, we have $\binom{p}{i} = 0$ in $GF(q)$.

Finally, if $p = 2$, $(\pm\beta)^2 = \pm\beta^2$, and if p is odd, $(\pm\beta)^p = \pm\beta^p$. This completes the proof of the theorem. □

Theorem 4.6.11. *Let m be a positive integer and p a prime. Then the smallest splitting field of the polynomial $g(x) = x^{p^m} - x$ regarded as a polynomial over $GF(p)$ has p^m elements.*

Proof: Every polynomial over $GF(p)$ has a smallest splitting field. Let $GF(Q)$ be the smallest splitting field of $g(x) = x^{p^m} - x$. Then in $GF(Q)$, $g(x)$ has p^m zeros, some possibly repeated. We will show that the p^m zeros are distinct and form a field. Consequently, we will conclude that $GF(Q)$ has p^m elements.

To prove that the set of zeros of $g(x)$ is a field, it suffices to prove that the set is closed under addition and multiplication, and that it contains inverses of all nonzero elements. Suppose α and β are zeros. Then by Theorem 4.6.10,

$$(\alpha \pm \beta)^{p^m} = \alpha^{p^m} \pm \beta^{p^m} = \alpha \pm \beta,$$

and thus $(\alpha \pm \beta)$ is a zero and the set is closed under addition. Next,

$$(\alpha\beta)^{p^m} = \alpha^{p^m}\beta^{p^m} = \alpha\beta,$$

and thus $\alpha\beta$ is a zero and the set is closed under multiplication. Next, note that $-\alpha$ is the additive inverse of α, and thus every element has an additive inverse. Similarly, α^{-1} is easily checked to be a zero whenever α is a zero.

Finally, we check that the p^m zeros of $x^{p^m} - x$ are distinct. This follows by examining the formal derivative:

$$\frac{d}{dx}\left[x^{p^m} - x\right] = ((p^m))x^{p^m - 1} - 1 = -1$$

because $((p)) = 0$ in $GF(Q)$. Hence $x^{p^m} - x$ can have no multiple zeros. This completes the proof. □

We now have a converse to Corollary 4.6.9.

Corollary 4.6.12. *For every prime p and positive integer m, there is a finite field with p^m elements.*

Finally, we will show that even if q is not a prime, but is a prime power, then $GF(q^m)$ can be constructed as an extension field of $GF(q)$. To do this, it suffices to show that a prime polynomial of degree m exists over $GF(q)$.

Theorem 4.6.13. *For every finite field $GF(q)$ and positive integer m, there exists at least one prime polynomial over $GF(q)$ of degree m.*

Proof: Because q is a prime power, then so is q^m. By Corollary 4.6.12, a field exists with q^m elements. This field has a primitive element α, and by Theorem 4.6.8, the minimal polynomial of α over $GF(q)$ is a prime polynomial of degree m. □

Corollary 4.6.14. *For every finite field $GF(q)$ and positive integer m, there exists at least one primitive polynomial over $GF(q)$ of degree m.*

Proof: Let α be a primitive element of $GF(q^m)$. Let $f(x)$ be the minimal polynomial of α over $GF(q)$. Then in the field of polynomials modulo $f(x)$, the primitive element $\alpha = x$ is a zero of $f(x)$, and thus the polynomial x represents a primitive element of the field. □

To close this chapter, we will describe square roots in a finite field.

Theorem 4.6.15. *Every element of the finite field $GF(2^m)$ has a square root in $GF(2^m)$. Half of the nonzero elements of $GF(p^m)$ – with p an odd prime – have a square root in $GF(p^m)$. The other nonzero elements of $GF(p^m)$ have a square root in the extension field $GF(p^{2m})$, but not in $GF(p^m)$.*

Proof: The zero element is its own square root in any field, so we need only to consider nonzero elements. First, consider a field $GF(2^m)$ of characteristic 2 and with primitive element α. Then α has odd order n, and any element β can be written as α^i for some i. Then $\sqrt{\beta} = \alpha^{i/2}$ if i is even, and $\sqrt{\beta} = \alpha^{(i+n)/2}$ if i is odd. In either case, $\sqrt{\beta}$ is an element of $GF(2^m)$.

Next, consider a field $GF(q)$ whose characteristic p is an odd prime and with primitive element $\alpha = \gamma^{q+1}$ where γ is a primitive element of the extension field $GF(q^2)$ of order $q^2 - 1 = (q+1)(q-1)$. Because q is a power of an odd prime, $q+1$ is even. Any element β can be written as α^i or as $\gamma^{(q+1)i}$ for some i. Then if i is even, $\sqrt{\beta} = \alpha^{i/2}$, which is an element of $GF(q)$. If i is odd, $\sqrt{\beta} = \gamma^{i(q+1)/2}$, which is an element of $GF(q^2)$, but not of $GF(q)$ because $i(q+1)/2$ is then not a multiple of $q+1$. □

Problems

4.1 Over $GF(2)$, let $p_1(x) = x^3 + 1$, and let $p_2(x) = x^4 + x^3 + x^2 + 1$.
  a. Find $\text{GCD}[p_1(x), p_2(x)]$.

b. Find $A(x)$ and $B(x)$ that satisfy

$$\text{GCD}[p_1(x), p_2(x)] = A(x)p_1(x) + B(x)p_2(x).$$

4.2 a. How many distinct second-degree monic polynomials of the form

$$x^2 + ax + b \quad b \neq 0$$

are there over $GF(16)$?

b. How many distinct polynomials of the form

$$(x - \beta)(x - \gamma) \quad \beta, \gamma \neq 0$$

are there over $GF(16)$?

c. Does this prove that irreducible second-degree polynomials exist? How many second-degree prime polynomials over $GF(16)$ are there?

4.3 Prove Theorem 4.1.2 by relating both sides to the division algorithm. Prove Theorem 4.3.4 in the same way.

4.4 a. Use the euclidean algorithm to find GCD(1573, 308).

b. Find integers A and B that satisfy

$$\text{GCD}(1573, 308) = 1573A + 308B.$$

4.5 Over $\mathbf{Z}/(15)$, the ring of integers modulo 15, show that the polynomial $p(x) = x^2 - 1$ has more than two zeros. Such a polynomial over a field can only have two zeros. Where does the proof of the theorem fail for a ring?

4.6 How many distinct monic polynomials over $GF(2)$ divide $x^6 - 1$?

4.7 Construct $GF(5)$ by constructing an addition table and a multiplication table.

4.8 Construct addition and multiplication tables for $GF(8)$ and $GF(9)$.

4.9 a. Prove that $p(x) = x^3 + x^2 + 2$ is irreducible over $GF(3)$.

b. What are the possible (multiplicative) orders of elements in $GF(27)$?

c. What is the order of the element represented by x if $p(x)$ above is used to construct $GF(27)$?

d. In the field $GF(27)$, find $(2x + 1)(x^2 + 2)$, assuming that $p(x)$ above was used to construct the field.

4.10 Find $3^{100} \pmod 5$.

4.11 Prove that the quotient rings $\mathbf{Z}/\langle q \rangle$ and $GF(q)[x]/\langle p(x) \rangle$ are rings.

4.12 The polynomial $p(x) = x^4 + x^3 + x^2 + x + 1$ is irreducible over $GF(2)$. Therefore the ring of polynomials modulo $p(x)$ is $GF(16)$.

a. Show that the field element represented by x in this construction is not a primitive element.

b. Show that the field element represented by $x + 1$ is primitive.

c. Find the minimal polynomial of the field element $x + 1$.

4.13 a. Construct addition and multiplication tables for $GF(2^3)$, using the irreducible polynomial $x^3 + x^2 + 1$ over $GF(2)$.

b. Repeat 4.13a with polynomial $x^3 + x + 1$, and show that the two fields are isomorphic. That is, by relabeling the elements of the first field, show that the second field is obtained.

4.14 Addition and multiplication tables for $GF(2^4)$ can be constructed in at least two different ways:

(i) using an irreducible polynomial of degree 4 over $GF(2)$;
(ii) using an irreducible polynomial of degree 2 over $GF(4)$.

Construct these tables using approach (ii).

4.15 The polynomial $p(x) = x^{20} + x^3 + 1$ is a primitive polynomial over $GF(2)$ and can be used to construct $GF(1,048,576)$; the element represented by the polynomial x will be a primitive element α.

a. What are the subfields of this field?

b. How many of these subfields have no proper subfields of their own other than $GF(2)$?

c. In this field, evaluate the expression

$$ab^2 - 1$$

where

$$a = x^{12} + x^7 + 1,$$
$$b = x^{13} + x^5 + x^2.$$

4.16 For the formal derivative of polynomials, prove that

$$[r(x)s(x)]' = r'(x)s(x) + r(x)s'(x),$$

and that if $a(x)^2$ divides $r(x)$, then $a(x)$ divides $r'(x)$.

4.17 Prove that $-\beta = \beta$ for any element β in a field of characteristic 2.

4.18 Prove that if $\text{GCD}[g_1(x), \ldots, g_n(x)] = 1$, then there exist polynomials $b_1(x), \ldots, b_n(x)$ satisfying

$$b_1(x)g_1(x) + \cdots + b_n(x)g_n(x) = 1.$$

4.19 If q is divisible by two primes, p and p', then in $GF(q)$ there is an element β of order p under addition and an element of order p' under addition. Show that this implies that the order of 1 under addition divides both p and p'. Does this show that the finite field $GF(q)$ can exist only if q is a power of a prime?

4.20 Prove that the extension field $GF(q^m)$ can be regarded as a vector space over $GF(q)$.

Notes

The subject of this chapter is a standard in the mathematical literature. The properties of Galois fields are developed in any book on abstract algebra, for example those by Birkhoff and MacLane (1953) or Van der Waerden (1949, 1953). However, the standard treatments are formal, concerned primarily with abstract properties and not concerned with examples or applications. Berlekamp (1968) and McEliece (1987) concentrate on the more immediately useful properties of Galois fields.

5 Cyclic Codes

Cyclic codes are a subclass of the class of linear codes obtained by imposing an additional strong structural requirement on the codes. Because of this structure, the search for error-control codes has been most successful within the class of cyclic codes. Here the theory of Galois fields has been used as a mathematical searchlight to spot the good codes. Outside the class of cyclic codes, the theory of Galois fields casts a dimmer light. Most of what has been accomplished builds on the ideas developed for cyclic codes.

The cyclic property in itself is not important and cyclic codes do not necessarily have a large minimum distance. Cyclic codes are introduced because their structure is closely related to the strong structure of the Galois fields. This is significant because the underlying Galois-field description of a cyclic code leads to encoding and decoding procedures that are algorithmic and computationally efficient. Algorithmic techniques have important practical applications, in contrast to the tabular decoding techniques that are used for arbitrary linear codes.

This chapter gives a leisurely introduction to cyclic codes as a special class of linear codes. An alternative approach to the topic of cyclic codes, based on the Fourier transform, is given in Chapter 6.

5.1 Viewing a code from an extension field

The cyclic codes over $GF(q)$ are a class of linear codes that can be looked at in a way that will prove to be quite powerful. Although these codes have symbols in the field $GF(q)$, we will often see them more clearly if we step up into an extension field $GF(q^m)$. Just as one uses the theory of functions of a complex variable to learn more about the functions of a real variable, so one can learn more about the functions over $GF(q)$ by studying them as functions over $GF(q^m)$. The introduction of the extension field may appear artificial, but one should be patient – the rewards will come in time.

We have seen that a linear code over $GF(q)$ can be described in terms of a check matrix $\boldsymbol{H}$. A vector $\boldsymbol{c}$ over $GF(q)$ is a codeword if and only if $\boldsymbol{c}\boldsymbol{H}^{\mathrm{T}} = \boldsymbol{0}$. For example, the following is a check matrix for the binary (7, 4) Hamming code:

$$\boldsymbol{H} = \begin{bmatrix} 1 & 0 & 0 & 1 & 0 & 1 & 1 \\ 0 & 1 & 0 & 1 & 1 & 1 & 0 \\ 0 & 0 & 1 & 0 & 1 & 1 & 1 \end{bmatrix}.$$

This check matrix can be written compactly by working in an extension field, identifying the columns of $\boldsymbol{H}$ with elements of $GF(8)$. Use the top element as the coefficient of z^0 in the polynomial representation of a field element, the next element as the coefficient of z^1, and the bottom element as the coefficient of z^2. Then using the polynomial $p(z) = z^3 + z + 1$ to construct $GF(8)$ and α as the primitive element represented by z, the matrix $\boldsymbol{H}$ becomes

$$\boldsymbol{H} = [\, \alpha^0 \;\; \alpha^1 \;\; \alpha^2 \;\; \alpha^3 \;\; \alpha^4 \;\; \alpha^5 \;\; \alpha^6 \,].$$

Now the check matrix is a 1 by 7 matrix over the extension field $GF(8)$. Using this matrix, a codeword is defined as any vector over $GF(2)$ such that, in the extension field $GF(8)$, the product with $\boldsymbol{H}^{\mathrm{T}}$ is zero:

$$\boldsymbol{c}\boldsymbol{H}^{\mathrm{T}} = 0.$$

But this product is

$$\sum_{i=0}^{6} c_i \alpha^i = 0.$$

Now we come to the idea of representing codewords by polynomials. The codeword $\boldsymbol{c}$ is represented by the *codeword polynomial*

$$c(x) = \sum_{i=0}^{n-1} c_i x^i.$$

The operation of multiplying the codeword by the check matrix becomes the operation of evaluating the polynomial $c(x)$ at $x = \alpha$. The condition for $c(x)$ to represent a codeword becomes the condition $c(\alpha) = 0$. Thus a binary polynomial $c(x)$ is a codeword polynomial if and only if α is a zero of $c(x)$. The (7, 4) Hamming code is represented as the set whose elements are the polynomials over $GF(2)$ of degree at most 6 that have α as a zero in $GF(8)$.

The method that we used above to re-express the (7, 4) Hamming code as a set of polynomials can be used for a large class of linear codes. Suppose that the check matrix $\boldsymbol{H}$ of a linear code has n columns and that $n - k$, the number of rows, is divisible by m. Each group of m rows can be re-expressed as a single row of elements from $GF(q^m)$. That is, the first m rows gathered together become a single row $(\beta_{11} \ldots \beta_{1n})$, and the

$\boldsymbol{H}$ matrix becomes

$$\boldsymbol{H} = \begin{bmatrix} \beta_{11} & \ldots & \beta_{1n} \\ \vdots & & \vdots \\ \beta_{r1} & \ldots & \beta_{rn} \end{bmatrix},$$

with $r = (n-k)/m$. This is now an r by n matrix over $GF(q^m)$ rather than an $(n-k)$ by n matrix over $GF(q)$. Of course, we have not really changed anything by this reinterpretation; we have just made the representation more compact.

In this chapter we shall study the special case in which the check matrix can be written in the form

$$\boldsymbol{H} = \begin{bmatrix} \gamma_1^0 & \gamma_1^1 & \ldots & \gamma_1^{n-2} & \gamma_1^{n-1} \\ \gamma_2^0 & \gamma_2^1 & \ldots & \gamma_2^{n-2} & \gamma_2^{n-1} \\ \vdots & & & & \vdots \\ \gamma_r^0 & \gamma_r^1 & \ldots & \gamma_r^{n-2} & \gamma_r^{n-1} \end{bmatrix}$$

where $\gamma_j \in GF(q^m)$ for $j = 1, \ldots, r$, and $n = q^m - 1$. This check matrix over $GF(q^m)$, with $n = q^m - 1$ columns and r rows, can also be written as a check matrix over $GF(q)$ with $n = q^m - 1$ columns and mr rows by replacing each field element from $GF(q^m)$ by its coefficients when expressed as a polynomial over $GF(q)$. We prefer to work in the larger field, however, because the concise structure is useful.

Each codeword $\boldsymbol{c}$ is a vector over $GF(q)$ such that, in the extension field $GF(q^m)$,

$$\boldsymbol{c}\boldsymbol{H}^{\mathrm{T}} = \boldsymbol{0}.$$

Because of the special form of $\boldsymbol{H}$, this can be written as

$$\sum_{i=0}^{n-1} c_i \gamma_j^i = 0 \quad j = 1, \ldots, r.$$

This is just the statement that the codeword polynomial $c(x)$ has zeros at $\gamma_1, \ldots, \gamma_r$. The code is defined as the set whose elements are the polynomials $c(x) = \sum_{i=0}^{n-1} c_i x^i$ of degree at most $n-1$ that satisfy $c(\gamma_j) = 0$ for $j = 1, \ldots, r$.

We have now made a transition from a matrix formulation of linear codes to a polynomial formulation for a special subclass of linear codes. The reason for restricting attention to this subclass is that the polynomial formulation makes it easier to discover good codes and to develop encoders and decoders. For reasons that will become clear in the next section, such codes are known as *cyclic codes*.

For example, take any primitive element α of the field $GF(16)$, an extension of $GF(2)$. To form a check matrix for a code over $GF(2)$ of blocklength 15, choose $\gamma_1 = \alpha$ and $\gamma_2 = \alpha^3$ and write

$$\boldsymbol{H} = \begin{bmatrix} \alpha^0 & \alpha^1 & \alpha^2 & \alpha^3 & \alpha^4 & \alpha^5 & \alpha^6 & \alpha^7 & \alpha^8 & \alpha^9 & \alpha^{10} & \alpha^{11} & \alpha^{12} & \alpha^{13} & \alpha^{14} \\ \alpha^0 & \alpha^3 & \alpha^6 & \alpha^9 & \alpha^{12} & \alpha^{15} & \alpha^{18} & \alpha^{21} & \alpha^{24} & \alpha^{27} & \alpha^{30} & \alpha^{33} & \alpha^{36} & \alpha^{39} & \alpha^{42} \end{bmatrix}.$$

Other choices for γ_1 and γ_2 will give other check matrices, some good and some bad. This choice happens to give a good check matrix, but the explanation will not be given until Section 5.6.

If desired, by using the representation of $GF(16)$ in Section 4.5, we can change this matrix to a check matrix over $GF(2)$. Replace each power of α by a four-bit column with the top element given by the coefficient of z^0 in the polynomial representation of that field element, the second element given by the coefficient of z^1, and so on. The result is

$$\boldsymbol{H} = \left[\begin{array}{ccccccccccccccc} 1 & 0 & 0 & 0 & 1 & 0 & 0 & 1 & 1 & 0 & 1 & 0 & 1 & 1 & 1 \\ 0 & 1 & 0 & 0 & 1 & 1 & 0 & 1 & 0 & 1 & 1 & 1 & 1 & 0 & 0 \\ 0 & 0 & 1 & 0 & 0 & 1 & 1 & 0 & 1 & 0 & 1 & 1 & 1 & 1 & 0 \\ 0 & 0 & 0 & 1 & 0 & 0 & 1 & 1 & 0 & 1 & 0 & 1 & 1 & 1 & 1 \\ \hline 1 & 0 & 0 & 0 & 1 & 1 & 0 & 0 & 0 & 1 & 1 & 0 & 0 & 0 & 1 \\ 0 & 0 & 0 & 1 & 1 & 0 & 0 & 0 & 1 & 1 & 0 & 0 & 0 & 1 & 1 \\ 0 & 0 & 1 & 0 & 1 & 0 & 0 & 1 & 0 & 1 & 0 & 0 & 1 & 0 & 1 \\ 0 & 1 & 1 & 1 & 1 & 0 & 1 & 1 & 1 & 1 & 0 & 1 & 1 & 1 & 1 \end{array}\right].$$

Although it is not obvious by inspection, these rows are linearly independent, and therefore the check matrix defines a $(15, 7)$ binary code. It is also not obvious, but every set of four columns is linearly independent, so the minimum distance of this code is 5. This code can also be represented as the set of polynomials over $GF(2)$ of degree 14 or less such that each polynomial $c(x)$ satisfies $c(\alpha) = c(\alpha^3) = 0$ in the extension field $GF(16)$. In other words, codewords are polynomials with zeros at α and α^3.

The check matrix $\boldsymbol{H}$ over the extension field $GF(q^m)$ will prove to be much easier to deal with, and thus we shall soon work mostly in the larger field. Of course, by restricting ourselves to check matrices with such a special form, we eliminate from further study the many linear codes that do not have this form.

5.2 Polynomial description of cyclic codes

Let us begin anew. A linear code $\mathcal{C}$ over $GF(q)$ is called a *cyclic code* if, whenever $\boldsymbol{c} = (c_0, c_1, \ldots, c_{n-1})$ is in $\mathcal{C}$, then $\boldsymbol{c}' = (c_{n-1}, c_0, \ldots, c_{n-2})$ is also in $\mathcal{C}$. The codeword $\boldsymbol{c}'$ is obtained by cyclically shifting the components of the codeword $\boldsymbol{c}$ one place to the right. Thus the cyclic shift operation maps the code $\mathcal{C}$ onto itself. This is an example of an *automorphism* of a code, which is any permutation of codeword components that maps the code onto itself.

The Hamming code of Table 5.1 is an example of a cyclic code. Every linear code over $GF(q)$ of blocklength n is a subspace of $GF(q)^n$; a cyclic code is a special kind of subspace because it has this cyclic property.

Table 5.1. *Syndrome evaluator table*

$e(x)$	$s(x)$
1	$R_{g(x)}[1]$
x	$R_{g(x)}[x]$
x^2	$R_{g(x)}[x^2]$
$\vdots$	$\vdots$
$1+x$	$R_{g(x)}[1+x]$
$1+x^2$	$R_{g(x)}[1+x^2]$
$\vdots$	$\vdots$

Each vector in $GF(q)^n$ can be represented as a polynomial in x of a degree less than, or equal to, $n-1$. The components of the vector are identified by the coefficients of the polynomial. The set of polynomials has a vector space structure identical to that of $GF(q)^n$. This set of polynomials also has a ring structure, defined in Section 4.4, called $GF(q)[x]/\langle x^n-1\rangle$. As a ring, the set has a product

$$p_1(x) \cdot p_2(x) = R_{x^n-1}[p_1(x)p_2(x)].$$

Notice that there are two different kinds of multiplication in the expression. The product on the left side is a product in $GF(q)[x]/\langle x^n-1\rangle$ that is defined in terms of the product in $GF(q)[x]$ on the right.

A cyclic shift can be written as a multiplication within this ring:

$$x \cdot p(x) = R_{x^n-1}[xp(x)].$$

Hence if the codewords of a code are represented by polynomials, the code is a subset of the ring $GF(q)[x]/\langle x^n-1\rangle$. If the code is a cyclic code, then $x \cdot c(x)$ is a codeword polynomial whenever $c(x)$ is a codeword polynomial.

Theorem 5.2.1. *In the ring $GF(q)[x]/\langle x^n-1\rangle$, a subset $\mathcal{C}$ is a cyclic code if and only if it satisfies the following two properties.*[1]

(i) $\mathcal{C}$ is a subgroup of $GF(q)[x]/\langle x^n-1\rangle$ under addition.
(ii) If $c(x) \in \mathcal{C}$, and $a(x) \in GF(q)[x]/\langle x^n-1\rangle$, then $R_{x^n-1}[a(x)c(x)] \in \mathcal{C}$.

Proof: The ring $GF(q)[x]/\langle x^n-1\rangle$ forms a vector space under addition. Suppose the subset $\mathcal{C}$ satisfies the two given properties. Then it is closed under addition and closed under multiplication by a scalar. Hence it is a vector subspace. It is also closed under multiplication by any ring element, in particular, under multiplication by x. Hence it is a cyclic code.

[1] This subset is known as an *ideal* of the ring. In general, a subset I of a ring R is called an ideal of R if (1) I is a subgroup of the additive group of R, and (2) if $r \in R$ and $a \in I$, then $ar \in I$.

Now suppose that it is a cyclic code. Then it is closed under linear combinations and closed under multiplication by x. But then it is closed under multiplication by the powers of x, and by linear combinations of the powers of x. That is, it is closed under multiplication by an arbitrary polynomial. Hence it satisfies the two properties, and the theorem is proved. □

Choose a nonzero codeword polynomial of the smallest degree from $\mathcal{C}$, and denote its degree by $n - k$ (it must be less than n). Multiply by a field element to make it a monic polynomial. This must also be in code $\mathcal{C}$ because the code is linear. No other monic polynomial of degree $n - k$ is in the code because, otherwise, the difference of the two monic polynomials would be in the code and have a degree smaller than $n - k$, contrary to this choice of polynomial.

The unique nonzero monic polynomial of the smallest degree is called the *generator polynomial* of $\mathcal{C}$ and is denoted by $g(x)$.

Theorem 5.2.2. *A cyclic code consists of all multiples of the generator polynomial $g(x)$ by polynomials of degree $k - 1$ or less.*

Proof: All such polynomials must be in the code, by Theorem 5.2.1, because $g(x)$ is in the code. But if any polynomial $c(x)$ is in the code, then by the division algorithm,

$$c(x) = Q(x)g(x) + s(x)$$

where the degree of $s(x)$ is smaller than the degree of $g(x)$. Then

$$s(x) = c(x) - Q(x)g(x)$$

is a codeword polynomial because both terms on the right are codeword polynomials and the code is linear. But $s(x)$ has a degree smaller than $n - k$, which is the smallest degree of any nonzero polynomial in the code. Hence $s(x) = 0$, and $c(x) = Q(x)g(x)$. □

Theorem 5.2.3. *There is a cyclic code of blocklength n with generator polynomial $g(x)$ if and only if $g(x)$ divides $x^n - 1$.*

Proof: The division algorithm states that

$$x^n - 1 = Q(x)g(x) + s(x),$$

with $s(x)$ having a degree less than that of $g(x)$. Then

$$0 = R_{x^n-1}(x^n - 1) = R_{x^n-1}[Q(x)g(x)] + R_{x^n-1}[s(x)].$$

Therefore

$$0 = R_{x^n-1}[Q(x)g(x)] + s(x).$$

The first term on the right side is a codeword polynomial as a consequence of Theorem 5.2.1. Then $s(x)$ is a codeword polynomial with a degree less than that of $g(x)$. The only such codeword polynomial is $s(x) = 0$. Thus $g(x)$ divides $x^n - 1$. Further, every polynomial that divides $x^n - 1$ can be used as a generator polynomial to define a cyclic code. This completes the proof. □

Whenever $g(x)$ divides $x^n - 1$, then the reciprocal polynomial $\widetilde{g}(x) = x^r g(x^{-1})$, where r is the degree of $g(x)$, also divides $x^n - 1$. The polynomial $g(x)$ and the reciprocal polynomial generate equivalent codes. If $c(x)$ is a codeword in the cyclic code generated by $g(x)$, then $\widetilde{c}(x) = x^{n-1}c(x^{-1})$ is a codeword in the cyclic code generated by $\widetilde{g}(x)$.

By Theorem 5.2.3, for any cyclic code $\mathcal{C}$ with generator polynomial $g(x)$ of degree $n - k$, there is another polynomial $h(x)$ of degree k such that

$$x^n - 1 = g(x)h(x).$$

The polynomial $h(x)$ is called the *check polynomial* of $\mathcal{C}$. Every codeword $c(x)$ satisfies

$$R_{x^n-1}[h(x)c(x)] = 0$$

because for some $a(x)$,

$$h(x)c(x) = h(x)g(x)a(x) = (x^n - 1)a(x).$$

The data sequence is represented by a *data polynomial* $a(x)$ of degree $k - 1$.[2] The set of data polynomials may be mapped into the set of codeword polynomials in any convenient way. A simple encoding rule is

$$c(x) = g(x)a(x).$$

This encoder is nonsystematic because the data polynomial $a(x)$ is not immediately visible in $c(x)$. A systematic encoding rule is somewhat more complicated. The idea is to insert the data into the high-order coefficients of the codeword, and then to choose the check symbols so as to obtain a legitimate codeword. That is, the codeword is of the form

$$c(x) = x^{n-k}a(x) + t(x)$$

where $t(x)$ is chosen so that

$$R_{g(x)}[c(x)] = 0.$$

This requires that

$$R_{g(x)}[x^{n-k}a(x)] + R_{g(x)}[t(x)] = 0,$$

[2] Actually, we should say that $a(x)$ has a degree less than, or equal to, $k - 1$, but in cases where $a(x)$ is one of a set of polynomials, it is more convenient – though imprecise – to refer to the maximum degree in the set.

and the degree of $t(x)$ is less than $n-k$, the degree of $g(x)$. Hence

$$t(x) = -R_{g(x)}[x^{n-k}a(x)].$$

This encoding rule is a one-to-one map because the k high-order coefficients of the polynomial are unique. The systematic encoding rule and the nonsystematic encoding rule produce the same set of codewords, but the association between the $a(x)$ and the $c(x)$ is different.

Let $c(x)$ denote a transmitted codeword polynomial of a cyclic code; the coefficients of the polynomial $c(x)$ are the symbols of the transmitted vector. Denote the senseword by the *senseword polynomial* $v(x)$. Let $e(x) = v(x) - c(x)$. The polynomial $e(x)$ is called the *error polynomial*. It has nonzero coefficients in those locations where channel errors occurred.

We define the *syndrome polynomial* $s(x)$, which will be used for decoding, as the remainder of $v(x)$ under division by $g(x)$:

$$\begin{aligned} s(x) &= R_{g(x)}[v(x)] \\ &= R_{g(x)}[c(x) + e(x)] \\ &= R_{g(x)}[e(x)]. \end{aligned}$$

The syndrome polynomial depends only on $e(x)$, and not on $c(x)$ or on $a(x)$.

A summary of the polynomials is as follows:

Generator polynomial:	$g(x)$	$\deg g(x) = n-k.$
Check polynomial:	$h(x)$	$\deg h(x) = k.$
Data polynomial:	$a(x)$	$\deg a(x) = k-1.$
Codeword polynomial:	$c(x)$	$\deg c(x) = n-1.$
Senseword polynomial:	$v(x)$	$\deg v(x) = n-1.$
Error polynomial:	$e(x)$	$\deg e(x) = n-1.$
Syndrome polynomial:	$s(x)$	$\deg s(x) = n-k-1.$

Theorem 5.2.4. *Let $d_{\min}$ be the minimum distance of a cyclic code $\mathcal{C}$. Every error polynomial of Hamming weight less than $\frac{1}{2}d_{\min}$ has a unique syndrome polynomial.*

Proof: Suppose that $e_1(x)$ and $e_2(x)$ each has a Hamming weight less than $\frac{1}{2}d_{\min}$ and each has the same syndrome polynomial. Then

$$e_1(x) = Q_1(x)g(x) + s(x),$$

$$e_2(x) = Q_2(x)g(x) + s(x),$$

and

$$e_1(x) - e_2(x) = [Q_1(x) - Q_2(x)]g(x).$$

By assumption, $e_1(x)$ and $e_2(x)$ each has a weight less than $\frac{1}{2}d_{\min}$, and thus the difference has a weight less than $d_{\min}$. The right side is a codeword. If it is nonzero, it has a weight of at least $d_{\min}$, the minimum weight of the code. Hence the right side is zero, and $e_1(x)$ equals $e_2(x)$. This proves the theorem. □

It is straightforward to compute the syndrome polynomial $s(x)$ from the senseword polynomial $v(x)$. The decoding task is to find the unique $e(x)$ with the least number of nonzero coefficients, satisfying

$$s(x) = R_{g(x)}[e(x)].$$

This can be done by constructing a table if the number of entries is not too large. For each correctable $e(x)$, compute $s(x)$ and form the syndrome evaluator table shown in Table 5.1. The decoder finds the error polynomial $e(x)$ by computing $s(x)$ from $v(x)$ and then finding $s(x)$ in the syndrome evaluator table, thereby finding the corresponding $e(x)$. The syndrome evaluator table, if it is not too large, can be realized in a computer memory or in a combinatorial logic circuit.

For the moment we will be satisfied with this simple decoding scheme, although it is too complicated to use except for very small codes. In later chapters, we shall develop algorithms that compute $e(x)$ from $s(x)$, which replace this storing of a precomputed table.

5.3 Minimal polynomials and conjugates

We have seen that a cyclic code of blocklength n over $GF(q)$ exists for each polynomial $g(x)$ over $GF(q)$ that divides $x^n - 1$. We now wish to study such generator polynomials explicitly. First, we will find all possible generator polynomials for cyclic codes with blocklength n of the form $q^m - 1$. The most direct approach is to find the divisors of $x^n - 1$ by writing $x^n - 1$ in terms of its prime factors:

$$x^n - 1 = f_1(x)f_2(x)\cdots f_s(x)$$

where s is the number of prime factors. These factors are the distinct minimal polynomials of the elements of $GF(q^m)$, where $n = q^m - 1$. Any subset of these factors can be multiplied together to produce a generator polynomial $g(x)$ that divides $x^n - 1$, so there are $2^s - 2$ different nontrivial cyclic codes of blocklength n (excluding the trivial cases $g(x) = 1$ and $g(x) = x^n - 1$). Which of these, if any, gives a code with a large minimum distance is a question that we will answer only gradually.

Suppose that $g(x)$ is a generator polynomial. It divides $x^n - 1$, and thus $g(x) = \prod f_i(x)$ where the product is over some of the prime polynomials dividing $x^n - 1$. The cyclic code generated by $g(x)$ consists of all polynomials that are divisible by each of these $f_i(x)$. We can define $g(x)$ by stating all of the prime polynomials that divide it.

In this section we shall look at the relationship between the prime polynomials and their zeros in an extension field. In particular, we will learn how to find prime polynomials, and hence generator polynomials, that have specified zeros. Eventually, we shall design codes by choosing zeros in an extension field. Let us start with certain preferred values of n called *primitive blocklengths*.

Definition 5.3.1. *A blocklength n of the form $n = q^m - 1$ is called a primitive blocklength for a code over $GF(q)$. A cyclic code over $GF(q)$ of primitive blocklength is called a primitive cyclic code.*

The field $GF(q^m)$ is an extension of $GF(q)$. By the unique factorization theorem, the prime factorization

$$x^{q^m-1} - 1 = f_1(x) \cdots f_s(x)$$

is unique over the field $GF(q)$. Because $g(x)$ divides $x^{q^m-1} - 1$, it must be a product of some of these polynomials. On the other hand, every nonzero element of $GF(q^m)$ is a zero of $x^{q^m-1} - 1$. Hence we can also factor $x^{q^m-1} - 1$ in the extension field $GF(q^m)$ to get

$$x^{q^m-1} - 1 = \prod_j (x - \beta_j)$$

where β_j ranges over all nonzero elements of $GF(q^m)$. It follows that each factor $f_\ell(x)$ can be further factored in $GF(q^m)$ into a product of some of these polynomials of degree 1, and that each β_j is a zero of exactly one of the $f_\ell(x)$. This $f_\ell(x)$ is the minimal polynomial of β_j, and we call it $f_j(x)$ also.[3] It is the smallest-degree polynomial with coefficients in the base field $GF(q)$ that has β_j as a zero in the extension field. Now we can relate the definition of cyclic codes to the earlier treatment in Section 5.1.

Theorem 5.3.2. *Let $g(x)$, the generator polynomial of a primitive cyclic code, have zeros at $\beta_1, \ldots, \beta_r$ in $GF(q^m)$. A polynomial $c(x)$ over $GF(q)$ is a codeword polynomial if and only if*

$$c(\beta_1) = c(\beta_2) = \cdots = c(\beta_r) = 0,$$

where $c(\beta_j)$ is evaluated in $GF(q^m)$.

Proof: Every codeword polynomial can be expressed in the form $c(x) = a(x)g(x)$. Therefore $c(\beta_j) = a(\beta_j)g(\beta_j) = 0$.

Conversely, suppose that $c(\beta_j) = 0$, and write

$$c(x) = Q(x)f_j(x) + s(x)$$

where $\deg s(x) < \deg f_j(x)$, and $f_j(x)$ is the minimal polynomial of β_j. But then

$$0 = c(\beta_j) = Q(\beta_j)f_j(\beta_j) + s(\beta_j) = s(\beta_j)$$

requires that $s(x) = 0$. Then $c(x)$ must be divisible by the minimal polynomial $f_j(x)$ for $j = 1, \ldots, r$. Hence $c(x)$ is divisible by LCM$[f_1(x), f_2(x), \ldots, f_r(x)] = g(x)$. □

[3] We apologize for the ambiguity in using $f_\ell(x)$ for $\ell = 1, \ldots, s$ to denote the distinct minimal polynomials, and $f_j(x)$ for $j = 1, \ldots, q^m - 1$ to denote the minimal polynomial of β_j. The context will convey the usage.

The set of zeros $\{\beta_1, \ldots, \beta_r\}$ is called the *complete defining set* of the cyclic code $\mathcal{C}$, corresponding to generator polynomial $g(x)$. A set $\mathcal{A} \subset \{\beta_1, \ldots, \beta_r\}$, consisting of at least one zero of each minimal polynomial factor of $g(x)$, is called a *defining set* of the code $\mathcal{C}$.

As an example of these ideas, take $n = 15$. All binary cyclic codes of blocklength 15 can be found by factoring $x^{15} - 1$ into prime polynomials:

$$x^{15} - 1 = (x + 1)(x^2 + x + 1)(x^4 + x + 1)(x^4 + x^3 + 1)(x^4 + x^3 + x^2 + x + 1).$$

We can verify this factorization by multiplication, and verify by trial and error that every factor is irreducible. There are $2^5 = 32$ subsets of these prime polynomials and hence 32 generator polynomials for cyclic codes of blocklength 15. Of these generator polynomials, two are trivial ($g(x) = x^{15} - 1$ with $k = 0$ and $g(x) = 1$ with $k = n$), and the rest generate nontrivial cyclic codes. To obtain one of these as an example, take

$$\begin{aligned} g(x) &= (x^4 + x^3 + 1)(x^4 + x^3 + x^2 + x + 1) \\ &= x^8 + x^4 + x^2 + x + 1. \end{aligned}$$

In particular, $g(\alpha) = 0$ and $g(\alpha^3) = 0$. Hence this is the example seen earlier in Section 5.1. A defining set of this code is $\mathcal{A} = \{\alpha, \alpha^3\}$. The set of all zeros of $g(x)$ is the complete defining set $\{\alpha, \alpha^2, \alpha^3, \alpha^4, \alpha^6, \alpha^8, \alpha^9, \alpha^{12}\}$. Four of the elements of the complete defining set are zeros of $x^4 + x^3 + 1$, and four are zeros of $x^4 + x^3 + x + 1$. An easy way to find the complete defining set is to write, for each $\alpha^j \in \mathcal{A}$, the exponent j in binary form. Thus

$$\begin{matrix} 0 & 0 & 0 & 1 \\ 0 & 0 & 1 & 1 \end{matrix}$$

are the values of j for which α^j is in the defining set. Then all cyclic shifts of these binary numbers are the values of j for which α^j is in the complete defining set. The eight cyclic shifts are

$$\begin{matrix} 0 & 0 & 0 & 1 \\ 0 & 0 & 1 & 0 \\ 0 & 0 & 1 & 1 \\ 0 & 1 & 0 & 0 \\ 0 & 1 & 1 & 0 \\ 1 & 0 & 0 & 0 \\ 1 & 0 & 0 & 1 \\ 1 & 1 & 0 & 0. \end{matrix}$$

These are the binary representations of $1, 2, 3, 4, 6, 8, 9, 12$, so the complete defining set is easily found.

Because $g(x)$ has degree 8, $n - k = 8$ and $k = 7$. Because $g(x)$ has weight 5, the minimum distance is not larger than 5. In Section 5.6, we shall see that any binary cyclic code satisfying $g(\alpha) = 0$ and $g(\alpha^3) = 0$ has a minimum distance at least 5. Hence this code is a $(15, 7, 5)$ code, and it can correct two errors. The check operation can be represented by the following:

$$c(\alpha^i) = 0 \quad c(\alpha^j) = 0$$

where, in the extension field $GF(16)$, α^i is any zero of $x^4 + x^3 + 1$, and α^j is any zero of $x^4 + x^3 + x^2 + x + 1$.

Now suppose that we wish to construct a polynomial $g(x)$ that has $\beta_1, \beta_2, \ldots, \beta_r$ as zeros. Find the minimal polynomials of these field elements, denoting them as $f_1(x), f_2(x), \ldots, f_r(x)$. Then

$$g(x) = \text{LCM}[f_1(x), f_2(x), \ldots, f_r(x)].$$

Thus our task reduces to the problem of finding, for any element β, its minimal polynomial $f(x)$.

Let $f(x)$, the minimal polynomial of β, have degree m'. Then it must have m' zeros in $GF(q^m)$. Therefore $f(x)$ is also the minimal polynomial for all of these zeros. We can identify these additional zeros by using the following two theorems.

Theorem 5.3.3. *Let p be the characteristic of the field $GF(q)$. Then for any polynomial $s(x)$ over $GF(q)$ of degree I and any integer m,*

$$\left[\sum_{i=0}^{I} s_i x^i\right]^{p^m} = \sum_{i=0}^{I} s_i^{p^m} x^{ip^m}.$$

Proof: We start with $m = 1$ and use the same reasoning as was used in the proof of Theorem 4.6.10. To proceed in small steps, define $s_1(x)$ by $s(x) = s_1(x)x + s_0$. Then in any field of characteristic p,

$$[s(x)]^p = \sum_{\ell=0}^{p} \binom{p}{\ell} [s_1(x)x]^{\ell} s_0^{p-\ell}.$$

But

$$\binom{p}{l} = \frac{p!}{l!(p-l)!} = \frac{p(p-1)!}{l!(p-l)!},$$

and p is a prime that does not appear in the denominator unless $l = 0$ or $l = p$. Hence except for $l = 0$ or p, $\binom{p}{\ell}$ is a multiple of p and equals zero modulo p. Then

$$[s(x)]^p = [s_1(x)]^p x^p + s_0^p.$$

Now apply the same reasoning to $s_1(x)$ and continue in this way. Eventually, this gives

$$[s(x)]^p = \sum_{i=0}^{I} s_i^p x^{ip},$$

as was to be proved for $m = 1$. Further,

$$[s(x)]^{p^2} = [[s(x)]^p]^p = \left[\sum_{i=0}^{I} s_i^p x^{ip}\right]^p = \sum_{i=0}^{I} s_i^{p^2} x^{ip^2}.$$

This can be repeated any number of times, and thus the theorem is true with p replaced by any power of p. □

Theorem 5.3.4. *Suppose that $f(x)$ is the minimal polynomial over $GF(q)$ of β, an element of $GF(q^m)$. Then $f(x)$ is also the minimal polynomial of β^q.*

Proof: Because q is a power of the field characteristic p, Theorem 5.3.3 gives

$$[f(x)]^q = \sum_{i=0}^{\deg f(x)} f_i^q (x^q)^i.$$

But the coefficients f_i are elements of $GF(q)$, and all elements of $GF(q)$ satisfy $\gamma^q = \gamma$. Therefore

$$[f(x)]^q = \sum_{i=0}^{\deg f(x)} f_i (x^q)^i = f(x^q),$$

and because $f(\beta) = 0$, we have

$$0 = [f(\beta)]^q = f(\beta^q),$$

and hence β^q is a zero of $f(x)$. Because $f(x)$ is a prime polynomial, $f(x)$ is the minimal polynomial of β^q, as was to be proved. □

Definition 5.3.5. *Two elements of $GF(q^m)$ that share the same minimal polynomial over $GF(q)$ are called conjugates* (*with respect to $GF(q)$*).

Generally, a single element can have more than one conjugate – in fact, excluding itself, it can have as many as $m - 1$ conjugates. We should also mention that the conjugacy relationship between two elements depends on the base field. Two elements of $GF(16)$ might be conjugates with respect to $GF(2)$, but not with respect to $GF(4)$.

The q-ary conjugates of an element β are easily found by using Theorem 5.3.4. If $f(x)$ is the minimal polynomial of β, then it is also the minimal polynomial of β^q and, in turn, of β^{q^2}, and so forth. Hence the elements in the set

$$\left\{\beta, \beta^q, \beta^{q^2}, \beta^{q^3}, \ldots, \beta^{q^{r-1}}\right\}$$

are all conjugates, where r is the smallest integer such that $\beta^{q^r} = \beta$. (Because $\beta^{q^m} = \beta$, we can show that r is a divisor of m.) This set is called a *set of conjugates*. The conjugates are all zeros of $f(x)$, and the following theorem shows that $f(x)$ has no other zeros.

Theorem 5.3.6. *The minimal polynomial of β over $GF(q)$ is*

$$f(x) = (x - \beta)(x - \beta^q) \cdots \left(x - \beta^{q^{r-1}}\right).$$

Proof: Certainly the minimal polynomial of β must have all of these zeros, as stated by Theorem 5.3.4, and thus the minimal polynomial cannot have a smaller degree. All we need to show is that $f(x)$, thus constructed, has coefficients only in $GF(q)$. We shall use the fact that in the field $GF(q^m)$, the set of elements that are zeros of $x^q - x$ forms the subfield $GF(q)$. First, find $[f(x)]^q$:

$$\begin{aligned}[f(x)]^q &= (x-\beta)^q(x-\beta^q)^q\left(x-\beta^{q^2}\right)^q \cdots \left(x-\beta^{q^{r-1}}\right)^q \\ &= (x^q-\beta^q)\left(x^q-\beta^{q^2}\right)\left(x^q-\beta^{q^3}\right)\cdots(x^q-\beta)\end{aligned}$$

where the last line follows from Theorem 5.3.3, and from the fact that $\beta^{q^r} = \beta$. Therefore

$$[f(x)]^q = f(x^q) = \sum_i f_i x^{iq},$$

whereas Theorem 5.3.3 asserts that

$$[f(x)]^q = \left[\sum_i f_i x^i\right]^q = \sum_i f_i^q x^{iq}.$$

Therefore $f_i^q = f_i$ for each i, and hence every f_i is in the subfield $GF(q)$, as was to be proved. □

For example, take α as a primitive element in $GF(256)$. Then

$$\{\alpha, \alpha^2, \alpha^4, \alpha^8, \alpha^{16}, \alpha^{32}, \alpha^{64}, \alpha^{128}\}$$

is a set of binary conjugates. It terminates with α^{128} because $\alpha^{255} = 1$, and hence $\alpha^{256} = \alpha$, which already appears. The minimal polynomial of α is

$$f(x) = (x-\alpha)(x-\alpha^2)(x-\alpha^4)\cdots(x-\alpha^{64})(x-\alpha^{128}),$$

which when multiplied out, must have all coefficients in $GF(2)$. Similarly, the set of binary conjugates that includes α^7 is

$$\{\alpha^7, \alpha^{14}, \alpha^{28}, \alpha^{56}, \alpha^{112}, \alpha^{224}, \alpha^{193}, \alpha^{131}\},$$

and the minimal polynomial of α^7 is

$$f(x) = (x-\alpha^7)(x-\alpha^{14})(x-\alpha^{28})\cdots(x-\alpha^{193})(x-\alpha^{131}),$$

which when multiplied out, must have coefficients in $GF(2)$.

Instead of $GF(2)$, we now take $GF(4)$ as the base field; it also is a subfield of $GF(256)$. Then the set of four-ary conjugates of α^7 is

$$\{\alpha^7, \alpha^{28}, \alpha^{112}, \alpha^{193}\},$$

and over $GF(4)$, the minimal polynomial of α^7 is

$$f(x) = (x-\alpha^7)(x-\alpha^{28})(x-\alpha^{112})(x-\alpha^{193}).$$

When multiplied out, this polynomial is

$$f(x) = x^4 + \alpha^{170}x^2 + \alpha^{85}x + \alpha^{85},$$

which must have coefficients only in $GF(4)$. In order to recognize these coefficients as elements of $GF(4)$, however, one must identify the field elements of $GF(4)$ among those of $GF(256)$. In the notation of $GF(256)$, the subfield $GF(4)$ is given by $\{0, 1, \alpha^{85}, \alpha^{170}\}$ because α^{85} and α^{170} are the only elements of order 3.

We have seen that for primitive cyclic codes there is a close relationship between $g(x)$ and the field $GF(q^m)$, as described by the zeros of $g(x)$. Let us now turn to cyclic codes that are not primitive cyclic codes. These nonprimitive cyclic codes are codes of blocklength n other than $q^m - 1$. The following theorem is the basis for relating many of these codes to the earlier discussion.

Theorem 5.3.7. *If n and q are coprime, then $x^n - 1$ divides $x^{q^m-1} - 1$ for some m, and $x^n - 1$ has n distinct zeros in the field $GF(q^m)$.*

Proof: It is only necessary to prove that n divides $q^m - 1$ for some m because, once this is proved, we can use the general factorization

$$z^b - 1 = (z - 1)(z^{b-1} + z^{b-2} + z^{b-3} + \cdots + z + 1)$$

with $z = x^n$ to show that

$$\begin{aligned} x^{q^m-1} - 1 &= (x^n)^b - 1 \\ &= (x^n - 1)\left(x^{n(b-1)} + x^{n(b-2)} + \cdots + x^n + 1\right). \end{aligned}$$

Therefore $x^n - 1$ divides $x^{q^m-1} - 1$ and, consequently, has n distinct zeros in $GF(q^m)$ because all zeros of $x^{q^m-1} - 1$ are distinct.

To prove that n divides $q^m - 1$ for some m, use the division algorithm to write the following set of $n + 1$ equations:

$$\begin{aligned} q &= Q_1 n + s_1 \\ q^2 &= Q_2 n + s_2 \\ q^3 &= Q_3 n + s_3 \\ &\vdots \quad \vdots \quad \vdots \\ q^n &= Q_n n + s_n \\ q^{n+1} &= Q_{n+1} n + s_{n+1}. \end{aligned}$$

All remainders are between 0 and $n - 1$. Because there are $n + 1$ remainders, two must be the same, say s_i and s_j, with i smaller than j. Then

$$q^j - q^i = Q_j n + s_j - Q_i n - s_i,$$

or

$$q^i(q^{j-i} - 1) = (Q_j - Q_i)n.$$

Because n and q are coprime, n must divide $q^{j-i} - 1$. Setting $m = j - i$ completes the proof. □

Using this theorem, we can describe any cyclic code over $GF(q)$ of blocklength n in a suitable extension field $GF(q^m)$, provided that n and q are coprime. For a cyclic code of blocklength n, $g(x)$ divides $x^n - 1$ and $x^n - 1$ divides $x^{q^m-1} - 1$, and thus $g(x)$ also divides $x^{q^m-1} - 1$. We shall always use the smallest such m. Then $q^m - 1 = nb$. Let α be primitive in $GF(q^m)$, and let $\beta = \alpha^b$. Then all zeros of $x^n - 1$, and thus of $g(x)$, are powers of β. The prime factors of $x^n - 1$ have only these zeros.

In summary, if we use $\beta = \alpha^b$ in place of α and choose only powers of β as zeros of $g(x)$, then we obtain a nonprimitive cyclic code of blocklength $n = (q^m - 1)/b$. All cyclic codes with n coprime to q can be obtained in this way, though not those for which n and q have a common factor.

5.4 Matrix description of cyclic codes

We began this chapter with a discussion relating a check matrix to the zeros of polynomials in an extension field. Then we proceeded to develop the idea of a cyclic code, but without further mention of check matrices. Now it is time to relate a cyclic code to its generator matrix and check matrix. There are a number of ways to write out these matrices. First of all, we can write out a check matrix by working in the extension field, as we did in Section 5.1. If $\gamma_j \in GF(q^m)$ for $j = 1, \ldots, r$ are the zeros of $g(x)$, then

$$\sum_{i=0}^{n-1} c_i \gamma_j^i = 0 \quad j = 1, \ldots, r,$$

which can be written in matrix form as

$$\boldsymbol{H} = \begin{bmatrix} \gamma_1^0 & \gamma_1^1 & \cdots & \gamma_1^{n-1} \\ \gamma_2^0 & \gamma_2^1 & \cdots & \gamma_2^{n-1} \\ \vdots & & & \\ \gamma_r^0 & \gamma_r^1 & \cdots & \gamma_r^{n-1} \end{bmatrix}.$$

This r by n matrix over $GF(q^m)$ can be changed into an rm by n matrix over $GF(q)$ by replacing each matrix element β by a column vector based on the coefficients of β expressed as a polynomial over $GF(q)$. This gives a check matrix over $GF(q)$, but any rows that may be linearly dependent represent excess baggage in the matrix. Delete the least number of rows that are necessary to obtain a matrix with linearly independent rows. This gives a check matrix for the code.

While the above procedure makes clear the relationship between zeros in the extension field and the check matrix, it is somewhat cumbersome to follow through. One can find the desired matrices from the generator polynomial without ever entering the extension field. One way is to construct the generator matrix from the generator polynomial by inspection. Codewords are of the form $c(x) = a(x)g(x)$, and it is simple to translate this equation into matrix form:

$$\boldsymbol{G} = \begin{bmatrix} 0 & \dots & 0 & g_{n-k} & g_{n-k-1} & \dots & g_2 & g_1 & g_0 \\ 0 & & g_{n-k} & g_{n-k-1} & g_{n-k-2} & \dots & g_1 & g_0 & 0 \\ 0 & & g_{n-k-1} & g_{n-k-2} & g_{n-k-3} & \dots & g_0 & 0 & 0 \\ \vdots & & & & & & & \vdots & \\ g_{n-k} & \dots & & & & & & 0 & 0 \end{bmatrix}.$$

A check matrix is:

$$\boldsymbol{H} = \begin{bmatrix} 0 & 0 & 0 & \dots & & & & \dots & h_{k-1} & h_k \\ \vdots & & & & & & & & & \vdots \\ 0 & h_0 & h_1 & & h_{k-1} & h_k & 0 & & 0 & 0 \\ h_0 & h_1 & h_2 & \dots & h_k & 0 & 0 & \dots & 0 & 0 \end{bmatrix}$$

where $h(x)$ is the check polynomial of the cyclic code. To verify that $\boldsymbol{G}\boldsymbol{H}^{\mathrm{T}} = \boldsymbol{0}$, let

$$u_r = \sum_{i=0}^{r} g_{r-i} h_i,$$

which is zero for $0 < r < n$ because $h(x)g(x) = x^n - 1$. But then we have

$$\boldsymbol{G}\boldsymbol{H}^{\mathrm{T}} = \begin{bmatrix} u_{n-1} & u_{n-2} & \dots & u_k \\ u_{n-2} & u_{n-3} & \dots & u_{k-1} \\ \vdots & & & \vdots \\ u_{n-k} & u_{n-k-1} & \dots & u_1 \end{bmatrix} = \boldsymbol{0},$$

and thus $\boldsymbol{H}$, as defined, is indeed a check matrix. Now we can see that the dual code $\mathcal{C}^{\perp}$ of the code $\mathcal{C}$ generated by $g(x)$ is also a cyclic code because $\boldsymbol{H}$ is its generator matrix and has the form of a generator matrix for a cyclic code. The generator polynomial for the dual code is the reciprocal polynomial of $h(x)$, given by $\widetilde{h}(x) = x^k h(x^{-1})$. Sometimes in discussions of cyclic codes, the code generated by $h(x)$ is also called the dual code, but to be precise, one should say only that it is equivalent to the dual code.

One can also easily write a generator matrix in systematic form. For each i corresponding to a data place, use the division algorithm to write

$$x^{n-i} = Q_i(x)g(x) + s_i(x) \quad i = 1, \dots, k$$

where

$$s_i(x) = \sum_{j=0}^{n-k-1} s_{ji} x^j.$$

Because

$$x^{n-i} - s_i(x) = Q_i(x)g(x)$$

is a multiple of $g(x)$, $x^{n-i} - s_i(x)$ is a codeword. Using the coefficients of the left side as elements of the generator matrix gives

$$G = \begin{bmatrix} -s_{0,\,k} & \dots & -s_{(n-k-1),\,k} & 1 & 0 & \dots & 0 \\ -s_{0,\,k-1} & & -s_{(n-k-1),\,k-1} & 0 & 1 & & 0 \\ \vdots & & & \vdots & & \vdots & \\ -s_{0,\,1} & \dots & -s_{(n-k-1),\,1} & 0 & 0 & \dots & 1 \end{bmatrix}.$$

This is a systematic generator matrix with the indices of the data places running from $n-k$ to $n-1$. To write this matrix, it is only necessary to carry out the computation of the $s_i(x)$ as remainders of x^i divided by $g(x)$. The check matrix

$$H = \begin{bmatrix} 1 & 0 & \dots & 0 & s_{0,\,k} & \dots & s_{0,\,1} \\ 0 & 1 & & 0 & s_{1,\,k} & & s_{1,\,1} \\ \vdots & & & \vdots & & & \vdots \\ 0 & & \dots & 1 & s_{(n-k-1),\,k} & \dots & s_{(n-k-1),\,1} \end{bmatrix}$$

is then written, using the principles of Section 3.2.

5.5 Hamming codes as cyclic codes

The code used to start this chapter, the (7, 4) Hamming code, has a generator polynomial

$$g(x) = x^3 + x + 1.$$

This polynomial has a zero in $GF(8)$ at the primitive element α, and thus all codewords satisfy $c(\alpha) = 0$. Similarly, to get a binary Hamming code for any primitive blocklength $n = 2^m - 1$, take $g(x)$ of degree m with a zero at primitive element α in $GF(2^m)$. This means that $g(x)$ is equal to $p(x)$, the minimal polynomial of α, which is the primitive polynomial used to construct $GF(2^m)$. Then $c(x) = a(x)g(x)$ and $c(\alpha) = 0$ for every codeword. That is, with the check matrix

$$\boldsymbol{H} = [\,\alpha^0 \quad \alpha^1 \quad \dots \quad \alpha^{n-1}\,],$$

the binary Hamming codewords satisfy $\boldsymbol{c}\boldsymbol{H}^{\mathrm{T}} = \boldsymbol{0}$.

There are also Hamming codes over nonbinary alphabets. In Section 3.3, we saw that for every m, there is a Hamming code over $GF(q)$ with $n = (q^m - 1)/(q - 1)$ and

$k = [(q^m - 1)/(q - 1)] - m$. In this section, we shall show that many of these, but not all, are cyclic codes.

Let α be primitive in $GF(q^m)$, and let $\beta = \alpha^{q-1}$. Then $\beta^{(q^m-1)/(q-1)} = 1$, and thus β is a zero of the polynomial $x^{(q^m-1)/(q-1)} - 1$. Hence the minimal polynomial of β divides $x^{(q^m-1)/(q-1)} - 1$ and serves as a generator polynomial for a cyclic code of blocklength $n = (q^m - 1)/(q - 1)$. The check matrix for this code is $\boldsymbol{H} = [\ \beta^0\ \beta^1\ \ldots\ \beta^{n-1}\ ]$.

For $q = 2$, we see that $\beta = \alpha$, and it is easy to prove that the code is a single-error-correcting code by giving a simple procedure for decoding single errors. The senseword is a polynomial of degree $n - 1$, given by

$$v(x) = a(x)g(x) + e(x)$$

where $e(x)$ has at most one nonzero coefficient. That is,

$$e(x) = 0 \text{ or } x^i.$$

The integer i indexes the locations at which an error occurs. We also use the elements of $GF(2^m)$ to index the error locations. The field element α^i is assigned to index component i. Because $g(\alpha) = 0$, we have $v(\alpha) = \alpha^i$, and all powers of α from 0 to $2^m - 2$ are distinct. The error location i is immediately determined from α^i unless $v(\alpha) = 0$, in which case there is no error. Therefore the code is a single-error-correcting code over $GF(2)$ with $n = 2^m - 1$ and $k = n - m$; in fact, it is a Hamming code over $GF(2)$.

We have just proved the following theorem for the special case of $q = 2$. The general case in which q is not equal to two is more difficult because all terms of the form $\gamma\beta_i$ must be distinct, where γ is any nonzero element of $GF(q)$.

Theorem 5.5.1. *Let α be primitive in $GF(q^m)$. The cyclic code over $GF(q)$ of blocklength $n = (q^m - 1)/(q - 1)$ with check matrix $\boldsymbol{H} = [\ \beta^0\ \beta^1 \ldots \beta^{n-1}\ ]$, where $\beta = \alpha^{q-1}$, has a minimum distance at least 3 if and only if n and $q - 1$ are coprime.*

Proof: Given that $\beta = \alpha^{q-1}$, suppose that two columns of $\boldsymbol{H}$, say columns i' and i'', are linearly dependent. Then

$$\beta^{i'} = \gamma\beta^{i''}$$

where γ is a nonzero element of $GF(q)$. Let $i = i' - i''$. Then $\beta^i = \gamma$ is an element of $GF(q)$. But the nonzero elements of $GF(q)$ can be expressed in terms of α as the first $q - 1$ powers of $\alpha^{(q^m-1)/(q-1)}$. Therefore, $\gamma \in \{\alpha^{\ell(q^m-1)/(q-1)} \mid \ell = 0, \ldots, q - 2\}$. Then

$$\beta^i = \left(\alpha^{(q^m-1)/(q-1)}\right)^{\ell} = \alpha^{\ell n}$$

for some ℓ less than $q - 1$. Further, because $\beta = \alpha^{q-1}$, we can conclude that

$$i(q - 1) = \ell n.$$

This equation can be satisfied by an i that is less than n if and only if n and $q-1$ are not coprime; that is, $\boldsymbol{H}$ can have two linearly dependent columns if and only if n and $q-1$ are not coprime. □

Theorem 5.5.2. *Let $n=(q^m-1)/(q-1)$. Then n and $q-1$ are coprime if and only if m and $q-1$ are coprime.*

Proof: We know that

$$n=\frac{q^m-1}{q-1}=q^{m-1}+q^{m-2}+\cdots+1,$$

and

$$q^{m-j}=(q-1)s_j+1$$

for some s_j because $q^{m-j}-1$ is divisible by $q-1$. Therefore by summing over j, we obtain

$$n=m+(q-1)\sum_{j=1}^{m}s_j,$$

and so n and $q-1$ are coprime if and only if m and $q-1$ are coprime. □

Theorem 5.5.2 established that the Hamming codes of blocklength $n=(q^m-1)/(q-1)$ over $GF(q)$ are cyclic if m and $q-1$ are coprime. We saw some Hamming codes in Section 3.3, however, for which m and $q-1$ are not coprime. For these few codes, the cyclic construction does not work. The smallest example of such is the $(21, 18)$ Hamming code over $GF(4)$. This code cannot be expressed in terms of a generator polynomial. Most Hamming codes of interest, however, are cyclic.

As an example of a cyclic Hamming code, we will find the generator polynomial for an $(85, 81)$ Hamming code over $GF(4)$. This construction takes place in $GF(256)$ and consists of finding the minimal polynomial over $GF(4)$ of $\beta=\alpha^3$. In $GF(256)$, the elements α^{85} and α^{170} are the only two elements of order 3, so we see that

$$GF(4)=\{0,1,\alpha^{85},\alpha^{170}\}$$

when expressed in the terminology of $GF(256)$. The conjugates of β with respect to $GF(2)$ are

$$\{\beta,\ \beta^2,\ \beta^4,\ \beta^8,\ \beta^{16},\ \beta^{32},\beta^{64},\ \beta^{128}\}.$$

We need the conjugates with respect to $GF(4)$. These are $\{\beta,\beta^4,\beta^{16},\beta^{64}\}$. Hence the desired generator polynomial is

$$\begin{aligned}g(x)&=(x-\beta)(x-\beta^4)(x-\beta^{16})(x-\beta^{64})\\&=(x-\alpha^3)(x-\alpha^{12})(x-\alpha^{48})(x-\alpha^{192}).\end{aligned}$$

When multiplied out, this polynomial must have coefficients only in $GF(4)$. Thus far we have not needed to make use of the structure of $GF(256)$, but now we do. Using the primitive polynomial $p(x) = x^8 + x^4 + x^3 + x^2 + 1$, we have $\alpha^8 = \alpha^4 + \alpha^3 + \alpha^2 + 1$. We can make this substitution repeatedly to reduce $g(x)$, or alternatively, we can write out a log table for multiplication in $GF(256)$. In either case, within an hour's worth of manual computation, we find that

$$g(x) = x^4 + x^3 + \alpha^{170}x + 1.$$

Finally, we change notation to something suitable for the subfield $GF(4)$ because we are finished with the big field $GF(256)$. That is,

$$g(x) = x^4 + x^3 + 3x + 1.$$

This is the desired generator polynomial of the $(85, 81)$ Hamming code over $GF(4)$.

5.6 Cyclic codes for correcting double errors

The Hamming codes were described in the previous section as cyclic codes whose generator polynomials have a zero at a suitable element of the appropriate extension field. These codes correct a single error. Now we turn to double-error-correcting codes over the field $GF(2)$. Let the blocklength n be $2^m - 1$ for some m, and let α be a primitive element of $GF(2^m)$. We will consider those binary cyclic codes having α and α^3 as zeros of the generator polynomial, and we will show that the resulting code corrects double errors.

Let $g(x)$ be the polynomial over $GF(2)$ of the smallest degree, having α and α^3 as zeros in $GF(2^m)$. By exhibiting a decoding procedure that corrects all single and double errors, we will prove that the minimum distance of this code is at least 5.

The senseword is a polynomial of degree $n - 1$ given by

$$v(x) = a(x)g(x) + e(x)$$

where $e(x)$ has at most two nonzero coefficients because we are considering at most two errors. That is,

$$e(x) = 0 \text{ or } x^i \text{ or } x^i + x^{i'}.$$

The integers i and i' index the two locations at which errors occur. We will also use the elements of $GF(2^m)$ to index the error locations. The field element α^i is assigned to index the ith component. In this role, the field elements are called the *location numbers*. Define the field elements $X_1 = \alpha^i$ and $X_2 = \alpha^{i'}$. The error-location numbers X_1 and X_2 must be unique because n, the order of α, is also the blocklength. If only one error occurs, then X_2 is zero. If no errors occur, then both X_1 and X_2 are zero.

Let $S_1 = v(\alpha)$ and $S_3 = v(\alpha^3)$. These field elements, also known as *syndromes*, can be evaluated immediately from $v(x)$. Because $g(x)$ has zeros at α and α^3, $S_1 = e(\alpha)$ and $S_3 = e(\alpha^3)$. Suppose two errors occur:

$$S_1 = \alpha^i + \alpha^{i'}$$
$$S_3 = \alpha^{3i} + \alpha^{3i'}.$$

But this is just a pair of equations in $GF(2^m)$ involving the two unknowns, X_1 and X_2:

$$S_1 = X_1 + X_2$$
$$S_3 = X_1^3 + X_2^3.$$

Under the assumption that at most two errors will occur, S_1 equals zero if and only if no errors occur. The decoder needs to proceed only if S_1 is not zero. If the above pair of nonlinear equations can be solved uniquely for X_1 and X_2, the two errors can be corrected, and so the code must have a minimum distance of at least 5.

It is not directly apparent how to solve this pair of equations. One way is to introduce a new polynomial that is defined with the error location numbers as zeros:

$$(x - X_1)(x - X_2) = x^2 + (X_1 + X_2)x + X_1X_2.$$

If we can find the coefficients of the polynomial on the right side, then we can factor the polynomial to find X_1 and X_2. But, over extensions of $GF(2)$,

$$S_1^3 + S_3 = X_1^2X_2 + X_2^2X_1 = S_1X_1X_2.$$

Thus

$$(x + X_1)(x + X_2) = x^2 + S_1x + \frac{S_1^3 + S_3}{S_1},$$

and $S_1 \neq 0$ if one or two errors occur. We know the polynomial on the right because we know S_1 and S_3. The two error-location numbers are the two zeros of the polynomial. The zeros of any polynomial over a field are unique. Therefore the code is a double-error-correcting code.

One can use any convenient procedure for decoding, but one procedure is to find the zeros of the above quadratic equation over $GF(2^m)$. This can be done by trial and error because there are only 2^m possibilities. Other decoding schemes will be studied in later chapters.

The double-error-correcting codes of this section illustrate how cyclic codes in one field can be constructed by working with zeros in a larger field. This general principle will be discussed in Chapter 6 for an arbitrary symbol field $GF(q)$ and an arbitrary number of correctable errors.

5.7 Quasi-cyclic codes and shortened cyclic codes

Closely related to the cyclic codes are the quasi-cyclic codes and the shortened cyclic codes.

Definition 5.7.1. *An (n, k) quasi-cyclic code is a linear block code such that, for some b coprime with n, the polynomial $x^b c(x) (\text{mod } x^n - 1)$ is a codeword polynomial whenever $c(x)$ is a codeword polynomial.*

Some good quasi-cyclic codes have been discovered by computer search. For example, the 4 by 12 matrix

$$\boldsymbol{G} = \begin{bmatrix} 1 & 1 & 1 & 0 & 1 & 1 & 0 & 0 & 0 & 0 & 0 & 1 \\ 0 & 0 & 1 & 1 & 1 & 1 & 0 & 1 & 1 & 0 & 0 & 0 \\ 0 & 0 & 0 & 0 & 0 & 1 & 1 & 1 & 1 & 0 & 1 & 1 \\ 0 & 1 & 1 & 0 & 0 & 0 & 0 & 0 & 1 & 1 & 1 & 1 \end{bmatrix}$$

is the generator matrix for a binary $(12, 4, 6)$ quasi-cyclic code with $b = 3$ because the cyclic shift of any row of $\boldsymbol{G}$ by three places produces another row. No better binary code of blocklength 12 exists with sixteen codewords.

Any cyclic code of composite blocklength can be punctured to form a quasi-cyclic code simply by dropping every bth symbol, where b is a factor of n. This quasi-cyclic code will be a shortened code if none of the dropped symbols are check symbols because then all of the dropped symbols are data symbols and can be first set to zero. With regard to a generator matrix in systematic form, dropping data places from $\boldsymbol{G}$ reduces the rank by b. By dropping the rows of $\boldsymbol{G}$ corresponding to the dropped columns, the matrix is restored to full rank.

More generally, any cyclic code can be shortened by any integer b less than k – that is, changed from an (n, k) code to an $(n - b, k - b)$ code by setting any fixed set of b data symbols to zero and dropping them from each codeword. Although dropping every bth place may produce a shortened code that is also quasi-cyclic, it is not the customary way to shorten a cyclic code. It is more usual to require that the dropped b data symbols are consecutive.

Definition 5.7.2. *An (n, k) linear code is called a proper shortened cyclic code if it can be obtained by deleting b consecutive places from an $(n + b, k + b)$ cyclic code.*

Because the unused symbols in a shortened code are always set to zero, they need not be transmitted, but the receiver can reinsert them and decode just as if the code were not shortened. If the original cyclic code has minimum distance d_{min}, then the shortened cyclic code also has minimum distance at least d_{min}. Similarly, if the original code can correct burst errors of length t, as will be discussed in Section 5.9, then the shortened code also can correct burst errors of length at least t.

A shortened code is no longer cyclic because $R_{x^{n'}-1}[xc(x)]$ generally is not a codeword when $c(x)$ is a codeword. A proper shortened cyclic code, however, does have an algebraic structure as a subset of an appropriate ring. Whereas cyclic codes are ideals in the ring of polynomials modulo $x^n - 1$, the following theorem shows that a proper shortened cyclic code is an ideal in the ring of polynomials modulo $f(x)$ for some polynomial $f(x)$ of degree $n' = n - b$.

Theorem 5.7.3. *If C is a proper shortened cyclic code, then there exists a polynomial $f(x)$ such that whenever $c(x)$ is a codeword and $a(x)$ is any polynomial, then $[R_{f(x)}a(x)c(x)]$ is also a codeword.*

Proof: Let $g(x)$ be the generator polynomial of the original cyclic code of blocklength n, and let n' be the blocklength of the shortened cyclic code. Then, by the division algorithm, we can write

$$x^{n'} = g(x)Q(x) + s(x).$$

Because $n' = n - b > n - k = \deg g(x)$, the remainder $s(x)$ has a degree smaller than n'. Let $f(x) = x^{n'} - s(x)$; then $f(x)$ has a degree of n', and $g(x)$ divides $f(x)$. Now if $c(x)$ is a multiple of $g(x)$ and $a(x)$ is any polynomial, then, again by the division algorithm, we can write

$$a(x)c(x) = f(x)Q'(x) + r(x).$$

Because $g(x)$ divides both $c(x)$ and $f(x)$, $r(x)$ is clearly a multiple of $g(x)$; thus in the ring $GF(q)[x]/\langle f(x)\rangle$, $a(x)c(x) = r(x)$ is a multiple of $g(x)$, as was to be proved. □

5.8 The Golay code as a cyclic code

The binary Golay code occupies a unique and important position in the subject of data-transmission codes. Because the spheres of radius t about codewords are so neatly packed, it is not surprising that the Golay code has close ties to many mathematical topics, and one can use it as a doorway for passing from the subject of error-control coding to deep topics in group theory and other areas of mathematics. From a practical point of view, however, it is probably fair to say that, because of its short blocklength, the binary Golay code has been left far behind in most practical applications of codes.

To construct binary cyclic codes of blocklength 23, we can find the polynomial factors of $x^{23} - 1$. It is easy to verify by direct multiplication that the polynomial

$$g(x) = x^{11} + x^{10} + x^6 + x^5 + x^4 + x^2 + 1,$$

and its reciprocal,

$$\widetilde{g}(x) = x^{11} + x^9 + x^7 + x^6 + x^5 + x + 1,$$

satisfy

$$(x-1)g(x)\widetilde{g}(x) = x^{23} - 1.$$

Hence $g(x)$ is the generator polynomial of a (23, 12) cyclic code over $GF(2)$. This cyclic code is the binary cyclic Golay code. Furthermore, the reciprocal polynomial $\widetilde{g}(x)$ is the generator polynomial of another (23, 12) cyclic code over $GF(2)$. The code generated by $\widetilde{g}(x)$ is equivalent to the code generated by $g(x)$. Indeed, it is the reciprocal cyclic Golay code obtained by replacing each codeword polynomial $c(x)$ by its reciprocal polynomial $\widetilde{c}(x)$.

Because the polynomials $g(x)$ and $\widetilde{g}(x)$ are irreducible, these are the only two (23, 12) binary cyclic codes. This can be quickly seen by noting that the binary conjugacy class of β modulo $\beta^{23} - 1$ is $\{\beta, \beta^2, \beta^4, \beta^8, \beta^{16}, \beta^9, \beta^{18}, \beta^{13}, \beta^3, \beta^6, \beta^{12}\}$. Consequently, the minimal polynomial of some element β in the splitting field of $x^{23} - 1$ has degree 11, and so it is either $g(x)$ or $\widetilde{g}(x)$. The field $GF(2^m)$ is seen to be $GF(2^{11})$ by noting that $2^{11} - 1 = 23 \cdot 89$, and 23 divides $2^m - 1$ for no smaller m.

It remains to show that the minimum distance of the cyclic Golay code is 7. This requires considerable work, which we break up into the proofs of the following two lemmas.

Lemma 5.8.1. *The Hamming weight of every nonzero codeword in the binary Golay code is at least 5.*

Proof: The proof uses the fact that a Vandermonde matrix with distinct columns is nonsingular. Because β, β^2, β^3, and β^4 are zeros of $g(x)$, they are zeros of every codeword polynomial. Hence every codeword satisfies $\boldsymbol{c}\boldsymbol{H}^{\mathrm{T}} = \boldsymbol{0}$ where

$$\boldsymbol{H} = \begin{bmatrix} 1 & \beta & \beta^2 & \cdots & \beta^{22} \\ 1 & \beta^2 & \beta^4 & & \beta^{21} \\ 1 & \beta^3 & \beta^6 & & \beta^{20} \\ 1 & \beta^4 & \beta^8 & \cdots & \beta^{19} \end{bmatrix}.$$

But any four columns of $\boldsymbol{H}$ form a nonzero Vandermonde matrix, which has a nonzero determinant if the elements of the first row are distinct. Therefore $\boldsymbol{H}$ is the check matrix of a code whose minimum weight is at least 5. □

The next lemma says that no codeword has Hamming weight 2, 6, 10, 14, 18, or 22.

Lemma 5.8.2. *The Hamming weight of every nonzero codeword of even weight in the binary Golay code is divisible by four.*

Proof: Let $c(x) = a(x)g(x)$ be any codeword of even weight, and let $\widetilde{c}(x) = x^{22}c(x^{-1})$ be the corresponding reciprocal codeword. Clearly, $\widetilde{c}(x) = \widetilde{a}(x)\widetilde{g}(x)$. Because $c(x)$ has even weight, $c(1) = 0$. This means that $(x - 1)$ divides $c(x)$. But $(x - 1)$ does not divide

$g(x)$, so we can conclude that $c(x) = b(x)(x-1)g(x)$. Therefore

$$\begin{aligned} c(x)\widetilde{c}(x) &= b(x)\widetilde{a}(x)(x-1)g(x)\widetilde{g}(x) \\ &= b(x)\widetilde{a}(x)(x^{23}-1), \end{aligned}$$

from which we conclude that $c(x)\widetilde{c}(x) = 0 \pmod{x^{23}-1}$. The equation holds over $GF(2)$ and so uses modulo-2 arithmetic on the coefficients. This means that if, instead, the equation is regarded as a polynomial equation (mod $x^{23}-1$) over the integers, then every coefficient of the polynomial product must be an even integer:

$$c(x)\widetilde{c}(x) = 2p(x) \pmod{x^{23}-1}$$

for some nonzero polynomial $p(x)$ with nonnegative integer coefficients. Of the coefficients of $p(x)$, the coefficient p_{22} has the simple form $p_{22} = \sum_{i=0}^{22} c_i^2$. This means that we can write the equation as

$$c(x)\widetilde{c}(x) - \left(\sum_{i=0}^{22} c_i^2\right) x^{22} = 2s(x) + Q(x)(x^{23}-1)$$

where $s(x)$ is an integer polynomial of degree at most 21. Replacing x by x^{-1} and multiplying by x^{44} yields

$$x^{44}\left[c(x^{-1})\widetilde{c}(x^{-1}) - \left(\sum_{i=0}^{22} c_i^2\right) x^{-22}\right] = 2x^{44}\left[s(x^{-1}) + Q(x^{-1})(x^{-23}-1)\right],$$

which reduces to

$$\widetilde{c}(x)c(x) - \left[\sum_{i=0}^{22} c_i^2\right] x^{22} = 2x^{21}s(x^{-1}) + 2(x^{23}-1)x^{21}s(x^{-1}) - \widetilde{Q}(x)(x^{23}-1)$$

where $\widetilde{Q}(x) = x^{21}Q(x^{-1})$. The left side is the same as before, so we conclude that

$$x^{21}s(x^{-1}) = s(x) \pmod{x^{23}-1}.$$

But $\deg s(x) \leq 21$, so $x^{21}s(x^{-1}) = \widetilde{s}(x)$. Thus $s(x)$ is its own reciprocal polynomial. The significance of this conclusion is that, over the integers,

$$s(1) = \sum_{i=0}^{21} s_i = 2\sum_{i=0}^{10} s_i,$$

so we can write

$$c(1)\widetilde{c}(1) - \sum_{i=0}^{20} c_i^2 = 2s(1) = 4\sum_{i=0}^{10} s_i.$$

Because the binary component c_i is nonzero in w places, this becomes

$$w(w-1) = 4\sum_{i=0}^{10} s_i.$$

Because w is even, w is a multiple of four, which completes the proof of the lemma. □

Theorem 5.8.3. *The binary Golay code is a perfect triple-error-correcting code.*

Proof: The Hamming bound says that the minimum distance of a (23, 12) binary code cannot be greater than 7. Thus it is only necessary to prove that the minimum weight is at least 7. By Lemma 5.8.1 it is at least 5, and by Lemma 5.8.2 it is not 6. Thus we need only to show that there are no codewords of weight 5.

But $\widetilde{g}(x)g(x)$ is a codeword; in fact, $g(x)\widetilde{g}(x) = \sum_{k=0}^{n-1} x^k$ because $(x-1)g(x)\widetilde{g}(x) = x^n - 1 = (x-1)\sum_{k=0}^{n-1} x^k$. Thus the vector with all ones is a codeword. The all-ones codeword can be added to any codeword of weight w to obtain a codeword of weight $23 - w$. Then by Lemma 5.8.2, we conclude that there can be no codewords of weight 21, 17, 13, 9, 5, or 1, and in particular, there can be no codewords of weight 5. Thus the theorem is proved. □

The (23, 12, 7) Golay code can be extended by one additional check bit formed as a simple sum of the original twenty-three bits. This increases the weight of every odd-weight codeword by one so that all codewords of the extended code have even weight, and the minimum distance is 8. It is an easy consequence of the proof of Theorem 5.8.3 that the Golay codewords can only have weights of 0, 7, 8, 11, 12, 15, 16, and 23, and so the extended Golay codewords can only have weights of 0, 8, 12, 16, and 24. The number of codewords of each weight for the Golay code and the extended Golay code, evaluated by computer, is shown in Table 5.2.

Table 5.2. *Number of Golay codewords*

	Number of codewords	
Weight	(23, 12) Code	Extended (24, 12) code
0	1	1
7	253	0
8	506	759
11	1288	0
12	1288	2576
15	506	0
16	253	759
23	1	0
24	—	1
	4096	4096

Besides the binary Golay code, there is also a perfect ternary (11, 6, 5) Golay code. These two codes are the only nontrivial examples of perfect codes that correct more than one error.

5.9 Cyclic codes for correcting burst errors

A code that has a minimum distance $2t + 1$ can correct any random pattern of t errors. Some channels, however, are susceptible mostly to burst errors. If one needs only to correct patterns of t errors that occur within a short segment of the message, but not arbitrary patterns of t errors, then one may use this relaxed requirement to obtain a more efficient code – that is, a code of higher rate. In this section, we shall give some cyclic codes for correcting burst errors. Because of the cyclic nature of the codes, we get some extra performance that we do not normally need – the codes correct not only burst errors, but cyclic burst errors as well. Cyclic bursts are defined as follows.

Definition 5.9.1. *A cyclic burst of length t is a vector whose nonzero components are among t (cyclically) consecutive components, the first and the last of which are nonzero.*

We can describe a cyclic burst of length t as a polynomial $e(x) = x^i b(x) \pmod{x^n - 1}$ where $b(x)$ is a polynomial of degree $t - 1$, with coefficient b_0 nonzero. Thus $b(x)$ describes the burst pattern, and x^i describes the starting location of the burst pattern. The length of the burst pattern is $\deg b(x) + 1$.

A cyclic code for correcting burst errors must have syndrome polynomials $s(x)$ that are distinct for each correctable burst pattern. That is, if

$$s(x) = R_{g(x)}[e(x)]$$

is different for each polynomial $e(x)$, representing a cyclic burst of length at most t, then the code is capable of correcting all cyclic burst errors of length at most t.

For example, the polynomial

$$g(x) = x^6 + x^3 + x^2 + x + 1$$

is the generator polynomial for a binary cyclic code of blocklength 15. The cyclic error bursts of length 3 or less can be enumerated. They are

$$\begin{aligned}
e(x) &= x^i & & i = 0, \ldots, 14 \\
e(x) &= x^i(1 + x) & \pmod{x^{15} - 1} \quad & i = 0, \ldots, 14 \\
e(x) &= x^i(1 + x^2) & \pmod{x^{15} - 1} \quad & i = 0, \ldots, 14 \\
e(x) &= x^i(1 + x + x^2) & \pmod{x^{15} - 1} \quad & i = 0, \ldots, 14.
\end{aligned}$$

By direct enumeration, it is straightforward, though tedious, to verify that the fifty-six syndromes of these fifty-six error patterns are distinct, and thus the cyclic code generated by $g(x)$ can correct all burst errors of length 3.

Notice that a codeword plus a correctable burst error pattern cannot be equal to a different codeword plus a correctable burst error pattern. In particular, no burst pattern of length 6 can be a codeword. In general, a linear code that can correct all burst patterns of length t or less cannot have a burst of length $2t$ or less as a codeword, because if it did, a burst of length t could change the codeword to a burst pattern of length t, which also could be obtained by making a burst error of length t in the all-zero codeword.

The following theorem resembles the Singleton bound for random-error-correcting codes, although the derivation is quite different; it is not restricted to cyclic codes.

Theorem 5.9.2 (Rieger Bound). *A linear block code that corrects all burst errors of length t or less must have at least $2t$ check symbols.*

Proof: Suppose the code corrects all burst errors of length t or less. Then no codeword is a burst of length $2t$. Any two vectors that are nonzero only in their first $2t$ components must be in different cosets of a standard array. Otherwise, their difference would be a codeword, and we have seen that no codewords are bursts of length $2t$. Therefore the number of cosets is at least as large as the number of such vectors. There are q^{2t} vectors that are zero except in their first $2t$ components; and hence there are at least q^{2t} cosets, and hence at least $2t$ check symbols. □

Notice that the earlier example of a cyclic burst-error-correcting code satisfies the Rieger bound with equality.

The class of cyclic codes includes many good codes for correcting burst errors. For small t and moderate blocklength, a number of good cyclic codes over $GF(2)$ have been found by computer search. Some of these are listed in Table 5.3. A large collection of longer codes of various lengths can be constructed from the codes of Table 5.3 by the techniques of shortening and interleaving. To get a (jn, jk) code from an (n, k) code, take any j codewords from the original code and merge the codewords by interleaving the symbols. If the original code can correct any burst error of length t, it is apparent that the interleaved code can correct any burst error of length jt. For example,

Table 5.3. *Some binary burst-error-correcting cyclic codes*

Generator polynomial	Parameters	Burst error-correction length
$x^4 + x^3 + x^2 + 1$	(7, 3)	2
$x^5 + x^4 + x^2 + 1$	(15, 10)	2
$x^6 + x^5 + x^4 + x^3 + 1$	(15, 9)	3
$x^6 + x^5 + x^4 + 1$	(31, 25)	2
$x^7 + x^6 + x^5 + x^3 + x^2 + 1$	(63, 56)	2
$x^8 + x^7 + x^6 + x^3 + 1$	(63, 55)	3
$x^{12} + x^8 + x^5 + x^3 + 1$	(511, 499)	4
$x^{13} + x^{10} + x^7 + x^6 + x^5 + x^4 + x^2 + 1$	(1023, 1010)	4

by taking four copies of the (31, 25) binary code given in Table 5.9, and interleaving the bits, one obtains a (124, 100) binary code. Because each of the individual codes can correct a burst error of length 2, the new code can correct any burst error of length 8.

The technique of interleaving creates a (jn, jk) cyclic code from an (n, k) cyclic code. Suppose that $g(x)$ is the generator polynomial of the original code. Then $g(x^j)$, which divides $x^{jn} - 1$, is the generator polynomial of the interleaved code regarded as a cyclic code. To see this, notice that interleaving the symbols of several data polynomials and then multiplying by $g(x^j)$ give the same codeword as multiplying each data polynomial by $g(x)$ and then interleaving these (n, k) codewords. Specifically, let

$$\begin{aligned} c_1(x) &= a_1(x)g(x) \\ c_2(x) &= a_2(x)g(x) \\ &\vdots \\ c_j(x) &= a_j(x)g(x) \end{aligned}$$

be the codewords of the individual (n, k) cyclic codes. To form the interleaved codeword, the symbols in each of these codewords are spread out, with $j - 1$ zeros inserted after every symbol. Then the codewords are delayed and combined to give the interleaved codeword as follows

$$\begin{aligned} c(x) &= c_1(x^j) + xc_2(x^j) + \cdots + x^{j-1}c_j(x^j) \\ &= a_1(x^j)g(x^j) + xa_2(x^j)g(x^j) + \cdots + x^{j-1}a_j(x^j)g(x^j) \\ &= [a_1(x^j) + xa_2(x^j) + \cdots + x^{j-1}a_j(x^j)]g(x^j). \end{aligned}$$

The bracketed term that has been factored out is formed by interleaving the j datawords. It can be replaced just as well by any arbitrary dataword $a(x)$. Hence

$$c(x) = a(x)g(x^j).$$

Replacing $g(x)$ by $g(x^j)$ is equivalent to interleaving j copies of the code generated by $g(x)$.

5.10 The Fire codes as cyclic codes

The Fire codes are a large class of cyclic burst-error-correcting codes that are constructed analytically. These are considered to be the best single-burst-correcting codes of high rate known. In Section 8.4, we shall see that decoders for such codes are easy to construct. Table 5.4 shows the parameters of some binary Fire codes for correcting burst errors of length t. The Fire codes are of very high rate, and the redundancy $n - k$ is least when the parameter m equals t. Then the redundancy is $3t - 1$, which exceeds the Rieger bound by $t - 1$. By interleaving Fire codes, longer codes for correcting longer burst errors can be constructed.

Table 5.4. *Parameters of some binary Fire codes*

(n, k)	t	m
(9, 4)	2	2
(35, 27)	3	3
(105, 94)	4	4
(279, 265)	5	5
(693, 676)	6	6
(1651, 1631)	7	7
(3315, 3294)	7	8
(8687, 8661)	9	9
(19437, 19408)	10	10

Definition 5.10.1. *A Fire code of blocklength n is a cyclic code over $GF(q)$ with generator polynomial*

$$g(x) = (x^{2t-1} - 1)p(x)$$

where $p(x)$ is a prime polynomial over $GF(q)$ that does not divide $x^{2t-1} - 1$ and whose degree m is not smaller than t, and n is the smallest integer such that $g(x)$ divides $x^n - 1$.

Theorem 5.10.2. *A Fire code has blocklength $n = \text{LCM}(2t - 1, q^m - 1)$.*

Proof: The smallest integer e for which $p(x)$ divides $x^e - 1$ is $e = q^m - 1$. By definition, n is the smallest integer for which $(x^{2t-1} - 1)p(x)$ divides $x^n - 1$. Because both $2t - 1$ and e divide $\text{LCM}(2t - 1, e)$, so $x^{2t-1} - 1$ divides $x^{\text{LCM}(2t-1, e)} - 1$ and $p(x)$ divides $x^{\text{LCM}(2t-1, e)} - 1$ because $x^e - 1$ does. This can happen for no n smaller than $\text{LCM}(2t - 1, e)$. □

Theorem 5.10.3. *A Fire code can correct all burst errors of length t or less.*

Proof: The code can correct all burst errors of length t or less if no two such bursts $x^i b(x)$ and $x^j b'(x)$ appear in the same coset. Because the code is a cyclic code, without loss of generality, we can take i equal to zero.

Suppose that two distinct nonzero bursts of length t or less, $b(x)$ and $x^j b'(x)$, are in the same coset of the code. Then their difference is a codeword. Because the difference is a multiple of $g(x)$, it is a multiple of $x^{2t-1} - 1$. Then $b(x) - x^j b'(x) = 0 \pmod{x^{2t-1} - 1}$, so

$$b(x) = x^j b'(x) \pmod{x^{2t-1} - 1}.$$

This means that j must be a multiple of $2t - 1$, so

$$b(x) = b'(x)x^{\ell(2t-1)}$$

for some nonnegative integer l. Because $l(2t-1) < n$, l is less than $q^m - 1$. Therefore $(x^{\ell(2t-1)} - 1)b(x)$ is a codeword, so

$$(x^{\ell(2t-1)} - 1)b(x) = a(x)(x^{2t-1} - 1)p(x).$$

Because $b(x)$ has a degree less than the degree of $p(x)$, $p(x)$ cannot divide $b(x)$. Suppose l is not zero, then $p(x)$ also cannot divide $x^{\ell(2t-1)} - 1$ because l is a nonnegative integer less than $q^m - 1$, and by definition of m, $p(x)$ divides $x^{\ell(2t-1)} - 1$ for no positive l smaller than $q^m - 1$. Thus l equals zero, and so j equals zero. This means that the two bursts are the same, contrary to the assumption, so the theorem is proved. □

For an example of a Fire code, choose

$$g(x) = (x^{11} + 1)(x^6 + x + 1),$$

which gives a (693, 676) binary Fire code for correcting a single burst of length 6.

For a second example of a Fire code, choose $m = t = 10$, and choose $p(x)$ as a primitive polynomial of degree 10. Then $e = 2^m - 1$, and we have a (19437, 19408) Fire code for correcting a single burst of length 10.

5.11 Cyclic codes for error detection

Cyclic codes are widely used for error detection. In this role they are sometimes called *cyclic redundancy codes*, a term that is common in the applications to error detection. In many of these applications there are various standard choices of error-detection code. Generally, cyclic redundancy codes are often understood to use a generator polynomial of the form $g(x) = (x + 1)p(x)$ where $p(x)$ is a primitive polynomial of degree m. These codes, as cyclic codes, have blocklength $n = 2^m - 1$ and dimension $2^m - 1 - (m + 1)$, but can be shortened.

If $g(x)$ is divisible by $x + 1$, then all the codewords are divisible by $x + 1$, so they have even weight. Because the codewords are divisible by $p(x)$, they are Hamming codewords. We conclude that a cyclic redundancy code with $g(x) = (x + 1)p(x)$ consists of the even weight codewords of a Hamming code. It has minimum distance 4.

For examples of codes with $d_{\min} = 4$, the primitive polynomials

$$p(x) = x^{15} + x^{14} + x^{13} + x^{12} + x^4 + x^3 + x^2 + x + 1$$

and

$$p(x) = x^{15} + x + 1,$$

when multiplied by $x + 1$, can be used as generator polynomials

$$g(x) = x^{16} + x^{12} + x^5 + 1$$
$$g(x) = x^{16} + x^{15} + x^2 + 1$$

to encode up to $2^{15} - 17$ data bits into cyclic codewords, possibly shortened, of blocklength up to $2^{15} - 1$.

For an example of a code with $d_{\min} = 3$, the primitive polynomial

$$p(x) = x^{32} + x^{26} + x^{23} + x^{22} + x^{16} + x^{12} + x^{11} + x^{10} + x^8 + x^7 + x^5 + x^4 + x^2 + x + 1$$

can be used as a generator polynomial to encode up to $2^{32} - 33$ data bits into a cyclic code, possibly shortened, of blocklength up to $2^{32} - 1$ bits. This cyclic redundancy code is a binary Hamming code, possibly shortened.

A cyclic code for detecting burst errors must have syndrome polynomials $s(x)$ that are nonzero for each detectable error pattern. The error-detection computation is straightforward. Simply divide the senseword by the generator polynomial $g(x)$ and check the remainder for zero. If it is not zero, an error has been detected. The remainder will be zero only if the senseword is a codeword, which is true only if the error word is a codeword. No pattern of weight less than $d_{\min}$ is a codeword, and no burst of length less than $n - k$ is a codeword because if $e(x) = b(x)$ is a burst of length less than $n - k$, then the syndrome is nonzero. Therefore a binary cyclic code can detect all error patterns of weight less than $d_{\min}$ and all cyclic burst error patterns of length less than $n - k$. Because there are 2^n sensewords and 2^k codewords, the probability that a randomly selected senseword will be accepted as a valid codeword is $1/2^{n-k}$.

For example, if $n - k = 32$, then only one randomly selected senseword in 4×10^9 will be accepted as a valid codeword. Moreover, if $d_{\min} = 4$, then every pattern of one, two, or three errors will be detected as will every burst error of length at most 32.

Problems

5.1 The polynomial

$$g(x) = x^8 + x^7 + x^6 + x^4 + 1$$

is the generator polynomial for a cyclic code over $GF(2)$ with blocklength $n = 15$.

a. Find the check polynomial.

b. How many errors can this code correct?

c. How many erasures can this code correct?

d. Find a generator matrix in systematic form.

5.2 Find the minimal polynomial for each element of $GF(16)$.

5.3 Find the generator polynomial for a binary $(31, 21)$ double-error-correcting cyclic code.

5.4 Find the generator polynomial for a binary $(21, 12)$ double-error-correcting cyclic code.

5.5 The polynomial

$$g(x) = x^6 + 3x^5 + x^4 + x^3 + 2x^2 + 2x + 1$$

is the generator polynomial for a double-error-correcting cyclic code of blocklength 15 over $GF(4)$.

a. Find a generator matrix in systematic form.

b. Show that every codeword in the code of Problem 5.1 is also a codeword in this code.

5.6 Let $g(x)$ be the generator polynomial of a cyclic code over $GF(q)$ of blocklength n. Prove that if q and n are coprime, then the all-ones word is a codeword if and only if $g(1) \neq 0$.

5.7 Suppose that $g(x) = g_{n-k}x^{n-k} + \cdots + g_0$ is the generator polynomial for a cyclic code. Prove that g_0 is not zero. Prove that the reciprocal polynomial $\widetilde{g}(x) = g_0 x^{n-k} + g_1 x^{n-k-1} + \cdots + g_{n-k}x^0$ is the generator polynomial for an equivalent cyclic code.

5.8 Suppose a binary cyclic code has the property that whenever $c(x)$ is a codeword, then so is its reciprocal polynomial $\widetilde{c}(x)$. Prove that $g(x) = \widetilde{g}(x)$. What is the corresponding statement for nonbinary codes?

5.9 Prove that all conjugates of a primitive element are also primitive elements.

5.10 Prove the following statement: A binary cyclic code of blocklength n is invariant under the permutation $c(x) \rightarrow c(x^2)(\bmod\ x^n - 1)$.

5.11 Find the generator polynomial of the (9, 7) Hamming code over $GF(8)$.

5.12 Expand Table 5.4 to include all Fire codes based on a primitive polynomial $p(x)$ of degree m for $12 \geq m \geq t$.

5.13 Suppose that $g_1(x)$ and $g_2(x)$ are generator polynomials for two codes, $\mathcal{C}_1$ and $\mathcal{C}_2$, of the same blocklength over $GF(q)$. Prove that if all the zeros of $g_1(x)$ are also zeros of $g_2(x)$ (which implies that $g_1(x)$ divides $g_2(x)$), then $\mathcal{C}_2$ is a subcode of $\mathcal{C}_1$.

5.14 The polynomial over $GF(4)$

$$g(x) = x^6 + 3x^5 + x^4 + x^3 + 2x^2 + 2x + 1$$

is known to be the generator polynomial for a (15, 9) double-error-correcting code over $GF(4)$.

a. Is $v(x) = x^{10} + 3x^2 + x + 2$ a codeword in this code?

b. What is the syndrome polynomial of $v(x)$?

c. How many syndrome polynomials must be tabulated to cover all correctable error patterns?

5.15 Factor $x^8 - 1$ over $GF(3)$. How many cyclic codes over $GF(3)$ of blocklength 8 are there?

5.16 Find the generator polynomial for the (11, 6, 5) ternary Golay code. Prove that the minimum distance is 5.

5.17 What are the blocklength, dimension, and burst correction performance of a binary Fire code with generator polynomial $g(x) = (x^{15} - 1)(x^8 + x^4 + x^3 + x^2 + 1)$?

5.18 Construct the generator polynomial for a cyclic Fire code of blocklength $n = 1143$ over $GF(256)$ that will correct all burst errors of length 5 or less.

5.19 Is there a cyclic Hamming code over $GF(5)$ of blocklength 156?

5.20 Express the (6, 3, 3) shortened binary Hamming code as a ring modulo a polynomial $f(x)$.

5.21 Show that a shortened binary cyclic code with generator polynomial of degree $n - k$ can detect a fraction $1 - 2^{-(n-k)+1}$ of all burst error patterns of length $n - k + 1$.

Notes

The major thread running through this chapter is based on the far-ranging ideas of Prange (1957, 1958). Prange introduced the notion of a cyclic code. The relationship of cyclic codes to the ideals of algebra was studied independently by Prange, and by Peterson (1960) and Kasami (1960). These works were carried out at the end of the 1950s, and formed a foundation for the major jolts that came in the 1960s when it was realized that cyclic codes could be embedded in an extension field. The ideas of this chapter quickly followed; much of the present chapter was written by using the machinery of the extension field.

The cyclic codes are important because the strong structure of the polynomial rings leads to structure in the codes as well as to the tools used to study the codes. It is these properties that lead to the cyclic property which, in itself, is not very important. The cyclic structure of Hamming codes was studied independently by Abramson (1960) and Elspas (1960). The Golay (23, 12, 7) binary code and the Golay (11, 6, 5) ternary code were published by Golay in 1949. It was shown to be unique by Pless (1968).

The study of burst-error-correcting codes was begun by Abramson (1959). Many of the burst-error-correcting codes used in practice have been found by computer search, many by Kasami (1963). Table 5.3 is based on the compilations of Peterson and Weldon (1968). The Fire codes were published by Fire in 1959.

6 Codes Based on the Fourier Transform

Applications of the discrete Fourier transform in the complex field occur throughout the subject of signal processing. Fourier transforms also exist in the Galois field $GF(q)$ and can play an important role in the study and processing of $GF(q)$-valued signals, that is, of codewords. By using the Fourier transform, the ideas of coding theory can be described in a setting that is much closer to the methods of signal processing. Cyclic codes can be defined as codes whose codewords have certain specified spectral components equal to zero.

In this chapter we shall study cyclic codes in the setting of the Fourier transform. In the next chapter, we shall study decoding algorithms for cyclic codes. The most important classes of cyclic codes studied in this chapter are the Reed–Solomon codes and their subcodes, the Bose–Chaudhuri–Hocquenghem (BCH) codes.

The BCH codes form a large class of multiple-error-correcting codes that occupy a prominent place in the theory and practice of error correction. This prominence is due to at least four reasons. (1) Provided the blocklength is not excessive, there are good codes in this class (but generally, not the best of known codes). (2) Relatively simple and implementable encoding and decoding techniques are known (although, if simplicity were the only consideration, other codes may be preferable). (3) The popular and powerful class of nonbinary BCH codes known as *Reed–Solomon codes* has a certain strong optimality property, and these codes have a well-understood distance structure. (4) A thorough understanding of Reed–Solomon codes is a good starting point for the study of many other classes of algebraic codes.

6.1 The Fourier transform

In the complex field, the discrete Fourier transform of $\boldsymbol{v} = (v_0, v_1, \ldots, v_{n-1})$, a vector of real or of complex numbers, is a vector $\boldsymbol{V} = (V_0, V_1, \ldots, V_{n-1})$, given by

$$V_k = \sum_{i=0}^{n-1} e^{-j2\pi n^{-1} ik} v_i \quad k = 0, \ldots, n-1$$

where $j = \sqrt{-1}$. The Fourier kernel $\exp(-j2\pi/n)$ is an nth root of unity in the field of complex numbers. In the finite field $GF(q)$, an element ω of order n is an nth root of unity. Drawing on the analogy between $\exp(-j2\pi/n)$ and ω, we have the following definition.

Definition 6.1.1. *Let $\boldsymbol{v} = (v_0, v_1, \ldots, v_{n-1})$ be a vector over $GF(q)$, and let ω be an element of $GF(q)$ of order n. The Fourier transform of the vector $\boldsymbol{v}$ is the vector $\boldsymbol{V} = (V_0, V_1, \ldots, V_{n-1})$ with components given by*

$$V_j = \sum_{i=0}^{n-1} \omega^{ij} v_i \quad j = 0, \ldots, n-1.$$

It is natural to call the discrete index i *time*, taking values on the *time axis* $\{0, 1, \ldots, n-1\}$, and to call $\boldsymbol{v}$ the *time-domain function* or the *signal*. Also, we might call the discrete index j *frequency*, taking values on the *frequency axis* $\{0, 1, \ldots, n-1\}$, and to call $\boldsymbol{V}$ the *frequency-domain function* or the *spectrum*.

The Fourier transform in a Galois field closely mimics the Fourier transform in the complex field with one important difference. In the complex field an element ω of order n, for example, $e^{-j2\pi/n}$, exists for every value of n. In the field $GF(q)$, however, such an ω exists only if n divides $q-1$. Moreover, if for some value of m, n divides $q^m - 1$, then there will be a Fourier transform of blocklength n in the extension field $GF(q^m)$. For this reason, a vector $\boldsymbol{v}$ of blocklength n over $GF(q)$ will also be regarded as a vector over $GF(q^m)$; it has a Fourier transform of blocklength n over $GF(q^m)$. This is completely analogous to the Fourier transform of a real-valued vector: Even though the time-domain vector $\boldsymbol{v}$ has components only in the real field, the transform $\boldsymbol{V}$ has components in the complex field. Similarly, for the Galois-field Fourier transform, even though the time-domain vector $\boldsymbol{v}$ is over the field $GF(q)$, the spectrum $\boldsymbol{V}$ may be over the extension field $GF(q^m)$. Any factor of $q^m - 1$ can be used as the blocklength of a Fourier transform over $GF(q)$, but the most important values for n are the primitive blocklengths, $n = q^m - 1$. Then ω is a primitive element of $GF(q^m)$. In error-control applications, for example, all the decoding action really takes place in the big field $GF(q^m)$ – we just happen to start with a vector consistent with the channel input and output, that is, a vector over the small field $GF(q)$.

The Fourier transform has many strong properties, which are summarized in Figure 6.1 and proved in this section. We shall prove most of these properties to make it quite clear that we are not restricted to any particular field when using them.

Theorem 6.1.2 (Inverse Fourier Transform). *The vector $\boldsymbol{v}$ is related to its spectrum $\boldsymbol{V}$ by*

$$v_i = \frac{1}{n} \sum_{j=0}^{n-1} \omega^{-ij} V_j$$

where n is interpreted as an integer of the field.

1. Additivity: $\lambda \boldsymbol{v} + \mu \boldsymbol{v}' \leftrightarrow \lambda \boldsymbol{V} + \mu \boldsymbol{V}'$.
2. Modulation: $(v_i \omega^{i\ell}) \leftrightarrow (V_{((j+\ell))})$.
3. Translation: $(v_{((i-\ell))}) \leftrightarrow (V_j \omega^{\ell j})$.
4. Inverse: $v_i = \frac{1}{n} \sum_{j=0}^{n-1} \omega^{-ij} V_j \quad i = 0, \ldots, n-1$

   where $n = 1 + 1 + \cdots + 1$ (n terms).
5. Convolution: $e_i = \sum_{\ell=0}^{n-1} f_{((i-\ell))} g_\ell \quad i = 0, \ldots, n-1$

   if and only if $E_j = F_j G_j \quad j = 0, \ldots, n-1$.
6. $v(x) = \sum_{i=0}^{n-1} v_i x^i$ has a zero at ω^j if and only if $V_j = 0$.
7. Decimation: $\{v_{bi}\} \leftrightarrow \{V_{Bj}\}$ where b is coprime to n, and $Bb = 1 (\text{mod } n)$.
8. Linear complexity: The weight of a sequence $\boldsymbol{v}$ is equal to the linear complexity of its inverse Fourier transform $\boldsymbol{V}$ (cyclically extended).

Figure 6.1. Properties of the Fourier transform. The double parentheses on subscripts indicate modulo n

Proof:

$$\sum_{j=0}^{n-1} \omega^{-ij} V_j = \sum_{j=0}^{n-1} \omega^{-ij} \sum_{k=0}^{n-1} \omega^{kj} v_k = \sum_{k=0}^{n-1} v_k \sum_{j=0}^{n-1} \omega^{(k-i)j}.$$

In any field,

$$x^n - 1 = (x-1)(x^{n-1} + x^{n-2} + \cdots + x + 1).$$

By the definition of ω, ω^r is a zero of the left side for all r. Hence for all $r \neq 0$ modulo n, ω^r is a zero of the last term. But this is equivalent to

$$\sum_{j=0}^{n-1} \omega^{rj} = 0 \quad r \neq 0 \; (\text{mod } n),$$

whereas if $r = 0$ modulo n,

$$\sum_{j=0}^{n-1} \omega^{rj} = n \; (\text{mod } p),$$

which is not zero if n is not a multiple of the field characteristic p. Because q is a power of p, $q^m - 1 = p^M - 1$ is a multiple of n, so n is not a multiple of p. Hence $n \neq 0 (\text{mod } p)$. Combining these facts proves the theorem. □

The *cyclic convolution*, denoted $\boldsymbol{f} * \boldsymbol{g}$, of two vectors, $\boldsymbol{f}$ and $\boldsymbol{g}$, of length n is defined as

$$e_i = \sum_{k=0}^{n-1} f_{((i-k))} g_k \quad i = 0, \ldots, n-1$$

where the double parentheses on the indices denote modulo n. An important and familiar link between the Fourier transform and the cyclic convolution is known as the *convolution theorem.*

Theorem 6.1.3 (Convolution Theorem). *The vector* $\boldsymbol{e}$ *is given by the cyclic convolution of the vectors* $\boldsymbol{f}$ *and* $\boldsymbol{g}$ *if and only if the components of the Fourier transforms satisfy:*

$$E_j = F_j G_j \quad j = 0, \ldots, n-1.$$

Proof: This holds because

$$\begin{aligned} e_i &= \sum_{k=0}^{n-1} f_{((i-k))} \left(\frac{1}{n} \sum_{j=0}^{n-1} \omega^{-jk} G_j \right) \\ &= \frac{1}{n} \sum_{j=0}^{n-1} \omega^{-ij} G_j \left(\sum_{k=0}^{n-1} \omega^{(i-k)j} f_{((i-k))} \right) = \frac{1}{n} \sum_{j=0}^{n-1} \omega^{-ij} G_j F_j. \end{aligned}$$

Because $\boldsymbol{e}$ is the inverse Fourier transform of $\boldsymbol{E}$, we conclude that $E_j = F_j G_j$. □

Theorem 6.1.4 (Modulation and Translation Properties). *If* $\{v_i\} \leftrightarrow \{V_j\}$ *is a Fourier transform pair, then the following are Fourier transform pairs:*

$$\begin{aligned} &\{\omega^i v_i\} \leftrightarrow \left\{V_{((j+1))}\right\} \\ &\left\{v_{((i-1))}\right\} \leftrightarrow \{\omega^j V_j\}. \end{aligned}$$

Proof: Immediate substitutions prove the theorem. □

Sometimes we represent a vector $\boldsymbol{v}$ by a polynomial $v(x)$. The polynomial

$$v(x) = v_{n-1}x^{n-1} + \cdots + v_1 x + v_0$$

can be transformed into a polynomial

$$V(x) = V_{n-1}x^{n-1} + \cdots + V_1 x + V_0$$

by means of the Fourier transform (where the coefficients of $V(x)$ are the components of the spectrum). The latter polynomial is called the *spectrum polynomial* of $v(x)$. Properties of the spectrum are closely related to the zeros of polynomials, as stated in the following theorem.

Theorem 6.1.5.

(i) *The polynomial* $v(x)$ *has a zero at* ω^j *if and only if the* j*th frequency component* V_j *equals zero.*

(ii) The polynomial $V(x)$ has a zero at ω^{-i} if and only if the ith time component v_i equals zero.

Proof: The proof of part (i) is immediate because

$$v(\omega^j) = \sum_{i=0}^{n-1} v_i \omega^{ij} = V_j.$$

The proof of part (ii) follows in the same way. □

Thus, in finite fields, when one speaks of zeros of polynomials or of spectral components equal to zero, one speaks of nearly the same thing, but the terminology and the insights are different. In the first formulation, one draws on insight into the properties of polynomials; in the second, one draws on an understanding of the Fourier transform.

If b and n are coprime, then $i \to bi \pmod n$ defines a permutation of a vector $\boldsymbol{c}$. The permutation $c'_i = c_{((bi))}$ is called *cyclic decimation* because every bth component, cyclically, is chosen. The following theorem describes how the spectrum of a cyclic decimation itself is a cyclic decimation of the original spectrum.

Theorem 6.1.6. *Let* $\text{GCD}(b, n) = 1$, *and let* $Bb = 1 (\text{mod } n)$. *The vector with components*

$$c'_i = c_{((bi))}$$

has a transform with components

$$C'_j = C_{((Bj))}$$

where all indices are interpreted modulo n.

Proof: The corollary to the euclidean algorithm states that

$$Bb + Nn = 1$$

for some integers B and N. Hence the required B exists. Therefore

$$\begin{aligned} C'_j &= \sum_{i=0}^{n-1} \omega^{ij} c'_i \\ &= \sum_{i=0}^{n-1} \omega^{(Bb+Nn)ij} c_{((bi))} \\ &= \sum_{i=0}^{n-1} \omega^{biBj} c_{((bi))}. \end{aligned}$$

The permutation $i' = bi$ rearranges the terms, but it does not affect the value of the sum. Thus

$$\begin{aligned} C'_j &= \sum_{i'=0}^{n-1} \omega^{i'Bj} c_{i'} \\ &= C_{((Bj))}, \end{aligned}$$

as was to be proved. □

A *linear recursion* over any field F is an expression of the form

$$V_k = -\sum_{j=1}^{L} \Lambda_j V_{k-j} \quad k = L, \ldots.$$

The linear recursion is characterized by the *length* L and the vector $\boldsymbol{\Lambda} = (\Lambda_1, \ldots, \Lambda_L)$ whose components are called the *connection weights*. It is denoted $(\boldsymbol{\Lambda}, L)$ or $(\Lambda(x), L)$. The *length* of the linear recursion is the length of the shortest subsequence from which the recursion can compute all subsequent terms of the sequence. The polynomial $\Lambda(x) = 1 + \sum_{j=1}^{L} \Lambda_j x^j$ is called a *connection polynomial*. Because Λ_L could be zero, the degree of $\Lambda(x)$ may be smaller than L. We say that the linear recursion $(\boldsymbol{\Lambda}, L)$ *produces* a sequence $\boldsymbol{V}$ that satisfies the above equation. Given a sequence V_0, $V_1, \ldots, V_{n-1}$ of elements of the field F, the length of the shortest linear recursion that will produce the sequence is called the *linear complexity* of $\boldsymbol{V}$.

If $\boldsymbol{V}$ has finite blocklength n (or is periodic with period n), then $\boldsymbol{V}$ can be regarded as the Fourier transform of a vector of blocklength n. We now examine the relationship between the linear complexity of $\boldsymbol{V}$ and the properties of its inverse Fourier transform $\boldsymbol{v}$. In this case, for reasons that we shall soon see, the connection polynomial for the shortest linear recursion that produces $\boldsymbol{V}$, cyclically repeated, is called the *locator polynomial*. Thus a locator polynomial is a connection polynomial of minimal length for a periodic sequence.

Consider a vector $\boldsymbol{v}$ of blocklength n and weight 1 with v_m as the single nonzero component of $\boldsymbol{v}$. Its transform is

$$V_j = v_m \omega^{jm}.$$

This spectrum can be generated by the recursion

$$V_j = \omega^m V_{j-1}$$

with the initial value $V_0 = v_m$. Thus the spectrum of a vector of weight 1 can be generated by a linear recursion of length 1. This is a special case of the linear complexity property given in the following theorem.

Theorem 6.1.7. *The linear complexity of the finite-length vector $\boldsymbol{V}$, cyclically repeated, is equal to the Hamming weight of its inverse Fourier transform $\boldsymbol{v}$.*

Proof: Let $i_1, i_2, \ldots, i_d$ denote the indices of the d nonzero components of $\boldsymbol{v} = (v_0, v_1, \ldots, v_{n-1})$. Consider the polynomial

$$\Lambda(x) = \prod_{\ell=1}^{d}(1 - x\omega^{i_\ell})$$
$$= \sum_{k=0}^{d} \Lambda_k x^k.$$

Let $\boldsymbol{\Lambda}$ be the vector whose components are the coefficients of $\Lambda(x)$. Let $\boldsymbol{\lambda}$ be the inverse Fourier transform of $\boldsymbol{\Lambda}$:

$$\lambda_i = \frac{1}{n}\sum_{k=0}^{n-1} \omega^{-ik}\Lambda_k$$
$$= \frac{1}{n}\Lambda(\omega^{-i})$$
$$= \frac{1}{n}\prod_{\ell=1}^{d}(1 - \omega^{-i}\omega^{i_\ell}).$$

Hence $\lambda_i = 0$ if and only if $i \in \{i_1, \ldots, i_d\}$. That is, $\lambda_i = 0$ if and only if $v_i \neq 0$. Consequently,

$$\lambda_i v_i = 0$$

for all i, so the convolution theorem implies that

$$\boldsymbol{\Lambda} * \boldsymbol{V} = 0,$$

or

$$\sum_{j=0}^{n-1} \Lambda_j V_{((k-j))} = 0.$$

But $\Lambda_0 = 1$ and $\Lambda_k = 0$ if $k > d$. Consequently, this becomes

$$V_k = -\sum_{j=1}^{d} \Lambda_j V_{((k-j))},$$

so the linear complexity is not larger than d. To prove that it is not smaller, suppose that

$$\sum_{j=0}^{L} \Lambda_j V_{((k-j))} = 0.$$

The inverse Fourier transform must satisfy

$$\lambda_i v_i = 0,$$

which implies that $\lambda_i = 0$ whenever $v_i \neq 0$. Because $\Lambda(x)$ has at most L zeros, we can conclude that L must be at least the weight of $\boldsymbol{v}$, and the theorem is proved. □

Corollary 6.1.8. *The Hamming weight of a vector* $\mathbf{V}$ *is equal to the linear complexity of its inverse Fourier transform* $\mathbf{v}$.

Corollary 6.1.9. *Let* n *and* q *be coprime. The linear complexity of a finite sequence of blocklength* n*, periodically repeated, is not changed by a cyclic shift, nor by a cyclic decimation by a step size coprime to the blocklength of the sequence.*

Proof: Theorem 5.3.7 implies that a Fourier transform of blocklength n exists. By the modulation property and the decimation property, the weight of the Fourier transform of the sequence remains the same under the operations of a cyclic shift or a cyclic decimation. Hence the theorem implies the corollary. □

6.2 Reed–Solomon codes

A *Reed–Solomon code* $\mathcal{C}$ of blocklength n over $GF(q)$, with n a divisor of $q-1$, is defined as the set of all words over $GF(q)$ of length n whose Fourier transform is equal to zero in a specified block of $d-1$ consecutive components, denoted $\{j_0, j_0+1, \ldots, j_0+d-2\}$. Because n divides $q-1$, the codeword components and the spectrum components are symbols in the same field, $GF(q)$.

If codeword $\boldsymbol{c}$ is cyclically shifted by one place, then its spectral component C_j is multiplied by ω^j. Because $C_j\omega^j$ is zero whenever C_j is zero, the cyclic shift of $\boldsymbol{c}$ is also a codeword. Hence $\mathcal{C}$ is a cyclic code. Because a Reed–Solomon code is a cyclic code, it has a generator polynomial, $g(x)$, that can be calculated. The minimal polynomial over $GF(q)$ of an element β in the same field is

$$f_\beta(x) = x - \beta.$$

Because the symbol field and the transform field are the same field, all minimal polynomials are of first degree. Thus $\{\omega^{j_0}, \omega^{j_0+1}, \ldots, \omega^{j_0+d-2}\}$ is the only defining set for the code. It is also the complete defining set.

One can choose any j_0 for a Reed–Solomon code. Then

$$g(x) = (x - \omega^{j_0})(x - \omega^{j_0+1}) \cdots (x - \omega^{j_0+d-2}).$$

This is always a polynomial of degree $d-1$. Hence a Reed–Solomon code satisfies

$$n - k = d - 1.$$

The blocklength of a Reed–Solomon code over $GF(q)$ is largest if ω has order $q-1$. Then ω is a primitive element, now denoted α, and the blocklength of the code is

$n = q - 1$. In this case, the Reed–Solomon code is called a *primitive Reed–Solomon code*.

The following theorem, proved by the fundamental theorem of algebra, asserts that Reed–Solomon codes are optimum in the sense of the Singleton bound. Thus a Reed–Solomon code with $d - 1$ consecutive zeros in the spectrum has a minimum distance satisfying $d_{\text{min}} = d$.

Theorem 6.2.1. *A Reed–Solomon code is a maximum-distance code, and the minimum distance satisfies*

$$d_{\text{min}} = n - k + 1.$$

Proof: It is sufficient to consider the defining set to be such that the last $d - 1$ components of $\boldsymbol{C}$ are equal to zero. Otherwise, use the modulation theorem to translate the defining set to the last $d - 1$ components thereby multiplying each codeword component by a power of ω, which does not change the weight of a codeword because components that were nonzero remain nonzero. Let

$$C(x) = \sum_{j=0}^{n-d} C_j x^j.$$

The corresponding codeword has components $c_i = \frac{1}{n} C(\omega^{-i})$. Because $C(x)$ is a polynomial of degree at most $n - d$, it can have at most $n - d$ zeros. Therefore, unless it is identically zero, $\boldsymbol{c}$ is nonzero in at least d components. The minimum distance d_{min} satisfies

$$d_{\text{min}} \geq d = n - k + 1$$

because $d - 1 = n - k$ for Reed–Solomon codes. But by the Singleton bound, for any linear code,

$$d_{\text{min}} \leq n - k + 1.$$

Hence $d_{\text{min}} = n - k + 1$, and $d_{\text{min}} = d$, as was to be proved. □

Theorem 6.2.1 tells us that for fixed n and k, no code can have a larger minimum distance than a Reed–Solomon code. Often, this is a strong justification for using a Reed–Solomon code. It should not be read, however, as saying more than it does. Another code having parameters (n', k') for which no Reed–Solomon code exists may be preferable to any existing (n, k) Reed–Solomon code. Reed–Solomon codes always have relatively short blocklengths compared to other cyclic codes over the same alphabet, and they are unacceptably short when the alphabet is small. A Reed–Solomon code over the binary alphabet $GF(2)$ has blocklength 1, and so is uninteresting.

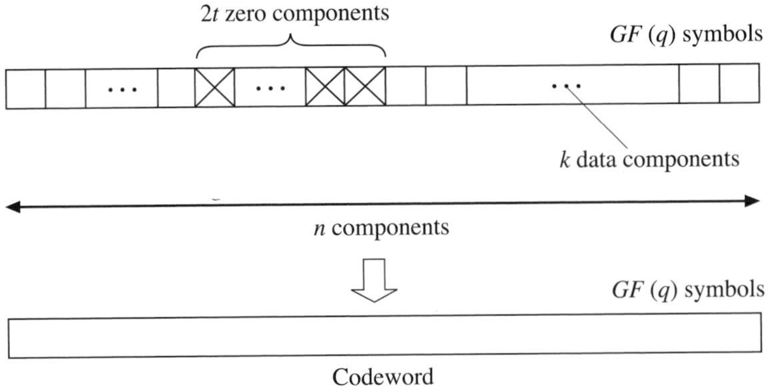

Figure 6.2. Encoding a Reed–Solomon code in the transform domain

An encoder for a Reed–Solomon code can be either systematic or nonsystematic, as discussed in Section 5.2. One may encode in the natural way using a generator polynomial. We call this encoder a *time-domain encoder* or a *code-domain encoder.* Alternatively, one may choose to encode the Reed–Solomon code directly in the transform domain by using the data symbols to specify spectral components. We call this a *frequency-domain encoder* or a *transform-domain encoder.* Every spectrum consistent with the spectral constraints yields a codeword. Encoding is as follows. Some set of $d-1$ cyclically consecutive frequencies, indexed by $j = j_0, \ldots, j_0+d-2$, is chosen as the set of spectral components constrained to zero. The $n-d+1$ unconstrained components of the spectrum are filled with data symbols from $GF(q)$. The inverse Fourier transform then produces a nonsystematic codeword, as shown in Figure 6.2. Because there are $n-d+1$ spectral components that can take on data values, we obtain a codeword of an $(n, n-d+1)$ Reed–Solomon code.

For a t-error-correcting Reed–Solomon code, choose $d-1 = 2t$. Then with $j_0 = 1$, which is a common choice for j_0, the generator polynomial is

$$g(x) = (x-\omega)(x-\omega^2)\cdots(x-\omega^{2t}).$$

This always is a polynomial of degree $2t$.

As an example, we shall find $g(x)$ for a $(7, 5)$ Reed–Solomon code over $GF(8)$ with $t = 1$. Any j_0 will do. Choose $j_0 = 1$ and $\omega = \alpha$

$$\begin{aligned} g(x) &= (x-\alpha^1)(x-\alpha^2) \\ &= x^2 + (z^2+z)x + (z+1), \end{aligned}$$

where the field elements of $GF(8)$ are expressed as polynomials in z, using Table 6.1. A data polynomial is a sequence of five 8-ary (octal) symbols (equivalent to fifteen bits). Suppose that

Table 6.1. *Representations of $GF(2^3)$*

Exponential notation	Polynomial notation	Binary notation	Octal notation	Minimal polynomial
0	0	000	0	
α^0	1	001	1	$x+1$
α^1	z	010	2	x^3+x^2+1
α^2	z^2	100	4	x^3+x^2+1
α^3	z^2+1	101	5	x^3+x+1
α^4	z^2+z+1	111	7	x^3+x^2+1
α^5	$z+1$	011	3	x^3+x+1
α^6	z^2+z	110	6	x^3+x+1

$$a(x) = (z^2+z)x^4 + x^3 + (z+1)x^2 + zx + (z^2+1)$$

is the data polynomial. The nonsystematic codeword is

$$\begin{aligned} c(x) &= a(x)g(x) \\ &= (\alpha^4 x^4 + x^3 + \alpha^3 x^2 + \alpha^1 x + \alpha^6)(x^2 + \alpha^4 x + \alpha^3) \\ &= \alpha^4 x^6 + \alpha^3 x^5 + \alpha^2 x^4 + 0x^3 + \alpha^5 x^2 + \alpha^6 x + \alpha^2, \end{aligned}$$

which is a sequence of seven octal symbols. This codeword also can be written with the alternative notation

$$c(x) = (z^2+z)x^6 + (z+1)x^5 + z^2x^4 + 0x^3 + (z^2+z+1)x^2 + (z^2+1)x + z^2.$$

The coefficients of the powers of x are the octal code symbols and, within each code symbol, the coefficients of the powers of z index the three-bit positions of that octal symbol.

The five octal data symbols of $a(x)$ can be specified in $8^5 = 32,768$ ways. Therefore there are 32,768 codewords in the (7, 5) Reed–Solomon code. Some of these codewords are listed in Figure 6.3, but even in this small code, there are far too many codewords to list them all. The list has been ordered as if the codewords were generated by a systematic encoder; the first five symbols can be interpreted as data and the last two symbols as checks.

As a second example, we shall find $g(x)$ for a (7, 3) Reed–Solomon code over $GF(8)$ with $t = 2$. Any j_0 will do. Choose $j_0 = 4$ and $\omega = \alpha$

$$\begin{aligned} g(x) &= (x-\alpha^4)(x-\alpha^5)(x-\alpha^6)(x-\alpha^0) \\ &= x^4 + (z^2+1)x^3 + (z^2+1)x^2 + (z+1)x + z \end{aligned}$$

where the field elements of $GF(8)$ are expressed as polynomials in z. A data polynomial represents a sequence of three 8-ary (octal) symbols (equivalent to nine bits). Suppose

0	0	0	0	0	0	0
0	0	0	0	1	6	3
0	0	0	0	2	7	6
0	0	0	0	3	1	5
			⋮			
0	0	0	1	0	1	1
0	0	0	1	1	7	2
0	0	0	1	2	6	7
0	0	0	1	3	0	4
			⋮			
0	0	0	7	0	7	7
0	0	0	7	1	1	4
0	0	0	7	2	0	1
0	0	0	7	3	6	2
			⋮			
0	0	1	0	0	7	3
0	0	1	0	1	1	0
0	0	1	0	2	0	5
0	0	1	0	3	6	6
			⋮			
0	0	1	1	0	6	2
0	0	1	1	1	0	1
0	0	1	1	2	1	4
0	0	1	1	3	7	7
			⋮			

Figure 6.3. The (7, 5) Reed–Solomon code

that

$$a(x) = (z^2 + z)x^2 + x + (z + 1)$$

is the data polynomial. The nonsystematic codeword is

$$\begin{aligned} c(x) &= a(x)g(x) \\ &= (\alpha^4 x^2 + x + \alpha^3)(x^4 + \alpha^6 x^3 + \alpha^6 x^2 + \alpha^3 x + \alpha) \\ &= \alpha^4 x^6 + \alpha x^5 + \alpha^6 x^4 + 0x^3 + 0x^2 + \alpha^5 x + \alpha^4, \end{aligned}$$

which is a sequence of seven octal symbols.

As a third example, we shall find $g(x)$ for a $(15, 11)$ Reed–Solomon code over $GF(16)$ with $t = 2$. Any j_0 will do. Choose $j_0 = 1$ and $\omega = \alpha$

$$g(x) = (x - \alpha)(x - \alpha^2)(x - \alpha^3)(x - \alpha^4).$$

Table 6.2. *Representations of $GF(2^4)$*

Exponential notation	Polynomial notation	Binary notation	Decimal notation	Minimal polynomial
0	0	0000	0	
α^0	1	0001	1	$x+1$
α^1	z	0010	2	x^4+x+1
α^2	z^2	0100	4	x^4+x+1
α^3	z^3	1000	8	$x^4+x^3+x^2+x+1$
α^4	$z+1$	0011	3	x^4+x+1
α^5	z^2+z	0110	6	x^2+x+1
α^6	z^3+z^2	1100	12	$x^4+x^3+x^2+x+1$
α^7	z^3+z+1	1011	11	x^4+x^3+1
α^8	z^2+1	0101	5	x^4+x+1
α^9	z^3+z	1010	10	$x^4+x^3+x^2+x+1$
α^{10}	z^2+z+1	0111	7	x^2+x+1
α^{11}	z^3+z^2+z	1110	14	x^4+x^3+1
α^{12}	z^3+z^2+z+1	1111	15	$x^4+x^3+x^2+x+1$
α^{13}	z^3+z^2+1	1101	13	x^4+x^3+1
α^{14}	z^3+1	1001	9	x^4+x^3+1

Table 6.2 gives $GF(16)$ as an extension field of $GF(2)$ constructed with the primitive polynomial $p(z) = z^4 + z + 1$. Then

$$\begin{aligned} g(x) &= x^4 + (z^3 + z^2 + 1)x^3 + (z^3 + z^2)x^2 + z^3x + (z^2 + z + 1) \\ &= x^4 + \alpha^{13}x^3 + \alpha^6x^2 + \alpha^3x + \alpha^{10} \end{aligned}$$

where the field elements of $GF(16)$ are expressed as polynomials in z, using Table 6.2. Because $g(x)$ has degree 4, $n - k = 4$ and $k = 11$. A data polynomial is a sequence of eleven 16-ary (hexadecimal) symbols (equivalent to forty-four bits).

6.3 Conjugacy constraints and idempotents

The Fourier transform of blocklength n over $GF(q)$ generally takes values in an extension field $GF(q^m)$. If we start with an arbitrary n-vector over $GF(q^m)$ and take the inverse Fourier transform, generally we do not get a vector over $GF(q)$; the components may take values in the larger field. We shall want constraints on the spectrum that will ensure that the inverse Fourier transform is a time-domain vector over $GF(q)$.

Constraints of this sort are familiar in the field of complex numbers. Recall that over the complex field, a spectrum $V(f)$ has a real-valued inverse Fourier transform if and only if $V^*(-f) = V(f)$. The next theorem gives a constraint, known as a *conjugacy constraint*, that provides an analogous condition for a finite field.

Theorem 6.3.1. *Let V be a vector of blocklength n of elements of $GF(q^m)$ where n is a divisor of $q^m - 1$. The inverse Fourier transform v is a vector of elements of $GF(q)$ if and only if the following equations are satisfied:*

$$V_j^q = V_{((qj))} \quad j = 0, \ldots, n-1.$$

Proof: By definition,

$$V_j = \sum_{i=0}^{n-1} \omega^{ij} v_i \quad j = 0, \ldots, n-1.$$

For a field of characteristic p, $(a+b)^{p^r} = a^{p^r} + b^{p^r}$ for any integer r. Further, if v_i is an element of $GF(q)$, then $v_i^q = v_i$. Combining these statements gives

$$V_j^q = \left(\sum_{i=0}^{n-1} \omega^{ij} v_i \right)^q = \sum_{i=0}^{n-1} \omega^{qij} v_i^q = \sum_{i=0}^{n-1} \omega^{qij} v_i = V_{((qj))}.$$

Conversely, suppose that for all j, $V_j^q = V_{((qj))}$. Then

$$\sum_{i=0}^{n-1} \omega^{iqj} v_i^q = \sum_{i=0}^{n-1} \omega^{iqj} v_i \quad j = 0, \ldots, n-1.$$

Let $k = qj (\text{mod } n)$. Because q is coprime to $n = q^m - 1$, as j ranges over all values between 0 and $n-1$, the integer k also takes on all values between 0 and $n-1$. Hence

$$\sum_{i=0}^{n-1} \omega^{ik} v_i^q = \sum_{i=0}^{n-1} \omega^{ik} v_i \quad k = 0, \ldots, n-1,$$

and by uniqueness of the Fourier transform, $v_i^q = v_i$ for all i. Thus v_i is a zero of $x^q - x$ for all i, and such zeros are always elements of $GF(q)$. □

To apply the theorem, the integers modulo n are divided into a collection of sets, known as *q-ary conjugacy classes*, as follows:

$$\mathcal{B}_j = \{j, jq, jq^2, \ldots, jq^{m_j - 1}\}$$

where m_j is the smallest positive integer satisfying $jq^{m_j} = j$ modulo n. There must be such an m_j because the field is finite and q is coprime to n.

For example, when $q = 2$ and $n = 7$, the conjugacy classes are

$$\mathcal{B}_0 = \{0\}$$
$$\mathcal{B}_1 = \{1, 2, 4\}$$
$$\mathcal{B}_3 = \{3, 6, 5\}.$$

Modulo 7

{ 0 }
{ 1, 2, 4 }
{ 3, 6, 5 }

Modulo 15

{ 0 }
{ 1, 2, 4, 8 }
{ 3, 6, 12, 9 }
{ 5, 10 }

Modulo 31

{ 0 }
{ 1, 2, 4, 8, 16 }
{ 3, 6, 12, 24, 17 }
{ 5, 10, 20, 9, 18 }
{ 7, 14, 28, 25, 19 }
{ 11, 22, 13, 26, 21 }
{ 15, 30, 29, 27, 25 }

Modulo 21

{ 0 }
{ 1, 2, 4, 8, 16, 11 }
{ 3, 6, 12 }
{ 5, 10, 20, 19, 17, 13 }
{ 7, 14, 28, 9, 18 }
{ 9, 18, 15 }

Modulo 63

{ 0 }
{ 1, 2, 4, 8, 16, 32 }
{ 3, 6, 12, 24, 48, 33 }
{ 5, 10, 20, 40, 17, 34 }
{ 7, 14, 28, 56, 49, 35 }
{ 9, 18, 36 }
{ 11, 22, 44, 25, 50, 36 }
{ 13, 26, 52, 41, 19, 38 }
{ 15, 30, 60, 57, 51, 39 }
{ 21, 42 }
{ 23, 46, 29, 58, 53, 43 }
{ 27, 54, 45 }
{ 31, 62, 61, 59, 55, 47 }

Figure 6.4. Conjugacy classes

The structure of the conjugacy classes may be more evident if the elements are written as binary m-tuples. In the form of binary m-tuples,

$$\mathcal{B}_0 = \{000\}$$

$$\mathcal{B}_1 = \{001, 010, 100\}$$

$$\mathcal{B}_3 = \{011, 110, 101\}.$$

With this notation, all elements of the same class are represented by binary numbers that are cyclic shifts of one another.

The conjugacy class $\mathcal{B}_j$, regarded as a set of indices, specifies a set of frequencies in the spectrum.[1] Theorem 6.3.1 asserts that if the time-domain signal is in $GF(q)$, then the value of the spectrum at any frequency in a conjugacy class specifies the value of the spectrum at all other frequencies in that conjugacy class. In the next section we shall develop this line of thought to give a spectral description of cyclic codes.

Figure 6.4 tabulates the conjugacy classes modulo n for some small integers. We have included the conjugacy classes of the integers modulo 21 as a reminder that any integer coprime to q might be used as the modulus. Notice that the members of conjugacy

[1] The term *chord* may be used as a picturesque depiction of the set of frequencies in the same conjugacy class.

classes modulo 21 can be multiplied by three, and the classes then become some of the conjugacy classes modulo 63.

Definition 6.3.2. *The q-ary trace of an element β of $GF(q^m)$ is the sum*

$$trace(\beta) = \sum_{i=0}^{m-1} \beta^{q^i}.$$

Clearly, by Theorem 4.6.10, the qth power of the q-ary trace of β is equal to the q-ary trace of β, and thus the q-ary trace is an element of $GF(q)$. If β has m elements in its conjugacy class, then trace(β) is the sum of all the elements in that conjugacy class. Otherwise, if β has r elements in its conjugacy class, then r divides m, and each element is added into the trace m/r times.

It follows from the definition of the trace and Theorem 4.6.10 that

$$\text{trace}(\beta + \gamma) = \text{trace}(\beta) + \text{trace}(\gamma)$$

and that all conjugates have the same trace.

Theorem 6.3.3. *Over $GF(q^m)$, the q-ary trace takes on each value of $GF(q)$ equally often, that is, q^{m-1} times.*

Proof: Let γ be an element of $GF(q)$. Suppose β is an element of $GF(q^m)$ whose q-ary trace is γ. Then β is a zero of the polynomial

$$x^{q^{m-1}} + x^{q^{m-2}} + \cdots + x^q + x - \gamma.$$

This polynomial has degree q^{m-1}, and thus it has at most q^{m-1} zeros. But there are only q such polynomials, and every element of $GF(q^m)$ must be a zero of one of them. The theorem follows. □

Theorem 6.3.4. *Let a be an element of $GF(2^m)$. The quadratic equation*

$$x^2 + x + a = 0$$

has a root in $GF(2^m)$ if and only if the binary trace of a equals zero.

Proof: Let β be a root of the quadratic equation $x^2 + x + a = 0$. Then the binary trace of both sides of the equation $\beta^2 + \beta + a = 0$ gives

$$\text{trace}(\beta^2 + \beta + a) = \text{trace}(0) = 0.$$

The trace distributes across addition, and the traces of β and β^2 are the same elements of $GF(2)$. Hence

$$\text{trace}(a) = 0.$$

Conversely, every β is a zero of $x^2 + x + a$ for some a, namely, a equal to $-(\beta + \beta^2)$. There are 2^{m-1} such a with zero trace, and this is just enough to form 2^{m-1} equations, each with two roots. The proof is complete. □

In a Galois field $GF(2^m)$, the polynomial

$$p(x) = x^2 + x$$

has the surprising property that it gives a linear function:

$$p(\beta + \gamma) = p(\beta) + p(\gamma),$$

as a consequence of Theorem 4.6.10. A polynomial with this property is called a *linearized polynomial*.[2] A linearized polynomial over the field $GF(q^m)$ is a polynomial $p(x)$ of the form

$$p(x) = \sum_{i=0}^{h} p_{q^i} x^{q^i}$$

where $p_{q^i} \in GF(q^m)$. A useful property of a linearized polynomial is that for every $\beta_1, \beta_2 \in GF(q^m)$ and $\lambda_1, \lambda_2 \in GF(q)$,

$$p(\lambda_1\beta_1 + \lambda_2\beta_2) = \lambda_1 p(\beta_1) + \lambda_2 p(\beta_2)$$

as a consequence of Theorem 4.6.10, and because $\lambda_1^{q^i} = \lambda_1$ and $\lambda_2^{q^i} = \lambda_2$.

Theorem 6.3.5. *The zeros of a linearized polynomial $p(x)$ over $GF(q)$ form a vector subspace of the splitting field of $p(x)$.*

Proof: The splitting field is the smallest extension field that contains the zeros of $p(x)$. Hence it is only necessary to show that the set of zeros of $p(x)$ is closed under linear combinations, and so forms a vector subspace. This follows immediately from the equation prior to Theorem 6.3.5. □

Now suppose that we choose a conjugacy class, $\mathcal{B}_k$, and define the spectrum:

$$W_j = \begin{cases} 0 & j \in \mathcal{B}_k \\ 1 & j \notin \mathcal{B}_k. \end{cases}$$

By Theorem 6.3.1, the inverse Fourier transform of this spectrum is a vector over $GF(q)$, which can be represented by the polynomial $w(x)$. It has the special property that

$$w(x)^2 = w(x) \pmod{x^n - 1}$$

[2] A "linear polynomial" would be a more logical term except that this term often is used loosely to refer to a polynomial of degree 1.

because the convolution $w(x)^2$ transforms into the componentwise product W_j^2 in the frequency domain, and clearly $W_j^2 = W_j$. Any polynomial $w(x)$, satisfying $w(x)^2 = w(x)(\text{mod } x^n - 1)$, is called an *idempotent polynomial*, or an *idempotent*. Every idempotent polynomial can be obtained in the following way. Choose several conjugacy classes; set $W_j = 0$ if j is in one of the chosen conjugacy classes, and otherwise, set $W_j = 1$. The inverse Fourier transform gives a polynomial in the time domain that is an idempotent.

Any codeword polynomial of a cyclic code that is also an idempotent, $w(x)$, with the property that

$$c(x)w(x) = c(x) \quad (\text{mod } x^n - 1)$$

for every other codeword polynomial $c(x)$, is called a *principal idempotent* of the code.

Theorem 6.3.6. *Every cyclic code has a unique principal idempotent.*

Proof: Let $g(x)$ be the generator polynomial, and let

$$W_j = \begin{cases} 0 & \text{if } g(\omega^j) = 0 \\ 1 & \text{if } g(\omega^j) \neq 0. \end{cases}$$

Then $w(x)$ is an idempotent. It has the same zeros as $g(x)$, and so it is a codeword. In addition, $W_j G_j = G_j$ for all j, and therefore $w(x)g(x) = g(x)$. Next, $c(x)$ is a codeword polynomial if and only if $c(x) = a(x)g(x)$ for some $a(x)$, and thus

$$c(x)w(x) = a(x)w(x)g(x) = a(x)g(x) = c(x) \quad (\text{mod } x^n - 1),$$

which proves the theorem. □

6.4 Spectral description of cyclic codes

Each codeword of a cyclic code of blocklength n can be represented by a polynomial $c(x)$ of degree at most $n - 1$. A nonsystematic form of an encoder is written as $c(x) = a(x)g(x)$ where $a(x)$ is a data polynomial of degree at most $k - 1$. This is equivalent to a cyclic convolution in the time domain: $\boldsymbol{c} = \boldsymbol{a} * \boldsymbol{g}$. Therefore in the frequency domain, the encoding operation can be written as the product: $C_j = A_j G_j$. Any spectrum C_j with values in the extension field $GF(q^m)$ that satisfies this expression is a *codeword spectrum*, provided that all components in the time domain are $GF(q)$-valued. Because the data spectrum is arbitrary, the role of G_j is to specify those j where the codeword spectrum C_j is required to be zero. Thus we can define a cyclic code alternatively as follows. Given a set of spectral indices, $\mathcal{B} = \{j_1, \ldots, j_{n-k}\}$, whose elements are called *check frequencies*, the cyclic code $\mathcal{C}$ is the set of words over $GF(q)$ whose spectrum is zero in components indexed by $j_1, \ldots, j_{n-k}$. Any such spectrum $\boldsymbol{C}$ can be regarded as having components of the form $A_j G_j$, and so by the convolution theorem, has an inverse

Fourier transform of the form $c(x) = a(x)g(x)$. Clearly, the set of check frequencies $\mathcal{B} = \{j_1, \ldots, j_{n-k}\}$ corresponds to the defining set $\mathcal{A} = \{\omega^{j_1}, \ldots, \omega^{j_{n-k}}\}$.

Although each codeword in a cyclic code is a vector over $GF(q)$, the codeword spectrum is a vector over $GF(q^m)$. Hence the codewords of a cyclic code can be described as the inverse Fourier transforms of spectral vectors that are constrained to zero in certain prescribed components, provided that these inverse Fourier transforms are $GF(q)$-valued. Not every spectrum that is zero in the prescribed components will give a codeword; some of these may have inverse transforms with components that are not in $GF(q)$. Only spectra that satisfy the conjugacy constraints can be used.

In later sections, we shall specify codes with large minimum distances by placing appropriate constraints on the spectra of the codewords. These constraints can be motivated by several bounds on the Hamming weight, $w(\boldsymbol{v})$, of a vector $\boldsymbol{v}$. The following bound, which is an immediate consequence of Theorem 6.2.1, also can be proved directly from the fundamental theorem of algebra.

Theorem 6.4.1 (BCH Bound). *Let n be a factor of $q^m - 1$ for some m. The only vector in $GF(q)^n$ of weight $d - 1$ or less that has $d - 1$ consecutive components of its spectrum equal to zero is the all-zero vector.*

Proof: The linear complexity property says that the linear complexity of $\boldsymbol{V}$, cyclically repeated, is equal to the weight of $\boldsymbol{v}$. Therefore, for any vector $\boldsymbol{v}$ of weight $d - 1$ or less

$$V_j = -\sum_{k=1}^{d-1} \Lambda_k V_{((j-k))}.$$

If $\boldsymbol{V}$ has $d - 1$ consecutive components equal to zero, the recursion implies that $\boldsymbol{V}$ is everywhere zero, and so $\boldsymbol{v}$ must be the all-zero vector. □

For example, the double-error-correcting cyclic codes, defined in Section 5.6, have spectral zeros at components with indices $j = 1$ and 3, and at all conjugates of 1 and 3. Because the conjugates of 1 include 2 and 4, the four consecutive spectral components C_1, C_2, C_3, and C_4 are equal to zero, and the BCH bound implies that the minimum weight of that code is at least 5.

Other properties of the Fourier transform can be used as well to move around spectral zeros or to create new spectral zeros so that the BCH bound can be applied. To give one more general bound, the BCH bound can be combined with the cyclic decimation property of the Fourier transform, described in Theorem 6.1.6. The components of a spectrum $\boldsymbol{V}$ can be rearranged by cyclic translation and by cyclic decimation by b if $\text{GCD}(b, n) = 1$. The components of the codeword $\boldsymbol{v}$ are then rearranged by cyclic decimation by B where $Bb = 1 (\text{mod } n)$.

Corollary 6.4.2. *Let n be a factor of $q^m - 1$ for some m, and b an integer that is coprime with n. The only vector in $GF(q)^n$ of weight $d - 1$ or less that has $d - 1$ bth-consecutive components of its spectrum equal to zero is the all-zero vector.*

Proof: The assumption of the theorem is that, for some integer a, $V_j = 0$ for $j = a + b\ell (\text{mod } n)$ where $\ell = 0, \ldots, d-2$. Let $V'_j = V_{B(j-a)}$. This is a translation by a, followed by a cyclic decimation by B. Consequently, the translation property and the cyclic decimation property of the Fourier transform imply that $\boldsymbol{v}'$ has the same weight as $\boldsymbol{v}$. But $\boldsymbol{V}'$ has $d-1$ consecutive components equal to zero, so by the BCH bound, $w(\boldsymbol{v}) = w(\boldsymbol{v}') \geq d$. □

Any pattern of spectral zeros in $\boldsymbol{V}$ can be cyclically shifted without changing the weight of $\boldsymbol{v}$. Therefore it is enough in the next theorem to choose a pattern of spectral zeros beginning at index $j = 0$. This theorem says that the $d-1$ consecutive spectral zeros of the BCH bound can be replaced by a pattern of s uniformly spaced subblocks, each with $d-s$ consecutive spectral zeros, provided the spacing of the subblocks is coprime with n. The spectral zeros occur at $j = \ell_1 + \ell_2 b (\text{mod } n)$, where $\ell_1 = 0, \ldots d-s-1$ and $\ell_2 = 0, \ldots, s-1$. Although the roles of ℓ_1 and ℓ_2 may appear to be different in the statement of the theorem, they are really the same. To see this, cyclically decimate the components of $\boldsymbol{V}$ by B so that $j' = Bj$ where $Bb = 1 (\text{mod } n)$. Such a B exists because $\text{GCD}(b, n) = 1$. Then the spectral zeros will be at $j' = \ell_2 + \ell_1 B (\text{mod } n)$, where $\ell_2 = 0, \ldots, s-1$ and $\ell_1 = 0, \ldots, d-s-1$, so the roles of $d-s-1$ and $s-1$ are interchanged.

Theorem 6.4.3 (Hartmann–Tzeng Bound). *Let n be a factor of $q^m - 1$ for some m, and b an integer that is coprime with n. The only vector $\boldsymbol{v}$ in $GF(q)^n$ of weight $d-1$ or less whose spectral components V_j equal zero for $j = \ell_1 + \ell_2 b$ (mod n), where $\ell_1 = 0, \ldots, d-s-1$ and $\ell_2 = 0, \ldots, s-1$, is the all-zero vector.*

Proof: Let $\boldsymbol{v}$ be such a nonzero vector with Fourier spectrum $\boldsymbol{V}$ and, without loss of generality, assume that $s \leq d-s$. Clearly, by the BCH bound applied to index ℓ_1, the Hamming weight satisfies $w(\boldsymbol{v}) \geq d-s+1$, and by Corollary 6.4.2 applied to index ℓ_2, $w(\boldsymbol{v}) \geq s+1$.

The idea of the proof is to construct a nonzero spectrum, $\boldsymbol{V}'$, with $d-1$ bth-consecutive spectral zeros, to which the corollary to the BCH bound can be applied to conclude that $w(\boldsymbol{v}) \geq w(\boldsymbol{v}') \geq d$. Let $\boldsymbol{V}^{(k)}$ be the kth cyclic shift of $\boldsymbol{V}$, given by

$$V_j^{(k)} = V_{((j+k))}.$$

For $k = 0, \ldots, d-s-1$, the vector $\boldsymbol{V}^{(k)}$ satisfies $V_{\ell b}^{(k)} = 0$ for $\ell = 0, \ldots, s-1$. Accordingly, for any set of coefficients $\Lambda_0, \Lambda_1, \ldots, \Lambda_{d-s-1}$, not all zero, the linear combination

$$V'_j = -\sum_{k=0}^{d-s-1} \Lambda_k V_{((j+k))} \quad j = 0, \ldots, n-1,$$

satisfies $V'_{\ell b} = 0$ for $\ell = 0, \ldots, d-s-1$. Not all such V'_j can be zero because, otherwise, this equation could be used to form a linear recursion on $\boldsymbol{V}$ of length at most $d - s$, while the linear complexity property says that no such linear recursion can have a length smaller than $w(\boldsymbol{v}) \geq d - s + 1$.

Thus, the inverse Fourier transform of $\boldsymbol{V}'$, denoted $\boldsymbol{v}'$, is nonzero. Each $\boldsymbol{V}^{(k)}$ has an inverse Fourier transform, denoted $\boldsymbol{v}^{(k)}$, that is nonzero everywhere that $\boldsymbol{v}$ is nonzero, and $\boldsymbol{v}'$ is a linear combination of these $\boldsymbol{v}^{(k)}$. Hence

$$0 < w(\boldsymbol{v}') \leq w(\boldsymbol{v}).$$

The nonzero spectrum $\boldsymbol{V}'$ has $d - s$ bth-consecutive spectral zeros with indices $j = 0, \ldots, (d - s - 1)b$. Setting $j = (d - s)b, \ldots, (d - 2)b$ gives the matrix equation

$$\begin{bmatrix} V'_{(d-s)b} \\ V'_{(d-s+1)b} \\ \vdots \\ V'_{(d-2)b} \end{bmatrix} = \begin{bmatrix} V_{(d-s)b} & \cdots & V_{(d-s)b+d-s-1} \\ V_{(d-s+1)b} & & V_{(d-s+1)b+d-s-1} \\ \vdots & & \vdots \\ V_{(d-2)b} & \cdots & V_{(d-2)b+d-s-1} \end{bmatrix} \begin{bmatrix} \Lambda_0 \\ \Lambda_1 \\ \vdots \\ \Lambda_{d-s-1} \end{bmatrix}.$$

We will choose $\Lambda_0, \Lambda_1, \ldots, \Lambda_{d-s-1}$ so that $V'_j = 0$ for $j = sb, (s+1)b, \ldots, (d-1)b$. Thus we must solve

$$\begin{bmatrix} V_{(d-s)b} & \cdots & V_{(d-s)b+d-s-1} \\ V_{(d-s+1)b} & & V_{(d-s+1)b+d-s-1} \\ \vdots & & \vdots \\ V_{(d-2)b} & \cdots & V_{(d-2)b+d-s-1} \end{bmatrix} \begin{bmatrix} \Lambda_0 \\ \Lambda_1 \\ \vdots \\ \Lambda_{d-s-1} \end{bmatrix} = \begin{bmatrix} 0 \\ 0 \\ \vdots \\ 0 \end{bmatrix}$$

for the vector $\boldsymbol{\Lambda}$. Each row of the equation defines a $(d - s - 1)$-dimensional hyperspace in $GF(q)^{d-s}$. Therefore the matrix equation specifies the intersection of $s - 1$ such $(d - s - 1)$-dimensional hyperspaces, an intersection which must have a dimension at least 1 because $s \leq d - s$. Hence there is a nonzero solution for $\boldsymbol{\Lambda}$, so a nonzero $\boldsymbol{V}'$ can be constructed with the desired $d - 1$ bth-consecutive spectral zeros. Hence $w(\boldsymbol{v}') \geq d$. This completes the proof of the theorem because $w(\boldsymbol{v}) \geq w(\boldsymbol{v}')$. □

The next corollary says that the sequence of subblocks need not be consecutive. Again, the roles of ℓ_1 and ℓ_2 can be interchanged by cyclically decimating the components of $\boldsymbol{V}$ by B.

Corollary 6.4.4 (Roos Bound). *Let n be a factor of $q^m - 1$ for some m. Let $GCD(n, b) = 1$. The only vector in $GF(q)^n$ of weight $d - 1$ or less whose spectral components V_j equal zero for $j = \ell_1 + \ell_2 b \pmod n$, where $\ell_1 = 0, \ldots, d - s - 2$ and ℓ_2 takes at least $s + 1$ values in the range $0, \ldots, d - 2$, is the all-zero vector.*

Proof: The proof is the same as the proof of Theorem 6.4.3. The linear combination is chosen to make $V'_j = 0$ for $j = \ell b$ for the $r - 1$ missing values of ℓ, which need not be restricted to any particular subset of missing values. □

As an example of the Roos bound, consider the binary cyclic code of blocklength 127 with spectral zeros at $j = 5$ and 9. These indices have conjugacy classes

$$\mathcal{B}_5 = \{5, 10, 20, 40, 80, 33, 66\}$$
$$\mathcal{B}_9 = \{9, 18, 36, 72, 17, 34, 68\}.$$

Spectral zeros occur at all indices in $\mathcal{B}_5$ and $\mathcal{B}_9$. In particular, the indices of spectral zeros include

$$(9, 10), (17, 18), (33, 34),$$

which have been parenthesized to form three blocks of zeros to which the Roos bound can be applied to conclude that $d_{\min} \geq 5$. Indeed, the proof of the Roos bound constructs a new word, not necessarily binary, with zeros at 9, 17, 25, and 33. Because the spacing, 8, is coprime to 127, the BCH bound implies that $d_{\min} \geq 5$.

As a second example, consider the binary cyclic code of blocklength 127 with spectral zeros at $j = 1, 5$, and 9. Now the spectral zeros include the blocks with indices

$$(8, 9, 10), (16, 17, 18), (32, 33, 34),$$

and the Roos bound implies that $d_{\min} \geq 6$ because a linear combination of cyclic shifts can produce spectral zeros at indices 8, 16, 24, 32, 40, and then the BCH bound applies. This code actually has $d_{\min} \geq 7$, which follows from a side calculation that shows there is no codeword of weight 6. Indeed, suppose that a codeword has even weight. Then $C_0 = 0$, and the spectral zeros include the blocks with indices

$$(0, 1, 2), (8, 9, 10), (16, 17, 18), (32, 33, 34).$$

A linear combination of cyclic shifts produces spectral zeros at 0, 8, 16, 24, 32, 40, and the Roos bound implies that if the weight is even, it is larger than 7. Consequently, $d_{\min} \geq 7$.

6.5 BCH codes

Recall that a cyclic Reed–Solomon code of minimum distance d and blocklength n dividing $q - 1$ is the set of all words over $GF(q)$ whose spectra are zero in a block of $d - 1$ consecutive components specified by the defining set $\{\omega^{j_0}, \omega^{j_0+1}, \ldots, \omega^{j_0+d-2}\}$. A Reed–Solomon code has the property that the code symbols and the spectrum symbols are in the same field, $GF(q)$. A BCH code of *designed distance* d and blocklength n dividing $q^m - 1$ is the set of all words over $GF(q)$ of blocklength n whose spectra are zero in a specified block of $d - 1$ consecutive components. In general, the components of a BCH codeword spectrum will lie in the larger field $GF(q^m)$. It is an immediate consequence of the BCH bound that the minimum distance $d_{\min}$ of the BCH code is at least as large as the designed distance d. Sometimes, however, the minimum distance of a BCH code is larger than the designed distance d.

The spectrum of each codeword of a BCH code lies in the extension field $GF(q^m)$ and must equal zero in a block of $d - 1$ consecutive components specified by the defining set $\{\omega^{j_0}, \omega^{j_0+1}, \ldots, \omega^{j_0+d-2}\}$. Therefore the codewords in the BCH code over $GF(q)$ are also codewords in the Reed–Solomon code over $GF(q^m)$. They are those Reed–Solomon codewords that happen to take values only in $GF(q)$. Thus every BCH code is a subfield-subcode of a Reed–Solomon code.

The relationship between a Reed–Solomon code and its BCH subfield-subcode is illustrated in Figure 6.5. Within the (7, 5) Reed–Solomon code over $GF(8)$, some of

Reed–Solomon code	Subfield-subcode
0 0 0 0 0 0 0	0 0 0 0 0 0 0
0 0 0 0 1 6 3	
0 0 0 0 2 7 6	
0 0 0 0 3 1 5	
⋮	
0 0 0 1 0 1 1	0 0 0 1 0 1 1
0 0 0 1 1 7 2	
0 0 0 1 2 6 7	
0 0 0 1 3 0 4	
⋮	
0 0 0 7 0 7 7	
0 0 0 7 1 1 4	
0 0 0 7 2 0 1	
0 0 0 7 3 6 2	
⋮	
0 0 1 0 0 7 3	
0 0 1 0 1 1 0	0 0 1 0 1 1 0
0 0 1 0 2 0 5	
0 0 1 0 3 6 6	
⋮	
0 0 1 1 0 6 2	
0 0 1 1 1 0 1	0 0 1 1 1 0 1
0 0 1 1 2 1 4	
0 0 1 1 3 7 7	
⋮	

Figure 6.5. Extracting a subfield-subcode from a (7, 5) code

the codewords happen to have only components that are zero or one. One can list these codewords to form a binary code of blocklength 7. The minimum distance of the BCH code must be at least 3 because the minimum distance in the original Reed–Solomon code is 3.

We can describe this binary BCH code more directly by using Theorem 6.4.1 to constrain the spectrum appropriately. Each spectral component C_j must satisfy $C_j^2 = C_{2j}$ to ensure that the codeword components have binary values. So that a single-bit error can be corrected, spectral components C_1 and C_2 are chosen as check frequencies and constrained to equal zero. Then Theorem 6.4.1 requires C_4 to equal zero. Theorem 6.4.1 also requires that $C_0^2 = C_0$, which means that C_0 is either zero or one and that $C_5 = C_6^2 = C_3^4$, which means that C_3 can be an arbitrary element of $GF(8)$, but then C_6 and C_5 are determined by C_3. Because four bits describe the spectrum (one bit to specify C_0 and three bits to specify C_3), the code is a binary $(7, 4)$ BCH code. The codewords are the inverse Fourier transforms of the codeword spectra. These codewords are shown in Figure 6.6. The figure makes it clear that the single-error-correcting BCH code is the same as the $(7, 4)$ Hamming code.

To specify the sixteen codewords more directly in the time domain, we form a generator polynomial $g(x)$ that has the required spectral zeros. These zeros are at α^1, α^2, and α^4. The first two zeros are needed to satisfy the BCH bound, and the third zero

Frequency-domain codewords							Time-domain codewords						
C_6	C_5	C_4	C_3	C_2	C_1	C_0	c_6	c_5	c_4	c_3	c_2	c_1	c_0
0	0	0	0	0	0	0	0	0	0	0	0	0	0
α^0	α^0	0	α^0	0	0	0	1	1	1	0	1	0	0
α^2	α^4	0	α^1	0	0	0	0	0	1	1	1	0	1
α^4	α^1	0	α^2	0	0	0	0	1	0	0	1	1	1
α^6	α^5	0	α^3	0	0	0	1	1	0	1	0	0	1
α^1	α^2	0	α^4	0	0	0	0	1	1	1	0	1	0
α^3	α^6	0	α^5	0	0	0	1	0	0	1	1	1	0
α^5	α^3	0	α^6	0	0	0	1	0	1	0	0	1	1
0	0	0	0	0	0	1	1	1	1	1	1	1	1
α^0	α^0	0	α^0	0	0	1	0	0	0	1	0	1	1
α^2	α^4	0	α^1	0	0	1	1	1	0	0	0	1	0
α^4	α^1	0	α^2	0	0	1	1	0	1	1	0	0	0
α^6	α^5	0	α^3	0	0	1	0	0	1	0	1	1	0
α^1	α^2	0	α^4	0	0	1	1	0	0	0	1	0	1
α^3	α^6	0	α^5	0	0	1	0	1	1	0	0	0	1
α^5	α^3	0	α^6	0	0	1	0	1	0	1	1	0	0

Figure 6.6. The $(7, 4)$ Hamming code

is needed to satisfy the conjugacy constraint. Therefore

$$\begin{aligned} g(x) &= (x - \alpha^1)(x - \alpha^2)(x - \alpha^4) \\ &= x^3 + x + 1. \end{aligned}$$

Notice that because $(x - \alpha^4)$ is included as a factor, $g(x)$ has only binary coefficients even though its zeros are specified in $GF(8)$.

In the general case, the integers modulo n are divided into the conjugacy classes:

$$\mathcal{B}_j = \{j, jq, jq^2, \ldots, jq^{m_j - 1}\}.$$

If the spectral component C_j is specified, then every other spectral component whose index is in the conjugacy class of j must be a power of C_j, and hence cannot be separately specified. Further, if the conjugacy class of j has r members, then we must have

$$C_j^{q^r} = C_j$$

and

$$C_j^{q^r - 1} = 1.$$

Consequently, we are not free to choose any element of $GF(q^m)$ for C_j but only the zero element or those elements whose order divides $q^r - 1$. Therefore if the conjugacy class of j has r members, then $C_j \in GF(q^r)$.

To specify a code in the frequency domain, we break the first $q^m - 1$ integers into conjugacy classes and select one integer to represent each class. These representatives specify the uniquely assignable symbols. Every other symbol indexed from the same set of conjugates is not free; it takes a value that is determined by a value indexed by another element of that conjugacy class. To form a BCH code, $d - 1$ consecutive spectral components are chosen as check frequencies and set to zero. The remaining assignable symbols are data symbols, arbitrary except for occasional constraints on the order.

For example, for $n = 63$, the conjugacy classes break the spectrum into the following:

$$\begin{aligned}
&\{C_0\}, \\
&\{C_1, C_2, C_4, C_8, C_{16}, C_{32}\}, \\
&\{C_3, C_6, C_{12}, C_{24}, C_{48}, C_{33}\}, \\
&\{C_5, C_{10}, C_{20}, C_{40}, C_{17}, C_{34}\}, \\
&\{C_7, C_{14}, C_{28}, C_{56}, C_{49}, C_{35}\}, \\
&\{C_9, C_{18}, C_{36}\}, \\
&\{C_{11}, C_{22}, C_{44}, C_{25}, C_{50}, C_{37}\}, \\
&\{C_{13}, C_{26}, C_{52}, C_{41}, C_{19}, C_{38}\}, \\
&\{C_{15}, C_{30}, C_{60}, C_{57}, C_{51}, C_{39}\}, \\
&\{C_{21}, C_{42}\},
\end{aligned}$$

$\{C_{23}, C_{46}, C_{29}, C_{58}, C_{53}, C_{43}\}$,
$\{C_{27}, C_{54}, C_{45}\}$,
$\{C_{31}, C_{62}, C_{61}, C_{59}, C_{55}, C_{47}\}$.

We choose the first component of each set to be the representative symbol. Once this element is specified, the remaining elements in that set are determined. If we take C_1, C_2, C_3, C_4, C_5, and C_6 to be check frequencies, then we have a triple-error-correcting binary BCH code of blocklength 63. Then C_0, C_7, C_9, C_{11}, C_{13}, C_{15}, C_{21}, C_{23}, C_{27}, and C_{31} are the data symbols. The values of C_9 and C_{27} must be elements of the subfield $GF(8)$ because $C_9^8 = C_9$ and $C_{27}^8 = C_{27}$. The value of C_{21} must be an element of the subfield $GF(4)$ because $C_{21}^4 = C_{21}$. The value of C_0 must be an element of the subfield $GF(2)$. The values of C_7, C_{11}, C_{13}, C_{15}, C_{23}, and C_{31} are arbitrary elements of $GF(64)$. It requires a total of forty-five data bits to specify these symbols. Hence we have the (63, 45) triple-error-correcting BCH code. The set of 2^{45} codewords that one obtains by encoding in the frequency domain is the same as the set of 2^{45} codewords that one obtains by encoding in the time domain using a generator polynomial. Up until the point at which the data is recovered, the decoder need not care how the data was encoded. At the last step, however, when the data is extracted from the corrected codeword, the decoder must know how the data is stored in the codeword. If the data bits were encoded in the frequency domain, then they must be read out of the frequency domain.

To encode in the time domain, one needs to find the generator polynomial of the code. This is given by

$$g(x) = \Pi(x - \omega^j)$$

where the product is over all j that have a conjugate (including j itself) in the range $1, \ldots, d-1$. This is because the codewords then have the form

$$c(x) = a(x)g(x)$$

and

$$\begin{aligned} C_j &= c(\omega^j) \\ &= a(\omega^j)g(\omega^j) \\ &= 0 \qquad\qquad j = 1, \ldots, d-1 \end{aligned}$$

by construction of $g(x)$. Equivalently,

$$g(x) = \text{LCM}[f_1(x), \ldots, f_{d-1}(x)]$$

where $f_j(x)$ is the minimal polynomial of ω^j for $j = 1, \ldots, d-1$.

Table 6.2 also gives the minimal polynomials over $GF(2)$ of all field elements in $GF(16)$ where $\alpha = z$ is primitive. Notice that the minimal polynomials of even powers of α always appear earlier in the list. This is a consequence of Theorem 5.3.4, which says that over $GF(2)$ any field element β and its square have the same minimal polynomial. This observation slightly reduces the work required to find $g(x)$.

The generator polynomial for the double-error-correcting BCH code of blocklength 15 is obtained as follows:

$$\begin{aligned} g(x) &= \mathrm{LCM}[f_1(x), f_2(x), f_3(x), f_4(x)] \\ &= \mathrm{LCM}[x^4+x+1, x^4+x+1, x^4+x^3+x^2+x+1, x^4+x+1] \\ &= (x^4+x+1)(x^4+x^3+x^2+x+1) \\ &= x^8+x^7+x^6+x^4+1. \end{aligned}$$

Because $g(x)$ has degree 8, $n-k=8$. Therefore $k=7$, and we have the generator polynomial for the $(15, 7)$ double-error-correcting BCH code. Notice that BCH codes are designed from a specification of n and t. The value of k is not known until after $g(x)$ is found.

Continuing in this way, we can construct generator polynomials for other primitive BCH codes of blocklength 15.

Let $t=3$:

$$\begin{aligned} g(x) &= \mathrm{LCM}[f_1(x), f_2(x), f_3(x), f_4(x), f_5(x), f_6(x)] \\ &= (x^4+x+1)(x^4+x^3+x^2+x+1)(x^2+x+1) \\ &= x^{10}+x^8+x^5+x^4+x^2+x+1. \end{aligned}$$

This is the generator polynomial for a $(15, 5)$ triple-error-correcting BCH code.

Let $t=4$:

$$\begin{aligned} g(x) &= \mathrm{LCM}[f_1(x), f_2(x), f_3(x), f_4(x), f_5(x), f_6(x), f_7(x), f_8(x)] \\ &= (x^4+x+1)(x^4+x^3+x^2+x+1)(x^2+x+1)(x^4+x^3+1) \\ &= x^{14}+x^{13}+x^{12}+x^{11}+x^{10}+x^9+x^8+x^7+x^6+x^5+x^4+ \\ &\quad x^3+x^2+x+1. \end{aligned}$$

This is the generator polynomial for a $(15, 1)$ BCH code. It is a simple repetition code that can correct seven errors.

Let $t=5$, 6, and 7. Each of these cases results in the same generator polynomial as for $t=4$. Beyond $t=7$, the BCH code is undefined because the number of nonzero field elements is 15.

Table 6.3 gives $GF(16)$ as an extension field of $GF(4)$ constructed with the primitive polynomial $p(z)=z^2+z+2$. This table also gives the minimal polynomials over $GF(4)$ of all field elements in $GF(16)$ where $\alpha=z$ is primitive. The generator polynomial for the single-error-correcting BCH code over $GF(4)$ of blocklength 15 is obtained as follows:

$$\begin{aligned} g(x) &= \mathrm{LCM}[f_1(x), f_2(x)] \\ &= (x^2+x+2)(x^2+x+3) \\ &= x^4+x+1. \end{aligned}$$

Table 6.3. *Representations of $GF(4^2)$*

$GF(4)$

+	0	1	2	3
0	0	1	2	3
1	1	0	3	2
2	2	3	0	1
3	3	2	1	0

·	0	1	2	3
0	0	0	0	0
1	0	1	2	3
2	0	2	3	1
3	0	3	1	2

Exponential notation	Polynomial notation	Quaternary notation	Decimal notation	Minimal polynomial
0	0	00	0	
α^0	1	01	1	$x+1$
α^1	z	10	4	x^2+x+2
α^2	$z+2$	12	6	x^2+x+3
α^3	$3z+2$	32	14	x^2+3x+1
α^4	$z+1$	11	5	x^2+x+2
α^5	2	02	2	$x+2$
α^6	$2z$	20	8	x^2+2x+1
α^7	$2z+3$	23	11	x^2+2x+2
α^8	$z+3$	13	7	x^2+x+3
α^9	$2z+2$	22	10	x^2+2x+1
α^{10}	3	03	3	$x+3$
α^{11}	$3z$	30	12	x^2+3x+3
α^{12}	$3z+1$	31	13	x^2+3x+1
α^{13}	$2z+1$	21	9	x^2+2x+2
α^{14}	$3z+3$	33	15	x^2+3x+3

This is the generator polynomial for a (15, 11) single-error-correcting BCH code over $GF(4)$. It encodes eleven quaternary symbols (equivalent to twenty-two bits) into fifteen quaternary symbols. It is not a Hamming code.

Continuing, we can find the generator polynomials for other BCH codes over $GF(4)$ of blocklength 15.

Let $t = 2$:

$$\begin{aligned} g(x) &= \text{LCM}[f_1(x), f_2(x), f_3(x), f_4(x)] \\ &= (x^2+x+2)(x^2+x+3)(x^2+3x+1) \\ &= x^6+3x^5+x^4+x^3+2x^2+2x+1. \end{aligned}$$

This is the generator polynomial for a (15, 9) double-error-correcting BCH code over $GF(4)$.

Let $t = 3$:

$$g(x) = x^9+3x^8+3x^7+2x^6+x^5+2x^4+x+2.$$

This gives the (15, 6) triple-error-correcting BCH code over $GF(4)$.

Let $t = 4$:

$$g(x) = x^{11} + x^{10} + 2x^8 + 3x^7 + 3x^6 + x^5 + 3x^4 + x^3 + x + 3.$$

This gives the (15, 4) four-error-correcting BCH code over $GF(4)$.

Let $t = 5$:

$$g(x) = x^{12} + 2x^{11} + 3x^{10} + 2x^9 + 2x^8 + x^7 + 3x^6 + 3x^4 + 3x^3 + x^2 + 2.$$

This gives the (15, 3) five-error-correcting BCH code over $GF(4)$.

Let $t = 6$:

$$g(x) = x^{14} + x^{13} + x^{12} + x^{11} + x^{10} + x^9 + x^8 + x^7 + x^6 + x^5 + x^4 + x^3 + x^2 + x + 1.$$

This gives the (15, 1) six-error-correcting BCH code over $GF(4)$. It is a simple repetition code and actually can correct seven errors.

The formal definition of a BCH code is more general than that of the BCH codes of primitive blocklength constructed above. In general, the BCH code takes $2t$ successive powers of any ω, not necessarily a primitive element, as the zeros of $g(x)$. The blocklength of the code is the order of ω, that is, the smallest n for which $\omega^n = 1$.

Definition 6.5.1. *Let q and m be given, and let ω be any element of $GF(q^m)$ of order n. Then for any positive integer t and any integer j_0, the corresponding BCH code is the cyclic code of blocklength n with the generator polynomial*

$$g(x) = \text{LCM}[f_{j_0}(x), f_{j_0+1}(x), \ldots, f_{j_0+2t-1}(x)]$$

where $f_j(x)$ is the minimal polynomial of ω^j.

Often one chooses $j_0 = 1$, which is usually – but not always – the choice that gives a $g(x)$ of smallest degree. With this choice, the code is referred to as a *narrow-sense BCH code*. Usually, one desires a large blocklength, and thus ω is chosen as an element with the largest order, that is, as a primitive element. With this choice, the code is referred to as a *primitive BCH code*.

6.6 The Peterson–Gorenstein–Zierler decoder

The BCH codes can be decoded by any general technique for decoding cyclic codes. However, there are much better algorithms that have been developed specifically for decoding BCH codes. In this section we develop the Peterson algorithm for finding the locations of the errors, and the Gorenstein–Zierler algorithm for finding their magnitudes.

By proving that the Peterson–Gorenstein–Zierler decoder can correct t errors, we provide an alternative proof that the minimum distance of a BCH code is at least as large as the designed distance. To simplify the equations, we will take $j_0 = 1$, although an arbitrary j_0 could be carried along with no change in the ideas.

Suppose that a BCH code is constructed based on the field element ω, possibly nonprimitive. The error polynomial is

$$e(x) = e_{n-1}x^{n-1} + e_{n-2}x^{n-2} + \cdots + e_1 x + e_0$$

where at most t coefficients are nonzero. Suppose that ν errors actually occur, $0 \le \nu \le t$, and that they occur at the unknown locations $i_1, i_2, \ldots, i_\nu$. The error polynomial can be written

$$e(x) = e_{i_1}x^{i_1} + e_{i_2}x^{i_2} + \cdots + e_{i_\nu}x^{i_\nu}$$

where e_{i_ℓ} is the magnitude of the ℓth error ($e_{i_\ell} = 1$ for binary codes). We do not know the error indices $i_1, \ldots, i_\nu$, nor in general do we know the error magnitudes $e_{i_1}, \ldots, e_{i_\nu}$. All of these must be computed to correct the errors. In fact, we do not even know the value of ν. All that we know at the start is the received polynomial $v(x)$.

The *syndrome* S_1 is defined as the received polynomial $v(x)$ evaluated at ω, recalling that $c(\omega) = 0$:

$$\begin{aligned} S_1 &= v(\omega) = c(\omega) + e(\omega) = e(\omega) \\ &= e_{i_1}\omega^{i_1} + e_{i_2}\omega^{i_2} + \cdots + e_{i_\nu}\omega^{i_\nu}. \end{aligned}$$

Since this notation is clumsy, to streamline it, we define the error values $Y_\ell = e_{i_\ell}$ for $\ell = 1, \ldots, \nu$, and the error-location numbers $X_\ell = \omega^{i_\ell}$ for $\ell = 1, \ldots, \nu$ where i_ℓ indexes the location of the ℓth error, and the field element Y_ℓ is the error value at this location. Notice that the error-location number for each component of the error pattern must be distinct because ω is an element of order n.

With this notation, the syndrome S_1 is given by

$$S_1 = Y_1 X_1 + Y_2 X_2 + \cdots + Y_\nu X_\nu.$$

Similarly, we can evaluate the received polynomial at each power of ω used in the definition of $g(x)$. Define the syndromes for $j = 1, \ldots, 2t$ by

$$S_j = v(\omega^j) = c(\omega^j) + e(\omega^j) = e(\omega^j).$$

These have the form of components of the Fourier transform of the error pattern and might be called *frequency-domain syndromes* to distinguish them from the time-domain syndromes discussed in Chapter 5.

We now have the following set of $2t$ simultaneous equations in the unknown error locations $X_1, \ldots, X_\nu$, and the unknown error values $Y_1, \ldots, Y_\nu$:

$$\begin{aligned}
S_1 &= Y_1X_1 + Y_2X_2 + \cdots + Y_\nu X_\nu\\
S_2 &= Y_1X_1^2 + Y_2X_2^2 + \cdots + Y_\nu X_\nu^2\\
S_3 &= Y_1X_1^3 + Y_2X_2^3 + \cdots + Y_\nu X_\nu^3\\
&\ \ \vdots \qquad\qquad\qquad\qquad\qquad \vdots\\
S_{2t} &= Y_1X_1^{2t} + Y_2X_2^{2t} + \cdots + Y_\nu X_\nu^{2t}.
\end{aligned}$$

Our task is to find the unknowns from the known syndromes. The decoding problem has now been reduced to the problem of solving a system of nonlinear equations. The system of equations must have at least one solution because of the way in which the syndromes are defined. We shall see that the solution is unique. The method of finding the solution, given below, is valid for such a system of equations over any field.

The set of nonlinear equations is too difficult to solve directly. Instead, we judiciously define some intermediate variables that can be computed from the syndromes and from which the error locations can then be computed, a process known as *locator decoding*.

Consider the polynomial in x,

$$\Lambda(x) = \Lambda_\nu x^\nu + \Lambda_{\nu-1}x^{\nu-1} + \cdots + \Lambda_1 x + 1,$$

known as the *error-locator polynomial* or the *locator polynomial*, which is defined to be the polynomial with zeros at the inverse error locations X_ℓ^{-1} for $\ell = 1, \ldots, \nu$. That is,

$$\Lambda(x) = (1 - xX_1)(1 - xX_2)\cdots(1 - xX_\nu).$$

If we knew the coefficients of $\Lambda(x)$, we could find the zeros of $\Lambda(x)$ to obtain the error locations. Therefore let us first try to compute the locator coefficients $\Lambda_1, \ldots, \Lambda_\nu$ from the syndromes. If we can do this, the problem is nearly solved.

Multiply the two equations for $\Lambda(x)$ by $Y_\ell X_\ell^{j+\nu}$ and set $x = X_\ell^{-1}$. The second equation is clearly zero, and we have

$$0 = Y_\ell X_\ell^{j+\nu}\left(1 + \Lambda_1 X_\ell^{-1} + \Lambda_2 X_\ell^{-2} + \cdots + \Lambda_{\nu-1}X_\ell^{-(\nu-1)} + \Lambda_\nu X_\ell^{-\nu}\right),$$

or

$$Y_\ell\left(X_\ell^{j+\nu} + \Lambda_1 X_\ell^{j+\nu-1} + \cdots + \Lambda_\nu X_\ell^{j}\right) = 0.$$

Such an equation holds for each ℓ and each j. Sum these equations from $\ell = 1$ to $\ell = \nu$. For each j, this gives

$$\sum_{\ell=1}^{\nu} Y_\ell\left(X_\ell^{j+\nu} + \Lambda_1 X_\ell^{j+\nu-1} + \cdots + \Lambda_\nu X_\ell^{j}\right) = 0$$

or

$$\sum_{\ell=1}^{\nu} Y_\ell X_\ell^{j+\nu} + \Lambda_1 \sum_{\ell=1}^{\nu} Y_\ell X_\ell^{j+\nu-1} + \cdots + \Lambda_\nu \sum_{\ell=1}^{\nu} Y_\ell X_\ell^{j} = 0.$$

The individual sums are recognized as syndromes, and thus the equation becomes

$$S_{j+\nu} + \Lambda_1 S_{j+\nu-1} + \Lambda_2 S_{j+\nu-2} + \cdots + \Lambda_\nu S_j = 0.$$

Because $\nu \leq t$, the subscripts always specify known syndromes if j is in the interval $1 \leq j \leq 2t - \nu$. Hence we have the set of equations

$$\Lambda_1 S_{j+\nu-1} + \Lambda_2 S_{j+\nu-2} + \cdots + \Lambda_\nu S_j = -S_{j+\nu} \quad j = 1, \ldots, 2t - \nu.$$

This is a set of linear equations relating the syndromes to the coefficients of $\Lambda(x)$. We can write the first ν of these equations in matrix form:

$$\begin{bmatrix} S_1 & S_2 & S_3 & \ldots & S_\nu \\ S_2 & S_3 & S_4 & \ldots & S_{\nu+1} \\ S_3 & S_4 & S_5 & \ldots & S_{\nu+2} \\ \vdots & & & & \vdots \\ S_\nu & S_{\nu+1} & S_{\nu+2} & \ldots & S_{2\nu-1} \end{bmatrix} \begin{bmatrix} \Lambda_\nu \\ \Lambda_{\nu-1} \\ \Lambda_{\nu-2} \\ \vdots \\ \Lambda_1 \end{bmatrix} = \begin{bmatrix} -S_{\nu+1} \\ -S_{\nu+2} \\ -S_{\nu+3} \\ \vdots \\ -S_{2\nu} \end{bmatrix}.$$

This equation can be solved by inverting the matrix provided the matrix is nonsingular. We shall now prove that if there are ν errors, then the matrix is nonsingular. This gives a condition that can be used to determine the number of errors that actually occurred.

Theorem 6.6.1. *The matrix of syndromes*

$$\boldsymbol{M}_\mu = \begin{bmatrix} S_1 & S_2 & \ldots & S_\mu \\ S_2 & S_3 & \ldots & S_{\mu+1} \\ \vdots & & & \vdots \\ S_\mu & S_{\mu+1} & \ldots & S_{2\mu-1} \end{bmatrix}$$

is nonsingular if μ is equal to ν, the number of errors that actually occurred. The matrix is singular if μ is greater than ν.

Proof: Let $X_\mu = 0$ for $\mu > \nu$. Let $\boldsymbol{A}$ be the Vandermonde matrix

$$\boldsymbol{A} = \begin{bmatrix} 1 & 1 & \ldots & 1 \\ X_1 & X_2 & \ldots & X_\mu \\ \vdots & & & \vdots \\ X_1^{\mu-1} & X_2^{\mu-1} & \ldots & X_\mu^{\mu-1} \end{bmatrix}$$

with elements $A_{ij} = X_j^{i-1}$, and let $\boldsymbol{B}$ be the diagonal matrix

$$\boldsymbol{B} = \begin{bmatrix} Y_1X_1 & 0 & \dots & 0 \\ 0 & Y_2X_2 & \dots & 0 \\ \vdots & & & \vdots \\ 0 & 0 & \dots & Y_\mu X_\mu \end{bmatrix}$$

with elements $B_{ij} = Y_i X_i \delta_{ij}$ where $\delta_{ij} = 1$ if $i = j$, and otherwise is zero. Then the matrix product $\boldsymbol{ABA}^{\mathrm{T}}$ has elements

$$\begin{aligned}(\boldsymbol{ABA}^{\mathrm{T}})_{ij} &= \sum_{\ell=1}^{\mu} X_\ell^{i-1} \sum_{k=1}^{\mu} Y_\ell X_\ell \delta_{\ell k} X_k^{j-1} \\ &= \sum_{\ell=1}^{\mu} X_\ell^{i-1} Y_\ell X_\ell X_\ell^{j-1} \\ &= \sum_{\ell=1}^{\mu} Y_\ell X_\ell^{i+j-1},\end{aligned}$$

which is the ij element of the matrix $\boldsymbol{M}_\mu$. Therefore $\boldsymbol{M}_\mu = \boldsymbol{ABA}^{\mathrm{T}}$. Hence the determinant of $\boldsymbol{M}_\mu$ satisfies

$$\det(\boldsymbol{M}_\mu) = \det(\boldsymbol{A})\det(\boldsymbol{B})\det(\boldsymbol{A}).$$

If μ is greater than ν, then $\det(\boldsymbol{B}) = 0$. Hence $\det(\boldsymbol{M}_\mu) = 0$, and $\boldsymbol{M}_\mu$ is singular. If μ is equal to ν, then $\det(\boldsymbol{B}) \neq 0$. Further, the Vandermonde matrix $\boldsymbol{A}$ has a nonzero determinant if the columns are different and nonzero, which is true if μ is equal to ν. Hence $\det(\boldsymbol{M}_\mu) \neq 0$. This completes the proof of the theorem. □

The theorem provides the basis of the *Peterson algorithm*. First, find the correct value of ν as follows. Compute the determinants of $\boldsymbol{M}_t, \boldsymbol{M}_{t-1}, \dots$ in succession, stopping when a nonzero determinant is obtained. The actual number ν of errors that occurred is then known. Next, invert $\boldsymbol{M}_\mu$ to compute $\Lambda(x)$. Find the zeros of $\Lambda(x)$ to find the error locations.

If the code is binary, the error magnitudes are known to be one. If the code is not binary, the error magnitudes must also be computed. Return to the equations defining the syndromes:

$$\begin{aligned} S_1 &= Y_1X_1 + Y_2X_2 + \cdots + Y_\nu X_\nu \\ S_2 &= Y_1X_1^2 + Y_2X_2^2 + \cdots + Y_\nu X_\nu^2 \\ &\vdots \\ S_{2t} &= Y_1X_1^{2t} + Y_2X_2^{2t} + \cdots + Y_\nu X_\nu^{2t}. \end{aligned}$$

Because the error locations X_ℓ are now known, these are a set of $2t$ linear equations in the unknown error magnitudes Y_ℓ. The first ν equations can be solved for the error magnitudes if the determinant of the matrix of coefficients is nonzero. This is the *Gorenstein–Zierler algorithm*. But

$$\det\begin{bmatrix} X_1 & X_2 & \dots & X_\nu \\ X_1^2 & X_2^2 & \dots & X_\nu^2 \\ \vdots & \vdots & & \vdots \\ X_1^\nu & X_2^\nu & \dots & X_\nu^\nu \end{bmatrix} = (X_1X_2\cdots X_\nu)\det\begin{bmatrix} 1 & 1 & \dots & 1 \\ X_1 & X_2 & \dots & X_\nu \\ \vdots & \vdots & & \vdots \\ X_1^{\nu-1} & X_2^{\nu-1} & \dots & X_\nu^{\nu-1} \end{bmatrix}.$$

This Vandermonde matrix does have a nonzero determinant if ν errors occurred because $X_1, X_2, \dots, X_\nu$ are nonzero and distinct.

The decoding algorithm is summarized in Figure 6.7. Here we have made j_0 arbitrary, although the derivation has treated the special case with $j_0 = 1$. The derivation for arbitrary j_0 is the same.

Usually, because there are only a finite number of field elements to check, the simplest way to find the zeros of $\Lambda(x)$ is by trial and error, a process known as a *Chien search*. One simply computes, in turn, $\Lambda(\omega^j)$ for each j and checks for zero. A simple way to organize the computation of the polynomial $\Lambda(x)$ at β is by *Horner's rule*:

$$\Lambda(\beta) = (\dots(((\Lambda_\nu\beta + \Lambda_{\nu-1})\beta + \Lambda_{\nu-2})\beta + \Lambda_{\nu-3})\beta + \cdots + \Lambda_0).$$

Horner's rule needs only ν multiplications and ν additions to compute $\Lambda(\beta)$. Alternatively, to use multiple concurrent multipliers, one for each j, the computation could be written

$$\Lambda(\alpha^i) = \sum_{j=0}^{\nu}\left[\Lambda_j\alpha^{j(i-1)}\right]\alpha^j.$$

As an example of the decoding procedure, take the BCH (15, 5, 7) triple-error-correcting code with a primitive blocklength and the generator polynomial

$$g(x) = x^{10} + x^8 + x^5 + x^4 + x^2 + x + 1.$$

For the example, we will take the received polynomial to be $v(x) = x^7 + x^2$. Obviously, because of the condition that three or fewer errors took place, the codeword must be the all-zero word, and $v(x) = e(x)$. The decoder, however, does not make this observation. We will proceed through the steps of the decoding algorithm. First, compute the syndromes, using the arithmetic of $GF(16)$ based on the polynomial $x^4 + x + 1$, and $\omega = \alpha$:

$$\begin{aligned} S_1 &= \alpha^7 + \alpha^2 = \alpha^{12} \\ S_2 &= \alpha^{14} + \alpha^4 = \alpha^9 \\ S_3 &= \alpha^{21} + \alpha^6 = 0 \\ S_4 &= \alpha^{28} + \alpha^8 = \alpha^3 \\ S_5 &= \alpha^{35} + \alpha^{10} = \alpha^0 \\ S_6 &= \alpha^{42} + \alpha^{12} = 0. \end{aligned}$$

p. 86 $\oplus$ = modulo 2

Enter $v(x)$

Compute syndromes
$S_j = v(\omega^{j+j_0-1})$ $j = 1, \ldots, 2t$

$v = t$

$$M = \begin{bmatrix} S_1 \cdots S_v \\ \vdots \quad \vdots \\ S_v \cdots S_{2v-1} \end{bmatrix}$$

det M = 0   Yes → $v \leftarrow v - 1$

No

$$\begin{bmatrix} \Lambda_v \\ \Lambda_{v+1} \\ \vdots \\ \Lambda_1 \end{bmatrix} = M^{-1} \begin{bmatrix} -S_{v+1} \\ -S_{v+2} \\ \vdots \\ -S_{2v} \end{bmatrix}$$

Find error location
X_ℓ $(\ell = 1, \ldots, v)$
by finding zeros of $\Lambda(x)$

$$\begin{bmatrix} Y_1 \\ \vdots \\ Y_v \end{bmatrix} = \begin{bmatrix} X_1 \cdots X_v \\ \vdots \quad \vdots \\ X_1^v \cdots X_v^v \end{bmatrix}^{-1} \begin{bmatrix} S_1 \\ \vdots \\ S_v \end{bmatrix}$$

Halt

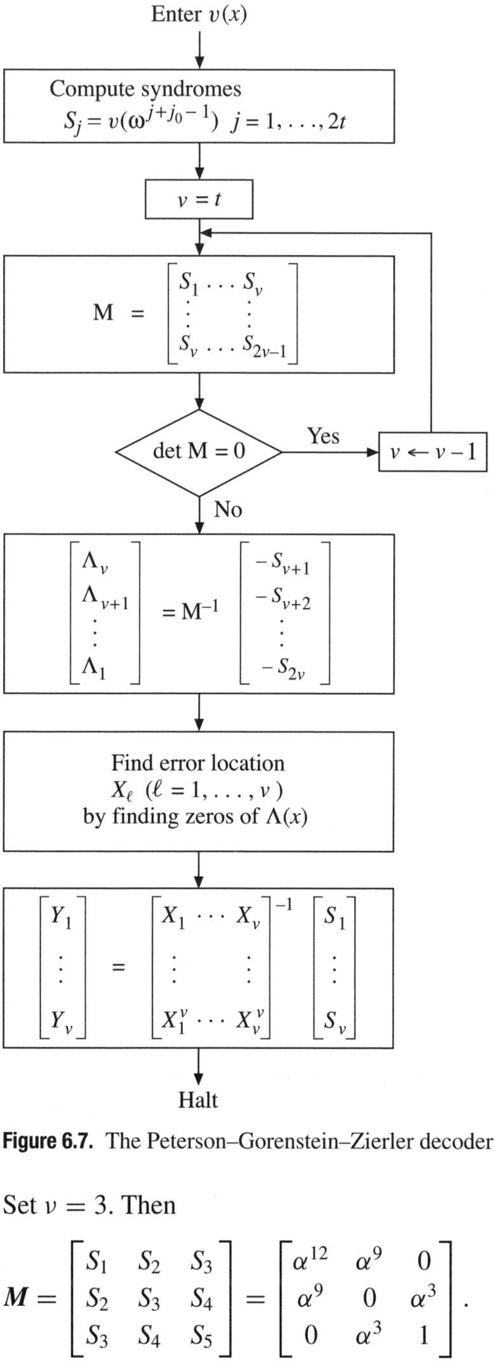

Figure 6.7. The Peterson–Gorenstein–Zierler decoder

Set $v = 3$. Then

$$M = \begin{bmatrix} S_1 & S_2 & S_3 \\ S_2 & S_3 & S_4 \\ S_3 & S_4 & S_5 \end{bmatrix} = \begin{bmatrix} \alpha^{12} & \alpha^9 & 0 \\ \alpha^9 & 0 & \alpha^3 \\ 0 & \alpha^3 & 1 \end{bmatrix}.$$

The determinant of M is zero; hence set $v = 2$. Then

$$M = \begin{bmatrix} S_1 & S_2 \\ S_2 & S_3 \end{bmatrix} = \begin{bmatrix} \alpha^{12} & \alpha^9 \\ \alpha^9 & 0 \end{bmatrix}.$$

The determinant is not zero; hence two errors occurred. Next,

$$\boldsymbol{M}^{-1} = \begin{bmatrix} 0 & \alpha^6 \\ \alpha^6 & \alpha^9 \end{bmatrix},$$

and

$$\begin{bmatrix} \Lambda_2 \\ \Lambda_1 \end{bmatrix} = \begin{bmatrix} 0 & \alpha^6 \\ \alpha^6 & \alpha^9 \end{bmatrix} \begin{bmatrix} 0 \\ \alpha^3 \end{bmatrix}$$

$$= \begin{bmatrix} \alpha^9 \\ \alpha^{12} \end{bmatrix}.$$

Hence

$$\Lambda(x) = \alpha^9 x^2 + \alpha^{12} x + 1.$$

By a Chien search, this factors as

$$\Lambda(x) = (\alpha^7 x + 1)(\alpha^2 x + 1).$$

The error-locator polynomial $\Lambda(x)$ has zeros at $x = \alpha^{-7}$ and at $x = \alpha^{-2}$, and the error locations are at the reciprocals of the zeros. Hence errors occurred in the second and seventh components. Because the code is binary, the error magnitudes are 1, and $e(x) = x^7 + x^2$.

6.7 The Reed–Muller codes as cyclic codes

The Reed–Muller codes are a class of binary codes for multiple error correction with a blocklength of the form 2^m. Two direct constructions for these codes will be given in Chapter 13. In this section, we give an indirect construction of these codes as cyclic codes to show that the Reed–Muller codes are subcodes of BCH codes. The Reed–Muller codes are easy to decode but, in general, the number of check symbols needed for a code of designed distance $d_{\min}$ is larger than for other known codes. For this reason, the Reed–Muller codes are not common in practice. In compensation, however, decoders for Reed–Muller codes can correct some error patterns that have more than t errors, or perform well with soft sensewords. We study the Reed–Muller codes here, not because of their applications, but because of their central place in the theory.

The cyclic Reed–Muller codes, of blocklength $2^m - 1$, which we describe in this section, are also called *punctured Reed–Muller codes* because the term Reed–Muller code is usually reserved for the extended code with blocklength 2^m. In Chapter 13, we shall see that the Reed–Muller codes, as well as the cyclic Reed–Muller codes, can be decoded by a simple method known as majority decoding.

Definition 6.7.1. *Let j be an integer with radix-2 representation*

$$j = j_0 + j_1 2 + j_2 2^2 + \cdots + j_{m-1} 2^{m-1}.$$

The radix-2 weight of j is

$$w_2(j) = j_0 + j_1 + \cdots + j_{m-1}$$

with addition as integers.

The radix-2 weight of j is simply the number of ones in a binary representation of j.

Definition 6.7.2. *A cyclic Reed–Muller code of order r and blocklength $n = 2^m - 1$ is the binary cyclic code whose defining set is $\mathcal{A} = \{\omega^j \mid 0 < w_2(j) < m - r\}$.*

The number of m-bit numbers with ℓ ones is $\binom{m}{\ell}$. It follows that a cyclic Reed–Muller code of blocklength $2^m - 1$ and order r has $n - k = \sum_{\ell=1}^{m-r-1} \binom{m}{\ell}$ check symbols. For example, a cyclic Reed–Muller code of blocklength $n = 2^4 - 1$ and order $r = 1$ has ten check symbols because $n - k = \binom{4}{1} + \binom{4}{2} = 10$. We shall see this cyclic code has minimum distance 5, so it is the (15, 5, 5) cyclic Reed–Muller code. When extended by a single check bit, it becomes the (16, 5, 6) Reed–Muller code.

A cyclic Reed–Muller code with $t = 1$ is a binary Hamming code because then ω^j is a zero of $g(x)$ whenever j is equal to one or a conjugate of one. The binary Hamming codes are the simplest examples of cyclic Reed–Muller codes.

Theorem 6.7.3. *The binary cyclic Reed–Muller code of order r and blocklength $n = 2^m - 1$ is a subcode of the binary BCH code of designed distance $d = 2^{m-r} - 1$.*

Proof: The radix-2 representation of $2^{m-r} - 1$ is an $(m - r)$-bit binary number consisting of all ones. Integers smaller than $2^{m-r} - 1$ are represented by binary numbers with fewer than $m - r$ ones. Therefore, for $j = 1, 2, \ldots, 2^{m-r} - 2$, $w_2(j) \le m - r - 1$, so the defining set for the Reed–Muller code contains the subset $\{\omega^j \mid j = 1, 2, \ldots, 2^{m-r} - 2\}$. This is the defining set for a BCH code so every codeword of the cyclic Reed–Muller code is also a codeword of the BCH code with this subset as the defining set. □

Corollary 6.7.4. *The minimum distance of a binary cyclic Reed–Muller code of order r and blocklength $n = 2^m - 1$ satisfies*

$$d_{\min} \ge 2^{m-r} - 1.$$

Proof: By the theorem, the code is contained in a BCH code of minimum distance $2^{m-r} - 1$. □

For example, take $m = 5$ and $r = 2$. This cyclic Reed–Muller code has blocklength 31. It has check frequencies corresponding to all nonzero j for which

$$w_2(j) \le 2,$$

that is, for which the binary representation of j has at most two ones. These are the binary numbers

0 0 0 0 1

0 0 0 1 1

0 0 1 0 1

and all cyclic shifts of these binary numbers. Thus check frequencies are at indices $j = 1$, 3, and 5, and at all elements of the conjugacy classes of 1, 3, and 5. This $(31, 16, 7)$ cyclic Reed–Muller code is identical to the $(31, 16, 7)$ binary BCH code.

As another example, take $m = 5$ and $r = 1$. This cyclic Reed–Muller code has blocklength 31 and can correct seven errors by majority decoding. Now the defining set is the set $\mathcal{A} = \{\omega^j \mid w_2(j) \leq 3\}$. These indices have binary representations

0 0 0 0 1

0 0 0 1 1

0 0 1 0 1

0 0 1 1 1

0 1 0 1 1

and all cyclic shifts of these binary numbers. Thus check frequencies correspond to indices $j = 1, 3, 5, 7$, and 11, and all elements of the conjugacy classes of these indices. This first-order cyclic Reed–Muller code is a $(31, 6, 15)$ code. Because 9 is in the conjugacy class of 5, and 13 is in the conjugacy class of 11, the $(31, 6, 15)$ cyclic Reed–Muller code is identical to the $(31, 6, 15)$ BCH code.

Thus we see that both the BCH $(31, 16, 7)$ code and the BCH $(31, 6, 15)$ code are also Reed–Muller codes and can be decoded by the majority decoding algorithms for Reed–Muller codes that are described in Chapter 13.

To see a different situation, take $m = 6$ and $r = 2$. This gives a $(63, 22, 15)$ second-order cyclic Reed–Muller code that can correct seven errors by majority decoding. The defining set of the code is $\mathcal{A} = \{\omega^j \mid w_2(j) \leq 3\}$. Thus ω^j is in the complete defining set if j has a binary representation of the form

0 0 0 0 0 1

0 0 0 0 1 1

0 0 0 1 0 1

0 0 0 1 1 1

0 0 1 0 0 1

0 0 1 0 1 1

0 0 1 1 0 1

0 1 0 1 0 1

and ω^j is in the complete defining set if the binary representation of j is a cyclic shift of one of these binary numbers. Thus the complete defining set is $\mathcal{A} = \{\omega^j \mid j = 1, 3, 5, 7, 9, 11, 13, 21\}$. This (63, 22, 15) second-order cyclic Reed–Muller code should be compared to a (63, 24, 15) BCH code, which has the defining set $\mathcal{A} = \{\omega^j \mid j = 1, 3, 5, 7, 9, 11, 13\}$. The Reed–Muller code has an inferior rate but can be decoded by majority logic. The extra zero at ω^{21} does not improve the minimum distance, but does simplify the decoding.

6.8 Extended Reed–Solomon codes

There is an orderly way to append two extra symbols to the codewords of a Reed–Solomon code. When a code is obtained by appending one or both of the extra symbols to each codeword, that code is called an *extended* Reed–Solomon code. If only one extension symbol is appended, the code is called a *singly extended* Reed–Solomon code. If both extension symbols are appended, the code is called a *doubly extended* Reed–Solomon code. The most important extended Reed–Solomon codes over a field of characteristic 2 are the singly extended primitive Reed–Solomon codes because such codes have a blocklength of the form 2^m, and this blocklength is convenient in many applications.

Each appended symbol can be viewed either as a data symbol or as a check symbol, that is, it may be used either to expand a code by increasing the rate, or to lengthen a code by increasing the minimum distance. The same extended code can be constructed either by expanding a Reed–Solomon code of minimum distance $d_{\min}$, or by lengthening a Reed–Solomon code of minimum distance $d_{\min} - 2$, so there is no need to describe both constructions.

The two new symbols must be identified in the extended codeword, and several index notations are in use. If the components of the unextended primitive codeword of blocklength $q - 1$ are indexed by the nonzero field elements, then the field element zero can be used to index one of the new components of the code. One additional symbol is needed to index the other new component. Generally, ∞ is used,[3] and the extended codeword of blocklength $q + 1$ is

$$\boldsymbol{c} = (c_0, c_{\omega^0}, c_{\omega^1}, c_{\omega^2}, \ldots, c_{\omega^{q-2}}, c_\infty).$$

If the components of the unextended code are indexed by the $q - 1$ exponents of the primitive element, then zero is not available to index a new symbol, and two new index

[3] It is a common and convenient practice to extend the field of real numbers by appending the extra symbols $+\infty$ and $-\infty$ and to define arithmetic operations for these. The extended reals do not form a field. One can do the same with a Galois field, appending the symbol ∞ and the operations $a + \infty = \infty$, $a \cdot \infty = \infty$. The set is no longer a field.

symbols are needed. We shall use $-$ and $+$ for these two new indices. Thus an extended codeword has blocklength $n = q + 1$ and is written

$$(c_-, c_0, c_1, c_2, \ldots, c_{q-3}, c_{q-2}, c_+),$$

with one new component at the beginning and one at the end of the codeword. The vector of blocklength $q - 1$ obtained by excluding c_- and c_+ is called the *interior*. When we speak of the spectrum of the codeword, we mean the spectrum of the interior.

Definition 6.8.1. *Let j_0 and t be arbitrary integers. A doubly extended Reed–Solomon code is a linear code over $GF(q)$ of blocklength $n = q + 1$ whose codeword $(c_-, c_0, c_1, \ldots, c_{q-2}, c_+)$ has a spectrum of blocklength $q - 1$, satisfying:*

1. $C_j = 0 \quad j = j_0 + 1, \ldots, j_0 + d - 3$
2. $C_{j_0} = c_-$
3. $C_{j_0+d-2} = c_+$.

The integer d is the designed distance of the extended Reed–Solomon code. The definition constrains $d - 3$ consecutive spectral components to equal zero, and the spectral components on either side of these $d - 3$ components are equal to c_- and c_+, respectively.

Theorem 6.8.2. *An extended Reed–Solomon code over $GF(q)$ is a $(q + 1, k)$ code with minimum distance $d = n - k + 1 = q - k + 2$.*

Proof: For simplicity, we may suppose that $j_0 = 0$. If j_0 were not equal to zero, the check matrix would be different only in that every element, except the elements in the first and last column, is multiplied by ω^{j_0}. However, this has no effect on subsequent arguments.

The check matrix is

$$\boldsymbol{H} = \begin{bmatrix} -1 & 1 & 1 & \ldots & 1 & 1 & 0 \\ 0 & \omega & \omega^2 & \ldots & \omega^{q-2} & \omega^{q-1} & 0 \\ 0 & \omega^2 & \omega^4 & \ldots & \omega^{2(q-2)} & \omega^{2(q-1)} & 0 \\ \vdots & & & & & \vdots & \\ 0 & \omega^{q-k-1} & \omega^{(q-k-1)2} & \ldots & \omega^{(q-k-1)(q-2)} & \omega^{(q-k-1)(q-1)} & 0 \\ 0 & \omega^{q-k} & \omega^{(q-k)2} & \ldots & \omega^{(q-k)(q-2)} & \omega^{(q-k)(q-1)} & -1 \end{bmatrix}.$$

The minimum distance of the code is at least d if every set of $d - 1$ columns of the check matrix forms a linearly independent set of vectors. If the first and last columns are deleted, then any set of $q - k + 1$ columns of the matrix forms a Vandermonde matrix with distinct columns, and so is nonsingular. Hence all the interior columns are linearly independent in sets of $q - k + 1$. But if instead we choose a set of $q - k + 1$ columns including the first column and the last column, then the determinant can be computed by expanding the matrix, in turn, about the one in the first column and then about the one in the last column. This chops off the first row and the last row, leaving

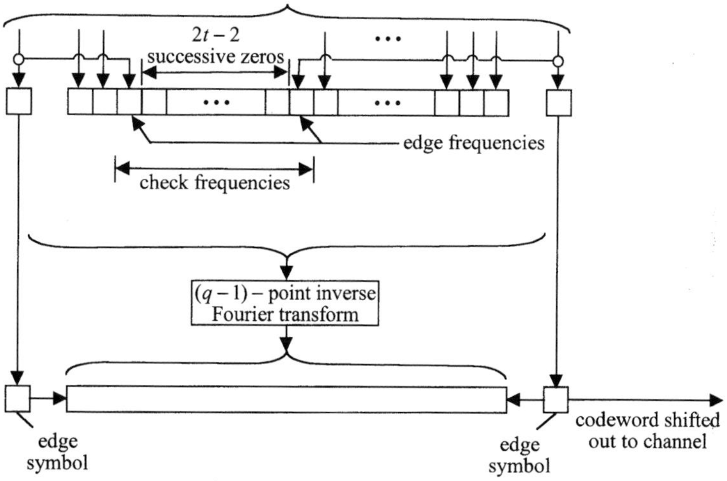

Figure 6.8. Encoding an extended Reed–Solomon code in the frequency domain

a Vandermonde matrix that again has a nonzero determinant. Hence any $q - k + 1$ columns are linearly independent, and thus the minimum distance is at least $q - k + 2$, as was to be proved. □

Figure 6.8 shows an encoder for an extended Reed–Solomon code. The encoder constrains to zero a block of $d - 3$ consecutive components of the spectrum. The remaining components of the spectrum contain arbitrary data symbols from $GF(q)$. The two values of the spectrum on the edges of the block of check frequencies are also arbitrary symbols from $GF(q)$, and these two symbols are appended to the codeword as c_+ and c_-. This gives an extended Reed–Solomon code with minimum distance $d_{\min}$ and blocklength $q + 1$.

To encode an (n, k) doubly extended Reed–Solomon code in the time domain, let $g(x)$ be the generator polynomial with zeros at $\omega^2, \ldots, \omega^{2t-1}$, and use $g(x)$ to encode k data symbols into $(c_0, c_1, \ldots, c_{n'-1})$. Then the extension symbols are defined by

$$c_- = \sum_{i=0}^{n'-1} c_i \omega^i \quad \text{and} \quad c_+ = \sum_{i=0}^{n'-1} c_i \omega^{2ti},$$

and these are attached to $(c_0, c_1, \ldots, c_{n'-1})$ to produce the codeword.

Any decoder for a Reed–Solomon code can be used to decode an extended Reed–Solomon code. Either the tails are correct and can be used to form two extra syndromes by

$$S_{j_0} = V_{j_0} - c_-$$
$$S_{j_0+d-2} = V_{j_0+d-2} - c_+,$$

or the tails contain at least one error. In the latter case, the interior contains at most $t - 1$ errors and the two extra syndromes are not needed. Only one codeword lies within a distance t of the senseword, so decoding with these two sets of syndromes cannot produce two different codewords. Other procedures for decoding an extended code will be described in Section 7.9.

6.9 Extended BCH codes

Any extended Reed–Solomon code over $GF(q^m)$ of blocklength $n = q^m + 1$ can be used to create a code of blocklength $n = q^m + 1$ over $GF(q)$ by taking the subfield-subcode of the extended Reed–Solomon code. Alternatively, the extended Reed–Solomon code of blocklength n can be used to create a code of blocklength mn by "tipping" each q^m-ary symbol into m q-ary symbols.

In this section, we shall describe a somewhat different construction of an extended code over $GF(q)$. The approach is to select those codewords from the extended Reed–Solomon code that have components only in $GF(q)$ in the $q^m - 1$ interior symbols. The two extended symbols are allowed to be arbitrary in $GF(q^m)$, but are represented in the codeword by leading and trailing "tails," each consisting of (up to) m q-ary symbols; each q^m-ary symbol is "tilted" into m q-ary symbols. These codes are called *extended BCH codes*.

A vector $\boldsymbol{C}$ over $GF(q^m)$ of length $q^m - 1$ is a valid codeword spectrum for a $GF(q)$-ary code if the conjugacy constraints

$$C_j^q = C_{((qj))}$$

are satisfied. To obtain an extended BCH code of designed distance d, choose $d - 1$ consecutive spectral components, starting at index j_0, and set

$$\begin{aligned} C_{j_0} &= c_- \\ C_j &= 0 \quad j = j_0 + 1, \ldots, j_0 + d - 3 \\ C_{j_0+d-2} &= c_+ \end{aligned}$$

where c_- and c_+ are elements of $GF(q^m)$. These symbols are arbitrary except for possible conjugacy constraints on the spectrum. The remaining spectral components are arbitrary, provided they satisfy the q-ary conjugacy constraints. The interior of the codeword is then the inverse Fourier transform of this spectrum. The interior is preceded by a tail consisting of at most m $GF(q)$-ary symbols representing c_-, and is followed by a similar tail representing c_+. It may be that conjugacy constraints force c_- or c_+ to have an order smaller than $q^m - 1$, in which case c_- or c_+ is represented by fewer than m q-ary data symbols. In such a case, that tail length will be a divisor of m.

Table 6.4. *Table of extended BCH codes, $j_0 = 0$*

d_{BCH}	$m = 4$	$m = 5$	$m = 6$	$m = 7$	$m = 8$
3	(20, 15)	(37, 31)	(70, 63)	(135, 127)	(264, 255)
5	(20, 11)	(37, 26)	(70, 57)	(135, 120)	(264, 247)
7	(18, 7)	(37, 21)	(70, 51)	(135, 113)	(264, 239)
9		(37, 16)	(70, 45)	(135, 106)	(264, 231)
11			(67, 39)	(135, 99)	(264, 223)
13		(37, 11)	(70, 36)	(135, 92)	(264, 215)
15			(70, 30)	(135, 85)	(264, 207)
17			(70, 24)	(135, 78)	(264, 199)
19					(264, 191)
21				(135, 71)	(264, 187)
23			(66, 18)	(135, 64)	(264, 179)
25			(70, 16)	(135, 57)	(264, 171)
27					(264, 163)
29				(135, 50)	(264, 155)

To illustrate the construction, we choose $j_0 = 0$ and construct some codes with spectra over $GF(64)$ by using the conjugacy classes modulo 63. All of the extended codes with odd minimum distances obtained from primitive BCH codes of blocklength n up to 255 and $j_0 = 0$ are shown in Table 6.4. Because $C_0 = C_0^2$, it must be either zero or one, and so c_- is a single binary symbol. The component C_{2t-1} (and hence c_+) is an arbitrary element of $GF(64)$ unless $2t - 1 = 9$, 18, 36, 27, 54, or 45, in which case c_+ contains three bits, or if $2t - 1 = 21$ or 42, in which case c_+ contains two bits. Further, $2t - 1$ must be the smallest integer in its conjugacy class because if there is another integer in its conjugacy class, that component of the spectrum is constrained to zero, which requires that C_{2t-1} must also be zero.

Let $t = 4$. Then $2t - 1 = 7$, and $C_0, C_1, \ldots, C_7$ are check frequencies, with $C_0 = c_-$ equal to any element of $GF(2)$ and $C_7 = c_+$ equal to any element of $GF(64)$. Further, C_0 corresponds to one check bit, and C_1, C_3, C_5, and C_7 each corresponds to six check bits. Hence $n = 63 + 1 + 6$, and we have a (70, 45) code whose minimum distance is at least 9.

Let $t = 8$. Then $2t - 1 = 15$, and $C_0, C_1, \ldots, C_{15}$ are check frequencies with $C_0 = c_-$ equal to any element of $GF(2)$, and $C_{15} = c_+$ equal to any element of $GF(64)$. Further, C_0 corresponds to one check bit, C_1, C_3, C_5, C_7, C_{11}, C_{13}, and C_{15} each corresponds to six check bits, and C_9 corresponds to three check bits. Hence we have a (70, 24) code whose minimum distance is at least 17.

There is another possible embellishment of a doubly extended binary BCH code. If the designed distance of the extended code is even, then the code can be further extended by an extra check bit as a parity check on one of the tails. This forces the minimum distance to be at least $d + 1$.

Table 6.5. *Table of extended BCH codes, $j_0 = -1$*

d_{BCH}	$m = 4$	$m = 5$	$m = 6$	$m = 7$	$m = 8$
3					
5	(24, 14)	(42, 30)	(76, 62)	(142, 126)	(272, 254)
7	(24, 10)	(42, 25)	(76, 56)	(142, 119)	(272, 246)
9		(42, 20)	(76, 50)	(142, 112)	(272, 238)
11		(42, 15)	(76, 44)	(142, 105)	(272, 230)
13			(73, 38)	(142, 98)	(272, 222)
15		(42, 10)	(76, 35)	(142, 91)	(272, 214)
17			(76, 29)	(142, 84)	(272, 206)
19			(76, 23)	(142, 77)	(272, 198)
21					(268, 190)
23				(142, 70)	(272, 186)
25			(72, 17)	(142, 63)	(272, 178)
27			(76, 15)	(142, 56)	(272, 170)
29				(142, 49)	(272, 162)

Proposition 6.9.1. *An extra check bit on one tail increases by one the minimum distance of a doubly extended binary BCH code whose defining set includes zero.*

Proof: The doubly extended codeword with the defining set $\{\omega^{j_0+1}, \ldots, \omega^{j_0+d-3}\}$ has the form

$$\boldsymbol{c} = (\overline{c}_-, c_0, c_1, \ldots, c_{n-1}, \overline{c}_+, c_{++})$$

where $c_0, \ldots, c_{n-1}$ are binary symbols, and each tail, $\overline{c}_-$ and $\overline{c}_+$, denotes an element of $GF(2^m)$ (or possibly of a subfield of $GF(2^m)$) that has been expressed in a binary representation, and the extra check bit c_{++} is the modulo-2 sum of the bits in $\overline{c}_+$.

The spectrum of the doubly extended code has $d - 3$ consecutive zeros. A codeword spectrum must satisfy the conjugacy constraints $C_j^2 = C_{2j}$. Because the consecutive spectral zeros include C_0, we conclude that $d - 3$ is odd and the interior contributes $d - 2$ to the codeword weight. Moreover, either $\overline{c}_+$ is zero and there are two more zeros at the end of the spectrum, or $(\overline{c}_+, c_{++})$ has weight at least two. In either case this tail contributes weight at least two to the codeword weight. Likewise either $\overline{c}_-$ is zero and there are two more zeros at the beginning of the spectrum, or $\overline{c}_-$ has weight at least one. Thus, this tail contributes weight at least one to the codeword. The total weight must be at least $d + 1$. □

To illustrate the construction, take $j_0 = -1$. With this choice of j_0, and if c_+ is nonzero, the minimum distance of the code is even because the upper check frequency must have an odd index. To make the minimum distance odd, we add one extra bit as a check on the bits of one tail. The extra bit allows the decoder to detect a single error in this tail.

All extended codes obtained from primitive BCH codes of blocklength n up to 255 and $j_0 = -1$ are shown in Table 6.5. We can also take other values for j_0, but with one exception – we will find no interesting new codes in the range $n \leq 500$, $d \leq 29$. The only exception is in $GF(64)$. With $j_0 = -9$, one can obtain a (70, 20, 21) code as an extended BCH code, which is better than any other known code of this n and k.

Problems

6.1 Prove the following standard properties of the Fourier transform, starting with the Fourier transform pair $\{c_i\} \leftrightarrow \{C_j\}$:

a. Linearity:

$$\{ac_i + bc_i'\} \leftrightarrow \{aC_j + bC_j'\}.$$

b. Cyclic shift:

$$\{c_{((i-1))}\} \leftrightarrow \{\omega^j C_j\}.$$

c. Modulation:

$$\{\omega^i c_i\} \leftrightarrow \left\{C_{((j+1))}\right\}.$$

6.2 Show that the Fourier transform of the vector with components $c_i = \omega^{ri}$ has a single nonzero spectral component. Which component is it if $r = 0$? Show that a vector that is nonzero only in a single component has a nonzero spectrum everywhere.

6.3 How many idempotents are there in the ring of polynomials modulo $x^{15} - 1$? List them.

6.4 The fundamental theorem of algebra says that a polynomial of degree d can have at most d zeros. Use the fundamental theorem of algebra to prove the BCH bound.

6.5 Find $g(x)$ for a binary double-error-correcting code of blocklength $n = 31$. Use a primitive α and the primitive polynomial $p(x) = x^5 + x^2 + 1$.

6.6 Find the codeword of the single-error-correcting Reed–Solomon code of blocklength 15 based on the primitive element α of $GF(2^4)$ that corresponds to the data polynomial $\alpha^5 x + \alpha^3$.

6.7 Show that the only Reed–Solomon codes over $GF(2)$ are the $(1, 1, 1)$ and the $(1, 0, \infty)$ binary codes. Show that the only extended Reed–Solomon codes over $GF(2)$ are the $(2, 2, 1)$, the $(2, 1, 2)$, and the $(2, 0, \infty)$ binary codes.

6.8 a. Find the generator polynomial for a (15, 7, 7) BCH code over $GF(4)$.

b. Find the generator polynomial for a (63, 55, 5) code over $GF(8)$.

6.9 Show that every BCH code is a subfield-subcode of a Reed–Solomon code of the same designed distance. Under what conditions is the rate of the subfield-subcode the same as the rate of the Reed–Solomon code? For a nonsystematically encoded $(7, 5)$ Reed–Solomon code, describe the sixteen datawords that will produce codewords that are also codewords in the Hamming $(7, 4)$ code.

6.10 How many elements are in the conjugacy class of 2(mod 11)? Does this prove that eleven divides $2^{10} - 1$? Why?

6.11 Find the generator polynomial for a $(23, 12)$ double-error-correcting BCH code over $GF(2)$. Notice that the resulting code is the Golay code. This gives an example of a BCH code whose minimum distance is larger than the designed distance.

6.12 Show that an $(11, 6)$ cyclic code over $GF(3)$ must use either $x^5 - x^3 + x^2 - x - 1$, or its reciprocal, as the generator polynomial.

6.13 A ternary channel transmits one of three symbols at each symbol time: a sinusoidal pulse at 0° phase angle, a sinusoidal pulse at 120° phase angle, or a sinusoidal pulse at 240° phase angle. Represent the channel symbols with the set $\{0, 1, 2\}$. Errors are made randomly and with equal probability. Design a triple-error-correcting code of blocklength 80 for this channel. What is the rate of the code? How might this code be used to transmit blocks of binary data? (A primitive polynomial of degree 4 over $GF(3)$ is $p(x) = x^4 + x + 2$.)

6.14 Prove that if $\mathcal{C}_1$ and $\mathcal{C}_2$ are cyclic codes of blocklength n with principal idempotents $w_1(x)$ and $w_2(x)$, respectively, then $\mathcal{C}_1 \cap \mathcal{C}_2$ has principal idempotent $w_1(x)w_2(x)(\text{mod } x^n - 1)$.

6.15 Prove the following statement: A binary BCH code of blocklength n is invariant under the permutation $c(x) \to c(x^2)(\text{mod } x^n - 1)$. How does the spectrum of $\boldsymbol{c}$ change if the components of $\boldsymbol{c}$ are permuted in this way?

6.16 Let m be odd and $n = 2^m - 1$. Using the Hartmann–Tzeng bound, show that the binary cyclic code of blocklength n with spectral zeros at $j = \pm 1$ has a minimum distance equal to at least 5. Such a code is called a *Melas double-error-correcting code*. How many codewords are there?

6.17 Let m be even and $n = 2^m + 1$. A Fourier transform over $GF(2)$ of blocklength n lies in $GF(2^{2m})$. Using the Hartmann–Tzeng bound, show that the binary cyclic code of blocklength n with a spectral zero at $j = 1$ has a minimum distance equal to at least 5. Such a code is called a *Zetterberg double-error-correcting code*. How many codewords are there?

6.18 Let m be odd (and at least equal to three) and $n = 2^m - 1$. Let $r = 2^{(m-3)/2} + 1$ and $s = 2^{(m-1)/2} + 1$. Prove that the binary cyclic code of blocklength n with spectral zeros at $j = 1, r$, and s has a minimum distance equal to at least 7.

6.19 Show that for every $(q + 1, k, d)$ doubly extended Reed–Solomon code over $GF(q)$, there is a BCH code over $GF(q)$ with the same parameters, but with defining set in the field $GF(q^2)$. (The two codes are actually equivalent.)

6.20 Show that the extended Reed–Solomon codes are maximum-distance codes.

6.21 The *q-ary norm* of an element β of $GF(q^m)$ is the product

$$\text{norm}(\beta) = \prod_{i=0}^{m-1} \beta^{q^i}.$$

Prove that if $\lambda \in GF(q)$

(i) $\text{norm}(\beta) \in GF(q)$;
(ii) $\text{norm}(\beta\gamma) = \text{norm}(\beta)\text{norm}(\gamma)$;
(iii) $\text{norm}(\lambda\gamma) = \lambda^m \text{norm}(\gamma)$.

6.22 Suppose that n' and n'' are coprime. A cyclic code of blocklength $n = n'n''$ can be formed as the product of two cyclic codes of blocklength n' and n''. Express the idempotent polynomial of the product code in terms of the idempotent polynomials of the two component codes.

6.23 Suppose that n' and n'' are coprime. Given a BCH code of designed distance d and blocklength n'' that has a minimum distance exactly d, prove that a BCH code of blocklength $n'n''$ and designed distance d has a minimum distance exactly d.

Notes

What we would now call a spectral description of an error-control code can be found in some early papers, although the relationship to the Fourier transform was not stated explicitly at that time. The original paper of Reed and Solomon (1960) introduced a spectral decoder for use in the proof of their code's minimum distance, but as that decoder was not practical, for many years the study of spectral decoding techniques was not continued. Mattson and Solomon (1961) described the spectral polynomial, which played a useful role in theoretical developments, but again, the essence of the spectral polynomial as a Fourier transform was not widely recognized, and the close relationship with the subject of signal processing was hidden. Mandelbaum (1979) developed similar ideas, although he used the language of interpolation.

Fourier transforms over a finite field were discussed by Pollard (1971). Their use in error-control codes was introduced by Gore (1973) and discussed further by Chien and Choy (1975) and Lempel and Winograd (1977). The relationship between the weight of a sequence and the linear complexity of its Fourier transform was developed by Blahut (1979). Schaub (1988) showed how the bounds on the minimum distance of cyclic codes and, hence, the theory of cyclic codes can be developed in a simple way by using the ideas of linear complexity. Chien (1972) had already made an observation in this direction.

The BCH codes were discovered independently by Hocquenghem (1959) and by Bose and Ray-Chaudhuri (1960). It was noticed quickly by Gorenstein and Zierler (1961) that the codes discovered earlier by Reed and Solomon (1960) are a special case of nonbinary BCH codes. The decoding algorithm, studied in this chapter, was first developed by Peterson (1960) for binary codes, and then extended by Gorenstein and Zierler (1961) for nonbinary codes. Reed–Solomon codes are attractive in applications and, when interleaved, are suitable for correcting burst errors as well as random errors.

Kasami and Tokura (1969) found an infinite subfamily of primitive BCH codes for which the BCH bound is strictly smaller than the true minimum distance; the first example is the (127, 43) BCH code with $d = 29$ and $d_{\min} = 31$. Chen (1970) found examples of cyclic codes that are better than BCH codes; an example is a (63, 28, 15) cyclic code.

The idea of an extended code came from many independent workers. Extended Reed–Solomon codes and their decoding were discussed by Wolf (1969). Constructions for extending BCH codes were discussed by Andryanov and Saskovets (1966) and by Sloane, Reddy, and Chen (1972). Kasahara *et al.* (1975) discussed the decoding of extended codes. The construction in the transform domain was discussed by Blahut (1980). The doubly extended Reed–Solomon code is equivalent to a BCH code.

7 Algorithms Based on the Fourier Transform

An error-control code must be judged not only by its rate and minimum distance, but also by whether a decoder can be built for it economically. Usually, there are many ways to decode a given code. A designer can choose among several decoding algorithms and their variations, and must be familiar with all of them so that the best one for a particular application can be chosen. The choice will depend not only on the code parameters, such as blocklength and minimum distance, but also on how the implementation is to be divided between hardware and software, on the required decoding speed, and even on the economics of available circuit components.

In this chapter we shall broaden our collection of decoding techniques by working in the frequency domain. Included here are techniques for decoding in the presence of both erasures and errors, and techniques for decoding beyond the designed distance of a code.

7.1 Spectral estimation in a finite field

In Section 6.6 we studied the task of decoding BCH codes, including Reed–Solomon codes, up to the designed distance. This task will now be reinterpreted from the point of view of the Fourier transform. Using the terminology of the frequency domain, we shall develop alternative decoding procedures. These procedures will take the form of spectral estimation in a finite field.

A senseword $\boldsymbol{v} = \boldsymbol{c} + \boldsymbol{e}$, with components $v_i = c_i + e_i$ for $i = 0, \ldots, n-1$, is the sum of a codeword $\boldsymbol{c}$ and an error word $\boldsymbol{e}$. The decoder must process the senseword so that the error word $\boldsymbol{e}$ is removed; the data is then recovered from the codeword. By construction of a BCH code of designed distance $2t+1$, there are $2t$ consecutive spectral components of $\boldsymbol{c}$ equal to zero:

$$C_j = 0 \quad j = j_0, \ldots, j_0 + 2t - 1$$

where $\{\omega^{j_0}, \omega^{j_0+1}, \ldots, \omega^{j_0+2t-1}\}$ is the defining set of the BCH code.

The Fourier transform of the senseword $\boldsymbol{v}$ has components $V_j = C_j + E_j$ for $j = 0, \ldots, n-1$. The (frequency-domain) syndromes are defined as those $2t$ components of this spectrum, from j_0 to $j_0 + 2t - 1$, corresponding to those spectral components where C_j is zero. It is convenient to index the syndromes starting with index one. Define

$$S_j = V_{j+j_0-1} = E_{j+j_0-1} \quad j = 1, \ldots, 2t.$$

The block of $2t$ syndromes gives us a window through which we can look at $2t$ of the n components of the error spectrum $\boldsymbol{E}$. But the BCH bound suggests that if the error pattern has a weight of at most t, then these $2t$ components of the error spectrum are enough to uniquely determine the error pattern.

Recall the linear complexity property of the Fourier transform, which says that if the error pattern $\boldsymbol{e}$ has weight ν, then the error spectrum satisfies the recursion

$$E_k = -\sum_{j=1}^{\nu} \Lambda_j E_{((k-j))} \quad k = 0, \ldots, n-1$$

for appropriate connection weights $\Lambda_1, \ldots, \Lambda_\nu$. To redevelop this equation here, suppose that there are ν errors in the senseword $\boldsymbol{v} = (v_0, \ldots, v_{n-1})$ at locations ω^{i_ℓ} for $\ell = 1, \ldots, \nu$. The error-locator polynomial

$$\Lambda(x) = \prod_{\ell=1}^{\nu} (1 - x\omega^{i_\ell})$$

defines a vector $\boldsymbol{\Lambda} = (\Lambda_0, \Lambda_1, \ldots, \Lambda_\nu, 0, \ldots, 0)$ with inverse Fourier transform $\boldsymbol{\lambda} = (\lambda_0, \lambda_1, \ldots, \lambda_{n-1})$ given by

$$\lambda_i = \frac{1}{n} \sum_{j=0}^{n-1} \Lambda_j \omega^{-ij} = \frac{1}{n} \Lambda(\omega^{-i}).$$

Clearly, $\Lambda(\omega^{-i})$ equals zero if and only if i is an error location. Thus $\lambda_i = 0$ whenever $e_i \neq 0$. Therefore $\lambda_i e_i = 0$ for all i. By the convolution theorem, the Fourier transform of this equation becomes a cyclic convolution in the frequency domain set equal to zero:

$$\sum_{j=0}^{\nu} \Lambda_j E_{((k-j))} = 0 \quad k = 0, \ldots, n-1$$

where the upper limit is ν because $\Lambda(x)$ is a polynomial of degree ν. Because Λ_0 equals one, this reduces to the recursion mentioned above. Furthermore, if $\nu \leq t$, then $\Lambda_j = 0$ for $j = \nu + 1, \ldots, t$, so the recursion then can be rewritten in the form:

$$E_k = -\sum_{j=1}^{t} \Lambda_j E_{((k-j))} \quad k = 0, \ldots, n-1$$

which may be preferred because the upper summation index t is now fixed. This system of n equations involves $2t$ known components of $\boldsymbol{E}$, given by the $2t$ syndromes, and $n - t$ unknowns, of which t are unknown coefficients of $\Lambda(x)$ and $n - 2t$ are unknown components of $\boldsymbol{E}$. Of the n equations, there are t equations that involve only the known

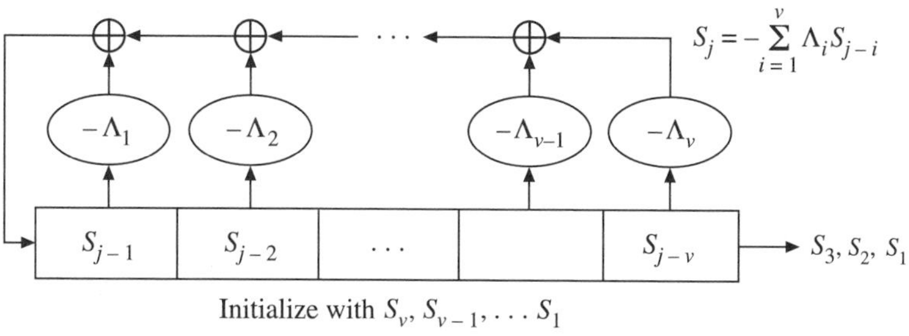

Figure 7.1. Error-locator polynomial as a shift-register circuit

components of $\boldsymbol{E}$, as expressed by the syndromes $S_j = E_{j+j_0-1}$ for $j = 1, \ldots, 2t$, and the t unknown components of $\boldsymbol{\Lambda}$, which are the coefficients of the unknown polynomial $\Lambda(x)$. That is, the t equations

$$S_k = -\sum_{j=1}^{t} \Lambda_j S_{((k-j))} \quad k = t+1, \ldots, 2t$$

involve only the syndromes, which are known, and the t unknown components of $\boldsymbol{\Lambda}$. This linear recursion, written in the form

$$S_k = -\sum_{j=1}^{\nu} \Lambda_j S_{k-j} \quad k = \nu+1, \ldots, 2t,$$

must be solved for $\boldsymbol{\Lambda}$ using the smallest possible value of ν. The recursion, denoted $(\boldsymbol{\Lambda}, \nu)$ or $(\Lambda(x), \nu)$, can be depicted as a *linear-feedback shift register*, shown in Figure 7.1. This system of equations is always solvable for $\boldsymbol{\Lambda}$ and ν. One way to solve it is by the Peterson algorithm, which was given in Section 6.6.

After $\boldsymbol{\Lambda}$ is computed, the remaining $n - 2t$ components of $\boldsymbol{S}$ can be obtained from $\Lambda(x)$ and the known components of $\boldsymbol{S}$ by the Gorenstein–Zierler algorithm. Alternatively, a procedure known as *recursive extension* can be chosen, using the linear recursion first to find S_{2t+1} from the known components of $\boldsymbol{S}$ and $\boldsymbol{\Lambda}$, then to find S_{2t+2}, and so on. Recursive extension can be described as the operation of the linear-feedback shift register with connection weights given by the components of $\boldsymbol{\Lambda}$ and initialized with $S_{t+1}, \ldots, S_{2t}$. In this way, S_j is computed for all j. Then E_j equals S_{j-j_0+1} and

$$C_j = V_j - E_j,$$

from which the codeword is easily recovered.

The decoder can be used with an encoder in the time domain, as shown in Figure 7.2, or an encoder in the frequency domain, as shown in Figure 7.3. If a time-domain encoder is used, then the inverse Fourier transform of the corrected codeword spectrum $\boldsymbol{C}$ must be computed to obtain the time-domain codeword $\boldsymbol{c}$, from which the data is recovered. If, instead, the encoder uses the data symbols in the frequency domain to specify the values of the spectrum, then the corrected spectrum gives the data symbols directly. That decoder does not compute an inverse Fourier transform.

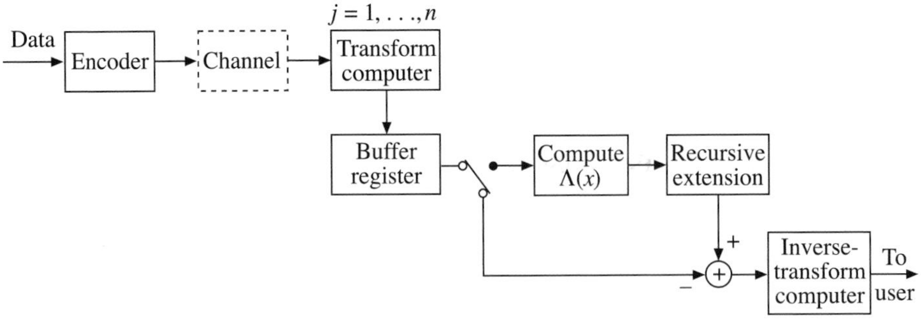

Figure 7.2. An encoder/decoder for BCH codes

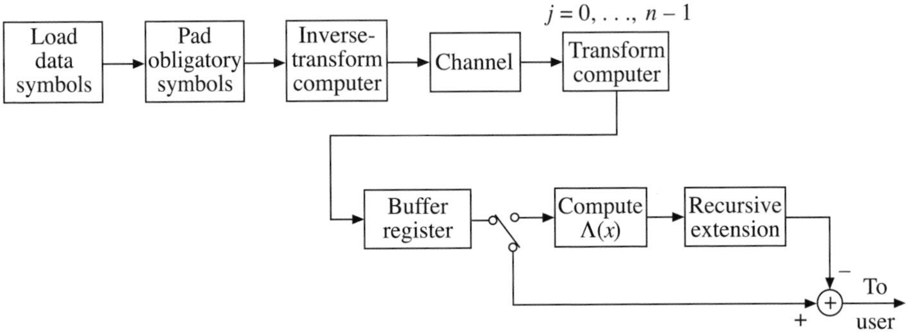

Figure 7.3. Frequency-domain encoder/decoder for BCH codes

Our main task in this chapter is to find efficient algorithms to compute a minimum-length linear recursion $(\Lambda(x), \nu)$ that produces the syndromes cyclically. This task can be reposed in the language of polynomials, the formulation of which we will find useful later. Define the (frequency-domain) *syndrome polynomial* as

$$S(x) = \sum_{j=1}^{2t} S_j x^j.$$

Let $\Lambda(x)$ be any polynomial of degree $\nu \leq t$. Define the *discrepancy polynomial* as $\Delta(x) = \Lambda(x)S(x)$. The discrepancy polynomial can always be decomposed as

$$\Delta(x) = \Lambda(x)S(x) = \Gamma(x) + x^{\nu+1}O(x) + x^{2t+1}\Theta(x)$$

where $\deg \Gamma(x) \leq \nu$ and $\deg x^{\nu+1}O(x) \leq 2t$. In this polynomial formulation, the coefficients of $x^{\nu+1}O(x)$ have the form $\sum_{j=0}^{\nu} \Lambda_j S_{k-j}$ for $k = \nu + 1, \ldots, 2t$, so $O(x) = 0$ if and only if the linear recursion $(\Lambda(x), \nu)$ produces the sequence $S_1, \ldots, S_{2t}$. Thus the decomposition

$$\Lambda(x)S(x) = \Gamma(x) + x^{2t+1}\Theta(x) \quad \deg \Gamma(x) \leq \nu$$

is just a way of saying that the discrepancy polynomial $\Delta(x) = \Lambda(x)S(x)$ has zero coefficients for $k = \nu + 1, \ldots, 2t$ if and only if the linear recursion $(\Lambda(x), \nu)$ produces the sequence $S_1, \ldots, S_{2t}$. Any algorithm for solving this polynomial equation is an algorithm for finding the locator polynomial $\Lambda(x)$.

7.2 Synthesis of linear recursions

To decode a BCH code or a Reed–Solomon code, as discussed in Section 7.1, one solves the following problem. Find the connection weights $(\Lambda_1, \ldots, \Lambda_\nu)$ for the smallest $\nu \leq t$ for which the system of equations

$$S_k = -\sum_{j=1}^{\nu} \Lambda_j S_{k-j} \quad k = \nu + 1, \ldots, 2t$$

has a solution. The Peterson algorithm solves for the linear recursion $\boldsymbol{\Lambda}$ by inverting a ν by ν matrix equation for various values of ν. For small ν, matrix inversion is not unreasonable; the number of multiplications necessary to invert a ν by ν matrix is proportional to ν^3. However, when ν is large, one should use a more efficient method of solution. The structure of the problem is used to find a method that is conceptually more intricate than direct matrix inversion, but computationally much simpler.

The *Berlekamp–Massey algorithm* solves the following modified problem. Find the connection weights $(\Lambda_1, \ldots, \Lambda_\nu)$ for the smallest $\nu \leq 2t$ for which the system of equations

$$S_k = -\sum_{j=1}^{\nu} \Lambda_j S_{k-j} \quad k = \nu + 1, \ldots, 2t$$

has a solution. In the modified problem, values of the integer ν as large as $2t$ are allowed, so for any arbitrary sequence $S_1, \ldots, S_{2t}$, the problem must have a solution. In particular, the value $\nu = 2t$, together with arbitrary values for $\Lambda_1, \ldots, \Lambda_{2t}$, will always satisfy the system of equations.

The modified problem is not the same as the original problem, which for an arbitrary sequence $S_1, \ldots, S_{2t}$ need not have a solution satisfying the constraint $\nu \leq t$. It is sufficient to solve the modified problem, however, because if the original problem has a solution, it is the same as the solution to the modified problem.

The modified problem can be viewed as the task of designing the shortest linear recursion $(\Lambda(x), \nu)$ that will produce the known sequence of syndromes. If, for this connection polynomial $\Lambda(x)$, either $\deg \Lambda(x) > t$ or $\Lambda_\nu = 0$, then $\Lambda(x)$ is not a legitimate error-locator polynomial. In this case, the syndromes cannot be explained by a correctable error pattern – a pattern of more than t errors has been detected.

The procedure that we will develop applies in any field and does not presume any special properties for the sequence $S_1, S_2, \ldots, S_{2t}$. The procedure is iterative. For each r, starting with $r = 1$, we will design a minimum-length linear recursion $(\Lambda^{(r)}(x), L_r)$ to produce $S_1, \ldots, S_r$. It need not be unique. At the start of iteration r, we will have already constructed a list of linear recursions

$$\begin{aligned}&\big(\Lambda^{(1)}(x), L_1\big),\\&\big(\Lambda^{(2)}(x), L_2\big),\\&\qquad\vdots\\&\big(\Lambda^{(r-1)}(x), L_{r-1}\big).\end{aligned}$$

The mth entry on the list produces the truncated sequence $S_1, \ldots, S_m$, and the last entry on the list produces the truncated sequence $S_1, \ldots, S_{r-1}$.

The main trick of the Berlekamp–Massey algorithm is to use these earlier iterates to compute a new minimum-length linear recursion $(\Lambda^{(r)}(x), L_r)$ that will produce the sequence $S_1, \ldots, S_{r-1}, S_r$. This is done by using as a trial candidate the linear recursion that produces the shorter sequence $S_1, \ldots, S_{r-2}, S_{r-1}$, and then possibly increasing its length and adjusting the connection weights.

At iteration r, compute the next output of the $(r-1)$th linear recursion:

$$\hat{S}_r = -\sum_{j=1}^{L_{r-1}} \Lambda_j^{(r-1)} S_{r-j}.$$

Subtract $\hat{S}_r$ from the desired output S_r to get a quantity Δ_r, known as the rth *discrepancy*:

$$\Delta_r = S_r - \hat{S}_r = S_r + \sum_{j=1}^{L_{r-1}} \Lambda_j^{(r-1)} S_{r-j}.$$

Equivalently,

$$\Delta_r = \sum_{j=0}^{L_{r-1}} \Lambda_j^{(r-1)} S_{r-j},$$

which is the rth coefficient of the polynomial product $\Lambda^{(r-1)}(x)S(x)$. If Δ_r is zero, then set $(\Lambda^{(r)}(x), L_r) = (\Lambda^{(r-1)}(x), L_{r-1})$, and the rth iteration is complete. Otherwise, the connection polynomial is modified as follows:

$$\Lambda^{(r)}(x) = \Lambda^{(r-1)}(x) + Ax^{\ell}\Lambda^{(m-1)}(x)$$

where A is an appropriate field element, ℓ is an integer, and $\Lambda^{(m-1)}(x)$ is one of the earlier connection polynomials. With the new connection polynomial, the discrepancy is

$$\begin{aligned}\Delta_r' &= \sum_{j=0}^{L_r} \Lambda_j^{(r)} S_{r-j}\\&= \sum_{j=0}^{L_r} \Lambda_j^{(r-1)} S_{r-j} + A\sum_{j=0}^{L_r} \Lambda_j^{(m-1)} S_{r-j-\ell}\end{aligned}$$

where L_r has not been specified, but it is at least as large as deg $\Lambda^{(r-1)}(x)$ and at least as large as deg $x^{\ell}\Lambda^{(m-1)}(x)$.

We are now ready to specify m, ℓ, and A. Choose an m smaller than r for which $\Delta_m \neq 0$, choose $\ell = r - m$, and choose $A = -\Delta_m^{-1}\Delta_r$. Then

$$\Delta_r' = \Delta_r - \frac{\Delta_r}{\Delta_m}\Delta_m = 0,$$

and also $\Delta_k = 0$ for $k < r$. Thus the new recursion $(\Lambda^{(r)}(x), L_r)$ produces the sequence $S_1, \ldots, S_{r-1}, S_r$.

We do not want just any such recursion, however; we want one of the shortest length. We still have not specified which of the m satisfying $\Delta_m \neq 0$ should be chosen. We now specify that m is the most recent iteration at which $L_m > L_{m-1}$, which ensures that we will get a shortest-length linear recursion at every iteration, but the proof of this statement will take some time to develop.

The computation is illustrated by Figure 7.4, with the recursions depicted as shift registers. The shift registers at the start of iterations m and r are shown in Figure 7.4(a). Iteration m is chosen such that, at iteration m, the shift register $(\Lambda^{(m-1)}(x), L_{m-1})$ failed to produce syndrome S_m, and $L_m > L_{m-1}$. Figure 7.4(b) shows the shift register $(\Lambda^{(m-1)}(x), L_{m-1})$ made into an auxiliary shift register by lengthening it, positioning it, and scaling its output so that it can compensate for the failure of $(\Lambda^{(r-1)}(x), L_{r-1})$ to produce S_r. The auxiliary shift register has an extra tap with a weight of one due to the term $\Lambda_0^{(m-1)}$. At the rth symbol time, a nonzero valve Δ_m comes from the auxiliary shift register. The coefficient A is selected to modify this nonzero value so that it can be added to the rth feedback of the main shift register, thereby producing the required syndrome S_r. Figure 7.4(c) shows the two shift registers merged into one shift register, which does not change the behavior. This gives the new shift register $(\Lambda^{(r)}(x), L_r)$. Sometimes $L_r = L_{r-1}$, and sometimes $L_r > L_{r-1}$. When the latter happens, we replace m with r for use in future iterations.

The following theorem gives the precise procedure for producing the shortest-length linear recursion with the desired property. The theorem uses the scratch polynomial $B(x)$ in place of $\Delta_m^{-1}\Lambda^{(m)}(x)$. The proof of the theorem occupies the remainder of the section.

Algorithm 7.2.1 (Berlekamp–Massey Algorithm). *In any field, let $S_1, \ldots, S_{2t}$ be given. With initial conditions $\Lambda^{(0)}(x) = 1$, $B^{(0)}(x) = 1$, and $L_0 = 0$, let the following set of equations for $r = 1, \ldots, 2t$ be used iteratively to compute $\Lambda^{(2t)}(x)$:*

$$\Delta_r = \sum_{j=0}^{n-1} \Lambda_j^{(r-1)} S_{r-j}$$

$$L_r = \delta_r(r - L_{r-1}) + \bar{\delta}_r L_{r-1}$$

$$\begin{bmatrix} \Lambda^{(r)}(x) \\ B^{(r)}(x) \end{bmatrix} = \begin{bmatrix} 1 & -\Delta_r x \\ \Delta_r^{-1}\delta_r & \bar{\delta}_r x \end{bmatrix} \begin{bmatrix} \Lambda^{(r-1)}(x) \\ B^{(r-1)}(x) \end{bmatrix}$$

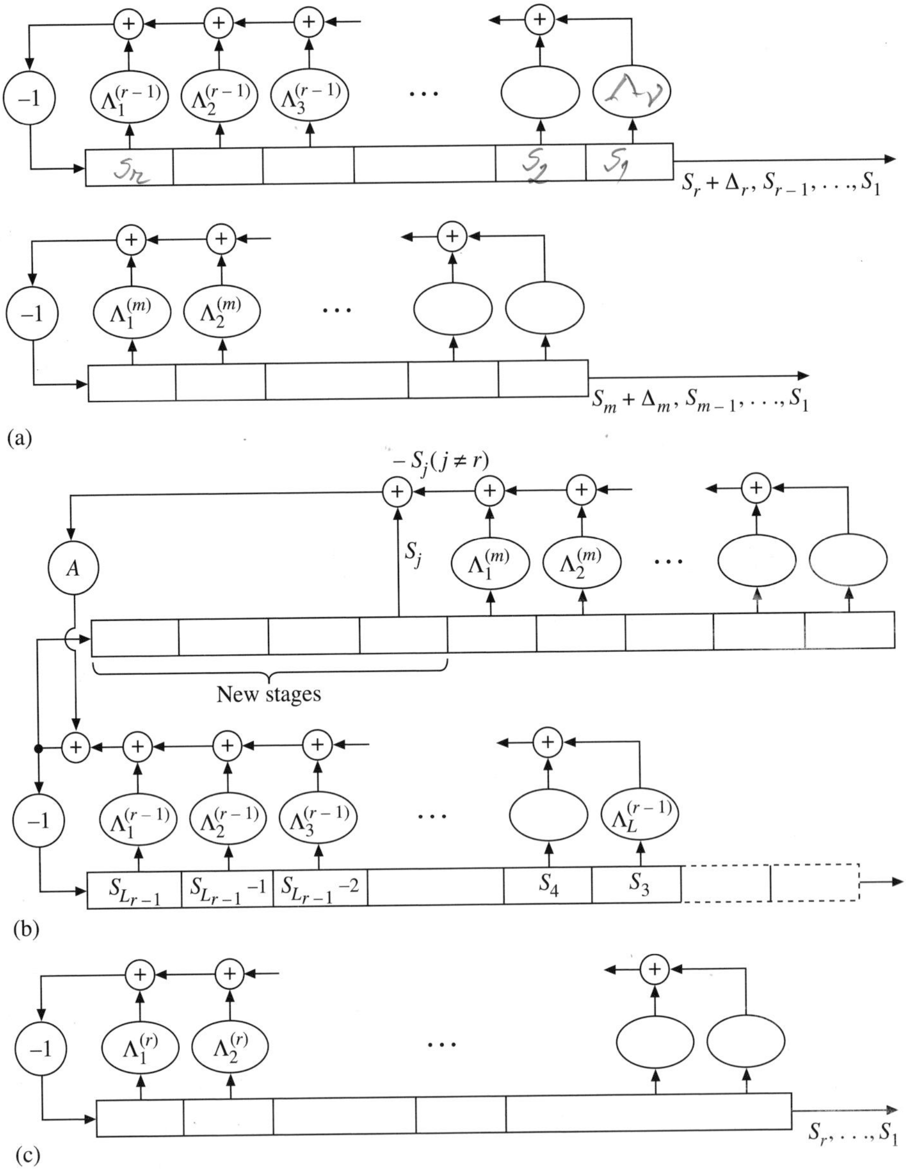

Figure 7.4. Berlekamp–Massey construction

where $\delta_r = (1 - \overline{\delta}_r) = 1$ *if both* $\Delta_r \neq 0$ *and* $2L_{r-1} \leq r - 1$*, and otherwise* $\delta_r = 0$. *Then* $(\Lambda^{(2t)}(x), L_{2t})$ *is a linear recursion of shortest length that produces* $S_1, \ldots, S_{2t}$.

Figure 7.5 restates the Berlekamp–Massey algorithm graphically in the form of a flow chart. In each of $2t$ iterations, the discrepancy Δ_r is computed. If the discrepancy is nonzero, $\Lambda(x)$ is changed by adding a multiple of $xB(x)$, which is a translated multiple

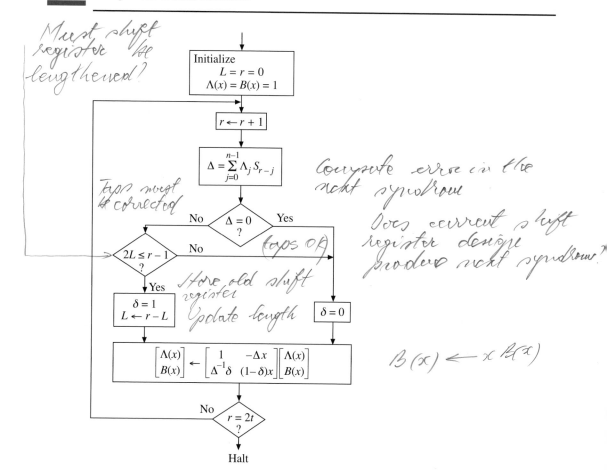

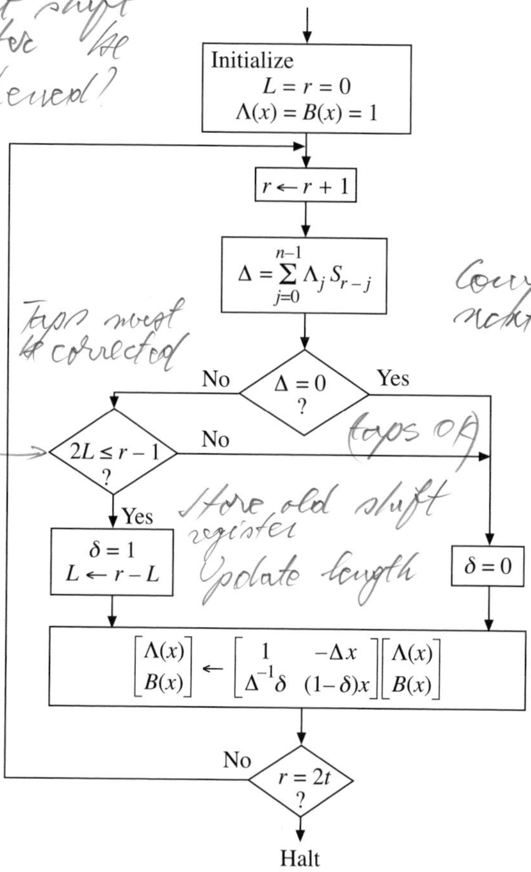

Figure 7.5. The Berlekamp–Massey algorithm

of a previous value of $\Lambda(x)$ that was stored when the length L_r was most recently changed. During each iteration when L_r is not changed, $B(x)$ is remultiplied by x so that a multiplication by x^{r-m} will be precomputed in time for the rth iteration. The decision to change L is based on the conditions $\Delta_r \neq 0$ and $2L \leq r - 1$, as required by the algorithm statement. In the computations, Δ_r may be zero, but only when δ_r is zero. Then the term $\Delta_r^{-1}\delta_r$ is defined to be zero. The matrix update requires at most $2t$ multiplications per iteration, and the calculation of Δ_r requires at most t multiplications per iteration. There are $2t$ iterations and so not more than $6t^2$ multiplications. In most instances, the algorithm is better than direct matrix inversion, which requires of the order of t^3 operations.

We shall illustrate the Berlekamp–Massey algorithm by decoding a binary (15, 5) BCH code in Section 7.3 and by decoding a (15, 9) Reed–Solomon code in Section 7.4. Both examples have a spectrum in $GF(16)$. The inner workings of the algorithm can be studied in these examples. One will notice that at the rth iteration, there may be more than one recursion of minimum length that produces the given sequence. During later iterations, whenever possible, another of the linear recursions of this length is selected. If no linear recursion of this length produces the next required symbol, then the linear recursion must be replaced by a longer one. The test to see if lengthening is required, however, uses the next symbol only to determine that $\Delta_r \neq 0$.

The proof of Algorithm 7.2.1 is broken into lemmas. In Lemma 7.2.3, which is based on Lemma 7.2.2, we find an inequality relationship between L_r and L_{r-1}. In Lemma 7.2.4, we use Algorithm 7.2.1 for the iterative construction of a linear recursion that produces the given sequence, and we conclude that the construction gives the shortest such linear recursion because it satisfies Lemma 7.2.3.

Lemma 7.2.2. *If the sequence $\boldsymbol{S} = (S_1, \ldots, S_r)$ has linear complexity L' and the sequence $\boldsymbol{T} = (T_1, \ldots, T_r)$ has linear complexity L'', then the sequence, $\boldsymbol{T} - \boldsymbol{S} = (T_1 - S_1, \ldots, T_r - S_r)$ has linear complexity not larger than $L' + L''$.*

Proof: Let L be the length of a linear recursion of minimum length that produces the sequence $\boldsymbol{T} - \boldsymbol{S} = (T_1 - S_1, \ldots, T_r - S_r)$. Let $(\Lambda'(x), L')$ be a linear recursion of minimum length that produces sequence $\boldsymbol{S}$. Let $(\Lambda''(x), L'')$ be a linear recursion of minimum length that produces sequence $\boldsymbol{T}$. Clearly, the jth coefficient of the polynomial product $\Lambda'(x)S(x)$ is equal to zero for $j = L' + 1, \ldots, r$, and the jth coefficient of the polynomial product $\Lambda''(x)T(x)$ is equal to zero for $j = L'' + 1, \ldots, r$. We can express these facts as a pair of polynomial relationships

$$\Lambda'(x)S(x) = \Gamma'(x) + x^{r+1}\Theta'(x)$$
$$\Lambda''(x)T(x) = \Gamma''(x) + x^{r+1}\Theta''(x)$$

for some $\Gamma'(x)$ and $\Gamma''(x)$ that satisfy $\deg \Gamma'(x) \leq L'$ and $\deg \Gamma''(x) \leq L''$, and for some $\Theta'(x)$ and $\Theta''(x)$.

Clearly, $L \leq r$, so whenever $L' + L'' > r$, we can conclude that $L < L' + L''$, and the lemma is true. Otherwise, $L' + L'' \leq r$. In this case, the two polynomial equations above imply that

$$\Lambda'(x)\Lambda''(x)[T(x) - S(x)] = [\Lambda'(x)\Gamma''(x) - \Lambda''(x)\Gamma'(x)] + x^{r+1}[\Lambda''(x)\Theta''(x) - \Lambda''(x)\Theta'(x)]$$

which has the form

$$\Lambda(x)[T(x) - S(x)] = \Gamma(x) + x^{r+1}\Theta(x)$$

with $\deg \Gamma(x) \leq \deg \Lambda(x) = L' + L''$, from which we observe that the linear recursion $(\Lambda'(x)\Lambda''(x), L' + L'')$ produces the sequence $\boldsymbol{T} - \boldsymbol{S}$. We conclude that the inequality $L \leq L' + L''$ must always be satisfied. □

Lemma 7.2.3. *Suppose that $(\Lambda^{(r-1)}(x), L_{r-1})$ is a linear recursion of the shortest length that produces $S_1, \ldots, S_{r-1}$; $(\Lambda^{(r)}(x), L_r)$ is a linear recursion of the shortest* *length that produces $S_1, \ldots, S_{r-1}, S_r$; and $\Lambda^{(r)}(x) \neq \Lambda^{(r-1)}(x)$. Then*

$$L_r \geq \max\left[L_{r-1}, r - L_{r-1}\right].$$

Proof: The inequality to be proved is a combination of two inequalities:

$$L_r \geq L_{r-1},$$
$$L_r \geq r - L_{r-1}.$$

The first inequality is obvious because, if a linear recursion produces a sequence, it must also produce any beginning segment of the sequence.

Let $\boldsymbol{\Delta} = (\Delta_1, \ldots, \Delta_r) = (0, 0, \ldots, 0, \Delta_r)$ where $\Delta_r \neq 0$. The sequence $\boldsymbol{\Delta}$ can be produced only by a linear recursion of length $L = r$. Suppose that $(\Lambda^{(r-1)}(x), L_{r-1})$ produces $\boldsymbol{S} + \boldsymbol{\Delta}$ and $(\Lambda^{(r)}(x), L_r)$ produces $\boldsymbol{S}$. Then with $\boldsymbol{T} = \boldsymbol{S} + \boldsymbol{\Delta}$ in the conclusion of Lemma 7.2.2, we have

$$r = L \leq L_{r-1} + L_r.$$

Consequently,

$$L_r \geq r - L_{r-1},$$

as was to be proved. □

If we can find a linear recursion that produces the given sequence and satisfies the inequality of Lemma 7.2.3 with equality, then it must be the shortest such linear recursion. The following lemma shows that Algorithm 7.2.1 accomplishes this.

Lemma 7.2.4. *Suppose that $(\Lambda^{(i)}(x), L_i)$, with $i = 1, \ldots, r-1$, is a sequence of minimum-length linear recursions such that $\Lambda^{(i)}(x)$ produces $S_1, \ldots, S_i$. If $(\Lambda^{(r-1)}(x), L_{r-1})$ does not produce $S_1, \ldots, S_r$, then any minimum-length linear recursion of length L_r that produces $S_1, \ldots, S_r$ has length*

$$L_r = \max\left[L_{r-1}, r - L_{r-1}\right].$$

Algorithm 7.2.1 gives such a recursion.

Proof: By Lemma 7.2.3, L_r cannot be smaller than $\max[L_{r-1}, r - L_{r-1}]$, so any linear recursion of length L_r that produces the required sequence must be a minimum-length linear recursion. We give a construction for a linear recursion of length L_r, assuming that we have constructed such linear recursions for each $i \leq r - 1$. For each i from 0 to $r - 1$, let $(\Lambda^{(i)}(x), L_i)$ be the minimum-length linear recursion that produces $S_1, \ldots, S_i$. For the induction argument, assume that

$$L_i = \max[L_{i-1}, i - L_{i-1}] \quad i = 1, \ldots, r-1$$

whenever $\Lambda^{(i)}(x) \neq \Lambda^{(i-1)}(x)$. This assumption is true for i such that S_i is the first nonzero syndrome because then $L_0 = 0$ and $L_i = i$. Let m denote the value of i at the most recent iteration step that required a change in the recursion length. At the end of iteration $r-1$, m is that integer such that $L_{r-1} = L_m > L_{m-1}$.

We now have

$$S_j + \sum_{i=1}^{L_{r-1}} \Lambda_i^{(r-1)} S_{j-i} = \sum_{i=0}^{L_{r-1}} \Lambda_i^{(r-1)} S_{j-i} = \begin{cases} 0 & j = L_{r-1}, \ldots, r-1 \\ \Delta_r & j = r. \end{cases}$$

If $\Delta_r = 0$, then the linear recursion $(\Lambda^{(r-1)}(x), L_{r-1})$ also produces the first r components. Thus

$$L_r = L_{r-1} \quad \text{and} \quad \Lambda^{(r)}(x) = \Lambda^{(r-1)}(x).$$

If $\Delta_r \neq 0$, then a new linear recursion must be designed. Recall that a change in length occurred at iteration $i = m$. Hence

$$S_j + \sum_{i=1}^{L_{m-1}} \Lambda_i^{(m-1)} S_{j-i} = \begin{cases} 0 & j = L_{m-1}, \ldots, m-1 \\ \Delta_m \neq 0 & j = m, \end{cases}$$

and by the induction hypothesis,

$$\begin{aligned} L_{r-1} = L_m &= \max[L_{m-1}, m - L_{m-1}] \\ &= m - L_{m-1} \end{aligned}$$

because $L_m > L_{m-1}$. Now choose the new polynomial

$$\Lambda^{(r)}(x) = \Lambda^{(r-1)}(x) - \Delta_r \Delta_m^{-1} x^{r-m} \Lambda^{(m-1)}(x),$$

and let $L_r = \deg \Lambda^{(r)}(x)$. Then because $\deg \Lambda^{(r-1)}(x) \leq L_{r-1}$ and

$$\deg[x^{r-m} \Lambda^{(m-1)}(x)] \leq r - m + L_{m-1},$$

we conclude that

$$L_r \leq \max[L_{r-1}, r - m + L_{m-1}] \leq \max[L_{r-1}, r - L_{r-1}].$$

Hence, recalling Lemma 7.2.2, if $\Lambda^{(r)}(x)$ produces $S_1, \ldots, S_r$, then $L_r = \max[L_{r-1}, r - L_{r-1}]$. It remains only to prove that the linear recursion $(\Lambda^{(r)}(x), L_r)$ produces the required sequence. But the discrepancy in the next term is

$$\begin{aligned} S_j + \sum_{i=1}^{L_r} \Lambda_i^{(r)} S_{j-i} &= S_j + \sum_{i=1}^{L_{r-1}} \Lambda_i^{(r-1)} S_{j-i} - \Delta_r \Delta_m^{-1} \left(S_{j-r+m} + \sum_{i=1}^{L_{m-1}} \Lambda_i^{(m-1)} S_{j-r+m-i} \right) \\ &= \begin{cases} 0 & j = L_r, L_r + 1, \ldots, r-1 \\ \Delta_r - \Delta_r \Delta_m^{-1} \Delta_m = 0 & j = r. \end{cases} \end{aligned}$$

Hence the linear recursion $(\Lambda^{(r)}(x), L_r)$ produces $S_1, \ldots, S_r$. In particular, $(\Lambda^{(2t)}(x), L_{2t})$ produces $S_1, \ldots, S_{2t}$, and the lemma is proved. □

7.3 Decoding of binary BCH codes

Most decoding algorithms for BCH codes can be used for codes over any finite field. When the field is $GF(2)$, however, it is necessary only to find the error locations; the error magnitude is always one (but the decoder might compute the error magnitude as a check). As an example, Table 7.1 shows the computations of the Berlekamp–Massey algorithm used to decode a noisy codeword of the (15, 5, 7) triple-error-correcting binary BCH code. The calculations can be followed by passing six times around the main loop of Figure 7.5 with the aid of a table of the arithmetic of $GF(16)$ (Table 6.2).

The possibility of simplification is suggested by the example of Table 7.1. In this example, Δ_r is always zero on even-numbered iterations. If this were always the case for binary codes, then even-numbered iterations can be skipped. The proof of this is based on the fact that, over $GF(2)$, the even-numbered syndromes are easily determined from the odd-numbered syndromes by the formula

$$S_{2j} = v(\omega^{2j}) = [v(\omega^j)]^2 = S_j^2,$$

as follows from Theorem 5.3.3. Let us calculate algebraic expressions for the coefficients of $\Lambda^{(r)}(x)$ for the first several values of r. Tracing the algorithm through the flow chart of Figure 7.5, and using the fact that $S_4 = S_2^2 = S_1^4$ for all binary codes, gives

$$\begin{aligned}
\Delta_1 &= S_1 & \Lambda^{(1)}(x) &= S_1 x + 1\\
\Delta_2 &= S_2 + S_1^2 = 0 & \Lambda^{(2)}(x) &= S_1 x + 1\\
\Delta_3 &= S_3 + S_1 S_2 & \Lambda^{(3)}(x) &= \left(S_1^{-1} S_3 + S_2\right)x^2 + S_1 x + 1\\
\Delta_4 &= S_4 + S_1 S_3 + S_1^{-1} S_2 S_3 + S_2^2 = 0. & &
\end{aligned}$$

This shows that, for any binary BCH code, Δ_2 and Δ_4 must always be zero. It is impractical to carry out the calculation in this explicit way to show that Δ_r is zero for all even r. Instead, we will formulate an indirect argument to show the general case.

Theorem 7.3.1. *In a field of characteristic 2, for any sequence $S_1, S_2, \ldots, S_{2\nu-1}$ satisfying $S_{2j} = S_j^2$ for $j \le \nu$, suppose that connection polynomial $\Lambda(x)$ of degree ν satisfies*

$$S_j = \sum_{i=1}^{\nu} \Lambda_i S_{j-i} \quad j = \nu, \ldots, 2\nu - 1.$$

Then the next term of the sequence,

$$S_{2\nu} = \sum_{i=1}^{n-1} \Lambda_i S_{2\nu-i},$$

satisfies

$$S_{2\nu} = S_\nu^2.$$

Table 7.1. *Sample Berlekamp–Massey computation*

BCH (15, 5) $t = 3$ code:

$$\begin{aligned}
g(x) &= x^{10} + x^8 + x^5 + x^4 + x^2 + x + 1 \\
a(x) &= 0 \\
v(x) &= x^7 + x^5 + x^2 = e(x) \\
S_1 &= \alpha^7 + \alpha^5 + \alpha^2 = \alpha^{14} \\
S_2 &= \alpha^{14} + \alpha^{10} + \alpha^4 = \alpha^{13} \\
S_3 &= \alpha^{21} + \alpha^{15} + \alpha^6 = 1 \\
S_4 &= \alpha^{28} + \alpha^{20} + \alpha^8 = \alpha^{11} \\
S_5 &= \alpha^{35} + \alpha^{25} + \alpha^{10} = \alpha^5 \\
S_6 &= \alpha^{42} + \alpha^{30} + \alpha^{12} = 1
\end{aligned}$$

r	Δ_r	$B(x)$	$\Lambda(x)$	L
0		1	1	0
1	α^{14}	α	$1+\alpha^{14}x$	1
2	0	αx	$1+\alpha^{14}x$	1
3	α^{11}	$\alpha^4+\alpha^3 x$	$1+\alpha^{14}x+\alpha^{12}x^2$	2
4	0	$\alpha^4 x+\alpha^3 x^2$	$1+\alpha^{14}x+\alpha^{12}x^2$	2
5	α^{11}	$\alpha^4+\alpha^3 x+\alpha x^2$	$1+\alpha^{14}x+\alpha^{11}x^2+\alpha^{14}x^3$	3
6	0	$\alpha^4 x+\alpha^3 x^2+\alpha x^3$	$1+\alpha^{14}x+\alpha^{11}x^2+\alpha^{14}x^3$	3

$$\begin{aligned}
\Lambda(x) &= 1 + \alpha^{14}x + \alpha^{11}x^2 + \alpha^{14}x^3 \\
&= (1+\alpha^7 x)(1+\alpha^5 x)(1+\alpha^2 x)
\end{aligned}$$

Proof: The proof consists of finding identical expressions for S_ν^2 and $S_{2\nu}$. First, we have

$$S_\nu^2 = \left(\sum_{i=1}^{n-1} \Lambda_i S_{\nu-i}\right)^2 = \sum_{i=1}^{n-1} \Lambda_i^2 S_{\nu-i}^2 = \sum_{i=1}^{n-1} \Lambda_i^2 S_{2\nu-2i}.$$

We also have

$$S_{2\nu} = \sum_{k=1}^{n-1} \Lambda_k S_{2\nu-k} = \sum_{k=1}^{n-1}\sum_{i=1}^{n-1} \Lambda_k \Lambda_i S_{2\nu-k-i}.$$

By symmetry, every term in the double sum with $i \neq k$ appears twice, and in a field of characteristic two, the two terms add to zero. This means that only the diagonal terms with $i = k$ contribute, so

$$S_{2\nu} = \sum_{i=1}^{n-1} \Lambda_i^2 S_{2\nu-2i},$$

which agrees with the expression for S_ν^2, and thus proves the theorem. □

We conclude that Δ_r is zero for even r, and we can analytically combine two iterations to give for odd r:

$$\Lambda^{(r)}(x) = \Lambda^{(r-2)}(x) - \Delta_r x^2 B^{(r-2)}(x),$$
$$B^{(r)}(x) = \delta_r \Delta_r^{-1} \Lambda^{(r-2)}(x) + (1 - \delta_r) x^2 B^{(r-2)}(x).$$

Using these formulas, iterations with even r can be skipped. This gives a faster decoder for binary codes. Notice that this improvement can be used even though the error pattern might contain more than t errors because only the conjugacy relations of the syndromes were used in the proof of the theorem – nothing was assumed about the weight of the binary error pattern. Therefore subsequent tests for more than t errors are still valid.

7.4 Decoding of nonbinary BCH codes

The Berlekamp–Massey algorithm can be used to compute the error-locator polynomial $\Lambda(x)$ from the $2t$ syndromes $S_1, \ldots, S_{2t}$ for any nonbinary BCH code such as a Reed–Solomon code. A sample computation of the error-locator polynomial for a $(15, 9, 7)$ Reed–Solomon triple-error-correcting code is given in Table 7.2. The calculations should be checked by tracing through the six iterations using the flow diagram of Figure 7.5.

For nonbinary codes, it is not enough to compute the error locations; the error values must be computed as well. There are a great many ways of organizing these computations. An alternative to recursive extension, and to the Gorenstein–Zierler algorithm, is to find the error locations from the zeros of $\Lambda(x)$ and to find the error values from yet another polynomial $\Gamma(x)$, called the *evaluator polynomial* or the *error-evaluator polynomial*, which is introduced for this purpose. Recall from Section 7.1 that the discrepancy polynomial $\Delta(x) = \Lambda(x)S(x)$ can always be decomposed as

$$\Lambda(x)S(x) = \Gamma(x) + x^{\nu+1} O(x) + x^{2t+1}\Theta(x)$$

where $O(x) = 0$ because $(\Lambda(x), \nu)$ produces the sequence $S_1, \ldots, S_{2t}$. Hence

$$\Lambda(x)S(x) = \Gamma(x) + x^{2t+1}\Theta(x)$$

and $\deg \Gamma(x) \leq \nu$. The polynomial $\Gamma(x)$ is the error-evaluator polynomial.

To decode, one computes the error-locator polynomial $\Lambda(x)$, possibly the error-evaluator polynomial $\Gamma(x)$, and the error-spectrum polynomial $E(x)$ or the error polynomial $e(x)$. An overview of selected alternatives is shown in Table 7.3. Some

Table 7.2. *Sample Berlekamp–Massey computation*

Reed–Solomon $(15, 9)$ $t = 3$ code:

$$\begin{aligned}
g(x) &= (x+\alpha)(x+\alpha^2)(x+\alpha^3)(x+\alpha^4)(x+\alpha^5)(x+\alpha^6) \\
&= x^6 + \alpha^{10}x^5 + \alpha^{14}x^4 + \alpha^4x^3 + \alpha^6x^2 + \alpha^9x + \alpha^6 \\
a(x) &= 0 \\
v(x) &= \alpha x^7 + \alpha^5x^5 + \alpha^{11}x^2 = e(x) \\
S_1 &= \alpha\alpha^7 + \alpha^5\alpha^5 + \alpha^{11}\alpha^2 = \alpha^{12} \\
S_2 &= \alpha\alpha^{14} + \alpha^5\alpha^{10} + \alpha^{11}\alpha^4 = 1 \\
S_3 &= \alpha\alpha^{21} + \alpha^5\alpha^{15} + \alpha^{11}\alpha^6 = \alpha^{14} \\
S_4 &= \alpha\alpha^{28} + \alpha^5\alpha^{20} + \alpha^{11}\alpha^8 = \alpha^{13} \\
S_5 &= \alpha\alpha^{35} + \alpha^5\alpha^{25} + \alpha^{11}\alpha^{10} = 1 \\
S_6 &= \alpha\alpha^{42} + \alpha^5\alpha^{30} + \alpha^{11}\alpha^{12} = \alpha^{11}
\end{aligned}$$

r	Δ_r	$B(x)$	$\Lambda(x)$	L
0		1	1	0
1	α^{12}	α^3	$1+\alpha^{12}x$	1
2	α^7	α^3x	$1+\alpha^3x$	1
3	1	$1+\alpha^3x$	$1+\alpha^3x+\alpha^3x^2$	2
4	1	$x+\alpha^3x^2$	$1+\alpha^{14}x$	2
5	α^{11}	$\alpha^4+\alpha^3x$	$1+\alpha^{14}x+\alpha^{11}x^2+\alpha^{14}x^3$	3
6	0	$\alpha^4x+\alpha^3x^2$	$1+\alpha^{14}x+\alpha^{11}x^2+\alpha^{14}x^3$	3

$$\begin{aligned}
\Lambda(x) &= 1+\alpha^{14}x+\alpha^{11}x^2+\alpha^{14}x^3 \\
&= (1+\alpha^7x)(1+\alpha^5x)(1+\alpha^2x)
\end{aligned}$$

of the algorithms listed there have been described previously, and some will be described in due course.

More directly, the error-evaluator polynomial $\Gamma(x)$ is defined from the error-locator polynomial for $S(x)$ as:

$$\Gamma(x) = \Lambda(x)S(x) \pmod{x^{2t+1}}.$$

Whereas the error-locator polynomial depends on the locations but not on the values of the errors, we shall see that the error-evaluator polynomial also depends on the error values.

The introduction of the error-evaluator polynomial leads to an alternative statement regarding the linear recursion that produces the error spectrum. The locator polynomial $\Lambda(x)$ for $\boldsymbol{e}$ satisfies

$$E(x)\Lambda(x) = 0 \pmod{x^n - 1},$$

Table 7.3. *A menu of algorithms*

Computation of error-locator polynomial $\Lambda(x)$
. Peterson algorithm
. Berlekamp–Massey algorithm
. Sugiyama algorithm
. Hybrid methods
Computation of error polynomial $e(x)$
. Gorenstein–Zierler algorithm
. Forney algorithm/Chien search
. Recursive extension
Computation of error-evaluator polynomial $\Gamma(x)$
. Polynomial multiplication $\Lambda(x)S(x)$
. Berlekamp algorithm
. Sugiyama algorithm

so the locator polynomial $\Lambda(x)$ is the connection polynomial for the periodic repetition of the sequence $E_0, E_1, \ldots, E_{n-1}$. Therefore,

$$E(x)\Lambda(x) = \Gamma(x)(x^n - 1).$$

The polynomial $\Gamma(x)$ that satisfies this equation is the error-evaluator polynomial, and its degree clearly satisfies $\deg \Gamma(x) \le \deg \Lambda(x)$.

We can reformulate the problem to portray the error-evaluator polynomial in a more central role. This formulation asks for the smallest integer $\nu \le t$ and for vectors $\mathbf{\Lambda} = (\Lambda_1, \ldots, \Lambda_\nu)$ and $\mathbf{\Gamma} = (\Gamma_1, \ldots, \Gamma_\nu)$ that satisfy the $2t$ equations

$$\Gamma_k = S_k + \sum_{j=1}^{\nu} \Lambda_j S_{k-j} \quad k = 1, \ldots, 2t$$

where $S_j = 0$ for $j \le 0$. Notice that k now runs over $2t$ values, so there are $2t$ equations and $2t$ unknowns. The solution is described by the two polynomials $\Lambda(x)$ and $\Gamma(x)$.

Theorem 7.4.1. *The error-evaluator polynomial can be written*

$$\Gamma(x) = \sum_{i=1}^{\nu} x Y_i X_i \prod_{j \ne i} (1 - X_j x).$$

Proof: Recall that the (frequency-domain) syndrome polynomial is given by

$$S(x) = \sum_{j=1}^{2t} S_j x^j = \sum_{j=1}^{2t} \sum_{\ell=1}^{\nu} Y_\ell X_\ell^j x^j$$

where $X_\ell = \omega^{i_\ell}$ is the error-location number of the ℓth error, and $Y_\ell = e_{i_\ell}$ is the value of the ℓth error. By definition of the terms in $\Gamma(x)$,

$$\Gamma(x) = \left[\sum_{j=1}^{2t}\sum_{i=1}^{\nu} Y_i X_i^j x^j\right]\left[\prod_{\ell=1}^{\nu}(1 - X_\ell x)\right] \pmod{x^{2t+1}}$$

$$= \sum_{i=1}^{\nu} x Y_i X_i \left[(1 - X_i x)\sum_{j=1}^{2t}(X_i x)^{j-1}\right]\prod_{\ell\neq i}(1 - X_\ell x) \pmod{x^{2t+1}}.$$

The term in square brackets is a factorization of $(1 - X_i^{2t}x^{2t})$. Therefore

$$\Gamma(x) = \sum_{i=1}^{\nu} x Y_i X_i\left(1 - X_i^{2t}x^{2t}\right)\prod_{\ell\neq i}(1 - X_\ell x) \pmod{x^{2t+1}}$$

$$= \sum_{i=1}^{\nu} x Y_i X_i \prod_{\ell\neq i}(1 - X_\ell x),$$

which is the expression that was to be proved. □

Now we are ready to give an expression for the error values that is much simpler than matrix inversion. It uses $\Lambda'(x)$, the formal derivative of $\Lambda(x)$.

Theorem 7.4.2 (Forney Algorithm). *The ℓth error value is given by*

$$Y_\ell = \frac{\Gamma\left(X_\ell^{-1}\right)}{\prod_{j\neq\ell}\left(1 - X_j X_\ell^{-1}\right)} = -X_\ell \frac{\Gamma\left(X_\ell^{-1}\right)}{\Lambda'\left(X_\ell^{-1}\right)}.$$

Proof: Theorem 7.4.1 states that

$$\Gamma(x) = \sum_{i=1}^{\nu} x Y_i X_i \prod_{j\neq i}(1 - X_j x).$$

Set $x = X_\ell^{-1}$. In the sum on i, every term will be zero except the term with $i = \ell$. Therefore

$$\Gamma\left(X_\ell^{-1}\right) = Y_\ell \prod_{j\neq\ell}\left(1 - X_j X_\ell^{-1}\right).$$

and

$$Y_\ell = \frac{\Gamma\left(X_\ell^{-1}\right)}{\prod\limits_{j\neq\ell}\left(1 - X_j X_\ell^{-1}\right)}.$$

Next we express the denominator in terms of the formal derivative of $\Lambda(x)$. Recall that the locator polynomial can be written in two ways

$$\Lambda(x) = \sum_{j=0}^{\nu}\Lambda_j x^j = \prod_{\ell=1}^{\nu}(1 - xX_\ell)$$

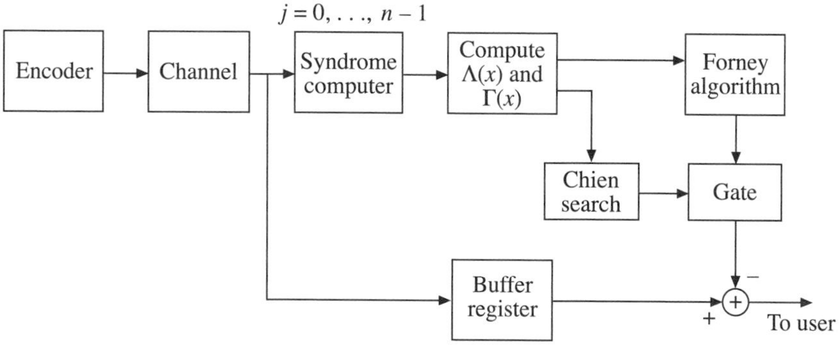

Figure 7.6. Another encoder/decoder scheme for BCH codes

so the formal derivative is

$$\Lambda'(x) = \sum_{j=0}^{\nu} j\Lambda_j x^{j-1} = -\sum_{\ell=1}^{\nu} X_i \prod_{j \neq \ell} (1 - xX_\ell).$$

Hence

$$\Lambda'\left(X_\ell^{-1}\right) = -X_\ell \prod_{j \neq \ell} \left(1 - X_j X_\ell^{-1}\right),$$

from which the theorem follows. □

The error-evaluator polynomial can be used with the Forney algorithm to compute the error values, as shown in Figure 7.6. This can be computationally faster than recursive extension, but the structure is not as simple. The Forney algorithm also is simpler computationally than the matrix inversion of the Gorenstein–Zierler algorithm, but it still requires division.

Figure 7.7 shows a flow diagram for a decoder that computes the frequency-domain codeword $\boldsymbol{C}$ from the frequency-domain senseword $\boldsymbol{V}$. The flow diagram must be preceded by a Fourier transform and followed by an inverse Fourier transform to provide a decoder that computes the codeword $\boldsymbol{c}$ from the time-domain senseword $\boldsymbol{v}$. The final $2t$ iterations of the flow diagram provide the recursive extension, but in the slightly modified form of the pair of equations

$$\Delta_r = \sum_{j=0}^{n-1} \Lambda_j S_{r-j}$$

$$S_r \leftarrow S_r - \Delta_r.$$

This restructuring allows the sum to be written with j starting at zero both for $r \leq 2t$ and for $r > 2t$, even though S_j for $j > 2t$ is not set to zero.

The decoder shown in Figure 7.7 has a simple structure for $r > 2t$, but requires $n - 2t$ such iterations. These iterations can be avoided by using the Forney algorithm, as shown in Figure 7.8, but this requires computation of the error-evaluator polynomial.

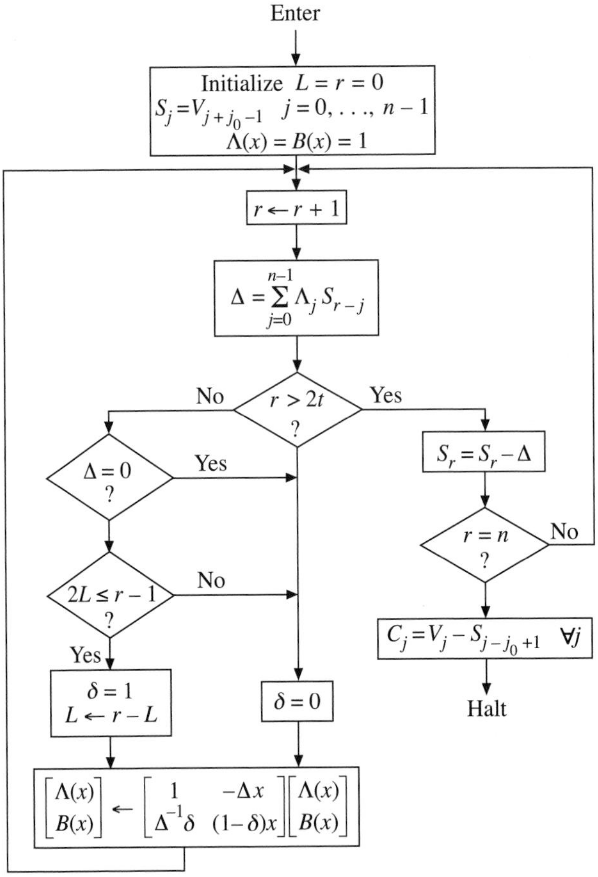

Figure 7.7. A frequency-domain decoder

A variation of the Berlekamp–Massey algorithm, which sometimes may be more attractive than the direct approach, is the *Berlekamp algorithm*. To compute the error-evaluator polynomial $\Gamma(x)$ from the defining equation

$$\Gamma(x) = \Lambda(x)S(x) \pmod{x^{2t+1}},$$

where

$$S(x) = \sum_{j=1}^{2t} S_j x^j,$$

requires a polynomial product. The Berlekamp algorithm replaces computation of the polynomial product $\Gamma(x) = \Lambda(x)S(x)$ with an iterative computation. It computes $\Gamma(x)$ in lockstep with the computation of $\Lambda(x)$ using a similar iteration. Define the iterate

$$\Gamma^{(r)}(x) = \Lambda^{(r)}(x)S(x) \pmod{x^{r+1}}$$

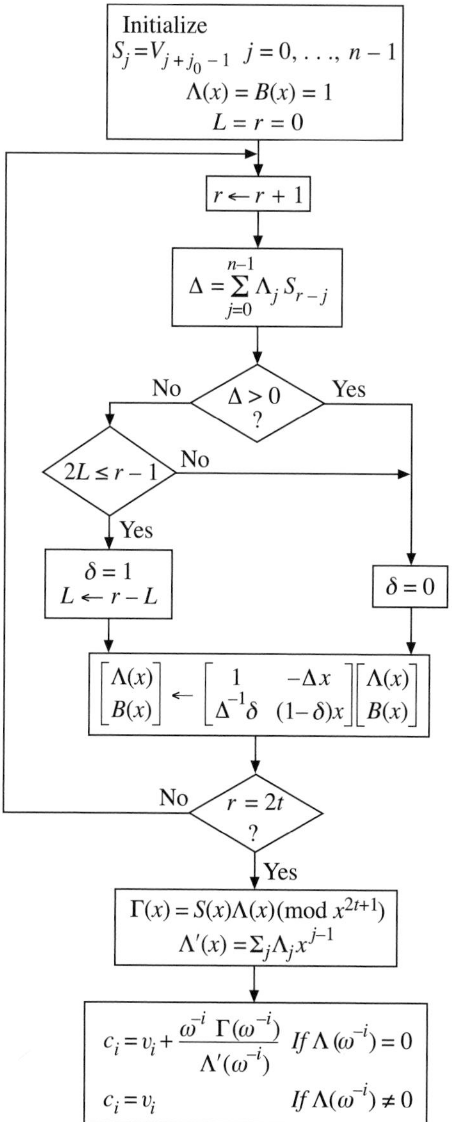

Figure 7.8. A decoder using the Forney algorithm

Clearly, the desired $\Gamma(x)$ is equal to $\Gamma^{(2t)}(x)$. Also define the scratch polynomial

$$A^{(r)}(x) = B^{(r)}(x)S(x) - x^r \pmod{x^{r+1}}.$$

The following algorithm iterates $\Gamma^{(r)}(x)$ and $A^{(r)}(x)$ directly to obtain $\Gamma(x)$ (with Δ_r and δ_r as before).

Algorithm 7.4.3. *Let* $\Gamma^{(0)}(x) = 0$, $A^{(0)}(x) = -1$, *and let*

$$\begin{bmatrix} \Gamma^{(r)}(x) \\ A^{(r)}(x) \end{bmatrix} = \begin{bmatrix} 1 & -\Delta_r x \\ \Delta_r^{-1}\delta_r & (1-\delta_r)x \end{bmatrix} \begin{bmatrix} \Gamma^{(r-1)}(x) \\ A^{(r-1)}(x) \end{bmatrix},$$

then $\Gamma^{(2t)}(x) = \Gamma(x)$.

Proof: Because

$$S(x)\Lambda^{(r)}(x) = \Gamma^{(r)}(x) \pmod{x^{r+1}},$$

we know that

$$S(x)\Lambda^{(r-1)}(x) = \Gamma^{(r-1)}(x) + \Delta x^r \pmod{x^{r+1}}$$

and, by definition of the iterates,

$$\begin{bmatrix} \Gamma^{(r)}(x) \\ A^{(r)}(x) \end{bmatrix} = \begin{bmatrix} S(x)\Lambda^{(r)}(x) \\ S(x)B^{(r)}(x) - x^r \end{bmatrix} \pmod{x^{r+1}}.$$

Using the iteration rule of Algorithm 7.2.1 to expand the right side leads to

$$\begin{aligned} \begin{bmatrix} \Gamma^{(r)}(x) \\ A^{(r)}(x) \end{bmatrix} &= \begin{bmatrix} 1 & -\Delta \\ \Delta^{-1}\delta & \overline{\delta} \end{bmatrix} \begin{bmatrix} S(x)\Lambda^{(r-1)}(x) \\ xS(x)B^{(r-1)}(x) \end{bmatrix} - \begin{bmatrix} 0 \\ x^r \end{bmatrix} \pmod{x^{r+1}} \\ &= \begin{bmatrix} 1 & -\Delta \\ \Delta^{-1}\delta & \overline{\delta} \end{bmatrix} \begin{bmatrix} \Gamma^{(r-1)}(x) + \Delta x^r \\ xA^{(r-1)}(x) + x^r \end{bmatrix} - \begin{bmatrix} 0 \\ x^r \end{bmatrix} \\ &= \begin{bmatrix} 1 & -\Delta \\ \Delta^{-1}\delta & \overline{\delta} \end{bmatrix} \begin{bmatrix} \Gamma^{(r-1)}(x) \\ xA^{(r-1)}(x) \end{bmatrix}, \end{aligned}$$

which is the iteration given in the theorem. It only remains to verify that the initialization gets the iterations off to the right start. The first iteration always gives

$$\begin{bmatrix} \Gamma^{(1)}(x) \\ A^{(1)}(x) \end{bmatrix} = \begin{bmatrix} 1 & -\Delta x \\ \Delta^{-1}\delta & \overline{\delta} x \end{bmatrix} \begin{bmatrix} 0 \\ -1 \end{bmatrix},$$

which, because $S_1 = \Delta_1$, reduces to

$$\Gamma^{(1)}(x) = S_1 x$$

and

$$A^{(1)}(x) = \begin{cases} -x & S_1 = 0 \\ 0 & S_1 \neq 0. \end{cases}$$

On the other hand, by the definition of $\Gamma^{(1)}(x)$ and $A^{(1)}(x)$, the first iteration should compute

$$\begin{aligned} \Gamma^{(1)}(x) &= S(x)(1 - \Delta x) \pmod{x^2} \\ &= S_1 x \end{aligned}$$

and

$$A^{(1)}(x) = \begin{cases} S(x)x - x \pmod{x^2} & \text{if } S_1 = 0 \\ S(x)S_1^{-1} - x \pmod{x^2} & \text{if } S_1 \neq 0 \end{cases}$$

$$= \begin{cases} -x & \text{if } S_1 = 0 \\ 0 & \text{if } S_1 \neq 0. \end{cases}$$

This agrees with the result of the first iteration, so the proof is complete. □

The iterations of Algorithm 7.2.1 and Algorithm 7.4.3 can be combined neatly into the following single statement:

$$\begin{bmatrix} \Lambda(x) & \Gamma(x) \\ B(x) & A(x) \end{bmatrix} \leftarrow \begin{bmatrix} 1 & -\Delta x \\ \Delta^{-1}\delta & \overline{\delta} x \end{bmatrix} \begin{bmatrix} \Lambda(x) & \Gamma(x) \\ B(x) & A(x) \end{bmatrix}.$$

A decoder based on this iteration is shown in Figure 7.9. Clearly both sides of this equation can be multiplied on the right side by any invertible matrix without invalidating the iteration.

7.5 Decoding with erasures and errors

Some channels make both errors and erasures. On such a channel the output alphabet consists of the channel input alphabet and a blank (or an equivalent symbol) denoting an erasure. A senseword is a vector of blocklength n of such symbols. The decoder must correct the errors and fill the erasures. Given a code with minimum distance $d_{\min}$, any pattern of ν errors and ρ erasures can be decoded, provided that

$$d_{\min} \geq 2\nu + 1 + \rho.$$

The largest ν that satisfies this inequality is denoted by t_ρ. To decode a senseword with ρ erasures, it is necessary to find a codeword that differs from the unerased portion of the senseword in at most t_ρ places. Any such codeword will do because, by the definition of t_ρ, such a codeword is unique. We need only to seek any solution because only one solution exists.

One way to decode is to guess the erased symbols and then apply any error-correction procedure. If the procedure finds that less than t_ρ of the unerased symbols are in error (and possibly some of the guesses for the erased symbols), then these errors can be corrected and the codeword restored. If the decoding fails, or if too many errors are found, then a new guess is made for the erased symbols. A trial-and-error procedure is impractical unless the number of combinations that must be tried for the erasures is small.

We will incorporate erasure filling into the decoding algorithm in a more central way. We will only treat BCH codes, and only those with $j_0 = 1$. Let v_i for $i = 0, \ldots, n-1$ be the components of the senseword. Suppose that erasures are made in locations

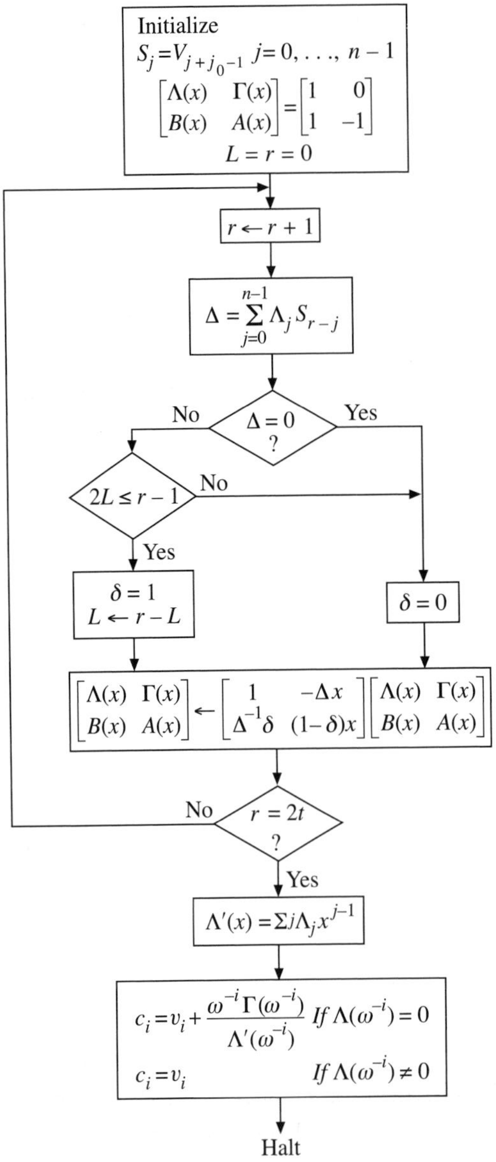

Figure 7.9. Another decoder using the Berlekamp algorithm and the Forney algorithm

$i_1, i_2, \ldots, i_\rho$. At these known locations, the senseword has blanks, which we may fill initially with any arbitrary symbols, but marking each by a special flag. We may choose an estimate of the codeword symbol by any appropriate means to fill the erasure. This symbol then is called a *readable erasure*.

Define the erasure vector as that vector of length n having component f_{i_ℓ} equal to the erased symbol for $\ell = 1, \ldots, \rho$, and in other components, f_i equals zero. Then

$$v_i = c_i + e_i + f_i \quad i = 0, \ldots, n-1.$$

Let ψ be any vector that is zero at every erased location, and otherwise nonzero. Then

$$\psi_i v_i = \psi_i(c_i + e_i + f_i) = \psi_i c_i + \psi_i e_i.$$

Define the modified senseword $v'_i = \psi_i v_i$, the modified error vector $e'_i = \psi_i e_i$, and the modified codeword $c'_i = \psi_i c_i$. The modified error vector $\boldsymbol{e}'$ has errors in the same locations as does $\boldsymbol{e}$. The equation becomes

$$v'_i = c'_i + e'_i.$$

Now the problem is to decode $\boldsymbol{v}'$ to find $\boldsymbol{e}'$, which we know how to do, in principle, if the weight of $\boldsymbol{e}'$ is smaller than $(d_{\min} - \rho)/2$.

The next step in the development is to choose ψ by working in the frequency domain. Let $U_\ell = \omega^{i_\ell}$ for $\ell = 1, \ldots, \rho$ denote the erasure locations. Define the erasure-locator polynomial:

$$\Psi(x) = \sum_{k=0}^{n-1} \Psi_k x^k = \prod_{\ell=1}^{\rho}(1 - xU_\ell).$$

This is defined so that the inverse Fourier transform of the vector $\boldsymbol{\Psi}$ has components ψ_i equal to zero whenever $f_i \neq 0$. Therefore $\psi_i f_i = 0$ for all i. In the frequency domain,

$$\boldsymbol{V}' = (\boldsymbol{\Psi} * \boldsymbol{C}) + \boldsymbol{E}'.$$

But $\boldsymbol{\Psi}$ is nonzero only in a block of length $\rho + 1$, and by the construction of a BCH code, $\boldsymbol{C}$ is zero in a block of length $2t$. Consequently, the convolution $\boldsymbol{\Psi} * \boldsymbol{C}$ is zero in a block of length $2t - \rho$. In this block, define the modified syndrome S'_j by $S'_j = V'_j$. Then

$$S'_j = (\boldsymbol{\Psi} * \boldsymbol{V})_j = E'_j,$$

which has enough syndrome components to compute the correctable patterns of the modified error vector $\boldsymbol{e}'$.

As in the errors-only case, we can find the error-locator polynomial $\Lambda(x)$ from these $2t - \rho$ known values of $\boldsymbol{E}'$, provided the number of errors ν is not more than $(2t - \rho)/2$. Once the error-locator polynomial is known, we can combine it with the erasure-locator polynomial and proceed as in the errors-only case. To do this, first define the *error-and-erasure-locator polynomial*.

$$\overline{\Lambda}(x) = \Psi(x)\Lambda(x).$$

The inverse Fourier transform of $\overline{\boldsymbol{\Lambda}}$ is zero at every error or erasure. That is, $\overline{\lambda}_i = 0$ if $e_i \neq 0$ or $f_i \neq 0$. Therefore $\overline{\lambda}_i(e_i + f_i) = 0$,

$$\overline{\boldsymbol{\Lambda}} * (\boldsymbol{E} + \boldsymbol{F}) = \boldsymbol{0},$$

and $\overline{\boldsymbol{\Lambda}}$ is nonzero in a block of length at most $\nu + \rho + 1$. Hence the $2t$ known values of $\boldsymbol{E} + \boldsymbol{F}$ can be recursively extended to n values by using this convolution equation and the known value of $\overline{\Lambda}$. Then

$$C_i = V_i - (E_i + F_i).$$

An inverse Fourier transform completes the decoding.

The step of computing the error-locator polynomial from the modified syndromes can use the Berlekamp–Massey algorithm. Starting with the initial estimates of $\Lambda^{(0)}(x) = 1$ and $B^{(0)}(x) = 1$ and proceeding through $2t - \rho$ iterations, the Berlekamp–Massey algorithm would compute $\Lambda(x)$ by a recursive procedure, using the modified syndromes

$$\Delta_r = \sum_{j=0}^{n-1} \Lambda_j^{(r-1)} S'_{r-j}$$

in the equation for Δ_r. After $2t - \rho$ iterations, the error-locator polynomial $\Lambda(x)$ is obtained.

It is possible to do much better, however, by combining several steps. Multiply the equations of the Berlekamp–Massey algorithm by $\Psi(x)$ to get the new equations

$$\Lambda^{(r)}(x)\Psi(x) = \Lambda^{(r-1)}(x)\Psi(x) - \Delta_r x B^{(r-1)}(x)\Psi(x)$$
$$B^{(r)}(x)\Psi(x) = (1 - \delta_r)x B^{(r-1)}(x)\Psi(x) + \delta_r \Delta_r^{-1} \Lambda^{(r-1)}(x)\Psi(x).$$

Thus if we start instead with the initial values $\Lambda^{(0)}(x) = B^{(0)}(x) = \Psi(x)$, we will iteratively compute the error-and-erasure-locator polynomial $\overline{\Lambda}(x)$, using the iterates $\overline{\Lambda}^{(r)}(x) = \Lambda^{(r)}(x)\Psi(x)$ and $\overline{B}^{(r)}(x) = B^{(r)}(x)\Psi(x)$. Even the computation of Δ_r works out simply because of the associativity property of the convolution. This is

$$\Delta_r = \sum_{j=0}^{n-1} \Lambda_{r-j}^{(r-1)} S'_j = \sum_{k=0}^{n-1} \overline{\Lambda}_{r-k}^{(r-1)} S_k = \sum_{j=0}^{n-1} \left(\sum_{k=0}^{n-1} \Lambda_k^{(r-1)} \Psi_{r-j-k} \right) S_j.$$

Therefore if we initialize the Berlekamp–Massey algorithm with $\Psi(x)$ instead of with one, the modified syndromes are computed implicitly and need not explicitly appear.

We now drop the notation $\overline{\Lambda}(x)$, replacing it with $\Lambda(x)$, which will now be called the error-and-erasure-locator polynomial. The Berlekamp–Massey algorithm initialized with $\Psi(x)$ produces recursively the error-and-erasure-locator polynomial, according to the equations

$$\Lambda^{(r)}(x) = \Lambda^{(r-1)}(x) - \Delta_r x B^{(r-1)}(x)$$
$$B^{(r)}(x) = (1 - \delta_r)x B^{(r-1)}(x) + \delta_r \Delta_r^{-1} \Lambda^{(r-1)}(x)$$
$$\Delta_r = \sum_{j=0}^{n-1} \Lambda_j^{(r-1)} S_{r-j},$$

which are exactly the same as for the case of errors-only decoding. Only the initialization is different.

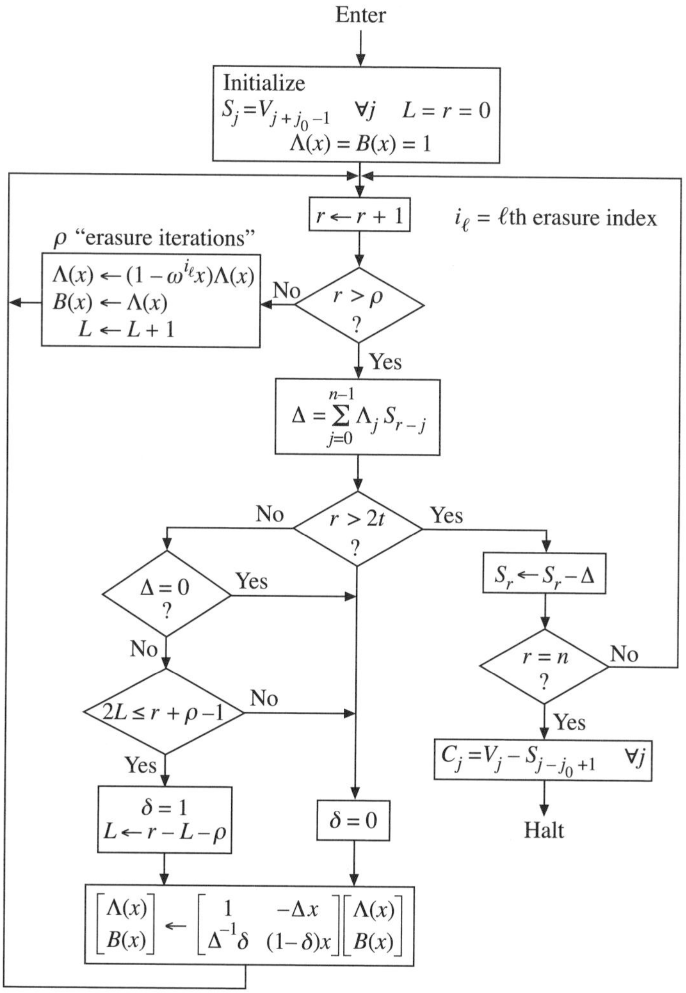

Figure 7.10. An error-and-erasure decoder for BCH codes

The Berlekamp–Massey algorithm, revised to handle erasures as well, is shown in Figure 7.10. This should be compared to Figure 7.7. Only the computation of the erasure-locator polynomial is new, which is a simple computation compared to other decoding computations. This consists of a special loop for the first ρ iterations, as shown in Figure 7.10. The index r counts out the first ρ iterations, which form the erasure polynomial, and then continues to count out the iterations of the Berlekamp–Massey algorithm, stopping when r reaches $2t$. The shift-register length L is increased once for each erasure, and thereafter the length changes according to the procedure of the Berlekamp–Massey algorithm, but modified by replacing r and L by $r - \rho$ and $L - \rho$, respectively.

7.6 Decoding in the time domain

The decoding procedure developed in Section 7.1, when stated in the language of spectral estimation, consists of a Fourier transform (syndrome computer), followed by a spectral analysis (Berlekamp–Massey algorithm), followed by an inverse Fourier transform (Chien search). In this section, we shall show how to transform the decoding procedure into the time domain, rather than transform the data into the frequency domain. This will eliminate both Fourier transforms from the decoder. The idea is to notice that the Berlekamp–Massey iteration is linear in the two updated polynomials. A time-domain counterpart to these equations can be found analytically by taking the inverse Fourier transform of the decoding equations. This gives a set of decoding equations that operates directly on the raw data, and the error correction is completed without the need to compute transforms.

To eliminate the need for computing a Fourier transform of the data, transform all variables in Algorithm 7.2.1. The transform of $\boldsymbol{\Lambda}$ is the time-domain error-locator vector $\boldsymbol{\lambda} = \{\lambda_i \mid i = 0, \ldots, n-1\}$. The transform of $\boldsymbol{B}$ is the time-domain scratch vector $\boldsymbol{b} = \{b_i \mid i = 0, \ldots, n-1\}$. The Berlekamp–Massey algorithm transforms into the iterative procedure in the time domain, described in the following theorem.

Algorithm 7.6.1. *Let $\boldsymbol{v}$ be at distance at most t from a codeword of a BCH code of designed distance $2t+1$. With initial conditions $\lambda_i^{(0)} = 1$, $b_i^{(0)} = 1$ for all i and $L_0 = 0$, let the following set of equations for $r = 1, \ldots, 2t$ be used iteratively to compute $\lambda_i^{(2t)}$ for $i = 0, \ldots, n-1$:*

$$\Delta_r = \sum_{i=0}^{n-1} \omega^{ir}\left(\lambda_i^{(r-1)} v_i\right)$$

$$L_r = \delta_r(r - L_{r-1}) + \overline{\delta}_r L_{r-1}$$

$$\begin{bmatrix} \lambda_i^{(r)} \\ b_i^{(r)} \end{bmatrix} = \begin{bmatrix} 1 & -\Delta_r \omega^{-i} \\ \Delta_r^{-1}\delta_r & \overline{\delta}_r \omega^{-i} \end{bmatrix} \begin{bmatrix} \lambda_i^{(r-1)} \\ b_i^{(r-1)} \end{bmatrix},$$

$r = 1, \ldots, 2t$, where $\delta_r = (1 - \overline{\delta}_r) = 1$ if both $\Delta_r \neq 0$ and $2L_{r-1} \leq r - 1$, and otherwise $\delta_r = 0$. Then $\lambda_i^{(2t)} = 0$ if and only if $e_i \neq 0$.

Proof: Take the Fourier transform of all vector quantities. Then the equations of Algorithm 7.2.1 are obtained, except for the equation

$$\Delta_r = \sum_{j=0}^{n-1} \Lambda_j^{(r-1)} V_{r-j},$$

which has $\boldsymbol{V}$ in place of $\boldsymbol{S}$. But $\Lambda_j^{(r-1)} = 0$ for $j > 2t$, and so $V_j = S_j$ for $j = 1, \ldots, 2t$.

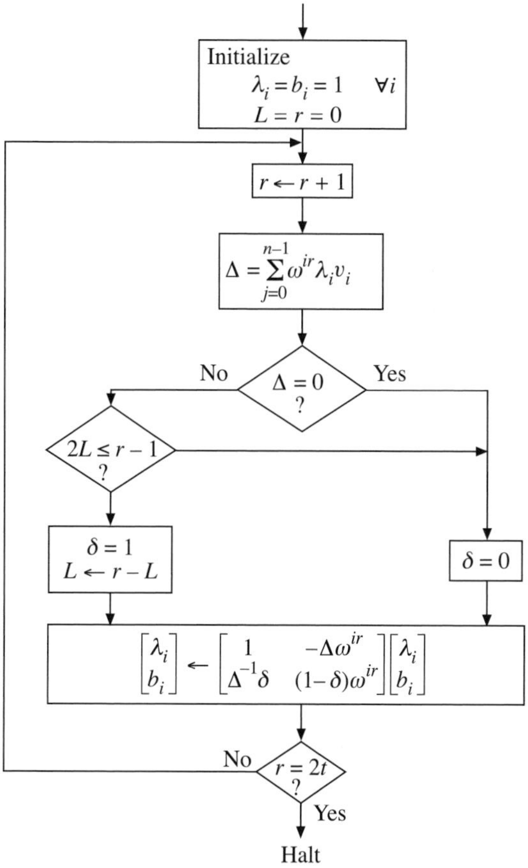

Figure 7.11. Time-domain Berlekamp–Massey algorithm

Therefore

$$\Delta_r = \sum_{j=0}^{n-1} \Lambda_j^{(r-1)} S_{r-j},$$

and the time-domain algorithm reduces exactly to the form of Algorithm 7.2.1. □

We call Algorithm 7.6.1 the *time-domain* (or the *code-domain*) *Berlekamp–Massey algorithm* which is given in flow-chart form in Figure 7.11. It computes the time-domain error-locator vector $\boldsymbol{\lambda}$ directly from the raw data v; no Fourier transform is needed. For binary codes, once the vector $\boldsymbol{\lambda}$ is known, correction is nearly complete because the senseword has an error in component i if and only if $\lambda_i = 0$. In Figure 7.12, a time-domain decoder for binary codes is shown.

For nonbinary codes, it does not suffice to compute only the error locations in the time domain. We must also compute the error values. In the frequency domain, all

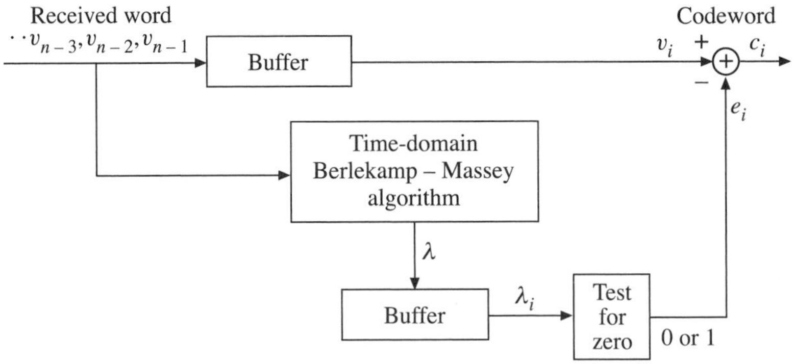

Figure 7.12. A time-domain decoder for a binary BCH code

components of $\boldsymbol{E}$ can be computed by the linear recursion

$$E_k = -\sum_{j=1}^{t} \Lambda_j E_{k-j} \quad k = 2t+1, \ldots, n-1.$$

To write the inverse Fourier transform of this equation, some restructuring is necessary. We will give the corresponding expression in the time domain in the next theorem.

Theorem 7.6.2. *Let $\boldsymbol{v} = \boldsymbol{c} + \boldsymbol{e}$ be a received noisy BCH codeword. Given the time-domain error locator $\boldsymbol{\lambda}$, the following set of equations,*

$$\Delta_r = \sum_{i=0}^{n-1} \omega^{ir} v_i^{(r-1)} \lambda_i$$

$$v_i^{(r)} = v_i^{(r-1)} - \Delta_r \omega^{-ri}$$

for $r = 2t+1, \ldots, n$, results in

$$v_i^{(n)} = e_i \quad i = 0, \ldots, n-1.$$

Proof: The equation of the recursive extension in the frequency domain is rewritten by separating it into two steps, starting with the senseword spectrum $\boldsymbol{V}$ and changing it to $\boldsymbol{E}$ – one component at a time:

$$\Delta_r = V_r - \left(-\sum_{j=1}^{n-1} \Lambda_j V_{r-j}\right) = \sum_{j=0}^{n-1} \Lambda_j V_{r-j}$$

$$V_j^{(r)} = \begin{cases} V_j^{(r-1)} & j \neq r \\ V_j^{(r-1)} - \Delta_r & j = r. \end{cases}$$

Because $V_j^{(2t)} = E_j$ for $j = 1, \ldots, 2t$ and $\Lambda_j = 0$ for $j > t$, the above equations are equivalent to

$$E_r = -\sum_{j=1}^{n-1} \Lambda_j E_{r-j}.$$

This equivalence proves the theorem. □

A flow chart in a field of characteristic 2 for a complete time-domain decoder, including erasure filling and a single extension symbol, is shown in Figure 7.13. The

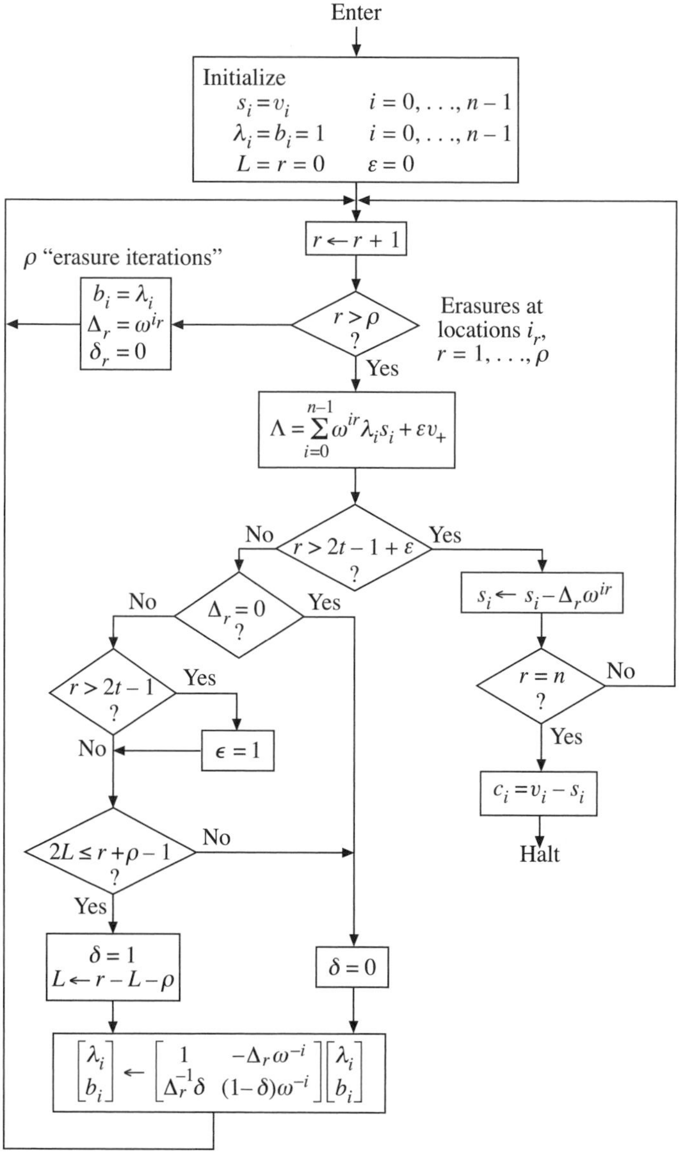

Figure 7.13. A time-domain BCH decoder

left side of the flow diagram is based on Algorithm 7.6.1, and the right side is based on Theorem 7.6.2. During the iterations on the left side of Figure 7.13, the inverse Fourier transform of the error-locator polynomial is being formed. During each pass on the right side, the senseword is changed into one component of the spectrum of the error word. Hence after the last iteration, the senseword has been changed into the error word.

The time-domain decoder can be attractive because there is only one major task in the computation. There is no syndrome computer or Chien search. This means that the decoder has a very simple structure, but the penalty is a longer running time. The number of major clocks required is n, which is independent of the number of errors corrected. During each such clock, vectors of length n are processed, and thus the complexity of the decoder is proportional to n^2. For high-rate codes of large blocklength, decoders with fewer computations can be found, but with a more complex structure. In some operations, structural simplicity may be at least as important as the number of arithmetic operations.

7.7 Decoding within the BCH bound

The decoding algorithms for BCH codes that have been developed in this chapter are bounded-distance decoders that will decode correctly whenever the number of errors does not exceed the packing radius of the code. When the distance from the senseword to the transmitted codeword is larger than the packing radius t, the senseword may lie within a distance t of another codeword, but usually the distance from the senseword to every codeword will be larger than t. Then the error pattern, though uncorrectable by a bounded-distance decoder, can be detected. This detection of an uncorrectable error pattern is called a *decoding failure*. It may be preferable to a *decoding error*, which occurs when the decoder produces the wrong codeword. In this section, we shall first discuss tests for detecting a decoding failure. Then we shall discuss making the decoding spheres smaller in order to decrease the probability of decoding error at the cost of a higher probability of decoding failure.

When the number of errors exceeds the packing radius, the various decoding algorithms can behave in different ways. In general, it is not true that the output of a decoding algorithm must always be a codeword. A particular decoding algorithm may fail to detect uncorrectable but detectable errors if this requirement was not imposed in the design of the decoding algorithm.

The most direct way of detecting uncorrectable errors is to check the decoder output, verifying that it is a true codeword and is within distance t of the decoder input. Alternatively, indirect tests can be embedded within an algorithm.

An error pattern in a BCH codeword that is uncorrectable with a bounded-distance decoder can be detected by noticing irregularities in the decoding computation. If the computed error-locator polynomial does not have a number of zeros in the error-locator field equal to its degree, or if the error pattern has components outside of the symbol field $GF(q)$, then the actual error pattern is not correctable. It is possible to catch these cues without completing all decoding calculations. They can be recognized by properties of the recursively computed error spectrum:

$$E_j = -\sum_{k=1}^{t} \Lambda_k E_{j-k}.$$

According to Theorem 6.4.1, the requirement that the error pattern be in $GF(q)$ can be recognized in the error spectrum over $GF(q^m)$ by the condition $E_j^q = E_{((qj))}$ for all j. (For a Reed–Solomon code, because $m = 1$, this test is useless.) According to the forthcoming Theorem 7.7.1, the requirement that $\deg \Lambda(x)$ should equal the number of zeros of $\Lambda(x)$ in $GF(q^m)$ can be recognized in the error spectrum over $GF(q^m)$ by the condition that $E_{((n+j))} = E_j$. If these two equations are tested as each E_j is computed, then an uncorrectable error pattern may be detected and the computation can be halted.

If either of the two conditions fails, that is, if for some j

$$E_j^q \neq E_{((qj))},$$

or

$$E_{((n+j))} \neq E_j,$$

then a pattern with more than t errors has been detected.

Theorem 7.7.1. *Let $n = q - 1$. Suppose $\Lambda(x)$ is a nonzero polynomial over $GF(q)$ of degree less than n. The number of distinct zeros of $\Lambda(x)$ in $GF(q)$ is equal to* $\deg \Lambda(x)$ *if and only if $E_{((n+j))} = E_j$ for any sequence $\mathbf{E}$ over $GF(q)$ for which the recursion*

$$E_j = -\sum_{k=1}^{\deg \Lambda(x)} \Lambda_k E_{j-k} \qquad j = 2t+1, \ldots, 2t+n$$

has minimum length.

Proof: Suppose ν, the number of distinct zeros of $\Lambda(x)$ in $GF(q)$, equals $\deg \Lambda(x)$. Then $\Lambda(x)$ divides $x^n - 1$, and there is a polynomial $\Lambda^0(x)$ with $\Lambda^0(x)\Lambda(x) = x^n - 1$. Because

$$\sum_{k=0}^{\nu} \Lambda_k E_{j-k} = 0,$$

with $\Lambda_0 = 1$, we have $\Lambda(x)E(x) = 0$, and thus $\Lambda^0(x)\Lambda(x)E(x) = (x^n - 1)E(x) = 0$. This implies that $E_j = E_{((n+j))}$ for all j.

Conversely, suppose that $E_j = E_{((n+j))}$ for all j. Then $(x^n - 1)E(x) = 0$. But we are given that $\Lambda(x)$ is a smallest-degree nonzero polynomial with the property that $\Lambda(x)E(x) = 0 \pmod{x^n - 1}$. By the division algorithm, write

$$x^n - 1 = \Lambda(x)Q(x) + r(x)$$

so that

$$(x^n - 1)E(x) = \Lambda(x)E(x)Q(x) + r(x)E(x).$$

Therefore

$$r(x)E(x) = 0 \pmod{x^n - 1}.$$

But $r(x)$ has a degree smaller than the degree of $\Lambda(x)$, and thus $r(x)$ must equal zero. Then $\Lambda(x)$ divides $x^n - 1$, and so all of its zeros are also zeros of $x^n - 1$. □

To reduce the probability of a false decoding, a bounded-distance decoder may be designed to correct only up to τ errors where $2\tau + 1$ is strictly smaller than $d_{\min}$. This must be so if $d_{\min}$ is even. Then a suitable decoder will be sure not to miscorrect a pattern if ν, the number of errors that actually occurred, satisfies

$$\tau + \nu < d_{\min}$$

because at least $d_{\min} - \tau$ errors are needed to move any codeword into an incorrect decoding sphere.

The Berlekamp–Massey algorithm is well-suited for such a reduced-distance decoder. The procedure is to perform 2τ iterations of the algorithm to produce $\Lambda(x)$ for an error pattern of up to τ errors. Then an additional $d_{\min} - 1 - 2\tau$ iterations are used to check if the discrepancy Δ_r is zero for each of these additional iterations. If Δ_r is nonzero for any of these additional iterations, the senseword is flagged as having more than τ errors. This procedure cannot fail to detect an error pattern unless more than $d_{\min} - 1 - \tau$ errors have occurred.

The following theorem provides the justification for this procedure.

Theorem 7.7.2. *Given syndromes S_j for $j = 1, \ldots, d_{\min} - 1$, let $(\Lambda^{(2\tau)}(x), L_{2\tau})$ be a minimum-length linear recursion with $L_{2\tau} \le \tau$ that will produce S_j for $j = 1, \ldots, 2\tau$. Suppose that*

$$\sum_{j=0}^{n-1} \Lambda_j^{(2\tau)} S_{k-j} = 0 \quad k = 2\tau + 1, \ldots, d_{\min} - 1.$$

Then either $\Lambda^{(2\tau)}(x)$ is the correct error-locator polynomial, or more than $d_{\min} - \tau - 1$ errors have occurred.

Proof: Let $d = d_{\min}$ and assume there are at most $d - \tau - 1$ errors. We are given syndromes S_j for $j = 1, \ldots, d-1$ and only used 2τ of them to form the locator polynomial, leaving $d - 2\tau$ unused. Suppose we are given $d - 2\tau$ extra syndromes S_j for $j = d, \ldots, 2d - 2\tau - 1$ by a genie or by a hypothetical side channel. Using the Berlekamp–Massey algorithm, we could compute a linear recursion for this longer sequence. But at iteration 2τ, $L_{2\tau} \leq \tau$, and by assumption, Δ_r equals zero for $r = 2\tau + 1, \ldots, d-1$, and thus L is not updated before iteration d. Hence by the rule for updating L,

$$\begin{aligned} L_{2\tau} &\geq d - L_{2\tau} \\ &\geq d - \tau, \end{aligned}$$

contrary to the assumption that there are at most $d - \tau - 1$ errors. □

7.8 Decoding beyond the BCH bound

Because complete decoders are computationally intractable, bounded-distance decoders are widely used. Sometimes, to improve performance, one may wish to continue the decoding beyond the designed distance. Techniques that decode a small distance beyond the BCH bound can be obtained by forcing the Berlekamp–Massey algorithm to continue past $2t$ iterations by introducing extra syndromes as unknowns. The decoder analytically continues its decoding algorithm, leaving the extra syndromes as unknowns, and the error polynomial is computed as a function of these unknowns. The unknowns are then selected by trial and error to obtain a smallest-weight error pattern in the symbol field of the code. Unfortunately, the complexity of these techniques increases very quickly as one passes beyond the designed distance, and thus only a limited penetration is possible.

If the code is in a subfield of the error-locator field, then procedures that first find the error-locator polynomial can lead to ambiguities. Some ambiguities may be resolved by rejecting those error polynomials with symbols not in the code symbol field.

The need for decoding beyond the BCH bound is illustrated by several examples. One example is a binary BCH code with its defining set a block of length $2t$ consecutive powers of ω, but with the actual minimum distance larger than $2t + 1$. (The Golay code is one example of such a code. Another is the $(127, 43, 31)$ BCH code, which has a designed distance of only 29.) A second example is a cyclic code with a defining set whose elements are nonconsecutive powers of ω. A third example is a decoder for some of the $(t+1)$-error patterns in a t-error-correcting Reed–Solomon code or BCH code.

Consider a Reed–Solomon code in $GF(q)$ of designed distance $2t+1$. Any polynomial $\Lambda(x)$ of degree τ, with τ distinct zeros in $GF(q)$, is a legitimate error-locator polynomial if

$$\sum_{j=0}^{n-1} \Lambda_j S_{r-j} = 0 \quad r = \tau+1, \ldots, 2t.$$

A smallest-degree polynomial $\Lambda(x)$ with $\Lambda_0 = 1$ produces a codeword of minimum weight. If it has a degree at most t, this polynomial is computed by the Berlekamp–Massey algorithm. Even when there are more than t errors, so τ is larger than t, the smallest-degree polynomial with these properties may be unique, and the senseword can be uniquely decoded whenever that polynomial can be found. If the smallest-degree polynomial is not unique, then there are several possible error patterns, all of the same weight τ, that agree with the senseword in $n-\tau$ places. A decoder that finds all of them is called a *list decoder*.

Suppose there are $t+1$ errors. Then the two unknown syndromes, S_{2t+1} and S_{2t+2}, if they were known, would be enough to find $\Lambda(x)$. Equivalently, the discrepancies Δ_{2t+1} and Δ_{2t+2}, if they were known, would be enough to find $\Lambda(x)$. Hence, we analytically continue the Berlekamp–Massey algorithm through two more iterations with these discrepancies as unknowns. Then we have

$$\begin{bmatrix} \Lambda^{(2t+2)}(x) \\ B^{(2t+2)}(x) \end{bmatrix} = \begin{bmatrix} 1 & -\Delta_{2t+2}x \\ \Delta_{2t+2}^{-1}\delta_{2t+2} & \overline{\delta}_{2t+2}x \end{bmatrix} \begin{bmatrix} 1 & -\Delta_{2t+1}x \\ \Delta_{2t+1}^{-1}\delta_{2t+1} & \overline{\delta}_{2t+1}x \end{bmatrix} \begin{bmatrix} \Lambda^{(2t)}(x) \\ B^{(2t)}(x) \end{bmatrix}$$

where the unknowns satisfy $\Delta_{2t+1}, \Delta_{2t+2} \in GF(q)$ and $\delta_{2t+1}, \delta_{2t+2} \in \{0, 1\}$. Everything else on the right side is known from the $2t$ syndromes.

The example of the Berlekamp–Massey algorithm used in Section 7.3 can be modified to give an example of decoding beyond the BCH bound, shown in Table 7.4. The code in this modified example is a (15, 7, 5) binary BCH code with spectral zeros at α, α^2, α^3, and α^4. The error pattern, however, has three errors. It is the same error pattern that was used with the (15, 5, 7) binary BCH code in the example in Section 7.3.

The first four iterations of the Berlekamp–Massey algorithm process syndromes S_1, S_2, S_3, and S_4. After these four iterations, because $L \neq \deg \Lambda(x)$, it is apparent that more than two errors occurred. Two more iterations are needed to find an error pattern with three errors. Assume an extra syndrome S_5, which is unknown, or equivalently an unknown discrepancy Δ_5. Because the code is a binary code, syndrome S_6 is known to equal S_3^2 and discrepancy Δ_6 is known to be zero. In terms of the unknown Δ_5, the locator polynomial is

$$\Lambda(x) = 1 + \alpha^{14}x + \alpha^{12}x^2 + \Delta_5(\alpha^4 x^2 + \alpha^3 x^3),$$

Table 7.4. *Example of decoding beyond the BCH bound*

BCH (15, 5) $t = 3$ code:

$$g(x) = x^{10} + x^8 + x^5 + x^4 + x^2 + x + 1$$
$$a(x) = 0$$
$$v(x) = x^7 + x^5 + x^2 = e(x)$$
$$S_1 = \alpha^7 + \alpha^5 + \alpha^2 = \alpha^{14}$$
$$S_2 = \alpha^{14} + \alpha^{10} + \alpha^4 = \alpha^{13}$$
$$S_3 = \alpha^{21} + \alpha^{15} + \alpha^6 = 1$$
$$S_4 = \alpha^{28} + \alpha^{20} + \alpha^8 = \alpha^{11}$$

r	Δ_r	$B(x)$	$\Lambda(x)$	L
0		1	1	0
1	α^{14}	α	$1+\alpha^{14}x$	1
2	0	αx	$1+\alpha^{14}x$	1
3	α^{11}	$\alpha^4 + \alpha^3 x$	$1+\alpha^{14}x+\alpha^{12}x^2$	2
4	0	$\alpha^4 x + \alpha^3 x^2$	$1+\alpha^{14}x+\alpha^{12}x^2$	2
5	Δ_5	$\Delta_5^{-1}(1+\alpha^{14}x+\alpha^{12}x^2)$	$\Lambda^{(4)}(x) + x\Delta_5 B^{(4)}(x)$	3
6	0	$\Delta_5^{-1}(x+\alpha^{14}x^2+\alpha^{12}x^3)$	$\Lambda^{(4)}(x) + x\Delta_5 B^{(4)}(x)$	3

and Δ_5 is a nonzero element of $GF(16)$. By trying each possible value for Δ_5, one finds three $\Lambda(x)$ that have three zeros in $GF(16)$. That is, by setting Δ_5 equal in turn to α^3, α^{11}, and α^{14}, we find three locator polynomials

$$\Lambda(x) = (1 - \alpha^0 x)(1 - \alpha^2 x)(1 - \alpha^7 x)$$
$$\Lambda(x) = (1 - \alpha^8 x)(1 - \alpha^{10} x)(1 - \alpha^{13} x)$$
$$\Lambda(x) = (1 - \alpha^5 x)(1 - \alpha^9 x)(1 - \alpha^{14} x),$$

and these determine the three error patterns

$$e(x) = 1 + x^8 + x^{13}$$
$$e(x) = x^2 + x^5 + x^7$$
$$e(x) = x + x^6 + x^{10}.$$

Then setting $c(x) = v(x) - e(x)$ leads to three possible codewords:

$$\mathbf{c} = (1, 0, 1, 0, 0, 1, 0, 1, 1, 0, 0, 0, 0, 1, 0)$$
$$\mathbf{c} = (0, 0, 0, 0, 0, 0, 0, 0, 0, 0, 0, 0, 0, 0, 0)$$
$$\mathbf{c} = (0, 1, 1, 0, 0, 1, 1, 1, 0, 0, 1, 0, 0, 0, 0),$$

each at distance three from the senseword. The decoder can form this list of codewords, but cannot reduce it to a single codeword. The decoding ambiguity cannot be resolved by the structure of the code; some additional information is needed.

One may modify this procedure to decode a binary code for which the defining set does not have consecutive elements. The (63, 28, 15) binary cyclic code, for example, has check frequencies $C_1, C_3, C_5, C_7, C_9, C_{11}$, and C_{21}. This code should be preferred to the (63, 24, 15) BCH code because it has a superior rate. With a little extra complexity, we can modify a BCH decoder to handle the (63, 28, 15) code. By introducing an unknown syndrome as above, all patterns of seven or fewer errors could be found that agree with the twelve consecutive check frequencies. Then S_{21} is computed for each of these candidates, possibly by recursive extension. Only one will agree with the observed value of S_{21}.

Alternatively, the procedure can be placed in the time domain. Transform the frequency-domain expression into the code domain:

$$\begin{bmatrix} \lambda_i^{(2t+2)}(x) \\ b_i^{(2t+2)}(x) \end{bmatrix} = \begin{bmatrix} 1 & -\Delta_{2t+2}\omega^{-i} \\ \Delta_{2t+2}^{-1}\delta_{2t+2} & \overline{\delta}_{2t+2}\omega^{-i} \end{bmatrix} \begin{bmatrix} 1 & -\Delta_{2t+1}\omega^{-i} \\ \Delta_{2t+1}^{-1}\delta_{2t+1} & \overline{\delta}_{2t+1}\omega^{-i} \end{bmatrix} \begin{bmatrix} \lambda_i^{(2t)} \\ b_i^{(2t)} \end{bmatrix}.$$

Now we must choose the unknowns Δ_{2t+1} and Δ_{2t+2}, if possible, so that the error pattern contains at most $t+1$ nonzero components. For a binary code Δ_{2t+2} is known to be zero. For a nonbinary code, both are unknown.

There are two cases:

1. $\lambda_i^{(2t+2)} = \lambda_i^{(2t)} - A_1\omega^{-i}b_i^{(2t)} - A_2\omega^{-2i}b_i^{(2t)}$
2. $\lambda_i^{(2t+2)} = \lambda_i^{(2t)} - A_1\omega^{-i}b_i^{(2t)} - A_2\omega^{-2i}\lambda_i^{(2t)}$

where A_1 and A_2 are constants that are not both zero, and A_1 is nonzero in the second equation. Each of these cases is to be searched over A_1 and A_2 for a solution, with $\lambda_i^{(2t+2)} = 0$ for exactly $t+1$ values of i. This locates $t+1$ errors, and the decoding can continue as for erasure decoding.

One can organize the search in a variety of ways. One possibility is a histogram approach. For case (2), for each value of A_1, prepare a histogram of $(\lambda_i^{(2t)} - A_1\omega^{-i}b_i^{(2t)})^{-1}(\omega^{-2i}\lambda_i^{(2t)})$. Any component of the histogram that takes the value $t+1$ corresponds to a possible $t+1$ error pattern.

7.9 Decoding of extended Reed–Solomon codes

To decode an extended Reed–Solomon code, divide the task into two cases: Either both extension symbols are correct, or at least one extension symbol is not correct. The message can be decoded once for each of these hypotheses. The doubly extended code has spectral components $C_{j_0+1}, \ldots, C_{j_0+2t-2}$ equal to zero, C_{j_0} equal to c_- and C_{j_0+2t-1} equal to c_+. Then $S_j = V_{j_0+j}$, for $j = 1, \ldots, 2t-2$ are the $2t-2$ interior syndromes, and $S_0 = V_{j_0} - v_-$ and $S_{2t-1} = V_{j_0+2t-1} - v_+$ are the extended syndromes. For the first hypothesis, the $2t-2$ interior syndromes $S_1, \ldots, S_{2t-2}$ are sufficient to find the entire spectrum of the error pattern in the interior because at most $t-1$ errors have occurred in the interior. For the second hypothesis, the $2t$ syndromes $S_0, \ldots, S_{2t-1}$ are sufficient to find the entire spectrum of the error pattern because it is certain that all errors are in the $n-2$ interior symbols. Although two trial decodings are made, only one pattern of t or fewer errors can be found because the code has a minimum distance of $2t+1$. Therefore either both decodings will give the same codeword or one of them will default.

For a singly extended code the computations can be combined in such a way that the decoding complexity is essentially the same as for a nonextended Reed–Solomon code. The singly extended code has spectral components $C_{j_0}, \ldots, C_{j_0+2t-2}$ equal to zero, and has the singly extended check symbol $c_+ = C_{j_0+2t-1}$. The $2t-1$ components equal to zero ensure that the code has distance $2t$ so that it can correct $t-1$ errors and detect t errors. The block of $2t$ syndromes begins with S_1 equal to V_{j_0} and ends with S_{2t} equal to $V_{j_0+2t-1} - v_+$, which is valid if v_+ is correct. Suppose that at most t errors have occurred; start the Berlekamp–Massey algorithm at syndrome S_1 and iterate out to syndrome S_{2t-1}. If $L_{2t-2} \le t-1$, and the discrepancy Δ_{2t-1} is zero, then (by Theorem 7.7.2) at most $t-1$ errors have occurred in the interior, and the error-locator polynomial is correct. Otherwise, t errors have occurred in the interior, and so v_+ is correct. The extra syndrome S_{2t} allows the Berlekamp–Massey algorithm to be continued through one more iteration for a total of $2t$ iterations, and hence a t-error-correcting error-locator polynomial is found.

Figure 7.14 shows a decoder for a singly extended Reed–Solomon code. The additional complexity needed for the single extension symbol is trivial. The binary variable ϵ simply controls whether or not iteration $2t$ of the Berlekamp–Massey algorithm will be executed.

7.10 Decoding with the euclidean algorithm

The euclidean algorithm can be used to develop algorithms for decoding BCH codes. We shall describe an algorithm that is easier to explain than the Berlekamp–Massey algorithm, but is less efficient because the polynomials that are iterated have a larger degree.

The euclidean algorithm for polynomials, given in Theorem 4.3.6, is a recursive procedure for calculating the greatest common divisor of two polynomials. The extended

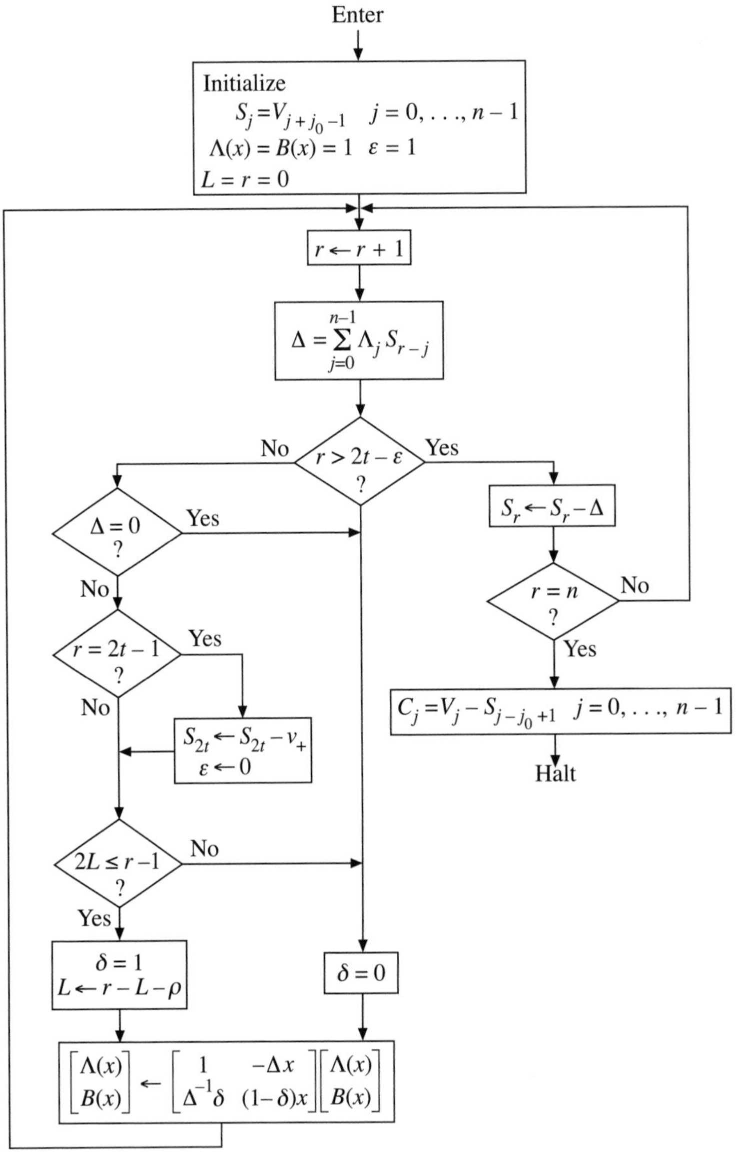

Figure 7.14. A decoder for a singly extended Reed–Solomon code

euclidean algorithm also computes the polynomials $a(x)$ and $b(x)$, satisfying

$$\mathrm{GCD}[s(x), t(x)] = a(x)s(x) + b(x)t(x).$$

We restate the extended euclidean algorithm in a convenient matrix form. For any polynomials $s(x)$ and $t(x)$ with $\deg s(x) \geq \deg t(x)$, let

$$Q^{(r)}(x) = \left\lfloor \frac{s^{(r)}(x)}{t^{(r)}(x)} \right\rfloor$$

denote the quotient polynomial. Let

$$\begin{bmatrix} s^{(r+1)}(x) \\ t^{(r+1)}(x) \end{bmatrix} = \begin{bmatrix} 0 & 1 \\ 1 & -Q^{(r)}(x) \end{bmatrix} \begin{bmatrix} s^{(r)}(x) \\ t^{(r)}(x) \end{bmatrix}$$

and

$$\mathbf{A}^{(r+1)}(x) = \prod_{\ell=r}^{0} \begin{bmatrix} 0 & 1 \\ 1 & -Q^{(\ell)}(x) \end{bmatrix}.$$

This can be rewritten in the form

$$\mathbf{A}^{(r+1)}(x) = \begin{bmatrix} 0 & 1 \\ 1 & -Q^{(r)}(x) \end{bmatrix} \mathbf{A}^{(r)}(x),$$

with an initial condition $\mathbf{A}^{(0)}(x) = \begin{bmatrix} 1 & 0 \\ 0 & 1 \end{bmatrix}$. Then the euclidean recursion can be written

$$\begin{bmatrix} s^{(r+1)}(x) \\ t^{(r+1)}(x) \end{bmatrix} = \mathbf{A}^{(r+1)}(x) \begin{bmatrix} s(x) \\ t(x) \end{bmatrix}.$$

Theorem 7.10.1 (Extended Euclidean Algorithm). *The euclidean recursion satisfies*

$$\begin{aligned} \gamma \mathrm{GCD}[s(x), t(x)] &= s^{(R)}(x) \\ &= A_{11}^{(R)}(x)s(x) + A_{12}^{(R)}(x)t(x), \end{aligned}$$

for some scalar γ, and where R is that r for which $t^{(R)}(x) = 0$.

Proof: The termination is well-defined because $\deg t^{(r+1)}(x) < \deg t^{(r)}(x)$, so eventually $t^{(R)}(x) = 0$ for some R. Therefore

$$\begin{bmatrix} s^{(R)}(x) \\ 0 \end{bmatrix} = \left\{ \prod_{r=R-1}^{0} \begin{bmatrix} 0 & 1 \\ 1 & -Q^{(r)}(x) \end{bmatrix} \right\} \begin{bmatrix} s(x) \\ t(x) \end{bmatrix},$$

and so any divisor of both $s(x)$ and $t(x)$ divides $s^{(R)}(x)$. On the other hand, the matrix inverse

$$\begin{bmatrix} 0 & 1 \\ 1 & -Q^{(r)}(x) \end{bmatrix}^{-1} = \begin{bmatrix} Q^{(r)}(x) & 1 \\ 1 & 0 \end{bmatrix}$$

leads to

$$\begin{bmatrix} s(x) \\ t(x) \end{bmatrix} = \left\{ \prod_{r=0}^{R-1} \begin{bmatrix} Q^{(r)}(x) & 1 \\ 1 & 0 \end{bmatrix} \right\} \begin{bmatrix} s^{(R)}(x) \\ 0 \end{bmatrix},$$

and so $s^{(R)}(x)$ must divide both $s(x)$ and $t(x)$. Thus $s^{(R)}(x)$ divides $\mathrm{GCD}[s(x), t(x)]$. Hence $\mathrm{GCD}[s(x), t(x)]$ divides $s^{(R)}(x)$ and is divisible by $s^{(R)}(x)$. Thus

$$s^{(R)}(x) = \gamma \mathrm{GCD}[s(x), t(x)].$$

Finally, because

$$\begin{bmatrix} s^{(R)}(x) \\ 0 \end{bmatrix} = \boldsymbol{A}^{(R)}(x) \begin{bmatrix} s(x) \\ t(x) \end{bmatrix},$$

we have

$$s^{(R)}(x) = A_{11}^{(R)}(x)s(x) + A_{12}^{(R)}(x)t(x).$$

This proves the last part of the theorem. □

Theorem 7.10.1 relates the matrix elements $A_{11}^{(R)}(x)$ and $A_{12}^{(R)}(x)$ to the greatest common divisor. The other two matrix elements also have a direct interpretation. We will need the inverse of the matrix $\boldsymbol{A}^{(r)}(x)$

$$\boldsymbol{A}^{(r)}(x) = \prod_{\ell=r-1}^{0} \begin{bmatrix} 0 & 1 \\ 1 & -Q^{(\ell)}(x) \end{bmatrix}.$$

From this it is clear that the determinant of $\boldsymbol{A}^{(r)}(x)$ is $(-1)^r$. Because the determinant is not a polynomial, the matrix $\boldsymbol{A}^{(r)}(x)$ does have an inverse. The inverse is

$$\begin{bmatrix} A_{11}^{(r)}(x) & A_{12}^{(r)}(x) \\ A_{21}^{(r)}(x) & A_{22}^{(r)}(x) \end{bmatrix}^{-1} = (-1)^r \begin{bmatrix} A_{22}^{(r)}(x) & -A_{12}^{(r)}(x) \\ -A_{21}^{(r)}(x) & A_{11}^{(r)}(x) \end{bmatrix}.$$

Corollary 7.10.2. *The polynomials $A_{21}^{(R)}(x)$ and $A_{22}^{(R)}(x)$ that are produced by the euclidean algorithm satisfy*

$$s(x) = (-1)^R A_{22}^{(R)}(x)\gamma GCD[s(x), t(x)],$$

$$t(x) = -(-1)^R A_{21}^{(R)}(x)\gamma GCD[s(x), t(x)].$$

Proof: Using the above expression for the inverse gives

$$\begin{bmatrix} s(x) \\ t(x) \end{bmatrix} = (-1)^R \begin{bmatrix} A_{22}^{(R)}(x) & -A_{12}^{(R)}(x) \\ -A_{21}^{(R)}(x) & A_{11}^{(R)}(x) \end{bmatrix} \begin{bmatrix} s^{(R)}(x) \\ 0 \end{bmatrix},$$

from which the corollary follows. □

Now, recall that the syndrome polynomial, the error-locator polynomial, and the error-evaluator polynomial are the unique polynomials that are related by the formula

$$\Gamma(x) = S(x)\Lambda(x) \pmod{x^{2t+1}},$$

and satisfy the conditions that $\deg \Lambda(x) \leq t$ and $\deg \Gamma(x) \leq t$. Let us look to the proof of the euclidean algorithm to see how this equation can be solved for $\Lambda(x)$ and $\Gamma(x)$. From this proof, we see that

$$\begin{bmatrix} s^{(r)}(x) \\ t^{(r)}(x) \end{bmatrix} = \begin{bmatrix} A_{11}^{(r)}(x) & A_{12}^{(r)}(x) \\ A_{21}^{(r)}(x) & A_{22}^{(r)}(x) \end{bmatrix} \begin{bmatrix} s(x) \\ t(x) \end{bmatrix},$$

and thus

$$t^{(r)}(x) = t(x)A_{22}^{(r)}(x) \pmod{s(x)},$$

which is the form of the equation that we wish to solve if we take $t(x) = S(x)$ and $s(x) = x^{2t+1}$. For each r, such an equation holds and gives a pair $(t^{(r)}(x), A_{22}^{(r)}(x))$ that satisfies the required equation. If for any r the polynomials have degrees that meet the constraints, then the problem is solved. Thus we need to find, if it exists, an r for which $\deg A_{22}^{(r)}(x) \leq t$ and $\deg t^{(r)}(x) \leq t$. Choose ρ as the first value of r satisfying

$$\deg t^{(\rho)}(x) \leq t.$$

Such a value ρ must exist because $\deg t^{(0)}(x) = 2t$ and the degree of $t^{(r)}(x)$ is strictly decreasing as r is increasing.

Algorithm 7.10.3. *Let $s^{(0)}(x) = x^{2t}$ and let $t^{(0)}(x) = S(x)$ be a polynomial of degree at most $2t$. Let $\mathbf{A}^{(0)}(x)$ be the two by two identity matrix, and*

$$Q^{(r)}(x) = \left\lfloor \frac{s^{(r)}(x)}{t^{(r)}(x)} \right\rfloor$$

$$\mathbf{A}^{(r+1)}(x) = \begin{bmatrix} 0 & 1 \\ 1 & -Q^{(r)}(x) \end{bmatrix} \mathbf{A}^{(r)}(x)$$

$$\begin{bmatrix} s^{(r+1)}(x) \\ t^{(r+1)}(x) \end{bmatrix} = \begin{bmatrix} 0 & 1 \\ 1 & -Q^{(r)}(x) \end{bmatrix} \begin{bmatrix} s^{(r)}(x) \\ t^{(r)}(x) \end{bmatrix}.$$

Then with ρ equal to the first value of r for which $\deg t^{(r)}(x) \leq t$, the polynomials

$$\Gamma(x) = t^{(\rho)}(x)$$
$$\Lambda(x) = A_{22}^{(\rho)}(x)$$

are solutions of

$$\Gamma(x) = S(x)\Lambda(x) \pmod{x^{2t+1}},$$

satisfying $\deg \Lambda(x) \leq t$, and $\deg \Gamma(x) \leq t$.

Proof: Let ρ be the first value of r for which the condition $\deg t^{(\rho)}(x) \leq t$ is satisfied. Then the condition $\deg \Gamma(x) \leq t$ is ensured by the definition of the stopping rule.

Furthermore,

$$t^{(\rho)}(x) = t(x)A_{22}^{(\rho)}(x) \pmod{x^{2t+1}}$$

is an immediate consequence of the equations of the euclidean algorithm.

The degree of $A_{22}^{(r)}(x)$ increases with r. We will show that

$$\deg A_{22}^{(\rho)}(x) \le t.$$

We will prove this by working with the inverse of the matrix $\boldsymbol{A}^{(r)}(x)$. First, recall that

$$\boldsymbol{A}^{(r)}(x) = \prod_{r'=0}^{r-1} \begin{bmatrix} 0 & 1 \\ 1 & -Q^{(r')}(x) \end{bmatrix}.$$

From this equation it is clear that $\deg A_{22}^{(r)}(x) > \deg A_{12}^{(r)}(x)$. Also recall that $\deg s^{(r)}(x) > \deg t^{(r)}(x)$. From these inequalities and the matrix equation

$$\begin{bmatrix} s(x) \\ t(x) \end{bmatrix} = (-1)^r \begin{bmatrix} A_{22}^{(r)}(x) & -A_{12}^{(r)}(x) \\ -A_{21}^{(r)}(x) & A_{11}^{(r)}(x) \end{bmatrix} \begin{bmatrix} s^{(r)}(x) \\ t^{(r)}(x) \end{bmatrix},$$

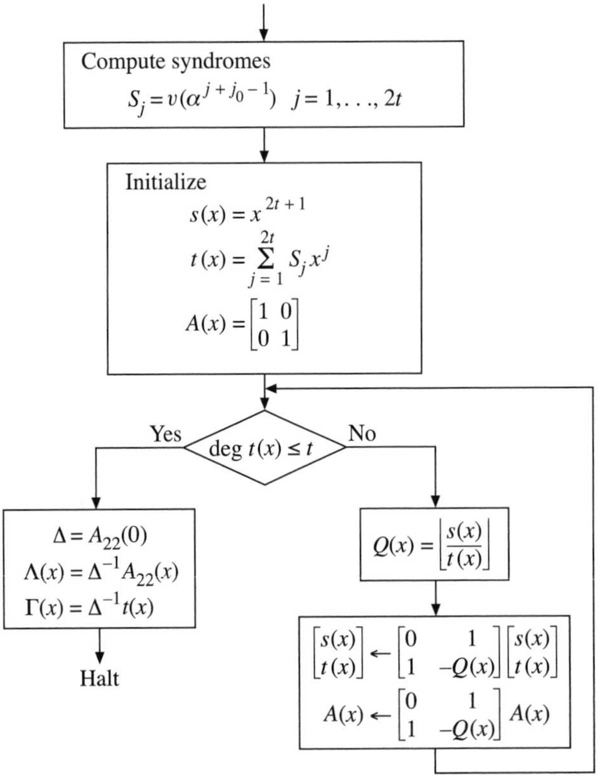

Figure 7.15. The Sugiyama algorithm

it is clear that $\deg s(x) = \deg A_{22}^{(r)}(x) + \deg s^{(r)}(x)$. Because $s^{(r)}(x) = t^{(r-1)}(x)$, and $\deg t^{(\rho-1)}(x) \geq t + 1$, this becomes

$$\begin{aligned}\deg A_{22}^{(\rho)}(x) &= \deg s(x) - \deg t^{(\rho-1)}(x) \\ &\leq 2t + 1 - (t + 1) = t,\end{aligned}$$

which completes the proof of the algorithm. □

If $S(x)$ is the syndrome polynomial of a t-error-correcting code in the presence of at most t errors, then Algorithm 7.10.3, sometimes called the *Sugiyama algorithm*, can be used to compute both the error-locator polynomial and the error-evaluator polynomial.

A flow chart of the Sugiyama algorithm for computing $\Lambda(x)$ and $\Gamma(x)$ is shown in Figure 7.15. To impose the condition $\Lambda_0 = 1$, which is not always necessary, one normalizes by writing

$$\Gamma(x) = \Delta^{-1} t^{(\rho)}(x)$$
$$\Lambda(x) = \Delta^{-1} A_{22}^{(\rho)}(x)$$

where $\Delta = A_{22}^{(p)}(0)$. The division by Δ ensures that $\Lambda_0 = 1$. Uniqueness is automatic because there is only one such solution for the syndrome of a t-error-correcting BCH code in the presence of at most t errors.

Problems

7.1 a. Find $g(x)$ for the double-error-correcting Reed–Solomon code based on the primitive element $\alpha \in GF(2^4)$ and $j_0 = 1$.
b. If $S_1 = \alpha^4$, $S_2 = 0$, $S_3 = \alpha^8$, and $S_4 = \alpha^2$, find the error-locator polynomial, the error locations, and the error values by using the Gorenstein–Zierler algorithm.
c. Repeat part (b) using the Berlekamp–Massey algorithm.
d. Repeat part (b) using the Sugiyama algorithm.

7.2 Suppose that a Reed–Solomon code has minimum distance $2t + 1$ and that ν is an integer smaller than t. We wish to correct the message if at most ν errors occur and to detect a bad message if more than ν and at most $2t - \nu$ errors occur. Describe how to accomplish this by using the Berlekamp–Massey algorithm. Repeat for a BCH code.

7.3 The polynomial $y^{20} + y^3 + 1$ is a primitive polynomial over $GF(2)$. The primitive element $\alpha = y$, with minimal polynomial $y^{20} + y^3 + 1$, is used to construct a binary double-error-correcting BCH code of blocklength $2^{20} - 1 = 1,048,575$.
a. A received message has syndromes

$$S_1 = y^4 \quad S_3 = y^{12} + y^9 + y^6.$$

Find the remaining syndromes.

b. Find the error-locator polynomial.

c. Find the location of the error or errors.

7.4 a. Find $g(x)$ for a binary BCH code of blocklength $n = 15$ with $d = 7$. Use a primitive α, $j_0 = 1$, and the irreducible polynomial $x^4 + x + 1$.

b. In the code of part (a), suppose a sequence $\boldsymbol{v}$ is received such that

$$S_1 = v(\alpha) = \alpha + \alpha^2,\ S_3 = v(\alpha^3) = \alpha^2,\ \text{and } S_5 = v(\alpha^5) = 1.$$

(i) Find S_2, S_4, and S_6.

(ii) Assuming that $\nu \leq 3$ errors took place, find ν.

(iii) Find the error-locator polynomial.

(iv) Find the error.

7.5 Show that the syndromes of a BCH code, given by

$$S_j = \sum_{j=0}^{n-1} \omega^{i(j+j_0)} v_i \quad j = 1, \ldots, 2t,$$

are a *sufficient statistic* for computing $\boldsymbol{e}$. That is, given the code $\mathcal{C}$, show that no information about $\boldsymbol{e}$ that is contained in $\boldsymbol{v}$ is lost by the map to syndromes.

7.6 Describe how an errors-only decoder for a binary code can be used twice to decode a senseword with $2t$ erasures. How can it be used to correct both errors and erasures?

7.7 A time-domain decoder with $2t$ clocks that uses Forney's algorithm can be developed if the inverse Fourier transforms of $\Gamma(x)$ and $\Lambda'(x)$ are available. Develop iterative equations similar to those of the Berlekamp–Massey algorithm to iteratively compute $\Gamma(x)$ and $\Lambda'(x)$. Use the inverse Fourier transforms of these equations to describe a time-domain decoder with $2t$ clocks.

7.8 Derive the Forney algorithm for arbitrary j_0.

7.9 What is the relationship between the syndrome polynomial $s(x)$ in the time domain, defined as $s(x) = R_{g(x)}[v(x)]$, and the syndrome polynomial in the frequency domain, defined as $S(x) = \sum_{j=1}^{2t} v(\omega^{j+j_0-1})x^j$. Give expressions for each in terms of the other.

7.10 Describe how a decoder for Reed–Solomon codes can be readily modified to decode any binary code up to the Hartmann–Tzeng bound. Can erasures be corrected as well as errors?

7.11 The polynomial $p(y) = y^3 + 2y + 1$ is a primitive polynomial over $GF(3)$. Minimal polynomials for selected elements of $GF(27)$ are as follows:

$$m_{\alpha^0}(x) = x - 1$$
$$m_{\alpha^1}(x) = x^3 - x + 1$$
$$m_{\alpha^2}(x) = x^3 + x^2 + x - 1$$
$$m_{\alpha^3}(x) = x^3 - x + 1$$

$$m_{\alpha^4}(x) = x^3 + x^2 - 1$$
$$m_{\alpha^5}(x) = x^3 - x^2 + x + 1$$
$$m_{\alpha^6}(x) = x^3 + x^2 + x - 1.$$

a. Without completing this list, state how many distinct minimal polynomials there are for elements of $GF(27)$.
b. Find the generator polynomial for a (26, 19) double-error-correcting BCH code over $GF(3)$.
c. What is the rate of this code? How many codewords are in the code? What can be said about the minimum distance of the code?
d. Given the senseword $v(x) = 2x^2 + 1$, prepare a table of iterates of the Berlekamp–Massey algorithm. Verify that the zeros of the computed error-locator polynomial point to the error locations.

7.12 The polynomial $p(y) = y^8 + y^4 + y^3 + y^2 + 1$ is a primitive polynomial for $GF(2)$. Based on this polynomial, a Reed–Solomon code is designed that has spectral zeros at $\alpha, \alpha^2, \alpha^3, \alpha^4$.

a. What is the code alphabet size? What is the rate of the code? What is the minimum distance?
b. A received message has syndromes

$$S_1 = y^4$$
$$S_2 = y^5 + y^4 + y^3 + y$$
$$S_3 = y^4 + y + 1$$
$$S_4 = y^6 + y^4 + y^3 + y.$$

Find the error locations and error magnitudes.

7.13 Given the Berlekamp iteration in the form

$$\begin{bmatrix} \Lambda^{(r)}(x) & \Gamma^{(r)}(x) \\ B^{(r)}(x) & A^{(r)}(x) \end{bmatrix} = \begin{bmatrix} 1 & -\Delta x \\ \Delta^{-1}\delta & (1-\delta)x \end{bmatrix} \begin{bmatrix} \Lambda^{(r-1)}(x) & \Gamma^{(r-1)}(x) \\ B^{(r-1)}(x) & A^{(r-1)}(x) \end{bmatrix},$$

with initialization

$$\begin{bmatrix} \Lambda^{(0)}(x) & \Gamma^{(0)}(x) \\ B^{(0)}(x) & A^{(0)}(x) \end{bmatrix} = \begin{bmatrix} 1 & 0 \\ 1 & -1 \end{bmatrix},$$

show that

$$\Gamma(x)B(x) - \Lambda(x)A(x) = x^{2t}.$$

Explain how this equation can be evaluated at the error locations to find the error values. This is a variation of an algorithm known as the *Horiguchi–Koetter algorithm*.

Notes

Locator decoding for BCH codes was introduced by Peterson (1960) for binary codes, and by Gorenstein and Zierler (1961) for nonbinary codes. Simplifications were found by Chien (1964) and Forney (1965). An iterative algorithm for finding the error-locator polynomial was discovered by Berlekamp (1968). Massey (1969) rederived and simplified the algorithm as a procedure for designing linear-feedback shift registers. Burton (1971) showed how to implement the Berlekamp–Massey algorithm without division, but division is still needed to compute the error magnitudes. The development of the Berlekamp–Massey algorithm using the linear complexity of sequences is due to Maurer and Viscardi (1984).

The decoding technique described by Reed and Solomon (1960) for their codes is somewhat similar to the frequency-domain techniques developed in this chapter, but thereafter techniques evolved primarily in terms of linear algebra. What amounts to a Fourier transform decoder was first proposed by Mandelbaum (1971), although in the terminology of the Chinese remainder theorem. Such a decoder was implemented in a computer program by Paschburg (1974).

The first explicit discussion of decoding from the spectral point of view can be found in work by Gore (1973), who pointed out that the data can be encoded in the frequency domain and also that the error spectrum can be obtained by recursive extension. The implementation of transform-domain decoders was also discussed by Michelson (1976). A general discussion of spectral decoding techniques can be found in Blahut (1979). The reflection of these decoders back into the time domain is from Blahut (1979, 1984).

The decoding algorithm that uses the euclidean algorithm as a subalgorithm was first discovered by Sugiyama *et al.* (1975). Welch and Scholtz (1979) derived a decoder using continued fractions that is quite similar to the use of the euclidean algorithm. Another way of using the euclidean algorithm to decode is due to Mandelbaum (1977). Another alternative to the Berlekamp–Massey algorithm is the Welch–Berlekamp algorithm (1983).

The version of the decoder for singly extended codes that appears in this chapter is from Blahut (1984). A decoder for doubly extended codes was described by Jensen (1995). Early work on extending decoding algorithms for BCH codes to both erasures and errors was done by Blum and Weiss (1960), and by Forney (1965). The technique of initializing the Berlekamp–Massey algorithm with the erasure-locator polynomial is due to Blahut (1979).

Decoding beyond the BCH bound can be found in Berlekamp (1968), Hartmann (1972), and Blahut (1979). A very different approach to coding beyond the BCH bound is the Sudan algorithm (1997). A complete algebraic decoding algorithm for BCH codes with a minimum distance equal to five was given by Berlekamp (1968), and a complete algebraic decoding algorithm for some BCH codes with a minimum distance equal to seven was given by Vanderhorst and Berger (1976). Except for these examples, complete decoding algorithms are rare. A comparison of the malfunction of various BCH decoders when the error is beyond the packing radius was discussed by Sarwate and Morrison (1990).

8 Implementation

In this chapter we shall turn our attention to the implementation of encoders and decoders. Elementary encoders and decoders will be based on shift-register circuits; more advanced encoders and decoders will be based on hardware circuits and software routines for Galois-field arithmetic.

Digital logic circuits can be organized easily into shift-register circuits that mimic the cyclic shifts and polynomial arithmetic used in the description of cyclic codes. Consequently, the structure of cyclic codes is closely related to the structure of shift-register circuits. These circuits are particularly well-suited to the implementation of many encoding and decoding procedures and often take the form of filters. In fact, many algorithms for decoding simple cyclic codes can be described most easily by using the symbolism of shift-register circuits. Studying such decoders for simple codes is worthwhile for its own sake, but it is also a good way to build up the insights and techniques that will be useful for designing decoders for large codes.

8.1 Logic circuits for finite-field arithmetic

Logic circuits are easily designed to execute the arithmetic of Galois fields, especially if q is a power of 2. We shall need circuit elements to store field elements and to perform arithmetic in the finite field.

A shift register, as shown in Figure 8.1, is a string of storage devices called *stages*. Each stage contains one element of $GF(q)$. The symbol contained in each storage device is displayed on an output line leaving that stage. Each storage device also has an input line that carries an element from $GF(q)$. When it is not designated otherwise, this input symbol is taken to be the zero element of the field. At discrete time instants known as *clock times*, the field element in the storage device is replaced by the field element on the input line. Modern electronic shift registers can be clocked at speeds in excess of a hundred million times per second.

In addition to shift-register stages, we shall use three other circuit elements, as shown in Figure 8.2. These are a *scaler*, an *adder*, and a *multiplier*. The scaler is a function of a single input variable. It multiplies an input field element of $GF(q)$ by a constant field

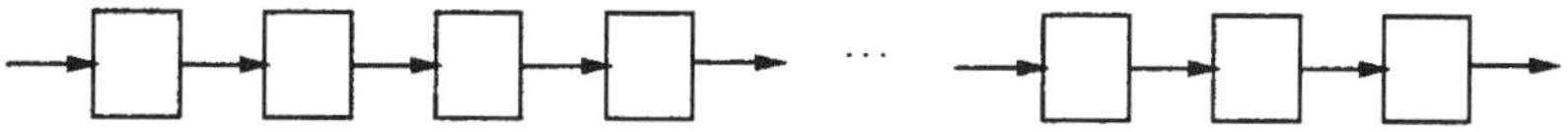

Figure 8.1. Shift register

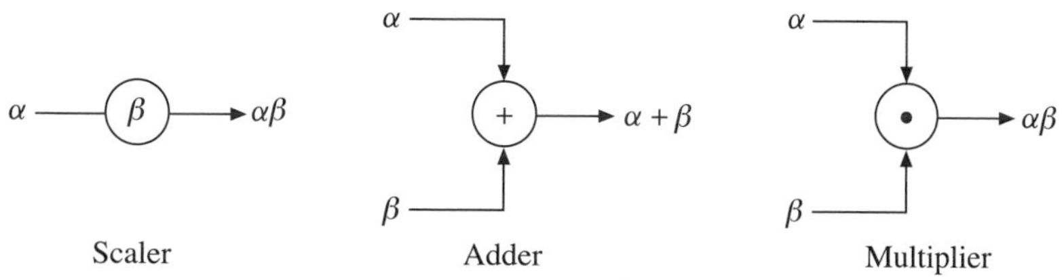

Figure 8.2. Circuit elements

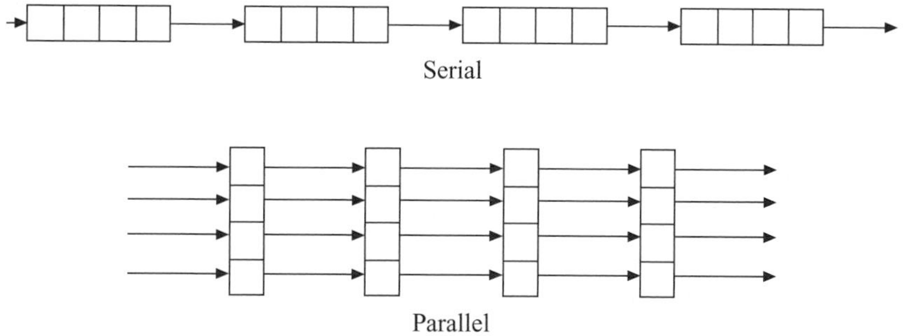

Figure 8.3. Hexadecimal shift registers made from binary components

element of $GF(q)$. The adder and multiplier are functions of two inputs from $GF(q)$. In the case of $GF(2)$, the adder is also known as an *exclusive-or gate*, and the multiplier is also known as an *and gate*.

A field element from $GF(2^m)$ can be represented by m bits and can be transferred within a circuit either serially (one bit at a time on a single wire), or in parallel (one bit on each of m parallel wires). Figure 8.3 shows the arrangement of binary shift-register stages for use with $GF(16)$. Each field element is represented by four bits in a four-bit shift register. A sequence of field elements is represented by a sequence of four-bit numbers in a chain of four-bit shift registers. It takes four clock times to shift a field element to the next stage.

Figure 8.4 shows a circuit for addition in $GF(16)$ with a serial data flow and a circuit for addition in $GF(16)$ with a parallel data flow. In each case, the addends are in shift registers at the start and, at the conclusion of the addition, the sum is in a third shift register. The circuit for serial addition requires four clock times to complete the addition; the circuit for parallel addition requires only one clock time but has more

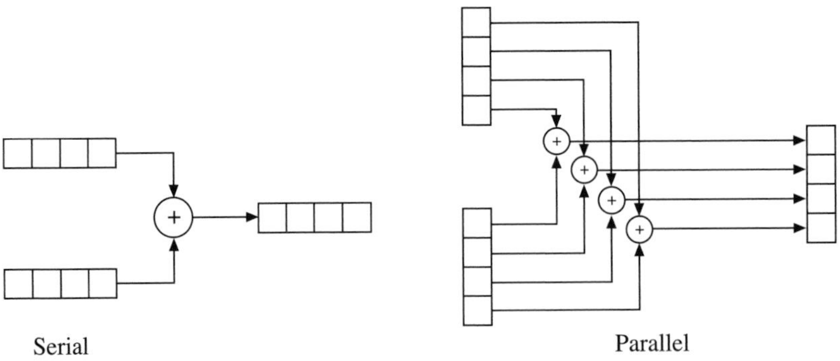

Figure 8.4. Addition of two field elements

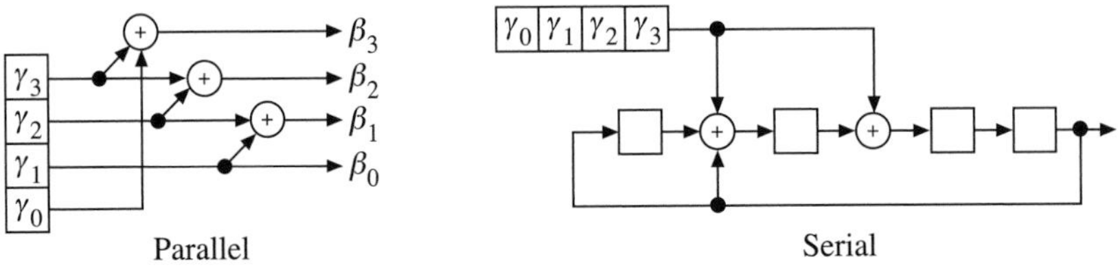

Figure 8.5. Multiplication by a constant field element $\beta = z^2 + z$

wires and modulo-2 adders. If desired, the addends could be fed back to the inputs of their own registers so that, at the end of the addition, the addends can be found in those registers for the next step of the computation.

A scaler in $GF(q)$ multiplies an arbitrary element of $GF(q)$ by a fixed element of $GF(q)$. The more difficult problem of multiplication of a variable by a variable is discussed in Sections 8.7 and 8.8. To explain a scaler, we design circuits for the multiplication of a field element by a fixed field element $\beta = z^5 = z^2 + z$ in $GF(16)$. We choose the specific representation of $GF(16)$ constructed with the primitive polynomial $p(z) = z^4 + z + 1$. Let $\gamma = \gamma_3 z^3 + \gamma_2 z^2 + \gamma_1 z + \gamma_0$ be an arbitrary field element. The computation of $\beta\gamma$ is as follows:

$$\begin{aligned}\beta\gamma &= \gamma_3 z^8 + \gamma_2 z^7 + \gamma_1 z^6 + \gamma_0 z^5 \pmod{z^4 + z + 1} \\ &= (\gamma_2 + \gamma_1)z^3 + (\gamma_3 + \gamma_1 + \gamma_0)z^2 + (\gamma_2 + \gamma_0)z + (\gamma_3 + \gamma_2).\end{aligned}$$

From this equation, the parallel multiplication circuit can be designed as in Figure 8.5. The serial circuit implements the two lines of the above equation. First, multiply by $z^2 + z$, then use a division circuit to reduce modulo $z^4 + z + 1$. The serial circuit takes four clock times to complete the multiplication by β after which the product $\beta\gamma$ is in the register.

Shift-register circuits can be used for the multiplication and division of polynomials over $GF(q)$. Consequently, they can be used in the design and construction of encoders and decoders for cyclic codes. Shift registers can also play a role as a quasi-mathematical notation that is an easily understood representation of some kinds of

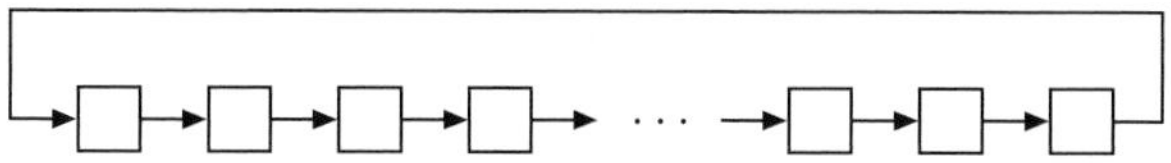

Figure 8.6. Circuit for cyclically shifting a polynomial

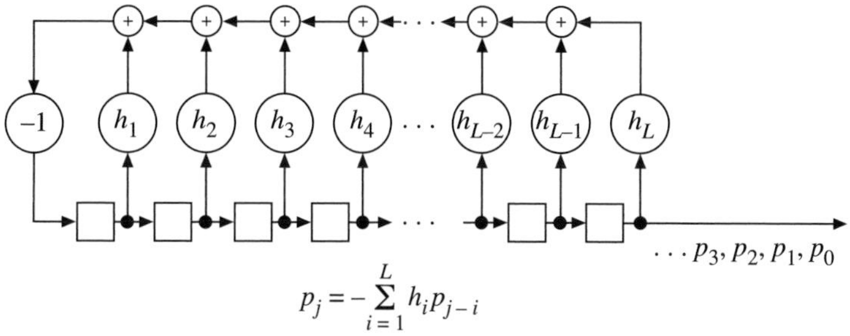

Figure 8.7. A linear-feedback shift register

polynomial manipulation. Thus, shift register circuits can represent polynomial relationships.

The n symbols contained in a shift register of length n may be interpreted as the coefficients of a polynomial of degree $n - 1$. Our usual convention will be that the shift-register circuits are shifted from left to right. Sometimes this requires that the polynomial coefficients appear in the shift register in descending order from right to left. Unfortunately, this is opposite to the most common convention used in writing polynomials in which the high-order coefficient is at the left.

A shift register connected in a ring can be used to cyclically shift a polynomial. Figure 8.6 shows an n-stage shift register used to cyclically shift a polynomial of degree $n - 1$. It computes $xv(x) (\bmod x^n - 1)$. This is the simplest example of a linear-feedback shift register.

A general linear-feedback shift register is shown in Figure 8.7. This circuit implements the recursion

$$p_j = -\sum_{i=1}^{L} h_i p_{j-i} \quad j \geq L.$$

When loaded initially with L symbols, $p_0, \ldots, p_{L-1}$, the shift-register output will be an unending sequence of symbols, starting at p_0 and satisfying the above recursion. Because of the feedback, this circuit is a member of a large class of filters known as *recursive filters*. The linear-feedback shift register is known as an *autoregressive filter* when used with an input, as in Figure 8.8. This circuit implements the equation

$$p_j = -\sum_{i=1}^{L} h_i p_{j-i} + a_j.$$

Instead of feeding back the shift-register tap outputs, one can feed them forward and use an externally generated sequence as input to the shift register. This

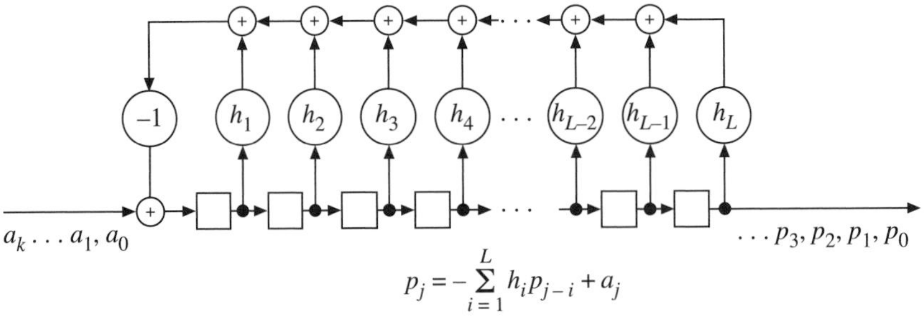

Figure 8.8. An autoregressive filter

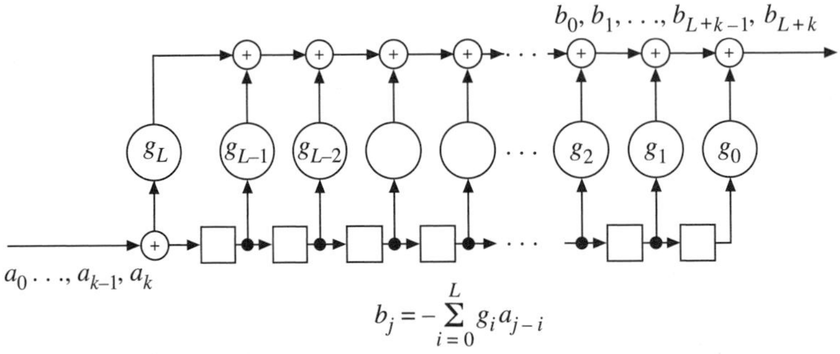

Figure 8.9. A linear-feedforward shift register

linear-feedforward shift register is shown in Figure 8.9. It is also called a *finite-impulse-response filter*, or a *nonrecursive filter*. Let the coefficients of the polynomial

$$g(x) = g_L x^L + \cdots + g_1 x + g_0$$

be equal to the tap weights of a feedforward shift register, and let the input and output sequences be represented by the polynomials

$$a(x) = a_k x^k + \cdots + a_1 x + a_0,$$
$$b(x) = b_{k+L} x^{k+L} + \cdots + b_1 x + b_0.$$

Then the polynomial product

$$b(x) = g(x)a(x)$$

is a representation of the operation of the shift register in Figure 8.9, with the understanding that the shift register contains zeros initially, and input a_0 is followed by L zeros. The coefficients of $a(x)$ and $g(x)$ are said to be convolved by the shift register because

$$b_j = \sum_{i=0}^{L} g_i a_{j-i}.$$

In terms of polynomials, the feedforward filter may be viewed as a device to multiply an arbitrary polynomial $a(x)$ by the fixed polynomial $g(x)$. We will also call it a multiply-by-$g(x)$ circuit.

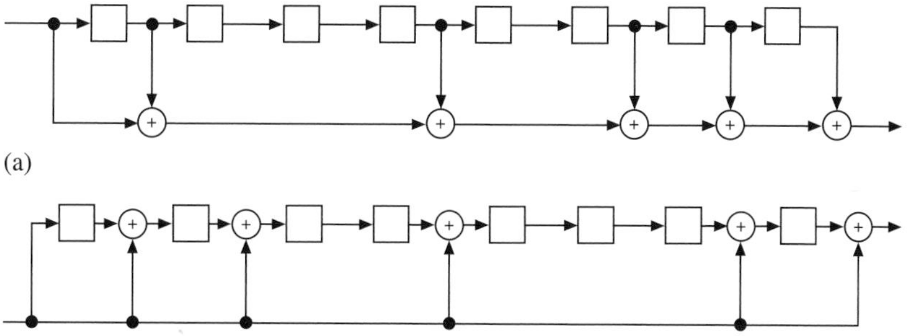

(a)

(b)

Figure 8.10. Two circuits for multiplication by $x^8 + x^7 + x^4 + x^2 + x + 1$

In Figure 8.10(a), we give an example of a multiply-by-$g(x)$ circuit in $GF(2)$ for the polynomial $g(x) = x^8 + x^7 + x^4 + x^2 + x + 1$. The circuit is a finite-impulse response filter over $GF(2)$. Notice that the internal stages of the shift register are read but not modified. It is possible to give an alternative arrangement in which the internal stages are modified, but this alternative is usually a more expensive circuit. In Figure 8.10(b), we show this alternative multiply-by-$g(x)$ circuit.

A shift-register circuit also can be used to divide an arbitrary polynomial by a fixed polynomial. The circuit that does this mimics the standard procedure for polynomial division. So that the divisor is a monic polynomial, a scalar can be factored out and handled separately. Long division proceeds as follows:

$$
\begin{array}{r|llll}
 & a_{n-1}x^{k-1} + & (a_{n-2} - a_{n-1}g_{n-k-1})x^{k-2} + \cdots & & \\
\hline
x^{n-k} + g_{n-k-1}x^{n-k-1} + \cdots + g_0 & a_{n-1}x^{n-1} + & a_{n-2}x^{n-2} + \cdots & & +a_1x + a_0 \\
 & a_{n-1}x^{n-1} + & a_{n-1}g_{n-k-1}x^{n-2} + \cdots & & \\
\cline{2-3}
 & & (a_{n-2} - a_{n-1}g_{n-k-1})x^{n-2} + \cdots & & \\
 & & (a_{n-2} - a_{n-1}g_{n-k-1})x^{n-2} + \cdots & & \\
\cline{2-3}
 & & \cdots & &
\end{array}
$$

This familiar procedure can be reformulated as a recursion on a pair of equations. Let $Q^{(r)}(x)$ and $R^{(r)}(x)$ be the trial quotient polynomial and the trial remainder polynomial, respectively, at the rth recursion, with initial conditions $Q^{(0)}(x) = 0$ and $R^{(0)}(x) = a(x)$. The recursive equations are:

$$
\begin{aligned}
Q^{(r)}(x) &= Q^{(r-1)}(x) + R_{n-r}^{(r-1)} x^{k-r}, \\
R^{(r)}(x) &= R^{(r-1)}(x) - R_{n-r}^{(r-1)} x^{k-r} g(x).
\end{aligned}
$$

The quotient and remainder polynomials are obtained after k iterations as $Q^{(k)}(x)$ and $R^{(k)}(x)$.

The circuit of Figure 8.11 is a circuit for dividing an arbitrary polynomial by the fixed polynomial $g(x)$. It may be understood by referring to either of the two descriptions of the long-division process described previously and noting that the circuit repeatedly subtracts multiples of $g(x)$. The only thing left out of the circuit is the subtraction of the term $R_{n-r}x^{n-r}$ from itself, which is unnecessary to compute because it is always

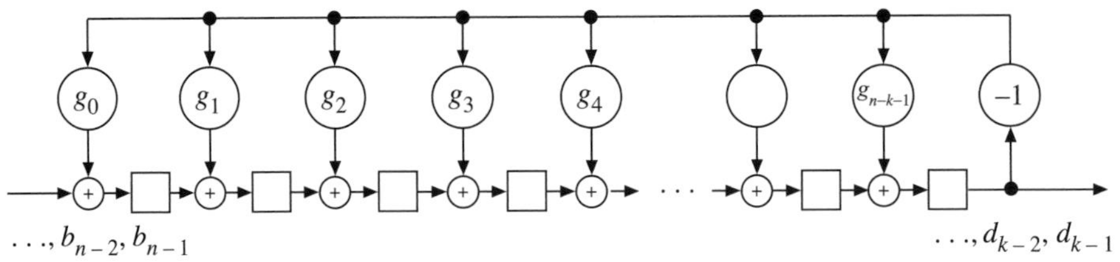

Figure 8.11. A circuit for dividing $b(x)$ by $g(x)$

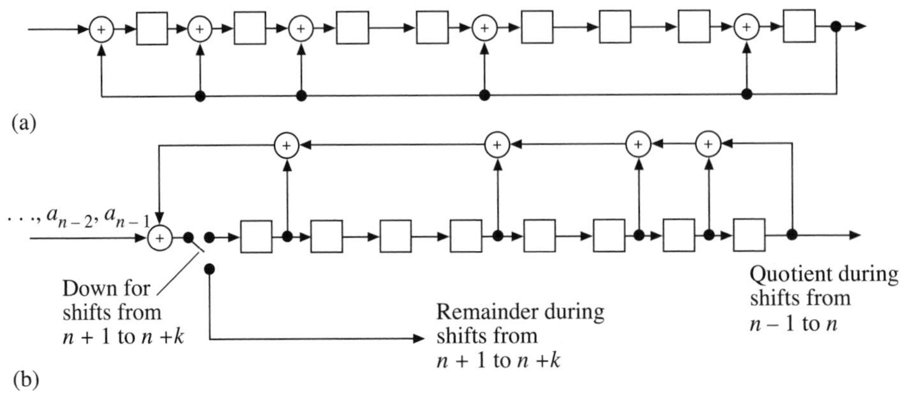

Figure 8.12. Two circuits for division by $x^8 + x^7 + x^4 + x^2 + x + 1$

zero. After n shifts, the quotient polynomial has been passed out of the shift register, and the remainder polynomial is found in the shift register.

In Figure 8.12(a), we give an example in $GF(2)$ of a divide-by-$g(x)$ circuit for the polynomial $g(x) = x^8 + x^7 + x^4 + x^2 + x + 1$.

Notice that the divide circuit requires additions to internal stages of the shift register. This is often an inconvenient circuit operation. It is possible to use an alternative circuit instead that only reads the internal shift-register stages, but does not modify them. To develop this circuit, we organize the work of polynomial division in another way. Refer to the longhand example that was written previously. The idea is to leave the subtractions pending until all subtractions in the same column can be done at the same time. To see how to do this, notice that we can write

$$R^{(r)}(x) = a(x) - Q^{(r)}(x)g(x),$$

and thus

$$R_{n-r}^{(r-1)} = a_{n-r} - \sum_{i=1}^{n-1} g_{n-r-i} Q_i^{(r-1)}$$

and

$$Q^{(r)}(x) = Q^{(r-1)}(x) + R_{n-r}^{(r-1)} x^{k-r}.$$

These equations can be implemented by a modification of a multiply-by-$g(x)$ circuit. Figure 8.12(b) shows such a circuit for dividing by the polynomial $g(x) = x^8 + x^7 + x^4 + x^2 + x + 1$.

8.2 Shift-register encoders and decoders

The structure of cyclic codes makes it attractive to use shift-register circuits for constructing encoders and decoders. Cyclic codes can be encoded nonsystematically by multiplication of the data polynomial $a(x)$ by the fixed polynomial $g(x)$ to obtain the codeword $c(x)$. This multiplication can be implemented with a feedforward shift register over $GF(q)$. A shift-register encoder for the (15, 11) Hamming code is shown in Figure 8.13. To encode a long stream of data bits into a sequence of (15, 11) Hamming codewords, simply break the datastream into eleven-bit blocks, insert a pad of four zeros after every data block, and pass the padded stream through the filter. This provides a sequence of noninteracting fifteen-bit Hamming codewords. The encoder, shown in Figure 8.14, is simple; the codewords, however, are not systematic.

If one wants a systematic codeword, a different encoder must be used. Insert the data into the high-order bits of the codeword and choose the check bits so that a legitimate codeword is obtained. The codeword is of the form

$$c(x) = x^{n-k}a(x) + t(x)$$

where

$$t(x) = -R_{g(x)}[x^{n-k}a(x)],$$

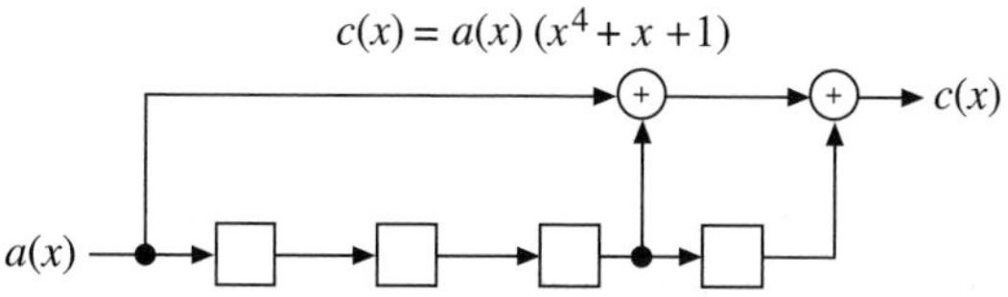

Figure 8.13. Nonsystematic (15, 11) encoder

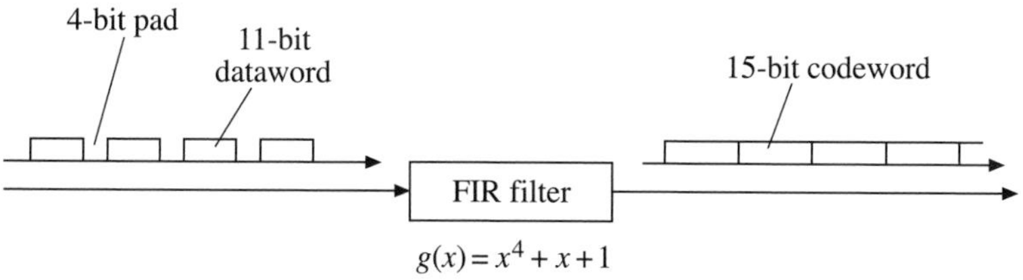

Figure 8.14. Encoding a long bit stream

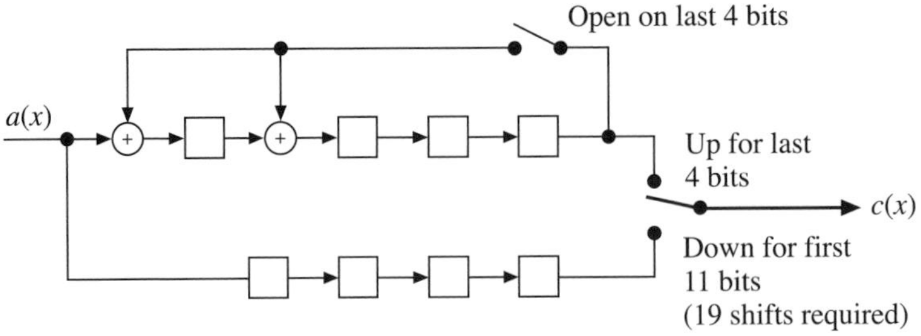

Figure 8.15. A systematic encoder for a (15, 11) Hamming code

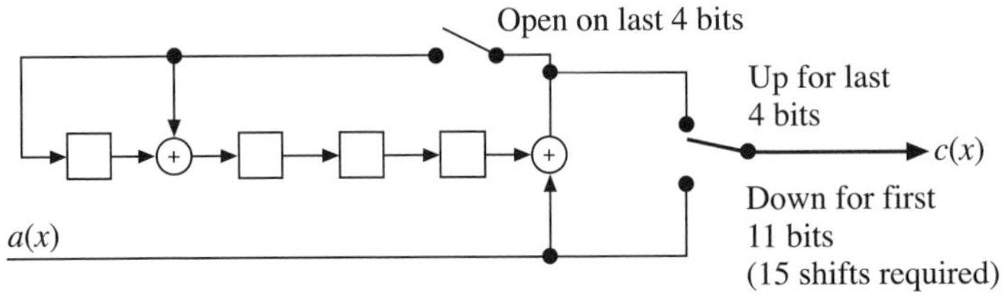

Figure 8.16. Another systematic encoder for a (15, 11) Hamming code

and thus

$$R_{g(x)}[c(x)] = 0.$$

The systematic encoder is implemented by using a divide-by-$g(x)$ circuit. For the (15, 11) Hamming code,

$$t(x) = R_{g(x)}[x^4 a(x)].$$

An implementation is shown in Figure 8.15. Eleven data bits, with the high-order bit first, are shifted into the divide-by-$(x^4 + x + 1)$ circuit from the left. Multiplication by x^4 is implicit in the timing of the circuit. The first bit that arrives is thought of as the coefficient of x^{15}. The division does not begin until four timing shifts are completed in order to position the first four bits in the register. Because of this, a four-stage buffer is included below the divide circuit so that the first bit is sent to the channel at the exact time the first step of the division occurs. After eleven steps of the division, all data bits have been sent to the channel, the division is complete, and the remainder, which provides the check bits, is ready to shift into the channel. During these last four shifts, the feedback path in the divide-by-$g(x)$ circuit is opened. Altogether it takes nineteen clock counts to complete the encoding.

It is possible to delete the initial four timing shifts by a slight change in the circuit. This is shown in Figure 8.16. To understand this circuit, one should notice that the incoming data bits need not immediately enter the divide-by-$g(x)$ circuit as long as they enter in time to form the feedback signal. The feedback in Figure 8.16 is the same as in Figure 8.15. Further, because the last bits of $x^4 a(x)$ are zero anyway, it will not matter if we fail to add them into the remainder. Thus Figure 8.16 computes the same

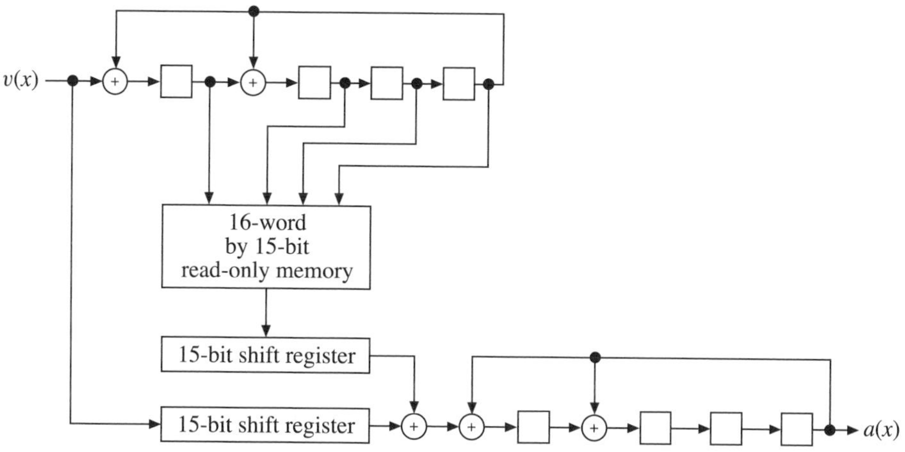

Figure 8.17. Syndrome decoder for a nonsystematic (15, 11) Hamming code

remainder as Figure 8.15, but it computes it and shifts it out in only fifteen clock counts, and so is much more convenient to use.

Now we will turn to the decoder. The channel transmits the coefficients of the polynomial $c(x)$. To this is added the error polynomial $e(x)$. The channel output is the received polynomial

$$v(x) = c(x) + e(x).$$

In Section 5.2, we described a table look-up decoding procedure that is conceptually simple. The senseword polynomial $v(x)$ is divided by the generator polynomial $g(x)$, and the remainder is the syndrome polynomial. The syndrome polynomial is used to retrieve the estimated error pattern from the syndrome evaluator table. For a binary code, the coefficients of the syndrome polynomial can be used directly as a memory address; the corresponding error pattern $e(x)$ is stored at that memory location.

Figure 8.17 shows a shift-register decoder for the nonsystematic (15, 11) Hamming code. There are four syndrome bits, and hence a memory with a four-bit address and a fifteen-bit word length is required. Such an approach may be practical for syndromes of small length, but we will see in the next section that other techniques can also be used. After the correct codeword $c(x)$ is recovered, the data bits are computed by using a divide-by-$g(x)$ circuit that implements the equation

$$a(x) = c(x)/g(x),$$

and this completes the decoder.

8.3 The Meggitt decoder

The most complicated part of the shift-register decoder described in the preceding section is the tabulation of the precomputed syndrome polynomials and their associated

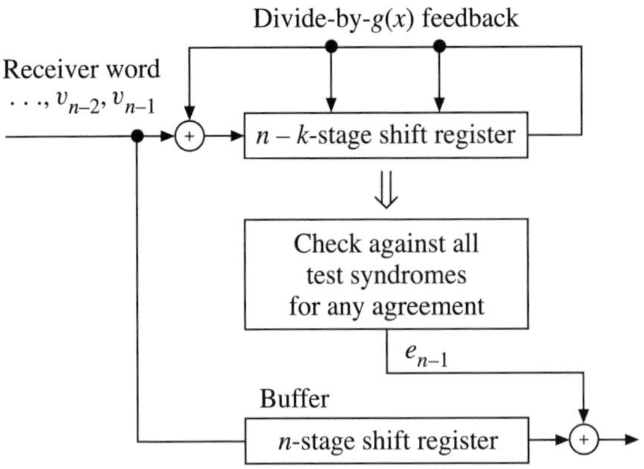

Figure 8.18. Meggitt decoder

error polynomials. This look-up decoder can be greatly simplified by using the strong algebraic structure of the code to find relationships between the syndromes. By relying on such relationships, it is necessary only to store the error polynomials associated with a few typical syndrome polynomials. Then simple computational procedures are used to compute other entries when they are needed.

Usually, when using shift-register circuits, one chooses to correct only a single error in the last stage of the shift register. This avoids the need to design the shift register such that all internal stages are directly accessible. The simplest decoder of this type, known as a *Meggitt decoder* and illustrated in Figure 8.18, tests the syndrome for all error patterns that have a symbol error in the high-order position. Consequently, the syndrome-evaluator table contains only those syndromes that correspond to an error pattern for which e_{n-1} is nonzero. If the measured syndrome is recognized by the syndrome evaluator, e_{n-1} is corrected. Otherwise, that symbol is correct. Errors in other positions are decoded later, using the cyclic structure of the code. The senseword is cyclically shifted by one position, and the process is repeated to test for a possible error in the next place ($e_{n-2} \neq 0$), which follows the high-order position. This process repeats component by component; each component, in turn, is tested for an error and is corrected whenever an error is found.

It is not necessary to actually recompute the syndrome for each cyclic shift of the senseword. The new syndrome can be computed easily from the original syndrome. The basic relationship is given by the following theorem.

Theorem 8.3.1 (Meggitt Theorem). *Suppose that* $g(x)h(x) = x^n - 1$, *and* $R_{g(x)}[v(x)] = s(x)$. *Then*

$$R_{g(x)}[xv(x)(\text{mod } x^n - 1)] = R_{g(x)}[xs(x)].$$

Proof: The proof consists of the combination of three simple statements.

(i) By definition:

$$v(x) = g(x)Q_1(x) + s(x),$$
$$xv(x) = xg(x)Q_1(x) + xs(x).$$

(ii) $xv(x)(\bmod\ x^n - 1) = xv(x) - (x^n - 1)v_{n-1}$.

(iii) By the division algorithm:

$$xs(x) = g(x)Q_2(x) + t(x)$$

uniquely, where $\deg t(x) < \deg g(x)$.

Combining lines (i), (ii), and (iii),

$$\begin{aligned} xv(x)(\bmod\ x^n - 1) &= [xQ_1(x) + Q_2(x) + v_{n-1}h(x)]g(x) + t(x) \\ &= Q_3(x)g(x) + t(x). \end{aligned}$$

Then $t(x)$ is unique because $\deg t(x) < \deg g(x)$ by the division algorithm, and

$$t(x) = R_{g(x)}[xv(x)(\bmod\ x^n - 1)] = R_{g(x)}[xs(x)],$$

as was to be proved. □

In particular, the theorem states that an error polynomial and its syndrome polynomial satisfies

$$R_{g(x)}[xe(x)(\bmod\ x^n - 1)] = R_{g(x)}[xs(x)].$$

If $e(x)$ is a correctable error pattern, then

$$e'(x) = xe(x)(\bmod\ x^n - 1)$$

is a cyclic shift of $e(x)$. Hence $e'(x)$ is a correctable error pattern with syndrome

$$s'(x) = R_{g(x)}[e'(x)] = R_{g(x)}[xs(x)].$$

This relationship shows how to compute syndromes of any cyclic shift of an error pattern whose syndrome is known. The computation can be achieved by a simple shift-register circuit, which often is much simpler than a syndrome look-up table.

Suppose that $s(x)$ and $e(x)$ are the measured syndrome polynomial and the true error polynomial, respectively, and we have precomputed that $\overline{s}(x)$ is the syndrome for the error polynomial $\overline{e}(x)$. We check whether $s(x) = \overline{s}(x)$, and if so, we know that $e(x) = \overline{e}(x)$. If not, we compute $R_{g(x)}[xs(x)]$ and compare this to $\overline{s}(x)$. If it agrees, then

$$\overline{e}(x) = xe(x)(\bmod\ x^n - 1),$$

which is a cyclic shift of $e(x)$; thus we know $e(x)$. Continue in this way. If $R_{g(x)}[xR_{g(x)}[xs(x)]]$ is equal to $\overline{s}(x)$, then $\overline{e}(x)$ is equal to $x^2e(x)(\bmod\ x^n - 1)$, and so on.

All syndromes corresponding to correctable error patterns with an error in the high-order position are tabulated as test syndromes. These test syndromes and the corresponding values of the error e_{n-1} are all that need to be stored. The true syndrome $s(x)$ is compared to all test syndromes. Then $R_{g(x)}[xs(x)]$ is computed and compared to all test syndromes. Repeating the process n times will find any correctable error pattern.

The decoder starts with all stages of the shift register set to zero. The first step in the shift-register decoder is to divide by $g(x)$. After this is completed, the divide-by-$g(x)$ shift register contains $s(x)$, and the buffer contains the senseword. Then the syndrome is compared to each test syndrome. If any agreement is found, the high-order symbol is corrected at the output of the buffer. Then the syndrome generator and the buffer are shifted once. This implements the operations of multiplication of the syndrome by x and division by $g(x)$. The content of the syndrome register has now been replaced by $R_{g(x)}[xs(x)]$, which is the syndrome of $xe(x)(\text{mod } x^n - 1)$. If this new syndrome is on the list of test syndromes, then there is an error in the position next to the high-order position, which now appears at the end of the buffer. This symbol is corrected, and the syndrome generator is shifted again; it is ready to test for an error in the next position. The process is repeated n times, at which time all error symbols have been corrected.

The use of the Meggitt theorem is easier to understand in terms of specific examples. Consider the (15, 11) Hamming code as defined by the generator polynomial $g(x) = x^4 + x + 1$. Because this is a single-error-correcting code, there is only one pattern with an error in the high-order bit. This error is

$$e(x) = x^{14},$$

and the associated syndrome is

$$s(x) = x^3 + 1.$$

The Meggitt decoder is shown in Figure 8.19. The syndrome is contained in the shift register after fifteen shifts. Of these, the first four shifts are used only to position the data in the register so that the division process can begin. If the syndrome is (1001),

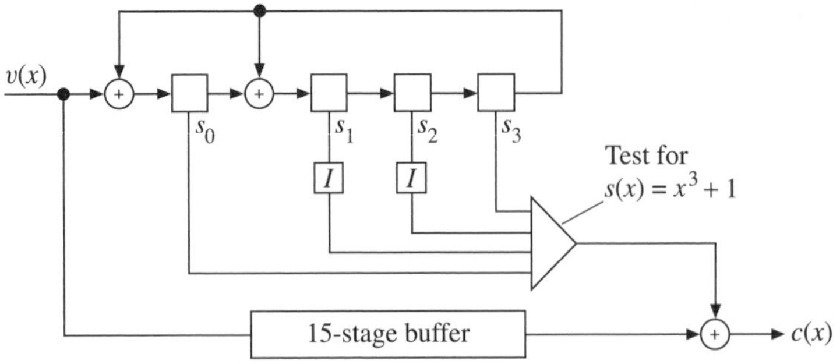

Figure 8.19. Meggitt decoder for a (15, 11) Hamming code

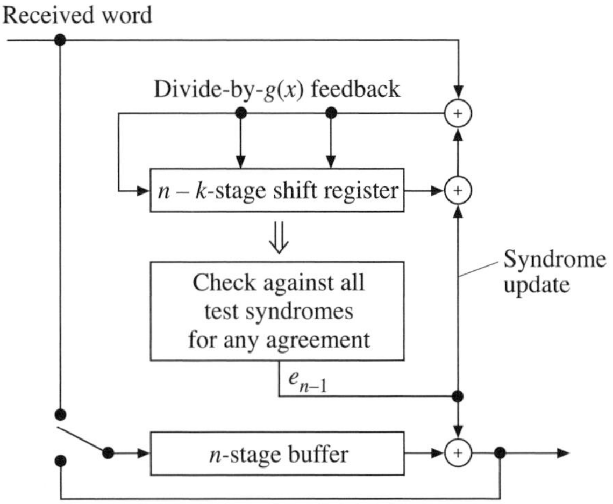

Figure 8.20. Another Meggitt decoder

then $e(x) = x^{14}$, and this bit is corrected. If after one shift the syndrome is (1001), then $e(x) = x^{13}$, and this bit is corrected. In this way, each of the fifteen bits, in turn, is available for correction. After fifteen such shifts, the error-correction process is complete. Hence thirty shifts are required to complete the correction process.

The Meggitt decoder often is seen in a slightly different form, as shown in Figure 8.20. The path labeled "syndrome update" is added, and the corrected symbols are fed back to the buffer. This allows the correction of errors to be spread over several passes.

The purpose of the syndrome update path is to remove from the circuit all traces of an error that has been corrected. Because of the Meggitt theorem, we need to treat only the case where the error is in the high-order symbol. All other error components will be taken care of automatically at the right time.

Let

$$e'(x) = e_{n-1}x^{n-1}.$$

Then the contribution to the syndrome from this error component is

$$s'(x) = R_{g(x)}[e_{n-1}x^{n-1}].$$

When e_{n-1} is corrected, it must be subtracted from the actual syndrome in the syndrome register, replacing $s(x)$ by $s(x) - s'(x)$. But $s'(x)$ can be nonzero in many places, and thus the subtraction affects many stages of the syndrome register and is a bother. To simplify the subtraction, change the definition of the syndrome to

$$s(x) = R_{g(x)}[x^{n-k}v(x)].$$

This is different from the original definition of the syndrome – but it is just as good – and we can do everything we did before with this modified definition. The advantage is that now, for a single error in the high-order place,

$$\begin{aligned} s'(x) &= R_{g(x)}[x^{n-k}e_{n-1}x^{n-1}] \\ &= R_{g(x)}[R_{x^n-1}[e_{n-1}x^{2n-k-1}]] \\ &= R_{g(x)}[e_{n-1}x^{n-k-1}] \\ &= e_{n-1}x^{n-k-1} \end{aligned}$$

because $g(x)$ has degree $n-k$. But now $s'(x)$ is nonzero only in a single place, the high-order position of $s(x)$. After e_{n-1} is corrected, in order to replace $s(x)$ by $s(x) - s'(x)$, it suffices to subtract e_{n-1} from the high-order component of the syndrome. The syndrome then has the contribution from the corrected error symbol removed from it.

The premultiplication of $e(x)$ by x^{n-k} is accomplished by feeding $v(x)$ into a different place in the syndrome generator.

For the (15, 11) Hamming code, already treated in Figure 8.19, a modified Meggitt decoder is shown in Figure 8.21. The received polynomial $v(x)$ is multiplied by x^4 and divided by $g(x)$, and thus the syndrome is

$$s(x) = R_{g(x)}[x^4 e(x)] = R_{g(x)}[R_{x^n-1}x^4[e(x)]].$$

The advantage of this modification is that the syndrome corresponding to $e(x) = x^{14}$ is

$$\begin{aligned} s(x) &= R_{x^4+x+1}[x^{18}] \\ &= R_{x^4+x+1}[x^3] \\ &= x^3, \end{aligned}$$

which can be corrected in the feedback signal.

The next example is a decoder for the (15, 7, 5) BCH double-error-correcting code, defined by the generator polynomial $g(x) = x^8 + x^7 + x^6 + x^4 + 1$. There are fifteen

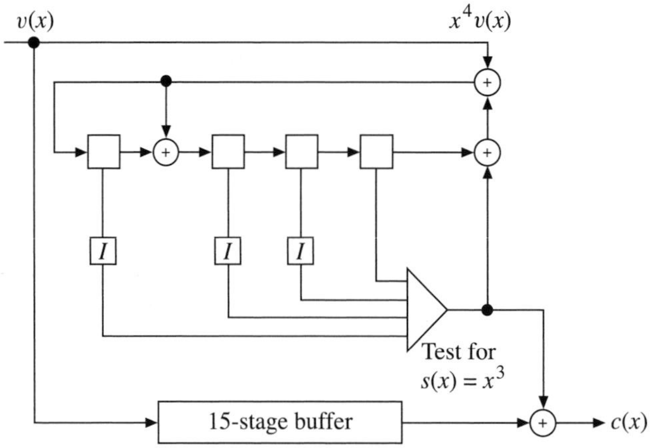

Figure 8.21. Modified Meggitt decoder for a (15, 11) Hamming code

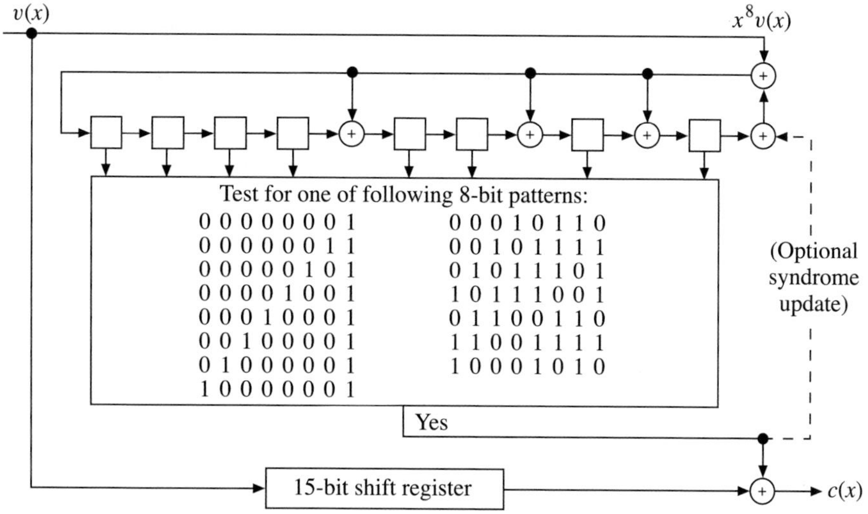

Figure 8.22. Meggitt decoder for (15, 7, 5) BCH code

correctable error patterns in which the high-order bit can be in error: one single-error pattern and fourteen double-error patterns.

The senseword is multiplied by x^8 and divided by $g(x)$. Hence if $e(x) = x^{14}$, then

$$s(x) = R_{g(x)}[x^8 \cdot x^{14}] = x^7.$$

Similarly, if $e(x) = x^{14} + x^{13}$, then

$$s(x) = R_{g(x)}[x^8 \cdot (x^{14} + x^{13})] = x^7 + x^6.$$

Proceeding in this way, the fifteen syndromes to be stored are computed. Hence the decoder must test for fifteen syndromes corresponding to these fifteen error patterns. Figure 8.22 shows the Meggitt decoder for the (15, 7, 5) BCH code. The content of the eight-bit syndrome register is compared to each of fifteen different eight-bit words to see if the high-order bit is in error. This is repeated after each of fifteen shifts so that each bit position, in turn, is examined for an error. A total of thirty shifts is required, fifteen to compute the syndrome and fifteen to locate the error.

This circuit can be further simplified to obtain the circuit of Figure 8.23. Notice that many of the fifteen correctable error patterns appear twice as the senseword is cyclically shifted. For example, (000000100000001) becomes (000000010000001) after eight shifts (one bit is underlined as a place marker). Each of these has a syndrome appearing in the table. If one of the two syndromes is deleted, the error pattern will still be caught eventually, after it is cyclically shifted. Hence one needs only to store the syndrome for one pattern of each such pair. There are eight syndromes then that need to be checked. These appear in the first column of the syndrome table in Figure 8.22. Now, however, the first error reaching the end of the buffer might not be corrected at this time, but at least one error will be corrected during fifteen shifts. It is necessary to execute two

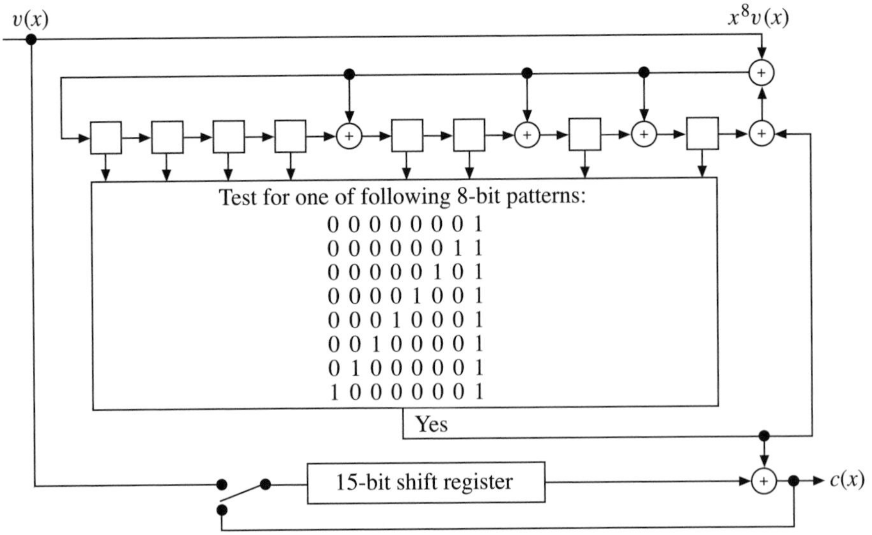

Figure 8.23. Another Meggitt decoder for (15, 7, 5) BCH code

cycles of shifts to correct all errors. This results in a total of forty-five shifts. The need for the syndrome update is now apparent because, otherwise, the second time an error reached the end of the buffer, the syndrome would again be unrecognized.

The Meggitt decoder for the (15, 7, 5) BCH code in Figure 8.23 is an example of an error-trapping decoder to be studied in the next section.

8.4 Error trapping

The Meggitt decoder described in Figure 8.23 has a syndrome table that upon examination is seen to contain only words of weight 1 and weight 2. In fact, these syndromes are exactly the same as correctable error patterns with the leftmost bits of the error pattern equal to zero and not shown. The syndrome acts as a window through which the error pattern can be seen. This kind of decoder is known as an *error-trapping decoder*, which is a modification of a Meggitt decoder that can be used for certain cyclic codes.

Suppose that all errors in a senseword occur close together. Then the syndrome, properly shifted, exhibits an exact copy of the error pattern. Let us define the length of an error pattern as the smallest number of sequential stages of a shift register that must be observed so that for some cyclic shift of the error pattern, all errors are within this segment of the shift register. Suppose the longest-length error pattern that must be corrected is not longer than the syndrome. Then for some cyclic shift of the error pattern, the syndrome is equal to the error pattern. For such codes, the syndrome generator is shifted until a correctable pattern is observed as the syndrome. The content of the

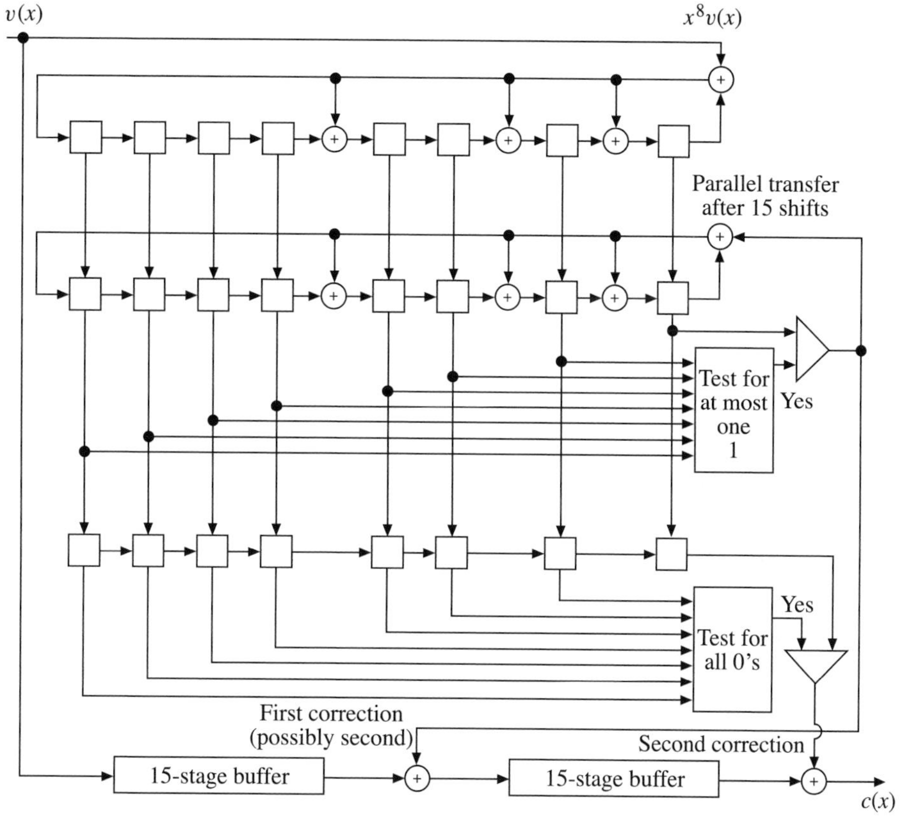

Figure 8.24. Pipelined version of the error-trapping decoder for a (15, 7, 5) BCH code

syndrome generator is subtracted from the cyclically shifted received polynomial and the correction is complete.

We have already seen one example of an error-trapping decoder in Figure 8.23; this one for a (15, 7, 5) BCH code. Forty-five shifts are required to correct any double-error pattern: fifteen shifts to generate the syndrome, fifteen shifts to correct the first error, and fifteen shifts to correct the second error. Figure 8.24 shows a pipelined version of the error-trapping decoder in which three syndrome generators are employed. This allows decoding of a nonending stream of fifteen-bit data blocks with the decoder shifting at the same speed as the incoming data. Figure 8.25 works through the forty-five cycles of the error-trapping decoder for a typical error pattern.

For a second example of error trapping, we will use the (7, 3, 5) Reed–Solomon code. This is a double-error-correcting code over $GF(8)$ with generator polynomial

$$g(x) = x^4 + (z + 1)x^3 + x^2 + zx + (z + 1)$$

where the field elements are expressed as polynomials in z. Alternatively,

$$g(x) = x^4 + \alpha^3 x^3 + x^2 + \alpha x + \alpha^3$$

$$a(x) = x^6 + 1$$
$$g(x) = x^8 + x^7 + x^6 + x^4 + 1$$
$$c(x) = x^{14} + x^{13} + x^{12} + x^{10} + x^8 + x^7 + x^4 + 1$$
$$e(x) = x^{11} + x$$
$$v(x) = x^{14} + x^{13} + x^{12} + x^{11} + x^{10} + x^8 + x^7 + x^4 + 1$$

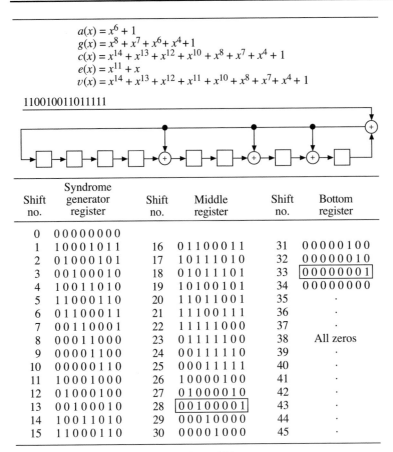

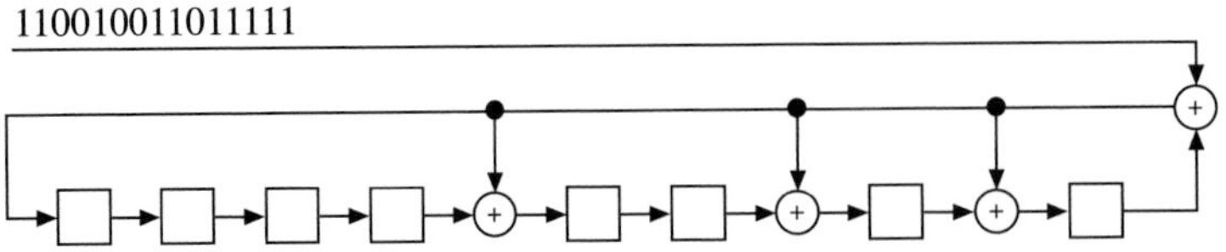

Shift no.	Syndrome generator register	Shift no.	Middle register	Shift no.	Bottom register
0	0 0 0 0 0 0 0 0 0				
1	1 0 0 0 1 0 1 1	16	0 1 1 0 0 0 1 1	31	0 0 0 0 0 1 0 0
2	0 1 0 0 0 1 0 1	17	1 0 1 1 1 0 1 0	32	0 0 0 0 0 0 1 0
3	0 0 1 0 0 0 1 0	18	0 1 0 1 1 1 0 1	33	0 0 0 0 0 0 0 1
4	1 0 0 1 1 0 1 0	19	1 0 1 0 0 1 0 1	34	0 0 0 0 0 0 0 0
5	1 1 0 0 0 1 1 0	20	1 1 0 1 1 0 0 1	35	·
6	0 1 1 0 0 0 1 1	21	1 1 1 0 0 1 1 1	36	·
7	0 0 1 1 0 0 0 1	22	1 1 1 1 1 0 0 0	37	·
8	0 0 0 1 1 0 0 0	23	0 1 1 1 1 1 0 0	38	All zeros
9	0 0 0 0 1 1 0 0	24	0 0 1 1 1 1 1 0	39	·
10	0 0 0 0 0 1 1 0	25	0 0 0 1 1 1 1 1	40	·
11	1 0 0 0 1 0 0 0	26	1 0 0 0 0 1 0 0	41	·
12	0 1 0 0 0 1 0 0	27	0 1 0 0 0 0 1 0	42	·
13	0 0 1 0 0 0 1 0	28	0 0 1 0 0 0 0 1	43	·
14	1 0 0 1 1 0 1 0	29	0 0 0 1 0 0 0 0	44	·
15	1 1 0 0 0 1 1 0	30	0 0 0 0 1 0 0 0	45	·

Note: Errors trapped at shifts number 28 and 33.

Figure 8.25. Error-trapping example for a (15, 7, 5) BCH code

where the field elements are expressed in terms of the primitive element $\alpha = z$. The error pattern $e(x)$ has at most two nonzero terms. It can always be cyclically shifted into a polynomial of degree 3 or less. Then, the syndromes are of degree 3 or less and error trapping can be applied.

The error-trapping decoder for the (7, 3, 5) Reed–Solomon code is shown in Figure 8.26. When implemented with binary circuits, the octal shift-register stages are three binary shift-register stages in parallel. All data paths are three-bits wide. The multiply-by-$(z + 1)$ (that is, by α^3) and multiply-by-z (that is, by α) circuits in the feedback path are simple three-input, three-output binary-logic circuits, which can be determined easily. Twenty-one shifts are necessary for this decoder to correct all errors. The first seven shifts compute the syndrome. The second seven shifts correct at least one error and sometimes both errors. The third seven shifts correct the second error if it is as yet uncorrected.

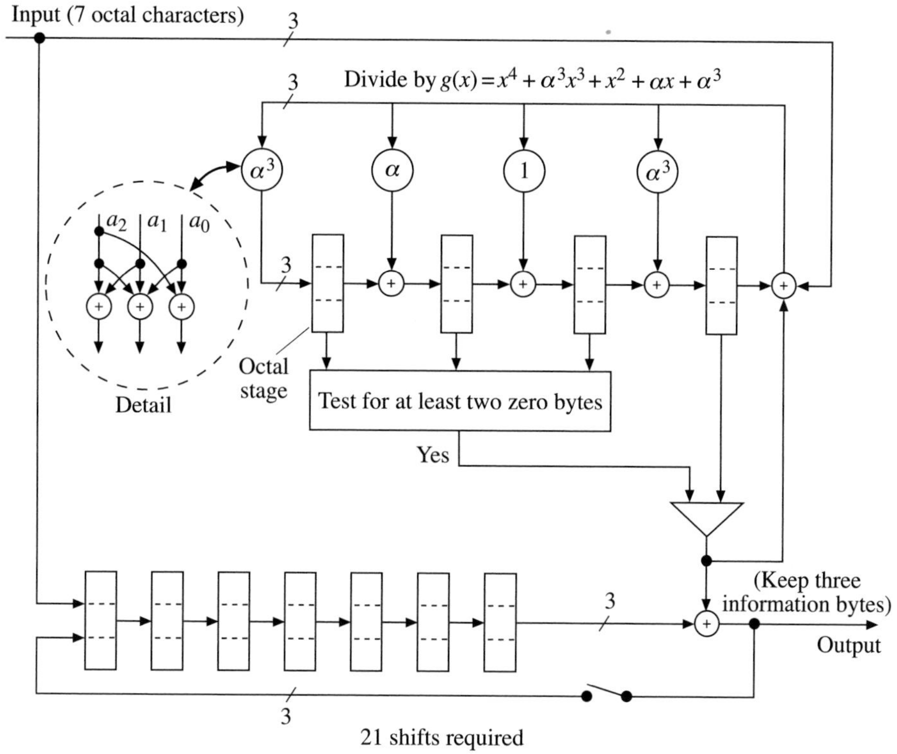

Figure 8.26. Error trapping for a (7, 3, 5) Reed–Solomon code

In Figure 8.27, we trace in detail a sample decoding operation for the (7, 3, 5) Reed–Solomon code. This illustrates error trapping and also gives a concrete example of the relationship between the abstract concept of a Galois field and the logic design of a shift-register circuit. For each shift, the contents of both the syndrome register and the data buffer are listed in the table. After seven shifts, the syndrome register contains the syndrome. Beginning with the eighth shift, the circuit will correct an error whenever it sees a trapped error pattern. This is a pattern with at most two nonzero symbols and with one of the nonzero symbols in the rightmost place. Such a pattern must always occur at least once in the second seven shifts if no more than two errors occur. In the example, it occurs at the thirteenth shift. Hence the error is trapped. Notice that the high-order syndrome symbols are normally to the right. When the error is trapped, we see a four-symbol segment of the cyclically shifted error pattern (e_4, e_5, e_6, e_0). Because the error is trapped, e_0 is corrected in the data buffer and set equal to zero (or subtracted from itself) in the syndrome register. After fourteen shifts, the syndrome register contains the syndrome of the remaining error pattern. The process repeats through a third set of seven shifts, after which the error correction is complete.

$\alpha^0 = 1$	$g(x) = x^4 + \alpha^3 x^3 + x^2 + \alpha x + \alpha^3$
$\alpha^1 = z$	$c(x) = \alpha^4 x^6 + \alpha^4 x^4 + \alpha^3 x^3 + \alpha^6 x + \alpha^6$
$\alpha^2 = z^2$	$e(x) = \alpha^4 x^4 + \alpha^3$
$\alpha^3 = z + 1$	$v(x) = \alpha^4 x^6 + \alpha^3 x^3 + \alpha^6 x + \alpha^4$
$\alpha^4 = z^2 + z$	
$\alpha^5 = z^2 + z + 1$	
$\alpha^6 = z^2 + 1$	

Shift	Syndrome register		Information register
0	$0\ 0\ 0\ 0$		$0\ 0\ 0\ 0\ 0\ 0\ 0$
1	$1\ \alpha^5 \alpha^4 1$		$\alpha^4 0\ 0\ 0\ 0\ 0\ 0$
2	$\alpha^3 \alpha^3 \alpha^4 \alpha^6$		$0\ \alpha^4 0\ 0\ 0\ 0\ 0$
3	$\alpha^2 \alpha\ \alpha^4 \alpha$		$0\ 0\ \alpha^4 0\ 0\ 0\ 0$
4	$\alpha^3 \alpha^4 \alpha^3 \alpha^6$		$\alpha^3 0\ 0\ \alpha^4 0\ 0\ 0$
5	$\alpha^2 \alpha\ \alpha^3 \alpha^5$		$0\ \alpha^3 0\ 0\ \alpha^4 0\ 0$
6	$\alpha^4 0\ 0\ \alpha^6$		$\alpha^6 0\ \alpha^3 0\ 0\ \alpha^4 0$
7	$\alpha^6 0\ \alpha^3 \alpha^6$		$\alpha^4 \alpha^6 0\ \alpha^3 0\ 0\ \alpha^4$
8	$\alpha^2 \alpha^2 \alpha^6 \alpha^5$		$\alpha^4 \alpha^4 \alpha^6 0\ \alpha^3 0\ 0$
9	$\alpha\ 1\ \alpha^3 \alpha^5$		$0\ \alpha^4 \alpha^4 \alpha^6 0\ \alpha^3 0$
10	$\alpha\ \alpha^5 \alpha^4 1$		$0\ 0\ \alpha^4 \alpha^4 \alpha^6 0\ \alpha^3$
11	$\alpha^3 0\ \alpha^4 \alpha^6$		$\alpha^3 0\ 0\ \alpha^4 \alpha^4 \alpha^6 0$
12	$\alpha^2 \alpha\ \alpha^6 \alpha$		$0\ \alpha^3 0\ 0\ \alpha^4 \alpha^4 \alpha^6$
13	$\boxed{\alpha^4 0\ 0\ \alpha^3}$	← Trapped error	$\alpha^6 0\ \alpha^3 0\ 0\ \alpha^4 \alpha^4$
14	$0\ \alpha^4 0\ 0$		$\alpha^6 \alpha^6 0\ \alpha^3 0\ 0\ \alpha^4$
15	$0\ 0\ \alpha^4 0$		$\alpha^6 \alpha^6 0\ \alpha^3 0\ 0\ \vert\ \alpha^4$
16	$\boxed{0\ 0\ 0\ \alpha^4}$	← Trapped error	$\alpha^6 \alpha^6 0\ \alpha^3 0\ \vert\ 0\ \alpha^4$
17	$0\ 0\ 0\ 0$		$\alpha^6 \alpha^6 0\ \alpha^3\ \vert\ \alpha^4 0\ \alpha^4$
18	$0\ 0\ 0\ 0$		$\alpha^6 \alpha^6 0\ \vert\ \alpha^3 \alpha^4 0$
19	$0\ 0\ 0\ 0$		$\alpha^6 \alpha^6\ \vert\ 0\ \alpha^3 \alpha^4$
20	$0\ 0\ 0\ 0$		$\alpha^6\ \vert\ \alpha^6 0\ \alpha^3$
21	$0\ 0\ 0\ 0$		$\vert\ \alpha^6 \alpha^6 0$

Figure 8.27. Error-trapping example

If the syndrome is nonzero initially, but no error pattern is trapped during the second set of seven shifts, then more than two errors occurred. Additional logic can be included easily to detect an uncorrectable error pattern.

For the next example of this section, we shall describe an error-trapping decoder for a burst-error-correcting code. In contrast to the situation for random-error-correcting codes, every burst-error-correcting code can be decoded by an error-trapping decoder. We will give a decoder for a (14, 6) code that corrects all burst errors of length 4. This code is obtained by interleaving two copies of the (7, 3) code from Table 5.3. The generator polynomial of the interleaved code is

$$g(x) = x^8 + x^6 + x^4 + 1.$$

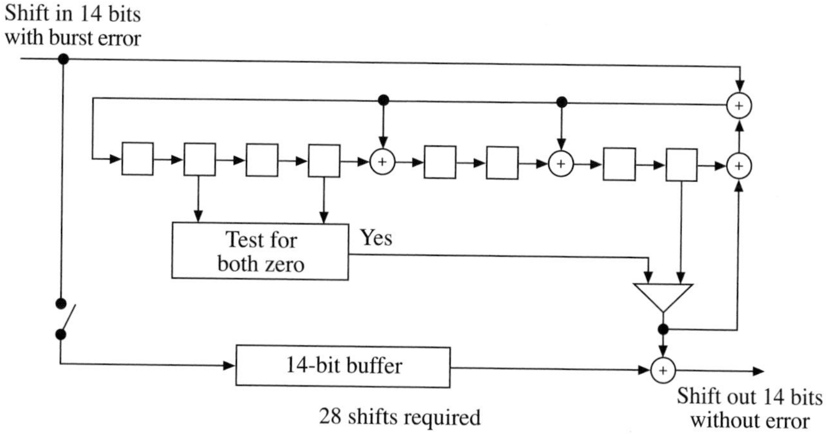

Figure 8.28. Error-trapping decoder for an interleaved burst-error-correcting code

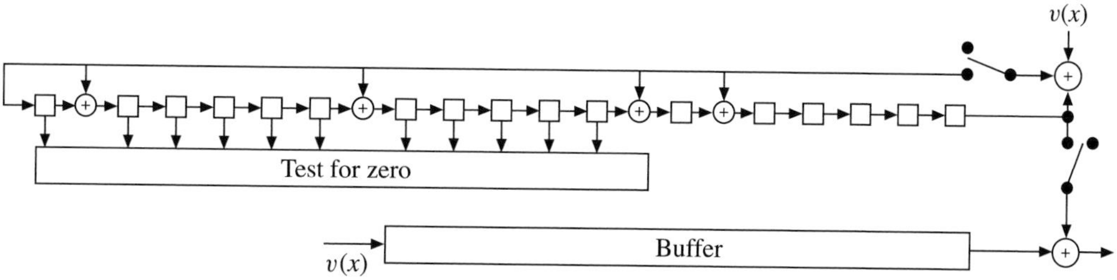

Figure 8.29. Error-trapping decoder for a Fire code

A burst error of length 4 or less is trapped if the four leftmost bits in the eight-bit syndrome register are all zero. We can improve this a little, however. Notice that the odd-numbered shift-register stages do not interact with the even-numbered stages. The circuit can be thought of as performing two independent syndrome calculations, one for each of the two underlying codes, but with the syndrome bit positions interleaved. Hence we can modify error trapping to obtain the circuit shown in Figure 8.28. This circuit will correct burst-error patterns of length 4 and also some other error patterns. Any error pattern consisting of a burst of length 2 in the even-numbered components and a burst of length 2 in the odd-numbered components will be corrected.

For the final example of this section, we shall describe a variant of an error-trapping decoder for a (693, 676) Fire code. Rather than use a standard error-trapping decoder, as shown in Figure 8.29, one can use the Chinese remainder theorem to replace the single syndrome by a pair of syndromes

$$s_1(x) = R_{x^{2t-1}-1}[v(x)] \quad s_2(x) = R_{p(x)}[v(x)].$$

The Chinese remainder theorem tells us that this pair of syndromes, $(s_1(x), s_2(x))$, is equivalent to the single syndrome $s(x)$. If the error $e(x)$ is equal to a correctable burst error $b(x)$ with nonzero b_0, then $s_1(x) = s_2(x) = b(x)$. Because $s_1(x)$ is equal to

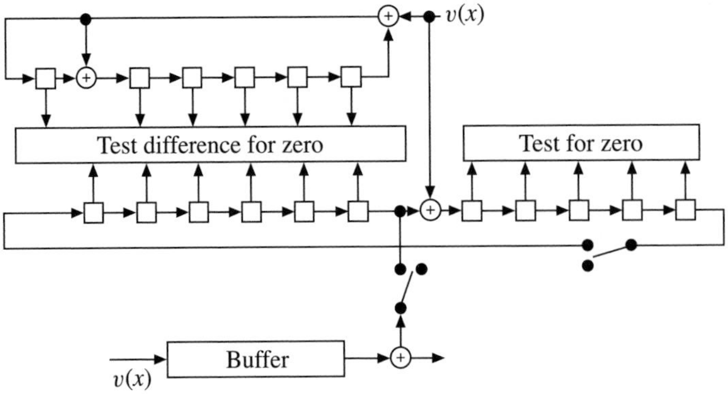

Figure 8.30. An alternative decoder for a Fire code

$R_{x^{2t-1}-1}[b(x)]$, $s_1(x)$ is a replica of the burst error that has been folded into a window of length $2t$.

Recall that a burst of length at most t is described by a polynomial $b(x)$ of degree at most $t-1$ with b_0 nonzero. This means that the burst error can be trapped. If $e(x) = b(x)$ is a burst of length at most t, then

$$s_1(x) = s_2(x) = b(x)$$

and the error is trapped. There are multiple cyclic shifts of $e(x)$ that will satisfy

$$s_1(x) = b(x),$$

but only one of these can satisfy $s_1(x) = s_2(x)$. An alternative error-trapping decoder for the (693, 676) Fire code is shown in Figure 8.30.

8.5 Modified error trapping

The idea of error trapping can be modified and augmented in many ways to accommodate the requirements of a particular code. In this section we shall look at several such modifications.

Sometimes, when error trapping cannot be used for a given code, it can be made to work by adding an extra circuit to remove a few troublesome error patterns. In general, this calls for a certain amount of ingenuity in order to obtain a satisfactory design. We shall illustrate such a decoder for the (23, 12) binary Golay code.

The error pattern in a Golay codeword is of length at most 23 and of weight at most 3. The syndrome register has a length of 11. If an error pattern cannot be trapped, then it cannot be cyclically shifted so that all three errors appear in the eleven low-order places. In this case, one can convince oneself (perhaps with a few sketches) that (with the twenty-three bits interpreted cyclically) one of the three error bits must have at least five zeros on one side of it, and at least six zeros on the other side. Therefore all error patterns can be cyclically shifted into one of the following three configurations (with bit positions numbered 0 through 22):

1. all errors are in the eleven high-order bits;
2. one error is in bit position 5 and the others are in the eleven high-order bits;
3. one error is in bit position 6 and the others are in the eleven high-order bits.

Therefore the decoder has the precomputed values $s_5(x) = R_{g(x)}[x^{n-k}x^5]$ and $s_6(x) = R_{g(x)}[x^{n-k}x^6]$. Then an error pattern is trapped if $s(x)$ has weight 3 or less, or if $s(x) - s_5(x)$ or $s(x) - s_6(x)$ has weight 2 or less. One can choose to correct all errors when this happens, or to initially correct only the two errors in the eleven low-order bits and to wait for the third error to be shifted down.

Dividing x^{16} and x^{17} by the generator polynomial

$$g(x) = x^{11} + x^{10} + x^6 + x^5 + x^4 + x^2 + 1$$

gives

$$s_5(x) = x^9 + x^8 + x^6 + x^5 + x^2 + x$$
$$s_6(x) = x^{10} + x^9 + x^7 + x^6 + x^3 + x^2.$$

Therefore if an error occurs in bit positions 5 or 6, the syndrome is (01100110110) or (11001101100), respectively. Two more errors occurring in the eleven high-order bit positions will cause two of these bits to be complemented at the appropriate locations.

The decoder looks for a syndrome pattern that differs from the all-zero syndrome in three or fewer places, or one that differs from either of the other two reference syndromes in two or fewer places.

Another way of modifying the method of error trapping to make it applicable to other cyclic codes is the method of *permutation decoding*. Binary cyclic codes are invariant under the permutation $i \to 2i$ modulo n; if $c(x)$ is a codeword, then $c(x^2)(\text{mod } x^n - 1)$ is also a codeword. Thus in addition to the cyclic shifts, permutations of the form $v(x) \to v(x^2)(\text{mod } x^n - 1)$ can be used to permute the errors into a pattern that can be trapped.

For example, every triple-error pattern $e(x)$ in a binary word of length 23 can be permuted into the eleven high-order bit positions by taking either $e(x)$, $e(x^2)$, $e(x^4)$, or one of their cyclic shifts (as can be verified by computer). Hence the binary (23, 12) code can be decoded by using an error-trapping decoder for the three permutations $v(x)$, $v(x^2)$, and $v(x^4)$. At least one of these will decode successfully.

Other automorphisms of a code also can be used with permutation decoding. However, the set of automorphisms of a code may be large, and there is no general procedure for finding a minimal set of permutations that will suffice to trap all errors. Permutation decoding is of limited attraction in applications because, except for the cyclic shifts, it is somewhat clumsy to implement the permutations with shift-register circuits.

Error trapping can also be used with shortened codes. A shortened cyclic code of blocklength $n' = n - b$ can be decoded by a Meggitt decoder designed for the original (n, k) code. However, the timing of this decoder is based on groups of n clock times, but the incoming codeword contains n' symbols. This mismatch in the timing may be an inconvenience in some applications. We shall describe a way to match the timing of the Meggitt decoder to groups of n' clock times. This speeds up the decoding of a shortened cyclic code.

We redefine the syndrome so that the shift-register clock times corresponding to unused symbols can be bypassed. When the b high-order symbols are dropped, the modified syndrome is defined as

$$s(x) = R_{g(x)}[x^{n-k+b}v(x)]$$

instead of as the remainder of $x^{n-k}v(x)$. To see the reason for this choice, suppose that the high-order bit of the shortened code is in error. That is, let

$$e(x) = e_{n'-1}x^{(n-b)-1}.$$

Then

$$\begin{aligned} s(x) &= R_{g(x)}[e_{n'-1}x^{2n-k-1}] \\ &= R_{g(x)}[R_{x^n-1}[e_{n'-1}x^{2n-k-1}]] \\ &= e_{n'-1}x^{n-k-1}. \end{aligned}$$

Hence $s(x)$ is nonzero only in the high-order bit. But now, by the rules of modulo computation, $s(x)$ can be written

$$s(x) = R_{g(x)}[a(x)v(x)]$$

where

$$a(x) = R_{g(x)}[x^{n-k+b}].$$

Thus all that needs to be done is to premultiply $v(x)$ by the fixed polynomial $a(x)$ prior to division by $g(x)$. This multiplication can be combined with the divide-by-$g(x)$ circuit by the method in which $v(x)$ is fed into the circuit, as shown in the following two examples.

The first example is a burst-correcting shortened cyclic code. Suppose the requirements of some application can be met by shortening the $(511, 499)$ burst-correcting code of Table 5.3 to obtain a $(272, 260)$ code. This code will correct all burst errors of length 4 or less. For this case, $g(x) = x^{12} + x^8 + x^5 + x^3 + 1$, $x^{n-k+b} = x^{251}$, and we need to compute

$$a(x) = R_{g(x)}[x^{251}].$$

One way of carrying out this calculation is by writing

$$x^{251} = (x^{12})^{16}(x^{12})^4(x^{11})$$

and

$$x^{12} = x^8 + x^5 + x^3 + 1.$$

The repeated squaring of x^{12} reduced modulo $x^{12} + x^8 + x^5 + x^3 + 1$ gives $(x^{12})^4$ and $(x^{12})^{16}$, and from these x^{251} modulo $x^{12} + x^8 + x^5 + x^3 + 1$ can be calculated quickly. Then

$$a(x) = x^{11} + x^9 + x^7 + x^3 + x^2 + 1.$$

Finally, the modified syndrome is

$$s(x) = R_{x^{12}+x^8+x^5+x^3+1}[(x^{11} + x^9 + x^7 + x^3 + x^2 + 1)v(x)].$$

This can be implemented with a single divide-by-$g(x)$ circuit if, at each iteration, the incoming coefficient of $v(x)$ is added into the proper positions of the current remainder to effect the multiplication by $a(x)$. A Meggitt decoder to decode the shortened code is shown in Figure 8.31. So that it will operate with a continuous stream of incoming bits, this decoder is pipelined into a syndrome computation and a syndrome tester.

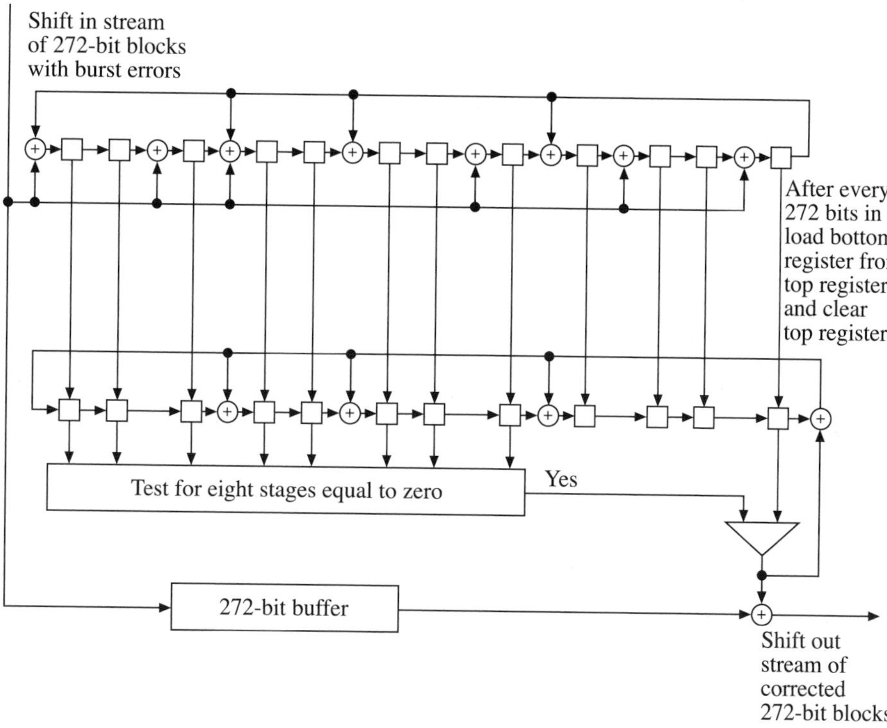

Figure 8.31. Error-trapping decoder for a (272, 260) burst-error-correcting code

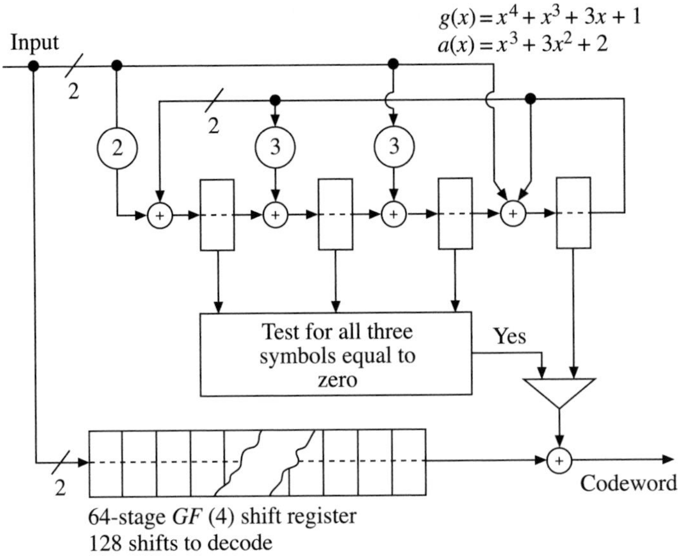

Figure 8.32. A Meggitt decoder for a (64, 60) shortened Hamming code over $GF(4)$

A second example of a decoder for a shortened cyclic code is a (64, 60) code over $GF(4)$, which is a shortened Hamming code. The (85, 81) Hamming code is a cyclic code, as was discussed in Section 5.5. The generator polynomial is $g(x) = x^4 + x^3 + 3x + 1$, and $x^{n-k+b} = x^{25}$. Then by carrying out long division, we have

$$a(x) = R_{x^4+x^3+3x+1}[x^{25}] = x^3 + 3x^2 + 2.$$

The modified syndrome is then computed by

$$s(x) = R_{x^4+x^3+3x+1}[(x^3 + 3x^2 + 2)v(x)].$$

The circuit of Figure 8.32 is a Meggitt decoder that will decode the shortened code in 128 clock times. It does this by multiplying the senseword polynomial by $x^3 + 3x^2 + 2$ and then dividing by $x^4 + x^3 + 3x + 1$ to produce the modified syndrome. Altogether, 64 clock times are used to compute the modified syndrome, and 64 clock times are used to correct the error pattern, making a total of 128 clock times. If desired, the decoder could be pipelined, or two decoders could be run in tandem, and then the decoder would run in step with the incoming data.

8.6 Architecture of Reed–Solomon decoders

Several algorithms and their variations were developed in Chapter 7 for decoding Reed–Solomon codes. These algorithms must be implemented by software routines or by hardware circuits to obtain a practical Reed–Solomon decoder. Just as there are many options in the choice of algorithm, there are many options in the choice of

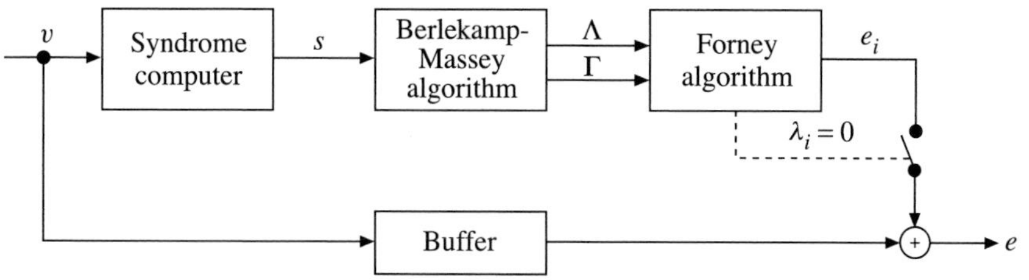

Figure 8.33. A decoder architecture

implementation. A good design will match the choice of algorithm to the choice of implementation.

The *architecture* of a decoder consists of an overview of its general structure without regard to the fine details. Figure 8.33 shows the architecture for a decoder that uses the Berlekamp algorithm together with the Forney algorithm. The decoder consists of three major functions that can be implemented separately. Indeed, three successive codewords can be at various stages of processing simultaneously within the decoder. While the most recent codeword is being processed in the syndrome computer, the next most recent codeword is being processed within the Berlekamp algorithm, and the least recent codeword is having its error symbols corrected by the Forney algorithm. All three computations must be finished when the next codeword arrives so that the output of each stage can be passed as the input to the next stage. A word is not completely decoded until three word times later. This kind of architecture is an example of a pipelined decoder.

The Berlekamp–Massey algorithm can be implemented with shift registers, as illustrated in Figure 8.34. A shift-register memory is provided for each of three polynomials $S(x)$, $\Lambda(x)$, and $B(x)$. Each memory cell is m bits wide, and stores one element of $GF(2^m)$. The data paths are m bits wide as well because the memory is to be used as an m bit wide shift register. It may, however, be physically realized with a random-access memory. Then an address register points to the end of the conceptual shift register in memory. Instead of moving data in the memory when the shift register is shifted, the address register is simply cyclically incremented modulo n.

The length of each shift register in the decoder is large enough to hold the largest degree of its polynomial, and possibly a little larger. Shorter polynomials are stored simply by filling out the register with zeros. Each of the $S(x)$ and $B(x)$ registers is one stage longer than needed to store the largest polynomial. The extra symbol in the syndrome register is set up so that the polynomials can be made to precess one position during each iteration. This supplies the multiplication of $B(x)$ by x and also offsets the index of the syndromes by r to provide S_{r-j}, which appears in the expression for Δ_r. To see this, refer to Figure 8.34; the syndrome register is shown as it is initialized. During each iteration, it will be shifted to the right by one symbol so that $S(x)$ will be timed properly for multiplication with $\Lambda(x)$ to compute the discrepancy. During one iteration, the Λ register is cycled twice, first to compute Δ, then to be updated.

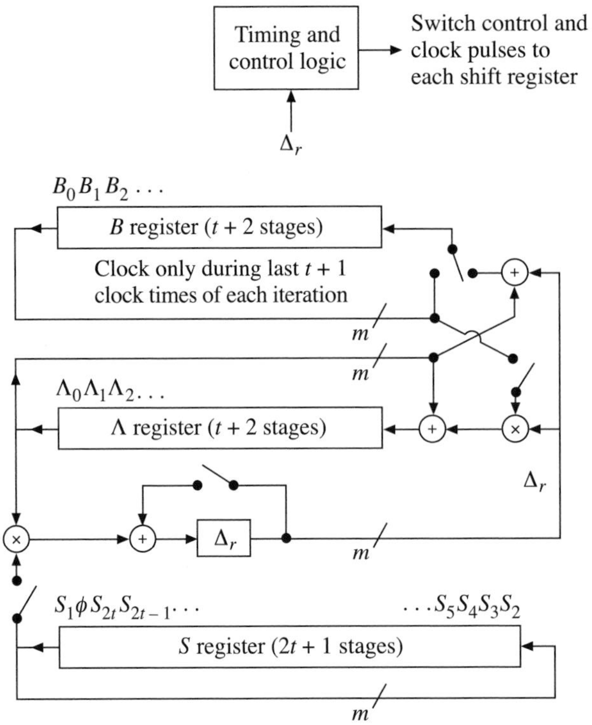

Figure 8.34. Outline of circuit for Berlekamp–Massey algorithm

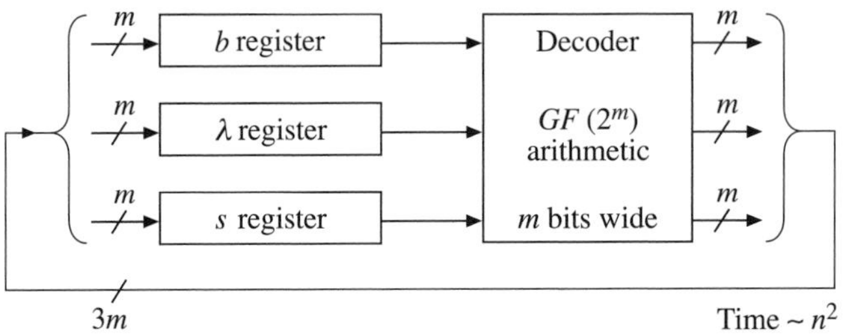

Figure 8.35. Architecture of a time-domain Reed–Solomon decoder

The Berlekamp–Massey implementation, shown in Figure 8.34, is only a portion of the decoder shown in Figure 8.33. It must be supported by the syndrome computer, the Forney algorithm, and some way to compute the error-evaluator polynomial. Furthermore the registers must be of variable length if t is a parameter to be chosen by the user.

The architecture of a much simpler decoder is the time-domain decoder, shown in Figure 8.35. This is a simple but slow decoder based on the methods of Section 7.6. The decoder uses three vectors $\boldsymbol{\lambda}$, $\boldsymbol{b}$, and $\boldsymbol{s}$, each of length n. The vector $\boldsymbol{s}$ is initialized equal to the senseword $\boldsymbol{v}$ and the iterations of Figure 7.13 change it to the corrected codeword $\boldsymbol{c}$. A total of n iterations is required: $2t$ iterations to correct the errors and $n - 2t$ iterations for the equivalent of recursive extension. In each of n iterations, the three vectors of length n are taken from memory, modified according to the flow of Figure 7.13, and returned to memory. Hence the decoding time is proportional to n^2. This is much slower than methods using the Forney algorithm. The advantage is the simplicity of the architecture and implementation.

The Reed–Solomon decoder of Figure 8.33 or Figure 8.34 is called a *universal decoder* whenever it is designed so that t can be selected by the user. In the decoder of Figure 8.35, t can be made selectable with no change in the complexity of the decoder. In addition, provisions for both erasure decoding and for decoding a single extension symbol can be included with very little extra complexity.

At the other extreme of the cost–complexity trade are very high-speed decoders in which the multiple components of the vectors are processed in space rather than time. A *systolic decoder* is a decoder based on a spatial array of identical or nearly identical computational cells. Usually, one imposes the design constraint that an individual cell can exchange most data only with neighboring cells, though it may be allowed to broadcast a small amount of global data to all cells. We will need to inspect and revise the equations of the Berlekamp–Massey algorithm to make them suitable for a systolic implementation.

The discrepancy equation

$$\Delta_r = \sum_{j=0}^{n-1} \Lambda_j S_{r-j},$$

is not satisfactory for a systolic implementation because Δ_r depends on all components of $\boldsymbol{\Lambda}$ and $\boldsymbol{S}$. If each cell holds only one component of $\boldsymbol{\Lambda}$ and of $\boldsymbol{S}$, then to compute Δ_r, data would need to flow from all cells to a single point. To avoid this, an alternative computation of Δ_r is used.

Multiply both sides of the Berlekamp–Massey recursion by $S(x)$ to give

$$\begin{bmatrix} \Lambda^{(r)}(x)S(x) \\ B^{(r)}(x)S(x) \end{bmatrix} \leftarrow \begin{bmatrix} 1 & -\Delta_r x \\ \Delta_r^{-1}\delta_r & \overline{\delta}_r x \end{bmatrix} \begin{bmatrix} \Lambda^{(r-1)}(x)S(x) \\ B^{(r-1)}(x)S(x) \end{bmatrix}.$$

The discrepancy Δ_r is the rth coefficient of the polynomial $\Delta^{(r)}(x) = \Lambda^{(r)}(x)S(x)$. Only the first $2t$ components are needed to compute the sequence of discrepancies, so define

$$G^{(r)}(x) = \Lambda^{(r)}(x)S(x) \pmod{x^{2t+1}}$$
$$F^{(r)}(x) = B^{(r)}(x)S(x) \pmod{x^{2t+1}}.$$

The recursion can now be written as

$$\begin{bmatrix} G(x) \\ F(x) \end{bmatrix} \leftarrow \begin{bmatrix} 1 & -\Delta_r x \\ \Delta_r^{-1}\delta_r & \overline{\delta}_r x \end{bmatrix} \begin{bmatrix} G(x) \\ F(x) \end{bmatrix} \pmod{x^{2t+1}},$$

where $G^{(r)}(x)$ and $F^{(r)}(x)$ are initialized to $S(x)$. Then $\Delta_r = G_r^{(r-1)}$.

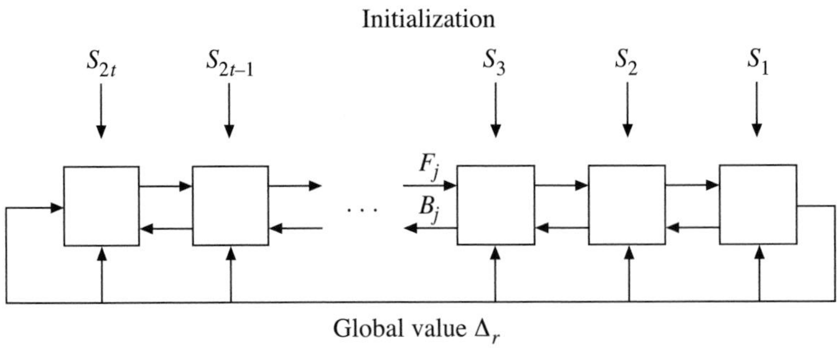

Figure 8.36. A systolic architecture

The update equation now can be rewritten as

$$\begin{bmatrix} \Lambda(x) & G(x) \\ B(x) & F(x) \end{bmatrix} \leftarrow \begin{bmatrix} 1 & -\Delta_r x \\ \Delta_r^{-1}\delta_r & \overline{\delta}_r x \end{bmatrix} \begin{bmatrix} \Lambda(x) & G(x) \\ B(x) & F(x) \end{bmatrix},$$

which can be broken into an expression for the jth component:

$$\begin{bmatrix} \Lambda_j^{(r)} & G_j^{(r)} \\ B_j^{(r)} & F_j^{(r)} \end{bmatrix} \leftarrow \begin{bmatrix} 1 & -\Delta_r \\ \Delta_r^{-1}\delta_r & \overline{\delta}_r \end{bmatrix} \begin{bmatrix} \Lambda_j^{(r-1)} & G_j^{(r-1)} \\ B_{j-1}^{(r-1)} & F_{j-1}^{(r-1)} \end{bmatrix}.$$

The jth cell uses data only from its own previous iteration, namely $\Lambda_j^{(r-1)}$, data from its neighbor's previous iteration, namely $B_{j-1}^{(r-1)}$, and the global variables Δ_r and δ_r. This equation is satisfactory for a systolic implementation. All cells can compute δ_r in lockstep. Cell r contains $\Delta_r = G_r^{(r-1)}$.

The systolic architecture of Figure 8.36 results from assigning one coefficient of each iteration to each cell. However, one small improvement has been made. Because we want to broadcast Δ_r from the rightmost cell, index $j = r$ is assigned to that cell for the $G(x)$ computation. Because we want Λ to be found in the leftmost cell at the end of the computation, index $j = 0$ is assigned to that cell in the $\Lambda(x)$ computation. Cell j performs the computation. Moreover, it is trivial to adjust the indices so that Δ_r is always found in a fixed cell. It is worth noting that this systolic architecture will also automatically compute $\Gamma(x)$.

8.7 Multipliers and inverters

A large decoder, such as one using the Berlekamp–Massey algorithm, primarily consists of devices for Galois-field addition, multiplication, and division, and data paths interconnecting these devices. Addition is much simpler than multiplication or division. Because there can be many multipliers; the major cost can be the cost of the multiplication of arbitrary elements of $GF(q)$. The cost of division may also be significant, but usually there are many fewer divisions than multiplications to perform, so the cost

Table 8.1. *Zech logarithms for $GF(8)$*

i	$Z(i)$
$-\infty$	0
0	$-\infty$
1	3
2	6
3	1
4	5
5	4
6	2

of multiplication dominates. Division can be performed by using an *inverse circuit* to compute b^{-1} followed by a multiplier to compute $b^{-1}a$.

Multiplication is easy if the field elements β and γ are represented as powers of the primitive element α. Thus the field elements

$$\beta = \alpha^i \quad \gamma = \alpha^j$$

are represented as the integers i and j. This requires a special symbol and special handling for the element zero because it is not a power of α. Multiplication, except for multiplication by zero, becomes integer addition modulo $q - 1$. Addition becomes more difficult. To add the field elements represented by exponents i and j, the elements α^i and α^j must be computed and then added to obtain α^k, from which k is obtained by taking the Galois-field logarithm to the base α. Stored log and antilog tables can facilitate these computations. The log table stores i at address α^i; the antilog table stores α^i at address i. If the data is normally stored in its natural form, multiplication consists of finding integers i and j in the log table at memory locations α^i and α^j, adding i and j modulo $q - 1$ to get k, then finding α^k in the antilog table at memory address k. If the data is normally stored in log form, then addition consists of finding field elements α^i and α^j in the antilog table at memory locations i and j, adding (by componentwise exclusive-or for fields of characteristic 2), then finding integer k in the log table at memory location α^k.

The look-up procedures can be simplified somewhat by a technique known as *Zech logarithms*. This technique allows all finite-field arithmetic to be performed in the logarithm domain – no antilog operation is needed. For each value of n, the Zech logarithm $Z(i)$ of i is defined by

$$\alpha^{Z(i)} = 1 + \alpha^i.$$

A table of Zech logarithms for $GF(8)$ is given in Table 8.1.

To use Zech logarithms, all field elements are represented in the logarithm domain. Thus α^i and α^j are represented by the indices i and j. Multiplication of α^i by α^j is implemented simply as the sum $i + j$ modulo $q - 1$. Addition of α^i and α^j is developed as follows

$$\begin{aligned}\alpha^i + \alpha^j &= \alpha^i(1 + \alpha^{j-i}) \\ &= \alpha^{i+Z(j-i)}.\end{aligned}$$

Thus the sum is represented by $i + Z(j - i)$, which is computed by a subtraction modulo $q - 1$, a look-up of $Z(j - i)$ in the Zech logarithm table, and a sum modulo $q - 1$.

Hardware implementations of general $GF(2^m)$ operations are quite different. Efficient logic structures for $GF(2^m)$ multiplication and inversion are in use. One measure of efficiency is the number of logic gates. Another measure is the regularity of the circuit structure.

Efficient logic circuits for multiplication do not use logarithms but work more closely with the structure of the field. Addition in $GF(2^m)$ is bit-by-bit modulo-2 addition, which is easily implemented in digital logic using exclusive-or gates, and all m bits of the sum can be computed in parallel. Multiplication in $GF(2^m)$, when the m bits are computed in parallel, requires circuits that are much more complicated. We shall describe one multiplier that has a regular structure and another multiplier that has a reduced number of gates.

To design a regular structure for the Galois-field product

$$c(x) = a(x)b(x) \pmod{p(x)},$$

recall that the modulo-$p(x)$ operation can be distributed across addition and multiplication by x, as is convenient. The equation can be rearranged as follows:

$$c(x) = R_{p(x)}[\ldots x R_{p(x)}[x R_{p(x)}[x b_{m-1} a(x)] + b_{m-2} a(x)] + \cdots] + b_0 a(x).$$

This suggests the cellular multiplier shown in Figure 8.37. It consists of the interconnection of m^2 standard cells, one of which is shown in Figure 8.38. In a standard cell, one bit of a and one bit of b are multiplied (an "and" gate), and the result is added in $GF(2)$ (an "exclusive-or" gate) to other terms and passed out the line on the diagonal, corresponding to multiplication by the position variable x. The other terms entering the modulo-2 sums are the overflow output of an earlier cell and the feedback from the modulo-$p(x)$ overflow. The coefficient p_i determines whether the feedback is used in the ith cell.

An alternative multiplier is one designed to reduce the number of gates. We will construct a multiplier for the field $GF(256)$ which will be regarded as an extension of $GF(16)$ formed using an irreducible polynomial $z^2 + Az + B$ over $GF(16)$. Every element of $GF(256)$ can be represented in the form $a_\ell + a_h z$ where $z^2 = Az + B$. To make this representation as simple as possible, choose $A = 1$ and $B = \omega$ where ω is a zero of $x^4 + x + 1$ in $GF(16)$. This polynomial is $z^2 + z + \omega$, which is irreducible in $GF(16)$, as can be verified easily.

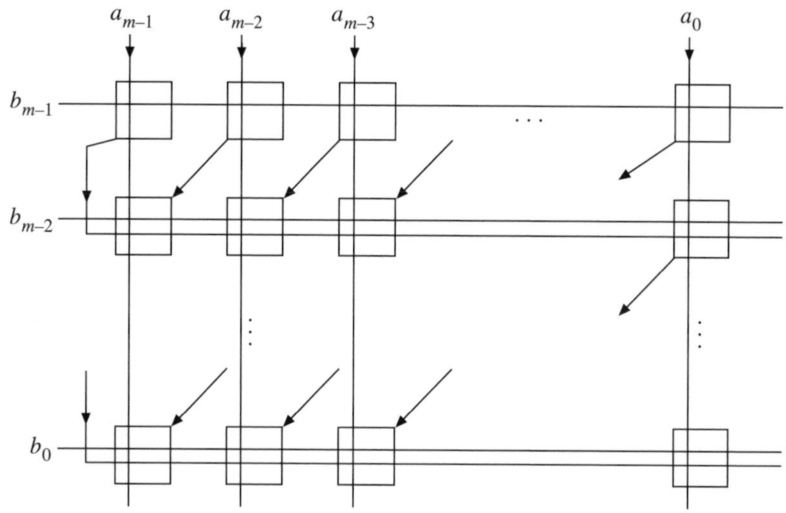

Figure 8.37. A cellular array for $GF(2^m)$ multiplication

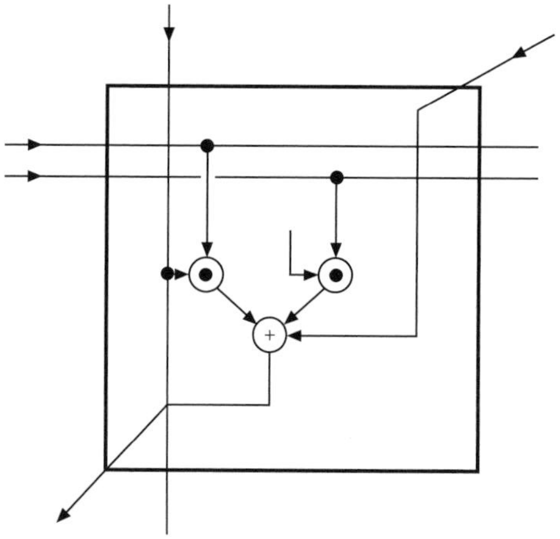

Figure 8.38. A single cell

Multiplication in $GF(256)$ can now be written out:

$$\begin{aligned} c = ab &= (a_\ell + a_h z)(b_\ell + b_h z) \\ &= (a_\ell b_\ell + a_h b_h \omega) + (a_\ell b_h + a_h b_\ell + a_h b_h)z. \end{aligned}$$

Therefore

$$\begin{aligned} c_\ell &= a_\ell b_\ell + a_h b_h \omega \\ c_h &= (a_\ell + a_h)(b_\ell + b_h) + a_\ell b_\ell. \end{aligned}$$

This form of multiplication in $GF(256)$ requires three general multiplications in $GF(16)$, plus a multiplication by ω and four additions in $GF(16)$. Multiplication by

$\omega = x$ in $GF(16)$ is trivial:

$$x(a_3x^3 + a_2x^2 + a_1x + a_0) = a_2x^3 + a_1x^2 + (a_3 + a_0)x + a_3$$

because $x^4 = x + 1$. This requires a single addition in $GF(2)$.

We now need an efficient general multiplier in $GF(16)$. The multiplication

$$\begin{aligned} c &= ab \\ &= (a_3x^3 + a_2x^2 + a_1x + a_0)(b_3x^3 + b_2x^2 + b_1x + b_0) \pmod{x^4 + x + 1} \end{aligned}$$

can be written as a matrix–vector product:

$$\begin{bmatrix} c_0 \\ c_1 \\ c_2 \\ c_3 \end{bmatrix} = \begin{bmatrix} a_0 & a_3 & a_2 & a_1 \\ a_1 & a_0 + a_3 & a_2 + a_3 & a_1 + a_2 \\ a_2 & a_1 & a_0 + a_3 & a_2 + a_3 \\ a_3 & a_2 & a_1 & a_0 + a_3 \end{bmatrix} \begin{bmatrix} b_0 \\ b_1 \\ b_2 \\ b_3 \end{bmatrix}.$$

All variables in this matrix–vector equation are elements of $GF(2)$ and all operations are $GF(2)$ operations. There are sixteen $GF(2)$ multiplications, which are implemented as *and* gates. There are fifteen $GF(2)$ additions – three before the multiplications and twelve after – and these are implemented as *exclusive-or* gates.

It is now straightforward to combine these calculations to construct a $GF(256)$ multiplier that uses forty-eight and gates and sixty-four exclusive-or gates. The pattern of interconnection between these gates does not have a regular structure but does have some repetition of sections. There may exist other multipliers with even fewer gates but with an even more irregular pattern of interconnection.

8.8 Bit-serial multipliers

A finite field $GF(q^m)$ can be regarded as a vector space over $GF(q)$. Then a basis for the vector space $GF(q^m)$ is any set of m linearly independent vectors over $GF(q)$:

$$\{\beta_0, \beta_1, \ldots, \beta_{m-1}\}.$$

Using a basis for $GF(2^m)$, any m-bit binary number $b = (b_{m-1}, b_{m-2}, \ldots, b_0)$ can be mapped to the field element

$$b = b_0\beta_0 + b_1\beta_1 + \cdots + b_{m-1}\beta_{m-1}.$$

There are a great many bases for $GF(2^m)$. The usual basis of $GF(2^m)$ is the *polynomial basis*

$$\{1, \alpha, \alpha^2, \ldots, \alpha^{m-1}\}.$$

Every field element is a linear combination over $GF(2)$ of these basis vectors, and this linear combination defines the map from m-bit binary numbers into field elements.

Although the polynomial basis for $GF(2^m)$ is perhaps the most natural, other bases are also in use. A *normal basis* for $GF(2^m)$ is a basis of the form

$$\left\{\alpha^{2^0}, \alpha^{2^1}, \alpha^{2^2}, \alpha^{2^3}, \ldots, \alpha^{2^{m-1}}\right\},$$

provided these elements are linearly independent over $GF(2)$. Then an element b has the form

$$b = \sum_{i=0}^{m-1} b_i \alpha^{2^i}$$

where $b_i \in GF(2)$. It is easy to add such numbers because addition is componentwise modulo-2 addition. It is also easy to square such numbers because $b_i^2 = b_i$ so

$$\begin{aligned} b^2 &= \sum_{i=0}^{m-1} b_i^2 \left(\alpha^{2^i}\right)^2 \\ &= \sum_{i=0}^{m-1} b_i \alpha^{2^{i+1}}. \end{aligned}$$

Thus

$$b^2 = b_{m-1}\alpha^{2^0} + b_0\alpha^{2^1} + \cdots + b_{m-2}\alpha^{2^{m-1}}.$$

Thus if $(b_0, b_1, \ldots, b_{m-1})$ is a representation of b with respect to the normal basis $\{\alpha^1, \alpha^2, \alpha^4, \ldots, \alpha^{2^{m-1}}\}$, then the cyclic shift $(b_{m-1}, b_0, \ldots, b_{m-2})$ is a representation of b^2. Squaring is not as important, however, as multiplication. Multiplication does not share the simple structure of squaring, but it can have an invariance under cyclic shifts.

The *Omura–Massey multiplier* is a bit-serial multiplier. It uses a normal basis as follows. Let

$$c = a \cdot b$$

where a, b, and c are represented with respect to the normal basis. Thus

$$\sum_{k=0}^{m-1} c_k \alpha^{2^k} = \left(\sum_{i=0}^{m-1} a_i \alpha^{2^i}\right)\left(\sum_{j=0}^{m-1} b_j \alpha^{2^j}\right).$$

Now suppose that we have constructed a circuit to compute the single coefficient c_{m-1} from $a = (a_0, \ldots, a_{m-1})$ and $b = (b_0, \ldots, b_{m-1})$. Then we can use this circuit to compute, in succession, the bits $c_{m-2}, c_{m-3}, \ldots, c_0$ as follows. Using the cyclic shift operator S, write $Sa = (a_{m-1}, a_0, \ldots, a_{m-2})$ and $Sb = (b_{m-1}, b_0, \ldots, b_{m-2})$. Hence if the inputs to the circuit are Sa and Sb, the output is c_{m-2}. Thus c_{m-1}, c_{m-2}, $c_{m-3}, \ldots, c_0$ are computed in succession by applying the inputs (a, b), (Sa, Sb), $(S^2a, S^2b), \ldots, (S^{m-1}a, S^{m-1}b)$.

To design the logic circuit, we now note that c_{m-1} is the coefficient of $\alpha^{2^{m-1}}$ in the normal basis representation of

$$c = a \cdot b = \sum_{i=0}^{m-1} \sum_{j=0}^{m-1} a_i b_j \alpha^{2^i + 2^j}.$$

Each of the m^2 powers of α on the right side is a field element, so it can be expressed in the form

$$\alpha^{2^i + 2^j} = \sum_{\ell=0}^{m-1} M_{ij\ell} \alpha^{2^\ell}$$

where $M_{ij\ell} \in GF(2)$. Then

$$\sum_{\ell=0}^{m-1} c_\ell \alpha^{2^\ell} = \sum_{\ell=0}^{m-1} \left[\sum_{i=0}^{m-1} \sum_{j=0}^{m-1} a_i M_{ij\ell} b_j \right] \alpha^{2^\ell}$$

Therefore c_{m-1} can be written in bilinear form

$$c_{m-1} = \sum_{i=0}^{m-1} \sum_{j=0}^{m-1} a_i M_{ij} b_j$$

where the third index on ℓ has been suppressed. The values of $M_{ij} \in GF(2)$ can be deduced by writing out the expansions in detail and identifying terms.

For example, in $GF(256)$ constructed by using the prime polynomial $p(x) = x^8 + x^7 + x^5 + x^3 + 1$, the elements M_{ij} form the matrix

$$\boldsymbol{M} = \begin{bmatrix} 0 & 0 & 0 & 1 & 0 & 0 & 0 & 0 \\ 0 & 0 & 0 & 0 & 0 & 0 & 1 & 1 \\ 0 & 0 & 0 & 0 & 1 & 0 & 1 & 0 \\ 1 & 0 & 0 & 0 & 0 & 1 & 1 & 1 \\ 0 & 0 & 1 & 0 & 0 & 1 & 0 & 0 \\ 0 & 0 & 0 & 1 & 1 & 0 & 0 & 0 \\ 0 & 1 & 1 & 1 & 0 & 0 & 0 & 1 \\ 0 & 1 & 0 & 1 & 0 & 0 & 1 & 1 \end{bmatrix}.$$

Then bit seven of c is

$$c_7 = \boldsymbol{a}\boldsymbol{M}\boldsymbol{b}^{\mathrm{T}},$$

which requires twenty-one "and" operations (because there are twenty-one ones in $\boldsymbol{M}$) and twenty "exclusive-or" operations. To compute other components of c,

$$c_{7-\ell} = (S^\ell \boldsymbol{a})^{\mathrm{T}} \boldsymbol{M} (S^\ell \boldsymbol{b})$$

where S is the cyclic shift operator on the m bits of $\boldsymbol{a}$ or $\boldsymbol{b}$.

Another approach is to use two different bases for the variables in a multiplication. Two bases $\{\mu_0, \mu_1, \ldots, \mu_{m-1}\}$ and $\{\lambda_0, \lambda_1, \ldots, \lambda_{m-1}\}$ are called *complementary bases* (or dual bases) if they satisfy the property

$$\text{trace}(\mu_i \lambda_k) = \delta_{ik} = \begin{cases} 1 & \text{if } i = k \\ 0 & \text{if } i \neq k. \end{cases}$$

Every basis has a unique complementary basis.

An m-bit binary number can be mapped into $GF(2^m)$ using any convenient basis, but the element of $GF(2^m)$ that results depends on which basis is used. Similarly, a fixed element of $GF(2^m)$ can be represented by an m-bit binary number by expressing the field element as a linear combination of basis elements – the binary number that results depends on which basis is used. Thus when two bases are used at the same time, care must be taken to distinguish situations where the field element is the fixed quantity and situations where the binary coefficients form the fixed quantity.

Theorem 8.8.1 (Projection Property). *Let $\{\mu_k\}$ and $\{\lambda_k\}$ be complementary bases of $GF(2^m)$. Any field element $\beta \in GF(2^m)$ has the representation*

$$\beta = \sum_{k=0}^{m-1} b_k \lambda_k$$

where $b_i = \text{trace}(\beta \mu_i)$.

Proof: Because $\{\lambda_k\}$ is a basis, every element $\beta \in GF(2^m)$ can be written as

$$\beta = \sum_{k=0}^{m-1} b_k \lambda_k$$

for some $b_0, b_1, \ldots, b_{m-1}$. But

$$\begin{aligned} \text{trace}(\beta \mu_i) &= \text{trace}\left(\mu_i \sum_{k=0}^{m-1} b_k \lambda_k\right) \\ &= \sum_{k=0}^{m-1} b_k \text{trace} \mu_i \lambda_k \\ &= \sum_{k=0}^{m-1} b_k \delta_{ik} \\ &= b_i, \end{aligned}$$

as was to be proved. □

The *Berlekamp bit-serial multiplier* uses the polynomial basis $\{1, \alpha, \alpha^2, \ldots, \alpha^{m-1}\}$ and the complementary polynomial basis $\{\lambda_0, \lambda_1, \ldots, \lambda_{m-1}\}$. The algorithm starts with the two field elements β and γ, one expressed in the polynomial basis and one expressed in the complementary polynomial basis, and computes the product σ expressed in the complementary polynomial basis. Let

$$\beta = \sum_{i=0}^{m-1} b_i \alpha^i$$

$$\gamma = \sum_{k=0}^{m-1} c_k \lambda_k.$$

Then

$$\begin{aligned}\sigma &= \beta\gamma \\ &= \sum_{i=0}^{m-1} \sum_{k=0}^{m-1} b_i c_k \alpha^i \lambda_k.\end{aligned}$$

The product is expressed in the complementary polynomial basis as

$$\sigma = \sum_{k=0}^{m-1} s_k \lambda_k$$

where, by Theorem 8.8.1,

$$\begin{aligned}s_k &= \text{trace}(\sigma \alpha^k) \\ &= \text{trace}(\beta \gamma \alpha^k).\end{aligned}$$

In particular,

$$\begin{aligned}s_0 &= \text{trace}(\beta\gamma) \\ &= \text{trace}\left(\sum_{i=0}^{m-1} \sum_{k=0}^{m-1} b_i c_k \alpha^i \lambda_k\right) \\ &= \sum_{i=0}^{m-1} \sum_{k=0}^{m-1} b_i c_k \text{trace} \alpha^i \lambda_k \\ &= \sum_{i=0}^{m-1} b_i c_i,\end{aligned}$$

which is a simple sum of products. Moreover, for each k, s_k can be obtained simply by first multiplying β by α^k, then repeating the same sum of products.

For example, let $p(z) = z^4 + z + 1$ be the primitive polynomial used to construct $GF(16)$. Then over $GF(2)$, $\{1, \alpha, \alpha^2, \alpha^3\}$ is the polynomial basis of $GF(2^4)$ and $\{\alpha^{14}, \alpha^2, \alpha, 1\}$ is the complementary polynomial basis. Let

$$\beta = b_0\alpha^0 + b_1\alpha^1 + b_2\alpha^2 + b_3\alpha^3$$

$$\gamma = c_0\alpha^{14} + c_1\alpha^2 + c_2\alpha^1 + c_3\alpha^0$$

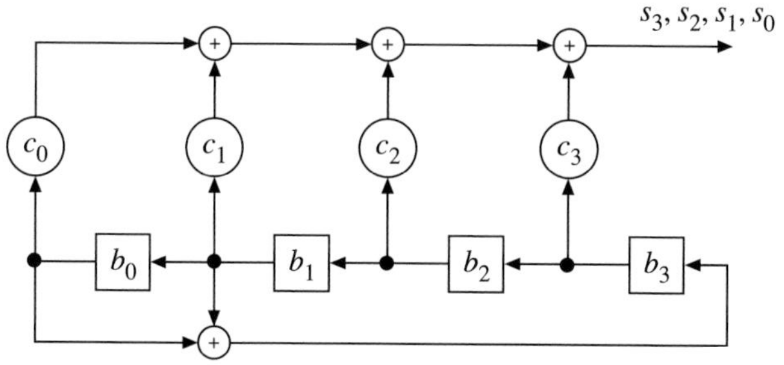

Figure 8.39. A Berlekamp bit-serial multiplier

and

$$\begin{aligned}\sigma &= \beta\gamma \\ &= s_0\alpha^{14} + s_1\alpha^2 + s_2\alpha^1 + s_3\alpha^0.\end{aligned}$$

Then

$$s_0 = b_0c_0 + b_1c_1 + b_2c_2 + b_3c_3,$$

and s_k can be obtained by the same expression if β is first replaced by $\beta\alpha^k$.

A circuit for implementing the Berlekamp bit-serial multiplier is shown in Figure 8.39. Compared with the parallel multiplier of Figure 8.37, the bit-serial multiplier has of the order of m fewer gates and of the order of m more clock cycles. The advantage of the circuit is the great reduction in the number of wires. The disadvantage is the need to use more than one representation of field elements. This may entail the need for the conversion of a variable from one basis to the other.

Problems

8.1 Over $GF(2)$, let $p(x) = x^2 + x + 1$, and let $q(x) = x^4 + x + 1$. Find the "Chinese polynomials" $a(x)$ and $b(x)$, satisfying $a(x)p(x) + b(x)q(x) = 1$.

8.2 How many errors can be corrected by the product of a $(7, 4)$ Hamming code with itself? Using a Hamming $(7, 4)$ decoder as an "off-the-shelf" building block, design a simple decoder that corrects to the packing radius of the product code.

8.3 Sketch an error-trapping decoder for a Reed–Solomon decoder that detects all detectable error patterns. How many syndromes correspond to correctable error patterns?

8.4 Sketch a circuit that will perform the permutation $i \rightarrow 2i \pmod{23}$. Can the $(23, 12, 7)$ Golay code be error-trapped by this permutation?

8.5 Design a modified error-trapping decoder for the ternary Golay code.

8.6 Design a shift register encoder for the (12, 4, 6) quasi-cyclic code with generator matrix

$$\boldsymbol{G} = \begin{bmatrix} 111 & 011 & 000 & 001 \\ 001 & 111 & 011 & 000 \\ 000 & 001 & 111 & 011 \\ 011 & 000 & 001 & 111 \end{bmatrix}.$$

8.7 Let a and b satisfy

$$an' + bn'' = 1$$

where n' and n'' are coprime. Given an (n', k') cyclic code with generator polynomial $g_1(x)$ and an (n'', k'') cyclic code with generator polynomial $g_2(x)$, prove that the cyclic product code with blocklength $n = n'n''$ has the generator polynomial

$$g(x) = \text{GCD}[g_1(x^{bn''}), g_2(x^{an'}), x^n - 1].$$

8.8 **(Inversion in $GF(16)$.)** Given nonzero $a \in GF(16)$, verify that $ba = 1$ in $GF(16)$ where

$$\begin{aligned} b_0 &= a_0 + a_1 + a_2 + a_3 + a_0a_2 + a_1a_2 + a_0a_1a_2 + a_1a_2a_3 \\ b_1 &= a_3 + a_0a_1 + a_0a_2 + a_1a_2 + a_1a_3 + a_0a_1a_3 \\ b_2 &= a_2 + a_3 + a_0a_1 + a_0a_2 + a_0a_3 + a_0a_2a_3 \\ b_3 &= a_1 + a_2 + a_3 + a_0a_3 + a_1a_3 + a_2a_3 + a_1a_2a_3. \end{aligned}$$

8.9 **(Inversion in $GF(256)$.)** An inverse in $GF(256)$ can be constructed from an inverse in $GF(16)$ as follows. Given nonzero $a \in GF(256)$ represented as $a = a_\ell + a_h z$ where $a_\ell, a_h \in GF(16)$, let $b = b_\ell + b_h z$ be the element that satisfies $a \cdot b = 1 \pmod{z^2 + z + \omega}$ where ω is primitive in $GF(16)$. Verify that

$$\begin{aligned} b_h &= \frac{a_h}{a_\ell(a_\ell + a_h) + \omega a_h^2} \\ b_\ell &= \frac{a_\ell + a_h}{a_\ell(a_\ell + a_h) + \omega a_h^2}. \end{aligned}$$

Describe a $GF(256)$ inverter with low gate count. How many gates are used? Describe a method for division in $GF(256)$.

8.10 **(Squaring.)** Design a circuit to compute squares in $GF(16)$. Design a circuit to compute squares in $GF(256)$.

8.11 A quarter-square multiplier is one that multiplies by squaring using the formula

$$ab = \tfrac{1}{4}[(a+b)^2 - (a-b)^2].$$

Can a quarter-square multiplier be used in a field of characteristic 2?

Notes

In this chapter the role of shift registers in encoders and decoders can now be seen clearly as a part of digital-filtering theory, but with a Galois field in place of the real or complex field. This was vaguely realized at the outset and became more obvious as the subject developed. The basic shift-register circuits are immediately apparent to most designers and have entered the literature without any fanfare. Shift-register circuits for encoding and decoding can be found in the work of Peterson (1960) and Chien (1964). The ideas appear in textbook form in the book by Peterson (1961). Meggitt published the design for his decoder in 1960 and 1961. The origin of the idea of error trapping is a little hazy, but credit is usually given to Prange (1957).

Kasami (1964) studied ways to augment an error-trapping decoder to handle correctable but untrappable error patterns (the modified error-trapping decoder for the Golay code, to be described in Section 10.5, uses the techniques of Kasami). The use of permutations other than cyclic shifts was studied by MacWilliams (1964). Other early work appears in the papers of Mitchell (1962) and Rudolph and Mitchell (1964).

Zech's logarithms were introduced in 1837 by Jacobi, as described in Lidl and Niederreiter (1984). The use of Zech's logarithms to factor polynomials of small degree was studied by Huber (1990). Willett (1980) discussed the use of alternative bases for multiplication, and Berlekamp (1982) introduced the idea of using a complementary pair of bases. This idea was developed further by Morii, Kasahara, and Whiting (1989). Imamura (1983) and Lempel (1988) discussed bases that are their own complements. The Omura–Massey multiplier (1986) can be found in a U.S. patent. The use of combinatorial logic circuits for Galois-field arithmetic was studied by Bartee and Schneider (1963); Laws and Rushforth (1971) revised such methods so that a cellular architecture could be used.

A discussion of pipelined Reed–Solomon decoders can be found in Shao *et al.* (1985). Sarwate and Shanbhag (2001) proposed an architecture that iterates the discrepancy polynomial instead of the evaluator polynomial. The tailoring of Reed–Solomon decoding algorithms for systolic architectures using the euclidean algorithm was studied by Citron (1986).

9 Convolutional Codes

To encode an infinite stream of data symbols with a block code, the datastream is broken into blocks, each of k data symbols, called *datawords*. The block code encodes each block of k data symbols into a block of n code symbols, called a *codeword*. The codewords are concatenated to form an infinite stream of code symbols.

In contrast to a block code is a *trellis code*. A trellis code also encodes a stream of data symbols into a stream of code symbols. A trellis code divides the datastream into blocks of length k, called *dataframes*, which are usually much smaller and are encoded into blocks of length n, called *codeframes*. In both cases, block codes and trellis codes, the datastream is broken into a sequence of blocks or frames, as is the codestream. When using a block code, a single block of the codestream depends only on a single block of the datastream, whereas when using a trellis code, a single frame of the codestream depends on multiple frames of the datastream.

The most important trellis codes are those known as *convolutional codes*. Convolutional codes are trellis codes that satisfy certain additional linearity and time-invariance properties. Although we introduce the general notion of a trellis code, we will be concerned mostly with the special class of convolutional codes.

The dataframes of a convolutional code are usually small and typically contain no more than a few data symbols. Each dataframe of length k corresponds to a codeframe of length n. Rather than coding a single dataframe into a single codeframe, however, a dataframe together with the previous m dataframes are encoded into a codeframe. Hence successive codeframes are linked by the encoding procedure, and the concatenated stream of such codeframes forms the infinitely long codeword.

9.1 Codes without a block structure

The basic notions of convolutional codes, and of other trellis codes, are introduced using the shift-register encoder of Figure 9.1. A sequence of data symbols is shifted into the encoder, beginning at time zero, and continuing indefinitely. The incoming datastream is broken into dataframes of k symbols. A dataframe may be as short as one data symbol. The encoder can store m dataframes, where m is called the *frame*

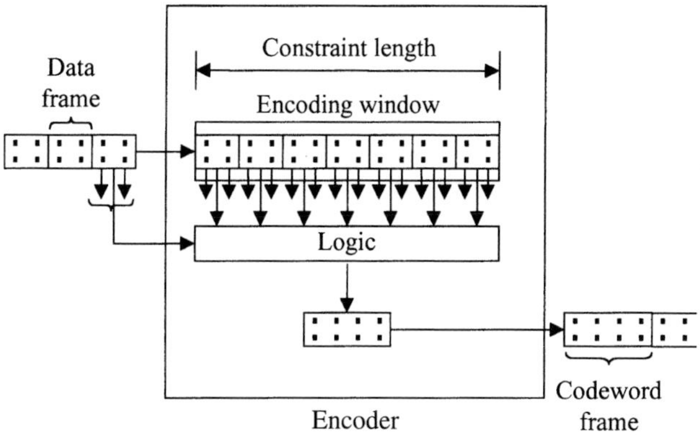

Figure 9.1. A shift-register encoder

memory of the encoder. At each clock time, a new dataframe is shifted into the shift register, and the oldest dataframe is shifted out and discarded. After each clock time, the encoder has stored the most recent m dataframes (a total of mk data symbols). At the start of the next clock time, from the next dataframe and the m stored dataframes, each of length k symbols, the encoder computes a single codeframe of length n symbols. This new codeframe is shifted out of the encoder as the next dataframe is shifted in. Hence the channel must transmit n code symbols for every k data symbols. The set of code sequences produced by such an encoder is called an (n, k) convolutional code or, more generally, an (n, k) trellis code. The *code rate* is the ratio k/n.

The code is the infinite set of all infinitely long codewords that one obtains by exciting the encoder with every possible input data sequence. Notice the careful distinction between the code and the encoder for the code; the code is a set of sequences, the encoder is the rule or device for computing one code sequence from one data sequence. The same code may be produced by many different encoders. The code, in principle, can be defined abstractly without reference to an encoder.

A convolutional code is required to have two properties, called *time invariance* and *linearity*. Time invariance means that if a datastream is delayed by a single dataframe (padded with a leading dataframe of zeros), then the codestream is delayed by a single codeframe (padded with a leading codeframe of zeros). Linearity means that any linear combination of two datastreams has a codeword that is the same linear combination of the codewords of the two datastreams. That is, if a_1 and a_2 are two datastreams with codewords $G(a_1)$ and $G(a_2)$, then $\beta a_1 + \gamma a_2$ has codeword

$$G(\beta a_1 + \gamma a_2) = \beta G(a_1) + \gamma G(a_2).$$

If these two properties are not required, the code is called a *trellis code*. Thus, a linear, time-invariant (n, k) trellis code is an (n, k) convolutional code.

Because of implementation considerations, practical convolutional codes use very small integers for k and n, often $k = 1$. This means that the choice for the code rate

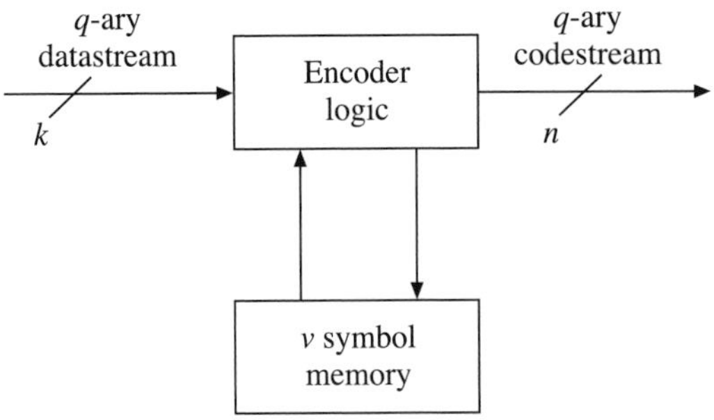

Figure 9.2. A general encoder for a trellis code

is limited. One cannot design a practical convolutional code with a code rate close to one, as can be conveniently done with a block code.

The number of q-ary memory cells in a minimal encoder for a convolutional code is an important descriptor of the code called the *constraint length* of the code and denoted by ν. This informal description of the constraint length will do for now, but a formal definition will be given later. Because the encoder, as shown in Figure 9.2, forms its output using the previous m dataframes, it is sufficient to store only these frames, so we can conclude that it suffices to store mk symbols. This implies that $\nu \leq mk$. This is not necessarily an equality because it may be that several patterns of these mk symbols are encoded the same, and so can be combined in some fashion in memory. Then the encoder could represent these mk symbols with fewer symbols. For example, the encoder of Figure 9.1 with $k = 3$ and $n = 5$, as shown, appears to have constraint length 21, but it may be possible to intermingle logic with the memory so that fewer symbols of memory suffice.

A systematically encoded (n, k) convolutional code is one in which each k-symbol dataframe appears unaltered as the first k symbols of the n symbols of the first codeframe that it affects. Systematic encoders for convolutional codes can be more satisfying because the data is visible in the encoded sequence and can be read directly if no errors are made. Because the code exists independently of the method of encoding, however, it is not correct to speak of systematic convolutional codes, only of systematic encoders.

Given an (n, k) convolutional code, a new convolutional code with the same parameters can be obtained by permuting the symbols of each frame in the same way. This, however, gives a code that is only trivially different from the original code. In general, two convolutional codes that are the same except for the order of components within a frame are called *equivalent* convolutional codes. Indeed, they are usually considered to be the same code.

Examples of encoders for two different convolutional codes over $GF(2)$ are shown in Figure 9.3 and Figure 9.4, both with $k = 1$ and $n = 2$. In each case, an arbitrary binary datastream is shifted into the encoder and the corresponding codestream is shifted out – two output code bits for each input data bit.

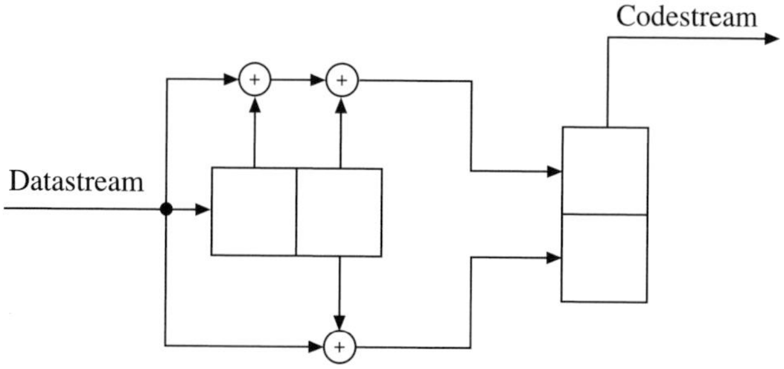

Figure 9.3. An encoder for a binary (2, 1) convolutional code

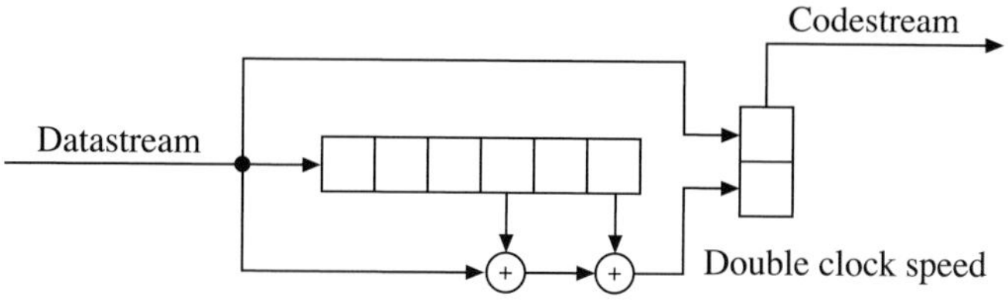

Figure 9.4. An encoder for another binary (2, 1) convolutional code

The example of Figure 9.3 is a nonsystematic encoder for a (2, 1) binary convolutional code of constraint length 2. The example of Figure 9.4 is a systematic encoder for a (2, 1) binary convolutional code of constraint length 5. In Figure 9.3, the input datastream is filtered twice by the upper and lower taps of the shift register. In Figure 9.4 the input datastream is filtered only by the lower taps. In either example, the two filter output sequences are interleaved in time by inserting the filter outputs into a buffer that is clocked out twice as fast as the incoming datastream is clocked in.

Only a few constructive classes of convolutional codes are known. Most of the best convolutional codes that are known and are in use today have been discovered by computer search; many such codes are tabulated in Figure 9.5 in terms of their generator polynomials, which are defined in Section 9.3. No classes with a satisfactory algebraic structure comparable to that of the t-error-correcting BCH codes are known. No satisfactory constructive method for finding a general class of good convolutional codes of long constraint length is known.

9.2 Trellis description of convolutional codes

A convolutional code can be described usefully in terms of a type of graph called a *trellis*. A trellis is a *directed graph* in which every node is at a well-defined depth with respect to a beginning node. A directed graph is one in which all branches are one-way branches; a node with a well-defined depth is one where all paths to it from a starting node have the same number of branches. We consider in this chapter only

$(mn,\ mk,\ d)$	Matrix of generator polynomials (in terms of polynomials coefficients)			
(6, 3, 5)	(x^2+1)	101	(x^2+x+1)	111
(8, 4, 6)	(x^3+x+1)	1011	(x^3+x^2+x+1)	1111
(10, 5, 7)		11001		10111
(12, 6, 8)		1101101		101111
(14, 7, 10)		11100101		1001111
(16, 8, 10)		100011101		10011111
(18, 9, 12)		1110111001		110101111
(20, 10, 12)		10111011001		10001101111
(22, 11, 14)		101110110001		110010111101
(24, 12, 15)		1101101010001		1000110111111
(26, 13, 16)		10111101110001		11001010011101
(28, 14, 16)				
(6, 3, 5)	101	111		111
(8, 4, 6)	1011	1011		1111
(10, 5, 7)	10101	11011		11111
(12, 6, 8)	111001	110101		101111
(14, 7, 10)	1101101	1010011		1011111
(16, 8, 10)	10101001	10011011		11101111
(18, 9, 12)	111101101	110011011		100100111
(20, 10, 12)	1111001001	101011101		110100111
(22, 11, 14)	11010111001	10011101101		10111110011
(24, 12, 15)	111011111001	110010111101		101011010011
(26, 13, 16)	1101101000001	1011110110001		1000110111111
(28, 14, 16)	10100101110001	10001101110111		11011010011111
(6, 3, 5)	101	111	111	111
(8, 4, 6)	1101	1011	1011	1111
(10, 5, 7)	10101	11011	10111	11111
(12, 6, 8)	110101	111011	100111	101111
(14, 7, 10)	1011101	101101	1110011	1100111
(16, 8, 10)	10111001	10111101	11010011	11110111
(18, 9, 12)	110011001	101110101	110110111	101001111
(20, 10, 12)	1111001001	1010111101	1101100111	1101010111
(22, 11, 14)	11101011001	11010111001	10011101101	10111110011
(24, 12, 15)	111011111001	110010111101	101011010011	10110001111
(26, 13, 16)	1010011001001	111110010101	1101111011011	1110101110111
(28, 14, 16)	11010010010001	1011110011001	1101010110111	11111010110111

Figure 9.5. Binary convolutional codes with maximum free distance

trellises semi-infinite to the right, with q^ν nodes in each column. The configuration of the branches connecting each column of nodes to the next column of nodes is the same for each column of nodes. A typical trellis for a binary code alphabet is shown in the diagram of Figure 9.6. Only nodes that can be reached by starting at the top left node and moving to the right are shown.

Each column of nodes of the trellis is indexed by an integer ℓ, called *time*. At time ℓ, every distinct state that can be assumed by the encoder is represented by a distinct node. An encoder of constraint length ν over a q-ary data alphabet can assume any of q^ν states, so there are q^ν nodes in a column depicting the set of states at one time;

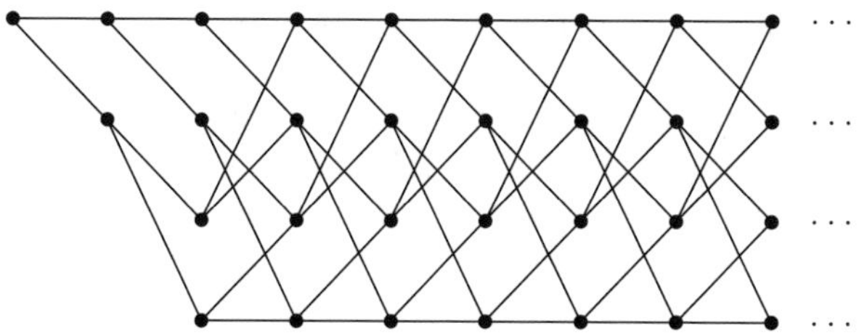

Figure 9.6. Trellis for a convolutional code

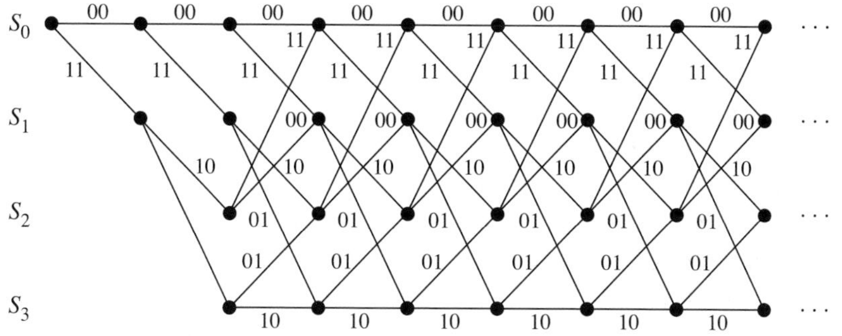

Figure 9.7. Trellis diagram for a convolutional code with constraint length 2

successive columns depict the set of states at successive times. If there are k data symbols per frame, then there must be q^k distinct branches that leave each node, one branch for each possible input dataframe. Usually two branches that leave the same node will terminate on different nodes, but this is not required.

A convolutional code is a set of codewords. The code should not be confused with an encoder for the convolutional code. There are many possible encoders. By using the trellis to specify the code, we describe the structure of the convolutional code without giving an encoder.

A simple example of a labeled trellis for a binary convolutional code with $k = 1$, $n = 2$, and $\nu = 2$ is shown in Figure 9.7. The convolutional code is the set of all semi-infinite binary words that may be read off by following any path through the trellis from left to right and reading the code bits written along that path. The encoding rule, however, is not given. The trellis in Figure 9.7 defines a linear code – and hence a convolutional code – because the $GF(2)$ sum of the symbol sequences along any pair of paths is a symbol sequence on another path. There are four states S_0, S_1, S_2, S_3 in Figure 9.7, corresponding to the four possible two-bit patterns in the encoder memory. Hence $\nu = 2$. Every node of the trellis has two branches leaving it so that one branch can be assigned to data bit zero, and one branch can be assigned to data bit one. Hence $k = 1$. Each branch leads to the new state caused by that input bit. Each branch is labeled with two code bits that will be generated if that branch is followed. Hence $n = 2$.

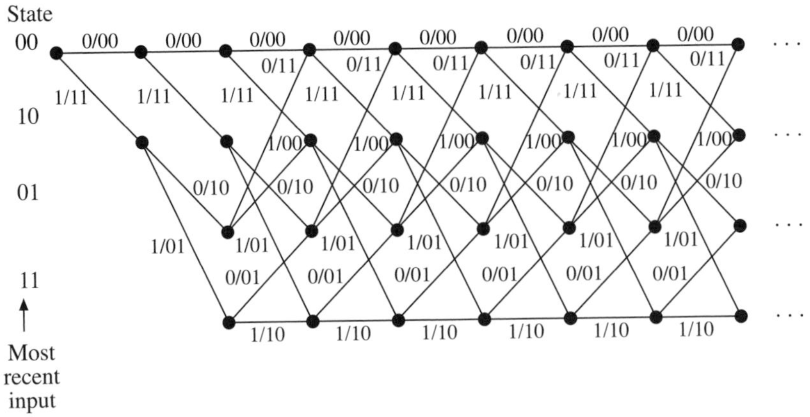

Figure 9.8. Trellis diagram annotated for a nonsystematic encoder

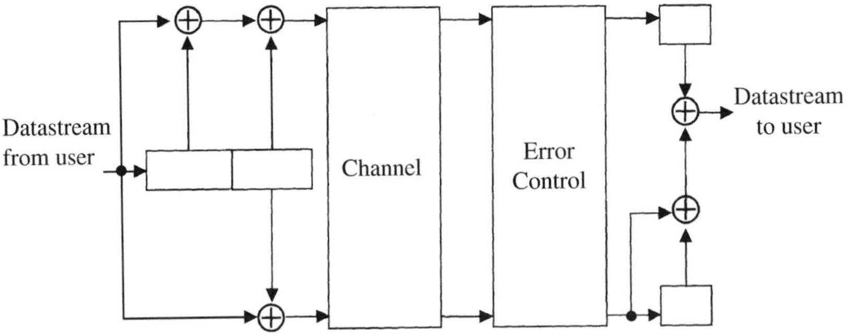

Figure 9.9. Application of a nonsystematic convolutional encoder

Use of the code of Figure 9.7 requires an encoder. This means that at each node one must specify which branch is assigned to represent a data bit zero, and which branch is assigned to represent a data bit one. In principle this assignment is completely arbitrary, but in practice, we want a rule that is easy to implement. We shall describe two encoders for this code: one that uses a two-bit feedforward shift register and is nonsystematic, and one that uses a two-bit feedback shift register and is systematic. In each case, the state corresponds to the two bits in the shift register, and the encoder produces the pair of code bits that label the branch of the trellis.

To design an encoder that uses a feedforward shift register, we must decide how to assign the states. We shall describe several options. For the first option, define the state to be the two most recent data bits, and so label the states, as in Figure 9.8, with two bits, the most recent on the left. The states are labeled with the contents of a two-bit memory in the form of a shift register, that stores the two most recent data bits, shown in Figure 9.9, so the state is equal to the two bits stored in this shift register. This assignment of states determines the state at the end of each branch and so implicitly determines the assignment of data bits to branches. The single incoming data bit determines the

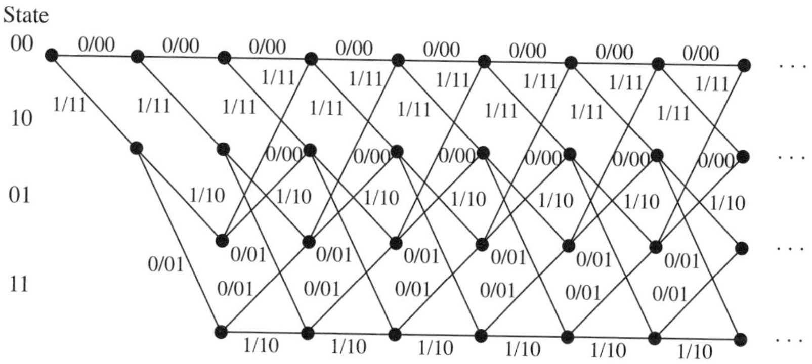

Figure 9.10. Trellis diagram annotated for a systematic encoder

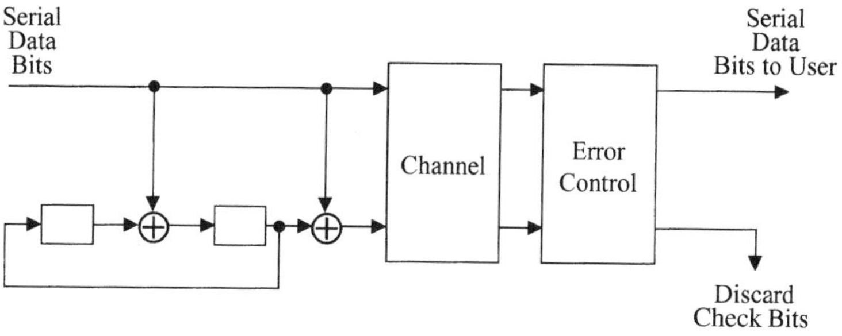

Figure 9.11. Application of a systematic convolutional encoder

next node and so determines a branch, and that data bit value is attached as another label to the branch in Figure 9.8. The code bits on the branch provide the output of the encoder, which is computed from the two state bits and the incoming data bit. A circuit that implements the required binary logic is shown on the left in Figure 9.9, which is the encoder we already saw in Figure 9.3. One kind of decoder, shown on the right in Figure 9.9, consists of two steps: one step to correct the errors and one step to invert the encoder. By inspection of Figure 9.9, one may verify that if all errors are removed in the first step, then every path through the circuit, except the path with no delay, is canceled by another path with the same delay, so the entire circuit is transparent to the datastream. Later we will see that this structure is easy to understand as a standard property of polynomials.

To design an alternative encoder for this code that is systematic, we use the trellis of Figure 9.7 from a different point of view. On each pair of branches coming from the same node, the first bit takes on both possible values. Therefore we now assign branches so that the first code bit is equal to the data bit. This leads to the trellis of Figure 9.10 and the encoder of Figure 9.11. The sets of codewords produced by the encoders of Figure 9.9 and Figure 9.11 are identical, but the way in which codewords represent data bits is different.

Each encoder of the example forms the same code, uses two stages of memory, and each encoder has four states in its trellis. An encoder for a convolutional code of constraint length ν would use ν stages of memory. There would be 2^ν states in the trellis for a binary code, and q^ν states in the trellis for a code over a q-ary alphabet. An (n, k) convolutional code can be encoded by n sets of finite-impulse-response filters, each set consisting of k feedforward shift registers. If the convolutional code is binary, then the filters are one bit wide and the additions and multiplications are modulo-2 arithmetic. The input to the encoder is a binary datastream at a rate of k bits per unit time, and the output of the encoder is a binary codestream to the channel at a rate of n bits per unit time.

In general, a systematic encoder that uses only feedforward shift registers may not exist for a particular convolutional code. However, a systematic encoder that also uses feedback shift registers always exists. We shall develop this statement in the next section by studying the structure of convolutional codes in a slightly more formal way.

9.3 Polynomial description of convolutional codes

A mathematical description of convolutional codes can be formulated in the language of polynomial rings, using the notation of polynomials instead of feedforward shift registers to describe the code. These polynomials, with coefficients in the field of the code, are called *generator polynomials* of the convolutional code. Every generator polynomial is an element of the ring $F[x]$.

For example, the encoder over $GF(2)$ shown in Figure 9.3 and Figure 9.9 has the pair of generator polynomials

$$g_0(x) = x^2 + x + 1$$
$$g_1(x) = x^2 + 1.$$

The encoder over $GF(2)$ shown in Figure 9.4 has the pair of generator polynomials

$$g_0(x) = 1$$
$$g_1(x) = x^5 + x^3 + 1.$$

The generator polynomials of an encoder for a convolutional code are commonly arranged into a k by n matrix of polynomials, called a *generator matrix*, and denoted $\boldsymbol{G}(x)$. For the encoder shown in Figure 9.3, the generator matrix is

$$\boldsymbol{G}(x) = [x^2 + x + 1 \quad x^2 + 1].$$

For the encoder shown in Figure 9.4, the generator matrix is

$$\boldsymbol{G}(x) = [1 \quad x^5 + x^3 + 1].$$

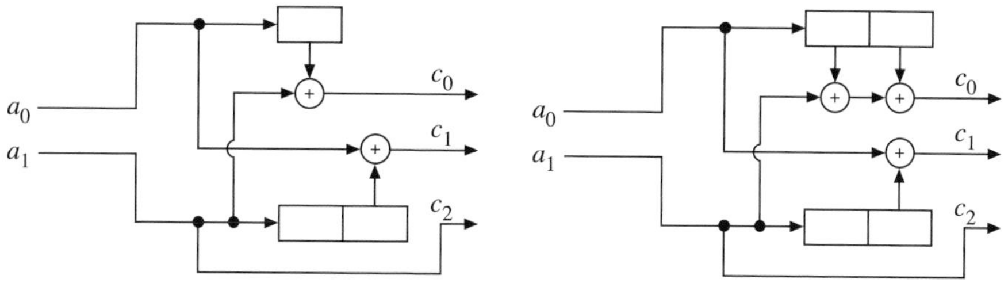

Figure 9.12. Encoders for two convolutional codes of rate 2/3

In general, an encoder for an (n, k) convolutional code is described by a total of kn generator polynomials, the longest of which has degree m. When k is larger than one, some of the generator polynomials may be the zero polynomial. Let $g_{ij}(x)$ for $i = 0, \ldots, k-1$ and $j = 0, \ldots, n-1$ be the set of generator polynomials. These polynomials can be put together into a k by n generator matrix, given by

$$\boldsymbol{G}(x) = [g_{ij}(x)].$$

We require that at least one generator polynomial does not have x as a factor. Such a polynomial is called a *delay-free polynomial*, and such a generator matrix is called a *delay-free generator matrix* This requirement is imposed because if all elements of $\boldsymbol{G}(x)$ did have x as a factor, then x could be factored out of $\boldsymbol{G}(x)$ and regarded as a simple frame delay. Thus, all of our generator matrices are delay-free generator matrices.

Two more examples of encoders for convolutional codes are shown in Figure 9.12. The first example has three stages of memory and so has constraint length three, the second example has four stages of memory and so has constraint length four. The first encoder has a generator matrix

$$\boldsymbol{G}(x) = \begin{bmatrix} x & 1 & 0 \\ 1 & x^2 & 1 \end{bmatrix}.$$

The second encoder has a generator matrix

$$\boldsymbol{G}(x) = \begin{bmatrix} x + x^2 & 1 & 0 \\ 1 & x^2 & 1 \end{bmatrix}.$$

Consider the input dataframe as k data symbols in parallel, and consider the sequence of input frames as k sequences of data symbols in parallel. These may be represented by k data polynomials $a_i(x)$ for $i = 0, 1, \ldots, k-1$, or as a row vector of such polynomials[1]

$$\boldsymbol{a}(x) = [a_0(x) \quad a_1(x) \quad \ldots \quad a_{k-1}(x)].$$

[1] It is convenient to call these polynomials even though they need not have finite degree.

If we break out the vector coefficients of the vector polynomial $\boldsymbol{a}(x)$, we can write this as

$$\boldsymbol{a}(x) = [a_{00}, a_{01}, \ldots, a_{0,k-1}] + [a_{10}, a_{11}, \ldots, a_{1,k-1}]x + [a_{20}, a_{21}, \ldots, a_{2,k-1}]x^2 + \cdots$$

Each bracketed term displays one dataframe of input symbols.

Similarly, the output codeword can be represented by n codeword polynomials $c_j(x)$ for $j = 0, 1, \ldots, n-1$, or as a vector of such polynomials

$$\boldsymbol{c}(x) = [c_0(x) \quad c_1(x) \quad \ldots \quad c_{n-1}(x)].$$

The ℓth codeframe consists of the set of ℓth coefficients from each of these n code polynomials. That is,

$$\boldsymbol{c}(x) = [c_{00}, c_{01}, \ldots, c_{0,n-1}] + [c_{10}, c_{11}, \ldots, c_{1,n-1}]x + [c_{20}, c_{21}, \ldots, c_{2,n-1}]x^2 + \cdots$$

The encoding operation now can be described compactly as a vector–matrix product

$$\boldsymbol{c}(x) = \boldsymbol{a}(x)\boldsymbol{G}(x).$$

Because a vector–matrix product involves only multiplications and additions, this equation is meaningful even though the elements of the arrays are polynomials. The encoding equation also suggests that the convolutional code can be made to have finite blocklength simply by restricting $\boldsymbol{a}(x)$ to have finite length. This convolutional code of finite length is actually a block code. It is called a *terminated convolutional code*. Another kind of block code, called a *truncated convolutional code*, is obtained by discarding terms $c_\ell x^\ell$ with ℓ larger than $n-1$.

A convolutional code can have many generator matrices. Two generator matrices that are related by column permutations and elementary row operations are called *equivalent generator matrices*. They will generate the same code or equivalent codes. A generator matrix for a convolutional code of the form

$$\boldsymbol{G}(x) = [\boldsymbol{I} \quad \boldsymbol{P}(x)],$$

($\boldsymbol{I}$ is a k by k identity matrix and $\boldsymbol{P}(x)$ is a k by $(n-k)$ matrix of polynomials) is called a *systematic generator matrix*. In contrast to a block code, for which a generator matrix can always be put into a systematic form by permutations of columns and elementary row operations, it is not possible, in general, to give a systematic polynomial generator matrix for a convolutional code. This is because the division of a polynomial by a polynomial, in general, is not a polynomial. However, if we allow the generator matrix to include rational functions of the form $a(x)/b(x)$ as elements, then we can always put the generator matrix into systematic form. The set of all rational functions of polynomials $\{a(x)/b(x) \mid b(x) \neq 0\}$ over the field F is itself a field, denoted $F(x)$, with an infinite number of elements. A matrix whose elements are from $F(x)$ is called a *rational matrix*. An element $a(x)/b(x)$ is called *realizable* if $b(x)$ is not divisible by x.

A rational generator matrix whose elements are all realizable is called a *realizable generator matrix*.

For example, the generator matrix

$$G(x) = [x^2 + 1 \quad x^2 + x + 1]$$

is equivalent to the systematic rational generator matrix

$$\begin{aligned} G'(x) &= \left[1 \quad \frac{x^2 + x + 1}{x^2 + 1} \right] \\ &= \left[1 \quad 1 + \frac{x}{x^2 + 1} \right]. \end{aligned}$$

What this means is that a systematic encoder exists, but it includes feedback in the encoder due to the denominator $x^2 + 1$. Inspection of the systematic generator matrix shows that the encoder of Figure 9.11 could have been designed by using the matrix entry

$$g_{01}(x) = 1 + \frac{x}{x^2 + 1}$$

to lay out the feedback shift register of the figure. Consequently, we have found a mathematical construction for passing between the two encoders shown in Figure 9.9 and Figure 9.11. In general, by using the feedback of polynomial division circuits, one can build a systematic encoder for any convolutional code.

In general, the same code $\mathcal{C}$ can be generated by many generator matrices, either of polynomials or of rational functions. Thus it is necessary to know when two matrices $G(x)$ and $G'(x)$ generate the same code. Two generator matrices $G(x)$ and $G'(x)$ generate the same code if and only if $G(x) = T(x)G'(x)$ where $T(x)$ is a rational k by k matrix with nonzero determinant, and so is invertible. Any rational matrix is equivalent to a rational matrix in systematic form and, in particular, to a rational systematic matrix in which the numerator and denominator of every element have no common polynomial factor. If $k = 1$, the denominators can be multiplied out to give a noncatastrophic, nonsystematic polynomial generator matrix, which generates the same code. If $k \neq 1$, it is more difficult to reach a similar conclusion – this general discussion is deferred until Section 9.10.

The notion of constraint length can be sharpened by expressing it in terms of the n by k matrix of generator polynomials $G(x)$. Because the constraint length is fundamentally a property of the code $\mathcal{C}$, not of the generator matrix, and $\mathcal{C}$ has many generator matrices, it is necessary to ensure that such a definition is consistent. The k by n matrix $G(x)$ has $\binom{n}{k}$ ways of forming a k by k submatrix by striking out $n - k$ columns. The determinants of these k by k submatrices are called *principal determinants*. Given the generator matrix of polynomials $G(x)$, the *internal degree* of the k by n matrix of polynomials $G(x)$, denoted $\deg_{\text{int}} G(x)$, is defined as the largest of the degrees of the k by k determinants of $\boldsymbol{G}(x)$. The *external degree* of the matrix $\boldsymbol{G}(x)$ is defined as

$$\deg_{\text{ext}} \boldsymbol{G}(x) = \sum_{i=0}^{k-1} \max_j [\deg g_{ij}(x)].$$

Each term of the sum, called a *row degree*, is the largest degree of any polynomial in that row.

Each term in the expansion of any k by k determinant is the product of k entries of $\boldsymbol{G}(x)$, one from each row. The entry from the ith row has degree at most $\max_j[\deg g_{ij}(x)]$ so each k by k determinant has degree not larger than $\deg_{\text{ext}} \boldsymbol{G}(x)$. Hence

$$\deg_{\text{int}} \boldsymbol{G}(x) \leq \deg_{\text{ext}} \boldsymbol{G}(x).$$

A *basic generator matrix* is a polynomial generator matrix for a convolutional code for which no polynomial generator matrix of smaller internal degree can be obtained by multiplication of the matrix by a rational invertible matrix $\boldsymbol{T}(x)$. The *constraint length* ν is defined as the internal degree of any basic generator matrix for $\mathcal{C}$. Recall that the frame memory of the convolutional code is

$$m = \max_i \max_j [\deg g_{ij}(x)]$$

from which we conclude that $\nu \leq km$.

Other ways of measuring the memory in the convolutional code can be defined but are less useful. The input span of the convolutional code is

$$K = k \max_{i,j} [\deg g_{ij}(x) + 1],$$

which measures the length of a subsequence of data symbols that can affect one code symbol. The output span of the convolutional code is

$$N = n \max_{i,j} [\deg g_{ij}(x) + 1],$$

which measures the length of a subsequence of code symbols that can be affected by a single data symbol. The encoders shown in Figure 9.3 and Figure 9.11 may be studied to illustrate these definitions.

9.4 Check matrices and inverse matrices

A convolutional code, which may be formed by a polynomial generator matrix $\boldsymbol{G}(x)$, is associated with two other kinds of matrices. These are *inverse matrices* and *check matrices*. The study of these matrices helps to clarify the structure of convolutional codes.

Definition 9.4.1. *A polynomial inverse matrix* $\boldsymbol{G}^{-1}(x)$ *for the polynomial matrix* $\boldsymbol{G}(x)$ *is an n by k matrix of polynomials such that*

$$\boldsymbol{G}(x)\boldsymbol{G}^{-1}(x) = \boldsymbol{I}$$

where $\boldsymbol{I}$ *is a k by k identity matrix.*

An inverse matrix may be useful to recover input data sequences from codewords whenever the encoder is nonsystematic, as in Figure 9.9.

Definition 9.4.2. *A polynomial check matrix* $\boldsymbol{H}(x)$ *is an* $(n-k)$ *by n matrix of polynomials that satisfies*

$$\boldsymbol{G}(x)\boldsymbol{H}(x)^{\mathrm{T}} = \boldsymbol{0}.$$

A check matrix can be used to compute syndromes and so has an essential role in syndrome decoding algorithms. The known syndrome decoding algorithms for convolutional codes, for most purposes, are inferior to graph-searching algorithms, which will be studied in Chapter 11. However, a graph-searching algorithm may be simpler if it processes syndromes rather than raw channel outputs.

If $\boldsymbol{G}(x)$ is a systematic polynomial generator matrix, a check matrix can be written immediately as

$$\boldsymbol{H}(x) = [-\boldsymbol{P}(x)^{\mathrm{T}} \quad \boldsymbol{I}]$$

where $\boldsymbol{I}$ here is an $(n-k)$ by $(n-k)$ identity matrix and $\boldsymbol{P}(x)$ is an $(n-k)$ by k matrix. It is straightforward to verify that

$$\boldsymbol{G}(x)\boldsymbol{H}(x)^{\mathrm{T}} = \boldsymbol{0}.$$

More generally, to find a check matrix corresponding to a generator matrix, first manipulate $\boldsymbol{G}(x)$ into a systematic form using permutations and elementary row operations, then write down the corresponding systematic check matrix as was done previously, then multiply out the denominators and undo the permutations.

The polynomial syndrome is a vector of polynomials, given by

$$\boldsymbol{s}(x) = \boldsymbol{v}(x)\boldsymbol{H}(x)^{\mathrm{T}}.$$

It is an $(n-k)$-component row vector of polynomials that is equal to the zero vector if and only if there are no correctable errors. The syndrome polynomials depend only on the errors and not on the encoded data. One way of designing a decoder is to use a tabulation of error patterns and the corresponding syndrome polynomials.

The encoder for a convolutional code must be invertible so that the data can be recovered after the errors are corrected. Moreover, the code must have the property that no uncorrectable error pattern of finite weight will destroy an infinite amount of data.

These properties of a polynomial generator matrix are closely related, and also related to the existence of a polynomial inverse matrix.

First, we shall discuss the special case where $k = 1$. In this case, the notation is simplified in that the generator polynomials have only a single index. Then

$$\boldsymbol{G}(x) = [g_0(x) \quad g_1(x) \quad \cdots \quad g_{n-1}(x)]$$

and

$$c_j(x) = a(x)g_j(x) \quad j = 0, \ldots, n-1.$$

Definition 9.4.3. *An* $(n, 1)$ *polynomial generator matrix, whose generator polynomials* $g_0(x), \ldots, g_{n-1}(x)$ *have a greatest common divisor satisfying*

$$\text{GCD}[g_0(x), \ldots, g_{n-1}(x)] = 1,$$

is called a noncatastrophic generator matrix. *Otherwise, it is called a catastrophic generator matrix.*

The catastrophic property is a property of the encoding, not of the code. The reason for discussing the notion of a catastrophic generator matrix, and the reason for dismissing such generator matrices from further consideration is that, for such an encoder, there are error patterns containing a finite number of channel errors that produce an infinite number of errors in the decoder output. Let

$$A(x) = \text{GCD}[g_0(x), \ldots, g_{n-1}(x)]$$

and let $a(x) = A(x)^{-1}$ by which we mean the infinite formal power series obtained by formally dividing one by $A(x)$. If $A(x)$ is not equal to one, $a(x)$ will have infinite weight. But for each j, $c_j(x) = a(x)g_j(x)$ has finite weight because $A(x)$, a factor of $g_j(x)$, cancels $a(x)$. If $a(x)$ is encoded and the channel makes a finite number of errors, one error in each place where one of the $c_j(x)$ is nonzero, the decoder will receive all zeros from the channel and conclude erroneously that $a(x) = 0$, thereby making an infinite number of errors.

For example, if $g_0(x) = x + 1$ and $g_1(x) = x^2 + 1$, then $\text{GCD}[g_0(x), g_1(x)] = x + 1$ over $GF(2)$. If $a(x) = 1 + x + x^2 + x^3 + \cdots$, then $c_0(x) = 1$, $c_1(x) = 1 + x$, and the codeword has weight 3. Three channel errors can change this codeword for the all-one dataword into the codeword for the all-zero dataword.

For a noncatastrophic generator matrix, the datastream $a(x)$ can be recovered from the codestream $\boldsymbol{c}(x)$ by using a corollary to the euclidean algorithm for polynomials to form $\boldsymbol{G}^{-1}(x)$. Corollary 4.3.7, generalized to n polynomials, states that for any set of polynomials $\{g_0(x), \ldots, g_{n-1}(x)\}$, if $\text{GCD}[g_0(x), \ldots, g_{n-1}(x)] = 1$, then polynomials $b_0(x), \ldots, b_{n-1}(x)$ exist, satisfying

$$b_0(x)g_0(x) + \cdots + b_{n-1}(x)g_{n-1}(x) = 1.$$

Therefore if the data polynomial $a(x)$ is encoded by

$$c_j(x) = a(x)g_j(x) \quad j = 0, \ldots, n-1,$$

we can recover $a(x)$ by

$$a(x) = b_0(x)c_0(x) + \cdots + b_{n-1}(x)c_{n-1}(x)$$

as is readily checked by simple substitution. Thus, the matrix

$$\boldsymbol{G}(x) = [g_0(x) \quad g_1(x) \quad \ldots \quad g_{n-1}(x)]$$

has inverse

$$\boldsymbol{G}^{-1}(x) = [b_0(x) \quad b_1(x) \quad \ldots \quad b_{n-1}(x)]^{\mathrm{T}}.$$

For example, in $GF(2)$,

$$\text{GCD}[x^5 + x^2 + x + 1, x^6 + x^2 + x] = 1.$$

Therefore polynomials $b_0(x)$ and $b_1(x)$ must exist such that, in $GF(2)$,

$$b_0(x)(x^5 + x^2 + x + 1) + b_1(x)(x^6 + x^2 + x) = 1.$$

These are

$$b_0(x) = x^4 + x^3 + 1$$
$$b_1(x) = x^3 + x^2 + 1$$

so that

$$\boldsymbol{G}^{-1}(x) = [x^4 + x^3 + 1 \quad x^3 + x^2 + 1].$$

A codeword composed of

$$c_0(x) = (x^5 + x^2 + x + 1)a(x)$$
$$c_1(x) = (x^6 + x^2 + x)a(x),$$

with generator polynomials $g_0(x) = x^5 + x^2 + x + 1$ and $g_1(x) = x^6 + x^2 + x$, can be inverted with the aid of $b_0(x)$ and $b_1(x)$ as follows:

$$\begin{aligned} c_0(x)b_0(x) + c_1(x)b_1(x) &= (x^5 + x^2 + x + 1)(x^4 + x^3 + 1)a(x) + \\ &\quad\ (x^6 + x^2 + x)(x^3 + x^2 + 1)a(x) \\ &= a(x). \end{aligned}$$

A shift-register circuit depicting this identity is shown in Figure 9.13.

In the general case, the matrix of generator polynomials $\boldsymbol{G}(x)$ has k rows. Then the definition of a noncatastrophic generator matrix is a little more elaborate, defined in terms of the principal determinants of $\boldsymbol{G}(x)$. If the k by k submatrices of $\boldsymbol{G}(x)$ are

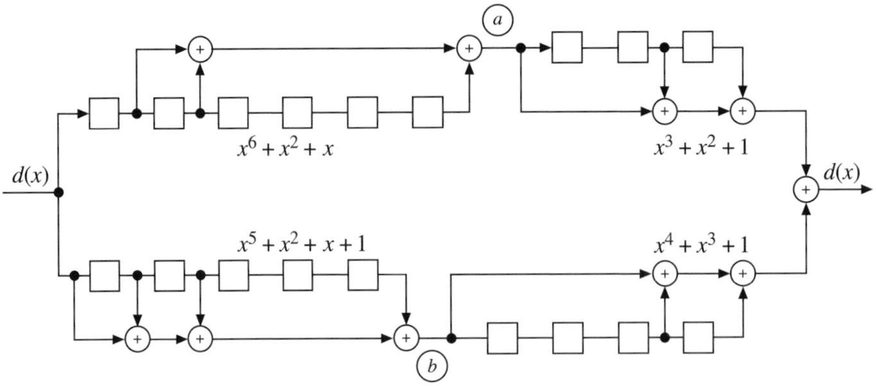

$1 = (x^3 + x^2 + 1)(x^6 + x^2 + x) + (x^4 + x^3 + 1)(x^5 + x^2 + x + 1)$

Figure 9.13. A transparent shift-register circuit

indexed by ℓ, and denoted $\boldsymbol{G}_\ell(x)$, then the determinant of the ℓth k by k submatrix will be denoted $|\boldsymbol{G}_\ell(x)|$.

Definition 9.4.4. *A polynomial generator matrix* $\boldsymbol{G}(x)$*, whose principal determinants* $|\boldsymbol{G}_\ell(x)|$ *satisfy*

$$\text{GCD}\left[|\boldsymbol{G}_\ell(x)|; \ell = 1, \ldots, \binom{n}{k}\right] = 1,$$

is called a noncatastrophic generator matrix. Otherwise it is called a catastrophic generator matrix.

A systematic generator matrix for a convolutional code will have a k by k submatrix equal to the identity matrix, and this submatrix has determinant one. Consequently, every systematic generator matrix is noncatastrophic.

A noncatastrophic polynomial generator matrix can be inverted because an n by k polynomial inverse matrix, $\boldsymbol{G}^{-1}(x)$, exists such that

$$\boldsymbol{G}(x)\boldsymbol{G}^{-1}(x) = \boldsymbol{I}$$

where $\boldsymbol{I}$ is the k by k identity matrix. The dataword $\boldsymbol{a}(x)$ is regenerated as $\boldsymbol{a}(x) = \boldsymbol{c}(x)\boldsymbol{G}^{-1}(x)$. The difficulty of finding the inverse $\boldsymbol{G}^{-1}(x)$ in the case where $k \neq 1$ is not much more difficult than that of finding the inverse via the euclidean algorithm in the case $k = 1$.

For example, let

$$\boldsymbol{G}(x) = \begin{bmatrix} 1 & x^2 & 1 \\ 0 & 1 & x + x^2 \end{bmatrix}.$$

There are three two by two submatrices, and the vector of the determinants of the three submatrices is

$$\Delta(x) = [1 \quad x + x^2 \quad 1 + x^3 + x^4].$$

The greatest common divisor of these three polynomials is one, so the inverse matrix $\boldsymbol{G}^{-1}(x)$ does exist. In fact, by computing $\boldsymbol{G}(x)\boldsymbol{G}^{-1}(x)$, one can easily verify that

$$\boldsymbol{G}^{-1}(x) = \begin{bmatrix} x^3 + x^4 & 1 + x^2 + x^3 + x^4 \\ x + x^2 & 1 + x + x^2 \\ 1 & 1 \end{bmatrix}$$

is an inverse for the generator matrix $\boldsymbol{G}(x)$. A procedure for computing the inverse matrix will be given in Section 9.10.

The inverse matrix $\boldsymbol{G}^{-1}(x)$ plays a role only in those decoders that first remove errors from the senseword to recover the corrected codeword. Then the datastream is recovered by using the matrix $\boldsymbol{G}^{-1}(x)$. Other kinds of decoders recover the datastream directly from the senseword. In such decoders $\boldsymbol{G}^{-1}(x)$ is not used.

9.5 Error correction and distance notions

When a convolutional codeword is passed through a channel, errors are made from time to time in the codeword symbols. The decoder must correct these errors by processing the senseword. The convolutional codeword is so long, however, that the decoder can remember only a part of the senseword at one time. Although the codeword is infinite in length, every decoding decision is made on a senseword segment of finite length. But, because of the structure of the code, no matter how one chops out a part of the senseword for the decoder to work with, there is some interaction with other parts of the senseword that the decoder does not see.

The number of senseword symbols that the decoder can store is called the *decoding window width* or the *decoding delay*. The decoding window width must be at least as large as the output span N, and usually it is several times larger. A decoder begins working at the beginning of the senseword and corrects errors, which are randomly and sparsely distributed along the codeword, as it comes upon them, using only the symbols in the decoding window. Occasionally, because of the nature of statistical fluctuations, there will be a cluster of errors that the code itself is not powerful enough to correct, and so the decoder will fail. It will be much more unlikely, if the decoding window width is large enough, that the decoder will fail because the decoding window is too small.

Our study of the decoding of convolutional codes first concentrates on the task of correcting errors in the first codeframe. If this first frame can be corrected and decoded, then the first dataframe is known. The effect of the first dataframe on the codeword can be computed and subtracted from the senseword. Then the task of decoding the second codeframe is the same as the task of decoding the first codeframe. Continuing in this way, if the first j codeframes can be corrected, then the task of decoding the $(j + 1)$th codeframe is the same as the task of decoding the first codeframe.

Occasionally, in any decoder, the first frame of the codeword will not be corrected properly because too many errors have occurred. An interval during which the decoder is not producing the correct codeword frame is called an *error event*. The error event is a decoding failure if the decoder detects this event and flags an uncorrectable error pattern. The error event is a decoding error if the decoder produces incorrect codeword symbols. After a decoding error or decoding failure occurs, a well-designed system will eventually return to a state of correct decoding. Otherwise, if the onset of a decoding error leads to a permanent state of error, we say that the decoder is subject to *infinite error propagation*. Infinite error propagation does not occur in a properly designed system.

A convolutional code has many minimum distances, determined by the length of the initial codeword segment over which the minimum distance is measured. The distance measure is defined in a way that is suitable for decoding only the first frame.

Definition 9.5.1. *The* ℓth *minimum distance* d_ℓ *of a convolutional code is equal to the smallest Hamming distance between any two initial codeword segments* ℓ *frames long that are different in the initial frame. The sequence* $d_1, d_2, d_3, \ldots$ *is called the distance profile of the convolutional code.*

Because a convolutional code is linear, one of the two codewords defining the minimum distance might just as well be the all-zero codeword. Thus, the ℓth minimum distance is equal to the ℓth minimum weight, which is the smallest-weight truncated codeword ℓ frames long that is nonzero in the first frame.

Suppose that a convolutional code has ℓth minimum distance d_ℓ. If at most t errors occur in the first ℓ frames, where $2t + 1 \leq d_\ell$, then those that occur in the first frame can be corrected as soon as the first ℓ frames of the senseword are received.

Definition 9.5.2. *The free Hamming distance of a convolutional code* $\mathcal{C}$ *is given by*

$$d_{\text{free}} = \max_{\ell} d_\ell.$$

The free Hamming distance is the smallest Hamming distance between any two distinct codewords. The distance profile of a convolutional code increases until it equals the free distance. That is,

$$d_1 \leq d_2 \leq \cdots \leq d_{\text{free}}.$$

The distance profile can be read from a labeled trellis. The convolutional code shown in Figure 9.7 has ℓth minimum distances $d_1 = 2$, $d_2 = 3$, $d_3 = 3$, $d_4 = 4$, and $d_i = 5$ for all i greater than four. The free distance of that code is five.

A convolutional code can correct any pattern of ν errors such that 2ν is less than d_{free}. Of course, a convolutional code can correct many other error patterns as well because many error patterns with ν errors may have 2ν larger than d_{free} and still have a unique closest codeword. Moreover, it can correct an indefinite number of such patterns provided they are spread sufficiently in the senseword. A convolutional codeword is infinitely long, in principle, and a channel with a nonzero probability of error, no matter how small, will make an infinite number of errors in an infinitely long time. This is why the decoder must be able to correct an infinite number of errors in an infinitely long codeword.

9.6 Matrix description of convolutional codes

A convolutional code consists of an infinite number of infinitely long codewords. It is linear and can be described by an infinite generator matrix over the field $GF(q)$. This description is an alternative to the description using a finite generator matrix over the ring $GF(q)[x]$. An infinite generator matrix over $GF(q)$ for a convolutional code, as described in this section, is cumbersome, so a polynomial description is usually preferred.

A given code has many generator matrices over $GF(q)$, but only a few of them are convenient to use. To obtain a generator matrix, the coefficients $g_{ij\ell}$ of the generator polynomials, given by

$$g_{ij}(x) = \sum_{\ell} g_{ij\ell} x^{\ell},$$

are arranged into matrices indexed by ℓ. For each ℓ, let $\boldsymbol{G}_\ell$ be the k by n matrix

$$\boldsymbol{G}_\ell = [g_{ij\ell}].$$

Then the generator matrix for the convolutional code has the infinitely recurring form

$$\boldsymbol{G} = \begin{bmatrix} \boldsymbol{G}_0 & \boldsymbol{G}_1 & \boldsymbol{G}_2 & \dots & \boldsymbol{G}_m & \boldsymbol{0} & \boldsymbol{0} & \boldsymbol{0} & \boldsymbol{0} & \dots \\ \boldsymbol{0} & \boldsymbol{G}_0 & \boldsymbol{G}_1 & \dots & \boldsymbol{G}_{m-1} & \boldsymbol{G}_m & \boldsymbol{0} & \boldsymbol{0} & \boldsymbol{0} & \dots \\ \boldsymbol{0} & \boldsymbol{0} & \boldsymbol{G}_0 & \dots & \boldsymbol{G}_{m-2} & \boldsymbol{G}_{m-1} & \boldsymbol{G}_m & \boldsymbol{0} & \boldsymbol{0} & \dots \\ \vdots & & & & & & & & & \end{bmatrix}$$

where the matrix continues indefinitely down and to the right and each $\boldsymbol{0}$ is a k by n matrix of zeros. Except for the diagonal band of m nonzero submatrices, all other entries are equal to zero.

When truncated to a block code of blocklength $(m+1)n$, the generator matrix for the convolutional code becomes

$$\boldsymbol{G}^{(m+1)} = \begin{bmatrix} \boldsymbol{G}_0 & \boldsymbol{G}_1 & \boldsymbol{G}_2 & \dots & \boldsymbol{G}_m \\ \boldsymbol{0} & \boldsymbol{G}_0 & \boldsymbol{G}_1 & \dots & \boldsymbol{G}_{m-1} \\ \boldsymbol{0} & \boldsymbol{0} & \boldsymbol{G}_0 & \dots & \boldsymbol{G}_{m-2} \\ \vdots & & & & \vdots \\ \boldsymbol{0} & \boldsymbol{0} & \boldsymbol{0} & & \boldsymbol{G}_0 \end{bmatrix}.$$

For a systematic convolutional code, the generator matrix can be written

$$G = \begin{bmatrix} I & P_0 & 0 & P_1 & 0 & P_2 & \dots & 0 & P_m & 0 & 0 & 0 & 0 & & \dots \\ 0 & 0 & I & P_0 & 0 & P_1 & \dots & 0 & P_{m-1} & 0 & P_m & 0 & 0 & & \dots \\ 0 & 0 & 0 & 0 & I & P_0 & \dots & 0 & P_{m-2} & 0 & P_{m-1} & 0 & P_m & & \dots \\ & & & & & & & & & 0 & P_{m-2} & 0 & P_{m-1} & & \\ & & & & & & & & & & & 0 & P_{m-2} & & \dots \end{bmatrix}$$

where the pattern repeats right-shifted in every row, and unspecified matrix entries are filled with zeros. Here I is a k by k identity matrix, 0 is a k by k matrix of zeros, and $P_0, \dots, P_m$ are k by $(n-k)$ matrices. The first row describes the encoding of the first dataframe into the first m codeframes. This matrix expression is a depiction of the shift-register description of the encoder.

A check matrix is any matrix H that satisfies

$$G^{(\ell)} H^{(\ell)\mathrm{T}} = 0 \quad \ell = 0, 1, 2, \dots$$

where $G^{(\ell)}$ and $H^{(\ell)}$ are the upper left submatrices of G and H, corresponding to ℓ frames. A check matrix H, which is an infinite-dimensional matrix, can be constructed from inspection of G. Inspection gives the matrix

$$H = \begin{bmatrix} P_0^{\mathrm{T}} & -I & & & & & & & & \dots \\ P_1^{\mathrm{T}} & 0 & P_0^{\mathrm{T}} & -I & & & & & & \\ P_2^{\mathrm{T}} & 0 & P_1^{\mathrm{T}} & 0 & P_0^{\mathrm{T}} & -I & & & & \\ \vdots & & & & & & & & & \\ P_m^{\mathrm{T}} & 0 & P_{m-1}^{\mathrm{T}} & 0 & P_{m-2}^{\mathrm{T}} & 0 & \dots & P_0^{\mathrm{T}} & -I & \dots \\ & & P_m^{\mathrm{T}} & 0 & P_{m-1}^{\mathrm{T}} & 0 & \dots & & & \\ & & & & P_m^{\mathrm{T}} & 0 & \dots & & & \end{bmatrix}$$

where, again, unspecified entries are all zeros.

For example, a systematic (2, 1) binary convolutional code, with $k = 1$ and $m = 1$, is described by the 1 by 1 matrices $P_0 = 1$ and $P_1 = 1$. Hence

$$G = \begin{bmatrix} 1 & 1 & 0 & 1 & & & & \\ & & 1 & 1 & 0 & 1 & & \\ & & & & 1 & 1 & 0 & 1 \\ & & & & & & \dots & \\ & & & & & & & \dots \end{bmatrix}$$

and

$$H = \begin{bmatrix} 1 & 1 & & & & \\ 1 & 0 & 1 & 1 & & \\ & & 1 & 0 & 1 & 1 \\ & & & \dots & & \\ & & & & \dots & \end{bmatrix}.$$

Another example is a (3, 1) systematic binary convolutional code, with $k = 1$ and $m = 3$, given by $\boldsymbol{P}_0 = [1\ 1]$, $\boldsymbol{P}_1 = [0\ 1]$, $\boldsymbol{P}_2 = [1\ 0]$, and $\boldsymbol{P}_3 = [1\ 1]$. Hence

$$\boldsymbol{G} = \left[\begin{array}{ccccccccccccccccccccc}
1 & 1 & 1 & 0 & 0 & 1 & 0 & 1 & 0 & 0 & 1 & 1 & & & & & & & & & \\
 & & & 1 & 1 & 1 & 0 & 0 & 1 & 0 & 1 & 0 & 0 & 1 & 1 & & & & & & \\
 & & & & & & 1 & 1 & 1 & 0 & 0 & 1 & 0 & 1 & 0 & 0 & 1 & 1 & & & \\
 & & & & & & & & & 1 & 1 & 1 & 0 & 0 & 1 & 0 & 1 & 0 & 0 & 1 & 1 \\
 & & & & & & & & & & & & & \cdots & & & & & & & \\
 & & & & & & & & & & & & & & & \cdots & & & & &
\end{array}\right]$$

and

$$\boldsymbol{H} = \left[\begin{array}{ccccccccccccccc}
1 & 1 & 0 & & & & & & & & & & & & \\
1 & 0 & 1 & & & & & & & & & & & & \\
0 & 0 & 0 & 1 & 1 & 0 & & & & & & & & & \\
1 & 0 & 0 & 1 & 0 & 1 & & & & & & & & & \\
1 & 0 & 0 & 0 & 0 & 0 & 1 & 1 & 0 & & & & & & \\
0 & 0 & 0 & 1 & 0 & 0 & 1 & 0 & 1 & & & & & & \\
1 & 0 & 0 & 1 & 0 & 0 & 0 & 0 & 0 & 1 & 1 & 0 & & & \\
1 & 0 & 0 & 0 & 0 & 0 & 1 & 0 & 0 & 1 & 0 & 1 & & & \\
 & & & 1 & 0 & 0 & 1 & 0 & 0 & 0 & 0 & 0 & 1 & 1 & 0 \\
 & & & 1 & 0 & 0 & 0 & 0 & 0 & 1 & 0 & 1 & 1 & 0 & 1 \\
 & & & & & & & \cdots & & & & & & & \\
 & & & & & & & & \cdots & & & & & &
\end{array}\right].$$

9.7 The Wyner–Ash codes as convolutional codes

A class of single-error-correcting binary convolutional codes, called *Wyner–Ash codes*, occupies a place among convolutional codes similar to the place occupied by the class of binary Hamming codes among block codes. The class of Wyner–Ash codes, however, has not played a major role, as has the class of Hamming codes.

For each positive integer m, there is a $(2^m, 2^m - 1)$ Wyner–Ash code defined in terms of the check matrix $\boldsymbol{H}'$ of the $(2^m, 2^m - m)$ extended binary Hamming code that has the all-ones row at the top. This matrix is an m by 2^m check matrix in which all 2^m columns are distinct and nonzero, and every column begins with a one. The check matrix for the Wyner–Ash code, in terms of $\boldsymbol{H}'$, is

$$\boldsymbol{H} = \left[\begin{array}{cccc}
\boldsymbol{H}' & & & \cdots \\
 & \boldsymbol{H}' & & \\
 & & \boldsymbol{H}' & \\
 & & \vdots & \\
 & & &
\end{array}\right],$$

where each successive copy of $\boldsymbol{H}'$ is moved down by only one row, and all other entries are zero. (Hence, in general, multiple rows of $\boldsymbol{H}'$ appear in each row of $\boldsymbol{H}$).

Theorem 9.7.1. *The Wyner–Ash code has a minimum distance $d_{\min}$ equal to 3; it is a single-error-correcting convolutional code.*

Proof: The code is linear so the minimum distance is equal to the minimum weight. Let $\boldsymbol{c}$ be a minimum weight codeword that is nonzero in the first frame. It must have an even number of ones in the first frame because the first row of $\boldsymbol{H}$ is all ones in the first frame and all zeros in every other frame. Because any two columns of $\boldsymbol{H}$ in the first frame are linearly independent, there must be at least one one in another frame, so the minimum weight is at least 3.

The minimum distance of the code is exactly 3, because there are two columns of $\boldsymbol{H}'$ that add to a column of the second frame, and so $\boldsymbol{H}$ has three dependent columns. □

For example, Figure 9.14 shows an encoder for the Wyner–Ash (4, 3) code constructed with $m = 2$ and the extended Hamming check matrix

$$\boldsymbol{H}' = \begin{bmatrix} 1 & 1 & 1 & 1 \\ 1 & 1 & 0 & 0 \\ 1 & 0 & 1 & 0 \end{bmatrix}.$$

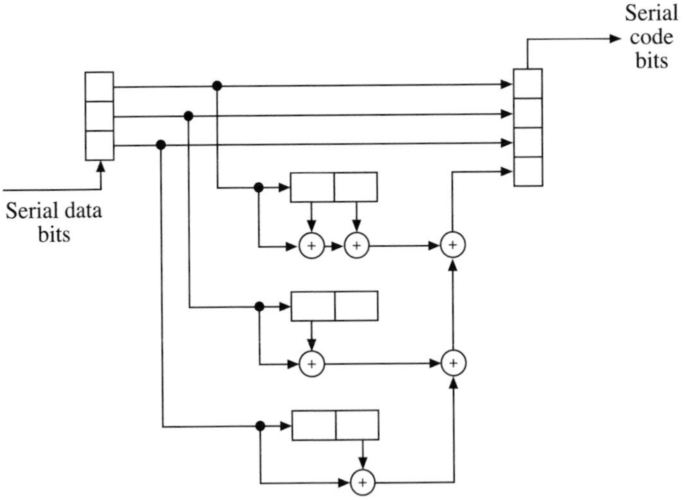

Figure 9.14. An encoder for the (4, 3) Wyner–Ash code

The check matrix for the convolutional code is

$$
\boldsymbol{H} = \begin{bmatrix}
1 & 1 & 1 & 1 \\
1 & 1 & 0 & 0 & 1 & 1 & 1 & 1 \\
1 & 0 & 1 & 0 & 1 & 1 & 0 & 0 & 1 & 1 & 1 & 1 \\
0 & 0 & 0 & 0 & 1 & 0 & 1 & 0 & 1 & 1 & 0 & 0 & 1 & 1 & 1 & 1 \\
0 & 0 & 0 & 0 & 0 & 0 & 0 & 1 & 0 & 1 & 0 & 1 & 1 & 0 & 0 & \dots \\
 & \vdots & & & \vdots & & & & \vdots & & & & \vdots
\end{bmatrix}.
$$

It is easy to see that the first, fourth, and sixth columns add to zero, so the code has minimum weight 3. The corresponding polynomial check matrix is

$$
\boldsymbol{H}(x) = [x^2 + x + 1 \quad x + 1 \quad x^2 + 1 \quad 1]
$$

and the polynomial generator matrix is

$$
\boldsymbol{G}(x) = \begin{bmatrix}
1 & 0 & 0 & x^2 + x + 1 \\
0 & 1 & 0 & x + 1 \\
0 & 0 & 1 & x^2 + 1
\end{bmatrix}.
$$

It is easy to check that $\boldsymbol{G}(x)\boldsymbol{H}(x)^{\mathrm{T}} = 0$. This matrix can be used to infer a generator matrix for the convolutional code

$$
\boldsymbol{G} = \begin{bmatrix}
1 & 0 & 0 & 1 & 0 & 0 & 0 & 1 & 0 & 0 & 0 & 1 \\
0 & 1 & 0 & 1 & 0 & 0 & 0 & 1 & 0 & 0 & 0 & 0 \\
0 & 0 & 1 & 1 & 0 & 0 & 0 & 0 & 0 & 0 & 0 & 1 \\
 & & & & 1 & 0 & 0 & 1 & 0 & 0 & 0 & 1 & 0 & 0 & 0 & 1 \\
 & & & & 0 & 1 & 0 & 1 & 0 & 0 & 0 & 1 & 0 & 0 & 0 & 0 \\
 & & & & 0 & 0 & 1 & 1 & 0 & 0 & 0 & 0 & 0 & 0 & 0 & 1 \\
 & & & & & & & & 1 & 0 & 0 & 1 & 0 & 0 & 0 & 1 & 0 & 0 & 0 & 1 \\
 & & & & & & & & 0 & 1 & 0 & 1 & 0 & 0 & 0 & 1 & 0 & 0 & 0 & 0 & \dots \\
 & & & & & & & & 0 & 0 & 1 & 1 & 0 & 0 & 0 & 0 & 0 & 0 & 0 & 1 \\
 & & \vdots & & & & & & \vdots
\end{bmatrix}.
$$

It is easy to check that $\boldsymbol{G}\boldsymbol{H}^{\mathrm{T}} = \boldsymbol{0}$. By inspection of $\boldsymbol{G}$, it is again clear that within the block of length 12, every nonzero codeword has a weight of 3 or greater. Consequently, the code can correct one error in a block of length 12.

9.8 Syndrome decoding algorithms

Suppose that the infinite length senseword $\boldsymbol{v}$ consists of a convolutional codeword and an error pattern:

$$\boldsymbol{v} = \boldsymbol{c} + \boldsymbol{e}.$$

Just as for block codes, the syndrome is given by

$$\boldsymbol{s} = \boldsymbol{v}\boldsymbol{H}^{\mathrm{T}} = \boldsymbol{e}\boldsymbol{H}^{\mathrm{T}}.$$

Now, however, the syndrome has infinite length, so it cannot be computed at once. It must be computed gradually, starting at the beginning, as the senseword symbols arrive. A syndrome decoder cannot look at the entire syndrome at once. A syndrome decoder works from the beginning of the senseword, computing components of $\boldsymbol{s}$ as it proceeds, correcting errors, and discarding components of $\boldsymbol{s}$ when they are too old. The decoder contains a table of syndrome segments and the corresponding segment of the error pattern. When the decoder sees a syndrome segment that is listed in the table, it corrects the initial segment of the senseword. We shall give examples of various syndrome decoders.

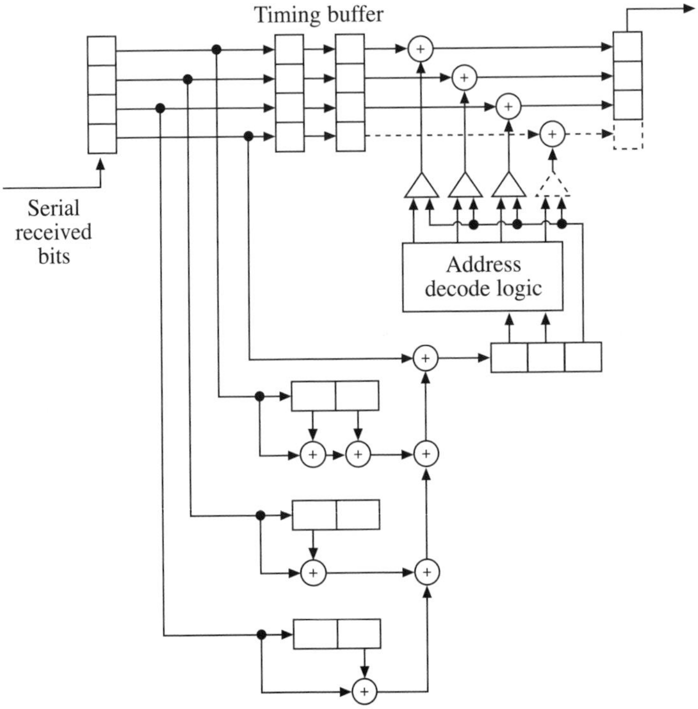

Figure 9.15. Syndrome decoder for the (4, 3) Wyner–Ash code

Error pattern				
Fourth frame	Third frame	Second frame	First frame	Syndrome
....	0000	0000	0001	111
	0000	0000	0010	011
	0000	0000	0100	101
	0000	0000	1000	001
	0000	0001	0000	110
	0000	0010	0000	110
	0000	0100	0000	010
	0000	1000	0000	010
	0000	0000	0000	100
	0001	0000	0000	100
	0010	0000	0000	100
	0100	0000	0000	100
	1000	0000	0000	100

Figure 9.16. Decoding table for (4, 3) Wyner–Ash code

First, consider the decoder, shown in Figure 9.15, for the (4, 3) Wyner–Ash code. The incoming serial bitstream is converted to n parallel lines, and the syndrome is computed by recomputing the check bit from the received data bits and subtracting it from the received check bit. To correct errors in the first frame, we need to know only the syndromes that are caused by a single error in the first frame. The possible patterns of a single error in the first frame and the first three syndrome bits are tabulated in Figure 9.16. The rightmost bit is a one if and only if an error has occurred in the first frame. The other two syndrome bits then specify the location of the error within the frame. Figure 9.16 also lists the first three bits of the syndrome for single errors in the second and third frames to illustrate that these syndromes are distinct from the syndromes for an error in the first frame.

The elements shown with dotted lines in the decoder of Figure 9.15 need to be included only to reconstruct the check bits of the codeword. Normally, the check bits are no longer needed so those elements of the decoder can be eliminated.

One more detail remains to be mentioned. Because syndromes are normally tabulated only for error patterns that have no errors in frames that come before the first frame, the syndrome must be updated after the leading frame is corrected so that the syndrome is correct in the next frame. To delete the effect of a corrected error from the syndrome, simply subtract the syndrome of the correction error pattern from the syndrome register.

As a second example, we take the (2, 1) double-error-correcting code with generator matrix

$$\boldsymbol{G}(x) = [x^2 + x + 1 \quad x^2 + 1].$$

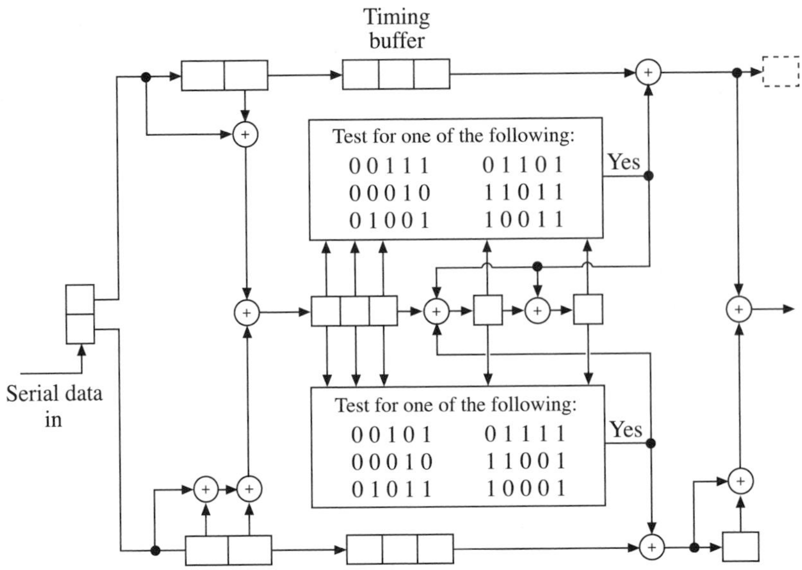

Figure 9.17. Syndrome decoder for a (2, 1) convolutional code

Error pattern				
Fourth frame	Third frame	Second frame	First frame	Syndrome
00	00	00	01	...000111
00	00	00	10	...000101
00	00	00	11	...000010
00	00	01	01	...001001
00	00	10	01	...001101
00	01	00	01	...011011
00	10	00	01	...010011
01	00	00	01	...111111
10	00	00	01	...101111
00	00	01	10	...001011
00	00	10	10	...001111
00	01	00	10	...011001
00	10	00	10	...010001
01	00	00	10	...111101
10	00	00	10	...101101

Figure 9.18. Decoding table for a (2, 1) convolutional code

An encoder for this code is shown in Figure 9.3. The syndrome decoder, shown in Figure 9.17, is based on the syndromes tabulated in Figure 9.18. This decoder circuit will correct up to two errors in the first frame provided there are not more than two errors in the first five frames. For each error that is corrected, the feedback removes from the syndrome register the contribution of that error to the syndrome. Any number of successive single-error or double-error patterns can be corrected provided their syndrome patterns do not overlap. Because the code is not a systematic code, the data bit must be recovered from the corrected codeword. The recovery circuit uses the relationship

$$\begin{aligned} 1 &= \text{GCD}[x^2 + x + 1, x^2 + 1] \\ &= x(x^2 + x + 1) + (x + 1)(x^2 + 1) \end{aligned}$$

to recover the data bits.

The decoder of Figure 9.17 is not a complete decoder because many syndromes exist that are not used. In principle, by tabulating other syndromes and expanding the length of the syndrome memory, one could decode some additional error patterns with more than two errors. However, selecting another code of longer blocklength and the same rate is probably a simpler way to improve performance.

The final example of this section is another systematic (2, 1) convolutional code whose encoder was shown in Figure 9.4. This code can correct two errors within a window of length 12. We describe a syndrome decoder for this code in this section, and a majority decoder in Section 13.6. A decoding table, given in Figure 9.19, lists syndromes above the gap for error patterns with an error in the rightmost data bit,

Error pattern	Syndrome
... 000000000001	... 111001
000000000011	111000
000000000101	001011
000000001001	111011
000000010001	011101
000000100001	111101
000001000001	110001
000010000001	110001
000100000001	101001
001000000001	110001
010000000001	011001
100000000001	011001
000000000010	000001
000000000010	110011
000000000110	000011
000000001010	100101
000000010010	000101
000000100010	000101
000001000010	001001
000010000010	001001
000100000010	010001
001000000010	010001
010000000010	100001
100000000010	100001

Figure 9.19. Decoding table for another (2, 1) convolutional code

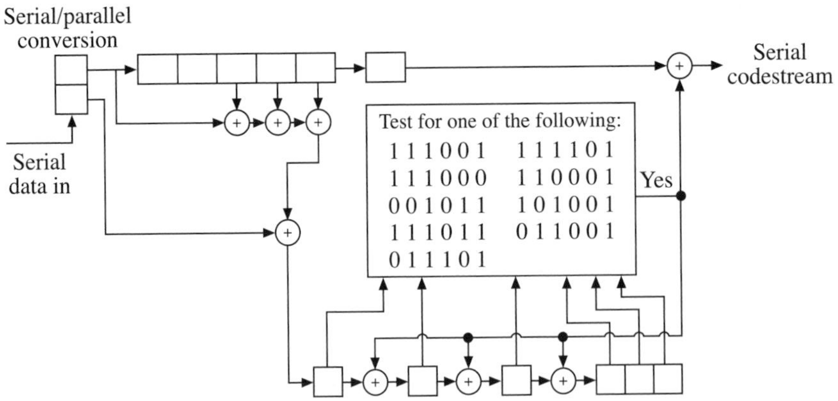

Figure 9.20. A decoder for another (2, 1) convolutional code

and lists syndromes below the line for error patterns with an error in the rightmost check bit and no error in the rightmost data bit. The decoder of Figure 9.20 only tests for syndromes above the horizontal line and thus provides correct data bits whenever the error pattern is within the design limits of the code. The decoder can detect some uncorrectable error patterns. This is a decoding failure and occurs when the syndrome is nonzero and does not appear anywhere in the table of Figure 9.19.

9.9 Convolutional codes for correcting error bursts

A burst of length t is a sequence of t symbols, the first and last of which are nonzero. A convolutional codeword is infinitely long, and many errors may occur in the senseword. We think of these errors as grouped into a number of bursts of various lengths. If the individual error burst occurs infrequently, then the decoder need contend with only one error burst at a time. A convolutional code for which a decoder can correct any single error burst of length t, provided that other error bursts are far enough away, is called a convolutional code with burst-correcting ability t.

Any t-error-correcting convolutional code, of course, will correct any error burst of length t. Convolutional codes for correcting longer bursts can be obtained by interleaving. To get a (jn, jk) convolutional code from an (n, k) convolutional code, take j copies of the encoder and merge the codewords by alternating the symbols. If the original code can correct any error burst of length t, it is apparent that the interleaved code can correct any error burst of length jt.

For example, the (2, 1) systematic convolutional code of constraint length 6 with generator polynomials

$$g_0(x) = 1$$
$$g_1(x) = x^6 + x^5 + x^2 + 1$$

can correct any two bit errors in any interval of length 14. By taking four copies of the (2, 1) code and interleaving the bits, one obtains a convolutional code that corrects any error burst of length 8.

The technique of interleaving creates a new convolutional code from a given convolutional code. If $g(x)$ is a generator polynomial of the given code, then $g(x^j)$ is a generator polynomial of the interleaved code. The generator polynomials of the above example become stretched into the new polynomials

$$g_0(x) = 1$$
$$g_1(x) = x^{24} + x^{20} + x^8 + 1.$$

This is the same behavior as was seen for interleaving cyclic block codes.

One can construct many burst-error-correcting convolutional codes by interleaving the short random-error-correcting convolutional codes tabulated in Figure 9.14. The interleaved codes will correct not only burst errors, but also many patterns of random errors.

A class of binary convolutional codes designed specifically for correcting burst errors is the class of *Iwadare codes*. An $(n, n-1)$ Iwadare code is defined in terms of a parameter λ and can correct any burst error of length $t = \lambda n$ or less. An unending sequence of burst errors can be corrected, provided successive burst errors are separated by a *guard space* of not less than $n(2n-1)(\lambda+1) - n$ bits.

Definition 9.9.1. *An $(n, n-1)$ Iwadare code is a binary convolutional code with a systematic check matrix of the form*

$$\boldsymbol{H}(x) = [1 \quad g_1(x)\; g_2(x) \quad \ldots \quad g_{n-1}(x)].$$

where

$$g_i(x) = x^{(\lambda+1)i-1} + x^{\lambda i+(\lambda+2)n-3}$$

and λ and n are any integers.

It is convenient to use only a single index for the check polynomials starting with $g_0(x) = 1$, followed by $g_1(x)$, which is the check polynomial of largest degree. The notation $g_i(x)$ has been chosen because the generator matrix is

$$\boldsymbol{G}(x) = \begin{bmatrix} g_1(x) & 1 & 0 & \ldots & 0 \\ g_2(x) & 0 & 1 & \ldots & 0 \\ \vdots & & & & \vdots \\ g_{n-1}(x) & 0 & 0 & \ldots & 1 \end{bmatrix}$$

so the check polynomials are also generator polynomials.

It is clear that an Iwadare code is an $(n, n-1)$ convolutional code where m, the frame memory, is

$$m = (\lambda + 1)(2n - 1) - 2.$$

A sequence of $m + 1$ zero frames will clear the encoder. This means that a sequence of $m + 1$ error-free frames will return the running syndrome register to all zeros. Hence a guard space of

$$g = n(\lambda + 1)(2n - 1) - n$$

error-free bits after a burst error will clear a syndrome decoder.

The purpose of an Iwadare code is to correct binary burst errors of length $t = \lambda n$. Suppose a burst error begins in frame zero and lasts for at most λn bits. If it does not begin in the first bit of frame zero, then it may last until frame λ and so occupy $\lambda + 1$ frames. However, for some positive integer r smaller than n, the first r bits of frame zero and the last $n - r$ bits of frame λ must be zero. The integer r cannot be equal to n because, if it were, the burst error, of length at most λn, would not start in frame zero. Because the final bit of the final frame of the burst is never in error, we can ensure that the check bits are in error in at most λ frames of a burst by requiring the bits within a frame to be in the order $e_{n-1}, \ldots, e_0$. We shall see later that this is essential to the behavior of the code.

An example of the Iwadare code is the (3, 2) code with $n = 3$ and $\lambda = 4$, which can correct all burst errors of length 12 or less. The systematic encoder is shown in Figure 9.21. The polynomials $c_1(x)$ and $c_2(x)$ are the polynomials of data symbols. The polynomial $c_0(x)$ is the polynomial of check symbols, given by

$$c_0(x) = (x^4 + x^{19})a_1(x) + (x^9 + x^{23})a_2(x).$$

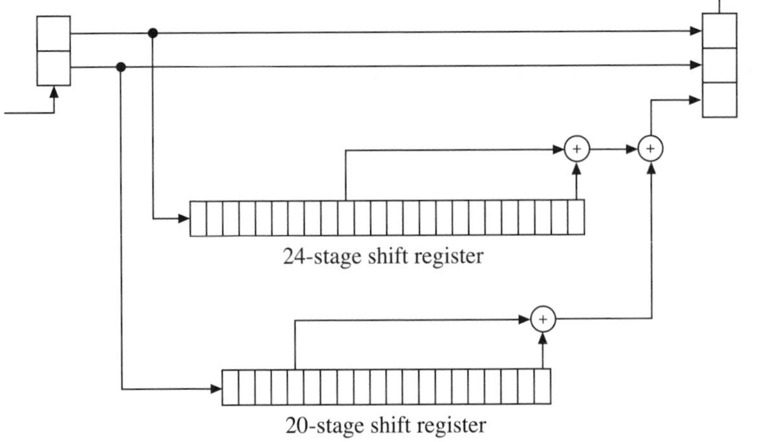

Figure 9.21. An encoder for a (3, 2) Iwadare code

The error polynomials are $e_0(x)$, $e_1(x)$, and $e_2(x)$. An error burst of length 12 that begins in frame zero cannot include the third bit of frame four. Therefore a burst of length 12 has the form

$$
\begin{aligned}
e_0(x) &= e_{00} + e_{01}x + e_{02}x^2 + e_{03}x^3 \\
e_1(x) &= e_{10} + e_{11}x + e_{12}x^2 + e_{13}x^3 + e_{14}x^4 \\
e_2(x) &= e_{20} + e_{21}x + e_{22}x^2 + e_{23}x^3 + e_{24}x^4
\end{aligned}
$$

with at least two edge bits equal to zero. This is because the error burst has a length of at most 12, so if e_{20} is not zero, then e_{24} and e_{14} are zero; and if e_{20} and e_{10} are not both zero, then e_{14} is zero.

The Iwadare generator polynomials have been defined so that every syndrome bit is affected by only a single bit of a correctable error burst. The syndrome polynomial is

$$s(x) = e_0(x) + (x^4 + x^{19})e_1(x) + (x^9 + x^{23})e_2(x).$$

Each nonzero bit of $e_0(x)$ produces a singlet in the syndrome; each nonzero syndrome bit due to $e_0(x)$ is followed by a zero after a delay of fourteen bits, and by another zero after a delay of fifteen bits. Each nonzero bit of $e_1(x)$ produces a doublet in the syndrome; the first occurrence is followed by an echo after a delay of fourteen bits. Similarly, each nonzero bit of $e_2(x)$ produces a doublet in the syndrome; the first occurrence is followed by an echo after a delay of fifteen bits. Thus, the syndrome polynomial begins with a single copy of $e_0(x)$, four bits long, followed by a doublet of copies of $e_1(x)$ due to $x^4 + x^{19}$, each five bits long, followed by a doublet of copies of $e_2(x)$ due to $x^9 + x^{23}$, each five bits long. The syndrome polynomial for a burst error beginning in frame zero is

$$
\begin{aligned}
s(x) = {} & e_{00} + e_{01}x + e_{02}x^2 + e_{03}x^3 + \\
& e_{10}x^4 + e_{11}x^5 + e_{12}x^6 + e_{13}x^7 + e_{14}x^8 + \\
& e_{20}x^9 + e_{21}x^{10} + e_{22}x^{11} + e_{23}x^{12} + e_{24}x^{13} + \\
& e_{10}x^{19} + e_{11}x^{20} + e_{12}x^{21} + e_{13}x^{22} + e_{14}x^{23} + \\
& e_{20}x^{23} + e_{21}x^{24} + e_{22}x^{25} + e_{23}x^{26} + e_{24}x^{27}.
\end{aligned}
$$

Although e_{14} and e_{20} are both coefficients of x^{23}, they cannot both be nonzero in a burst of length 12, so there will be no cancellation of coefficients of x^{23}.

If, instead, the error pattern begins in frame ℓ, then $s(x)$ is simply multiplied by x^ℓ.

Theorem 9.9.2. *A binary Iwadare code, when the bits within a frame are serialized in the order $e_{n-1}, \ldots, e_0$, corrects any sequence of burst errors, each of length $t = \lambda n$ or less, separated by not less than a guard space of $n(\lambda + 1)(2n - 1) - n$ bits.*

Proof: Because $n-k=1$ for an Iwadare code, there is only one syndrome polynomial, given by

$$s(x)=e_0(x)+\sum_{i=1}^{n-1} g_i(x)e_i(x)$$
$$=e_0(x)+\sum_{i=1}^{n-1}\left[x^{(\lambda+1)i-1}+x^{\lambda i+(\lambda+2)n-3}\right]e_i(x).$$

It is enough to show that every burst error starting in frame zero has a distinct syndrome that cannot be caused by a burst error starting in a later frame. Because the total burst length is at most λn, the position within its frame of the first bit error occurring in frame zero must come after the position within its frame of the last bit error occurring in frame λ. In particular, in frame λ, the last bit, $e_{0,\lambda-1}$, must equal zero. Otherwise the burst of length at most λn would not start in frame zero. Consequently $e_0(x)$ has a degree at most $\lambda-1$, and for $i\neq 0$, $e_i(x)$ has a degree at most λ. The syndrome is now written

$$s(x)=e_0(x)+x^{\lambda}e_1(x)+x^{2\lambda+1}e_2(x)+\cdots+x^{(n-1)\lambda+(n-2)}e_{n-1}(x)+$$
$$x^{(\lambda+2)n-3}\left[x^{\lambda}e_1(x)+x^{2\lambda}e_2(x)+\cdots+x^{(n-1)\lambda}e_{n-1}(x)\right].$$

Because $\deg e_0(x)\le\lambda-1$ and $\deg e_i(x)\le\lambda$ for $i\neq 0$, the syndrome explicitly displays all error components starting in frame zero grouped into nonoverlapping segments. Hence, no two burst errors starting in frame zero have the same syndrome.

To finish the proof, suppose that a burst error begins in a later frame. The bits in each syndrome doublet are spread by more than t bits so both ones of each syndrome doublet must be caused by the same error bit. We see by inspection of the form of $s(x)$ that s_0 cannot be zero. Likewise, $s(x)$ cannot have both a nonzero s_λ and a nonzero $s_{\lambda+(\lambda+2)n-3}$ because only $e_{0\lambda}$ could contribute to s_λ and $e_{0\lambda}$ does not contribute to $s_{\lambda+(\lambda+2)n-3}$. Likewise, $s(x)$ cannot have both a nonzero $s_{2\lambda+1}$ and a nonzero $s_{2\lambda+(\lambda+2)n-3}$ because only $e_{0,2\lambda+1}$ or $e_{1,\lambda+1}$ could contribute to $s_{2\lambda+1}$ and neither contributes to $s_{2\lambda+(\lambda+2)n-3}$. Continuing in this way, we conclude that no syndrome of a burst starting in a later frame is equal to a syndrome of a burst starting in frame zero. □

Now we are ready to describe the decoder shown in Figure 9.22 for the (3, 2) Iwadare code with $n=3$ and $\lambda=4$. The decoder tests syndrome positions s_i and s_{i-15} to determine if the rightmost bit in the $v_1(x)$ register is in error. If both are one, an appropriate correction is made to the datastream and to the syndrome register. Simultaneously, to find an error in $v_2(x)$ four frames later, the decoder tests syndrome positions s_i and s_{i-14}. If both are equal to one and s_4 is equal to zero, the rightmost bit in the $v_2(x)$ register is found to be in error and an appropriate correction is made to the datastream and to the syndrome register.

After a error burst of length 12 occurs, there must be an error-free guard space so that the correction process will not be disturbed. This will be the case if at least twenty-four

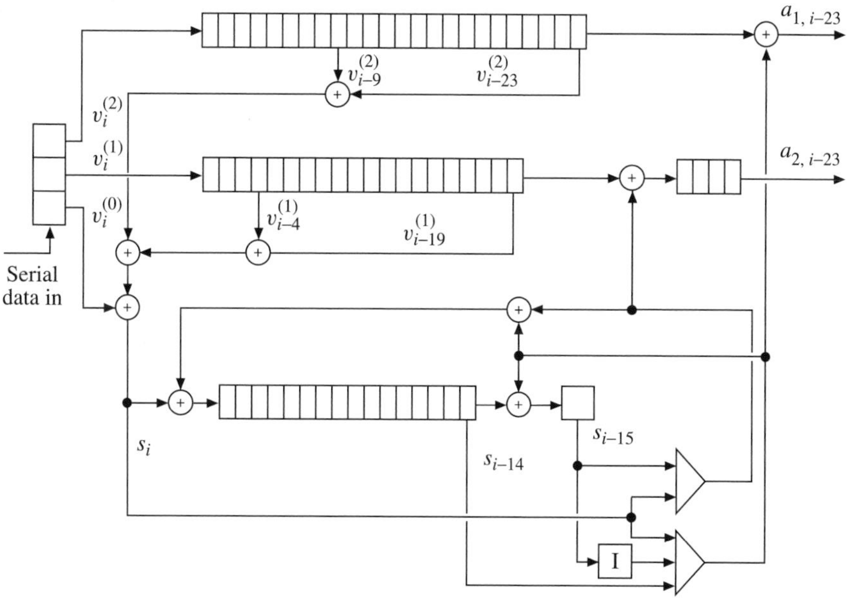

Figure 9.22. A decoder for the (3, 2) Iwadare code

error-free frames occur between error bursts. The decoder will successfully correct multiple error burst patterns; each burst must be of length 12 or less and successive bursts must be separated by at least seventy-two error-free bits.

9.10 Algebraic structure of convolutional codes

In this section, to conclude the chapter, the class of convolutional codes will be given a formal mathematical structure, which allows the properties of a code to be made precise. We shall study the structure of convolutional codes using both the ring of polynomials, denoted $F[x]$, and the field of rational forms of polynomials, denoted $F(x)$, and defined as

$$F(x) = \left\{ \frac{a(x)}{b(x)} \,\middle|\, a(x), b(x) \in F[x], b(x) \neq 0 \right\}$$

with the usual definitions of addition and multiplication of rationals. Whereas a generator matrix for a block code is a matrix over the field F, a generator matrix for a convolutional code is a matrix over the ring $F[x]$ or the field $F(x)$. The distance structure of the convolutional code, however, is defined in the base field F, not in the ring $F[x]$ or the field $F(x)$.

In place of the algebraic structure of a vector space, which was used to study block codes, the algebraic structure of convolutional codes uses the notion of a *module*.

A module is similar to a vector space, but a module is defined over a ring rather than a field. Hence a module does not have all the properties of a vector space.

Definition 9.10.1. *Let R be a ring. The elements of R will be called scalars. A module is a set M together with an operation called addition (denoted by $+$) on pairs of elements of M, and an operation called scalar multiplication (denoted by juxtaposition) on an element of R and an element of M to produce an element of M, provided the following axioms are satisfied.*

1. *M is an abelian group under addition.*
2. *Distributivity. For any elements m_1, m_2 of M, and any scalar r,*

   $$r(m_1 + m_2) = rm_1 + rm_2.$$
3. *Distributivity. For any element m of M, $1m = m$ and for any scalars r_1 and r_2,*

   $$(r_1 + r_2)m = r_1 m + r_2 m$$
4. *Associativity. For any element m of M, and any scalars r_1 and r_2,*

   $$(r_1 r_2)m = r_1(r_2 m).$$

If R is also a field, then the module M is also a vector space. Hence, a module is more general than a vector space, and has weaker properties.

Because it depends only on the operations of addition and multiplication, the Laplace expansion formula for determinants holds for a matrix over a polynomial ring. If $\boldsymbol{A}(x)$ is a square matrix of polynomials, then, for any i

$$\det(\boldsymbol{A}(x)) = \sum_{\ell=1}^{n} a_{i\ell}(x) C_{i\ell}(x)$$

where $a_{i\ell}(x)$ is an element of the matrix and $C_{i\ell}(x)$ is the cofactor of $a_{i\ell}(x)$. Except for sign, $C_{i\ell}(x)$ is a minor of the matrix $\boldsymbol{A}(x)$. It follows from the Laplace expansion formula that any common polynomial factor of all of the minors of $\boldsymbol{A}(x)$ is also a polynomial factor of $\det(\boldsymbol{A}(x))$. Furthermore, an i by i minor itself is a determinant and can be expanded in terms of its $i-1$ by $i-1$ minors. Hence the greatest common divisor of all determinants of i by i submatrices, denoted $\Delta_i(x)$, is a polynomial multiple of the greatest common divisor of all determinants of $i-1$ by $i-1$ submatrices, denoted $\Delta_{i-1}(x)$.

This latter remark holds even if the original matrix is not square. An n by k matrix has $\binom{n}{k}$ k by k square submatrices and $\binom{n}{k}$ principal determinants. The nonzero principal determinants have a greatest common divisor. This polynomial $\Delta_k(x)$ is a polynomial multiple of the greatest common divisor of all $i-1$ by $i-1$ submatrices.

Definition 9.10.2. *The invariant factors, denoted $\gamma_i(x)$ for $i = 1, \ldots, r$, of an n by k matrix of polynomials of rank r are*

$$\gamma_i(x) = \frac{\Delta_i(x)}{\Delta_{i-1}(x)}$$

where $\Delta_i(x)$ is the greatest common divisor of the set of all nonzero i by i minors of $G(x)$ and $\Delta_0(x) = 1$.

It is an immediate consequence of the definition of the invariant factors that

$$\Delta_i(x) = \prod_{\ell=1}^{i} \gamma_\ell(x).$$

Moreover, for the special case of a 1 by n polynomial matrix

$$A(x) = [g_0(x) \quad \cdots \quad g_{n-1}(x)],$$

the invariant factor has a simple interpretation. In this case the minors of $A(x)$ are the elements themselves, so the only invariant factor of a one by n matrix of polynomials is the greatest common divisor of its elements. Therefore the notion of an invariant factor is a generalization of the notion of the greatest common divisor of a set of polynomials. This observation suggests that there may exist an algorithm, analogous to the euclidean algorithm, for computing the invariant factors. Our next task is to develop this generalization of the euclidean algorithm, as well as a generalization of the extended euclidean algorithm.

A square polynomial matrix $A(x)$ whose determinant is equal to a nonzero field element is called a *unimodular matrix*. A unimodular matrix $A(x)$ has the property that $A^{-1}(x)$ exists as a polynomial matrix. Therefore, if $A(x)$ is a unimodular matrix, any polynomial matrix $M(x)$ for which the matrix product $M(x)A(x)$ is defined can be recovered from that matrix product by

$$M(x) = [M(x)A(x)]A^{-1}(x).$$

An n by n unimodular matrix is a unit in the ring of n by n matrices over $GF(q)[x]$.

Definition 9.10.3. *A k by n matrix of rank r is said to be factored in Smith form if*

$$G(x) = a(x)\Gamma(x)B(x)$$

where $A(x)$ and $B(x)$ are unimodular matrices and $\Gamma(x)$ is a k by n diagonal matrix with the r invariant factors of $G(x)$ as nonzero polynomial entries on the first r diagonal elements and zero elements on the diagonal thereafter.

For example, in the ring $GF(2)[x]$, the following is a factorization in the Smith form:

$$\begin{bmatrix} x+1 & x & 1 \\ x^2 & 1 & x^2+x+1 \end{bmatrix} =$$

$$\begin{bmatrix} 1 & 0 \\ x^2+x+1 & 1 \end{bmatrix} \begin{bmatrix} 1 & 0 & 0 \\ 0 & 1 & 0 \end{bmatrix} \begin{bmatrix} x+1 & x & 1 \\ x^3+x^2+1 & x^3+x^2+x+1 & 0 \\ x^2+x & x^2+x+1 & 0 \end{bmatrix}.$$

There are two invariant factors, each equal to one. The inverse of this matrix is:

$$\begin{bmatrix} x+1 & x & 1 \\ x^2 & 1 & x^2+x+1 \end{bmatrix}^{-1} = \begin{bmatrix} 0 & x^2+x+1 & x^3+x^2+x+1 \\ 0 & x^2+x & x^3+x^2+1 \\ 1 & x^2+1 & x^3+x+1 \end{bmatrix} \times$$

$$\begin{bmatrix} 1 & 0 \\ 0 & 1 \\ 0 & 0 \end{bmatrix} \begin{bmatrix} 1 & 0 \\ x^2+x+1 & 1 \end{bmatrix}$$

$$= \begin{bmatrix} x^4+x^2+1 & x^2+x+1 \\ x^4+x & x^2+x \\ x^4+x^3+x & x^2+1 \end{bmatrix}$$

which is easily computed by computing the inverse of each factor. It can be easily verified that the product of $\boldsymbol{G}(x)$ and $\boldsymbol{G}^{-1}(x)$ is the two by two identity matrix.

In general, because $\boldsymbol{A}(x)$ and $\boldsymbol{B}(x)$ are unimodular matrices, they each have an inverse over the polynomial ring. Consequently, it is now rather clear how to construct the inverse matrix for a full rank matrix $\boldsymbol{G}(x)$. Simply compute the Smith form and write $\boldsymbol{G}^{-1}(x) = \boldsymbol{B}^{-1}(x)\boldsymbol{\Gamma}^{-1}(x)\boldsymbol{A}^{-1}(x)$, where $\boldsymbol{\Gamma}^{-1}(x)$ is the diagonal matrix whose diagonal elements are $\gamma_i^{-1}(x)$. If the invariant factors are all scalars, the inverse matrix $\boldsymbol{G}^{-1}(x)$ is a polynomial matrix.

The following theorem asserts that a Smith form always exists. The proof of the theorem is based on the Smith algorithm. It consists of a sequence of elementary row and column operations.

Theorem 9.10.4. *Every k by n binary polynomial matrix $\boldsymbol{G}(x)$ with $k \le n$ can be factored in Smith form.*

Proof: This is a consequence of the Smith algorithm, which follows. □

The Smith algorithm, given next, may be regarded as a generalization of the euclidean algorithm. The extended Smith algorithm, which can be regarded as a generalization of the extended euclidean algorithm, can be obtained from the Smith algorithm simply by collecting, as they arise, the elementary matrices that are used in the Smith algorithm.

The function of the Smith algorithm is to find a matrix that is in the block form

$$\boldsymbol{G}'(x) = \begin{bmatrix} \gamma(x) & \boldsymbol{0} \\ \boldsymbol{0} & \boldsymbol{G}''(x) \end{bmatrix}$$

and is equivalent to $\boldsymbol{G}(x)$, where $\gamma(x)$ is a polynomial. The first invariant factor is obtained from this first diagonal element of the matrix $\boldsymbol{G}'(x)$. To find the next invariant factor and, in turn, all invariant factors, the Smith algorithm calls itself with $\boldsymbol{G}''(x)$ as the input. We may suppose that $\boldsymbol{G}(x)$ has at least one nonzero element because, otherwise, there is nothing to compute.

Algorithm 9.10.5 (Smith Algorithm). *If a given k by n matrix* $\mathbf{G}(x)$ *is equivalent to a matrix in the form*

$$\mathbf{G}'(x) = \begin{bmatrix} \gamma(x) & \mathbf{0} \\ \mathbf{0} & \mathbf{G}''(x) \end{bmatrix},$$

and every element is divisible by $\gamma(x)$*, then the element* $\gamma(x)$ *is the first invariant factor of* $\mathbf{G}(x)$*. Otherwise, if* $\mathbf{G}'(x)$ *does not have this form, then proceed through the three steps given next.*

Step 1. *By using row permutations and column permutations, rearrange the elements of* $\mathbf{G}(x)$ *so that the element* $g_{11}(x)$ *in the upper left corner is nonzero and has degree at least as small as the degree of any other nonzero element of* $\mathbf{G}(x)$.

Step 2. *If every nonzero element of the first row and the first column is divisible by* $g_{11}(x)$*, go to Step 3. Otherwise, reduce the degree of the element* $g_{11}(x)$ *as follows. Using the division algorithm for polynomials, write a nonzero element of the first row as*

$$g_{1j}(x) = Q_{1j}(x)g_{11}(x) + r_{1j}(x)$$

or write a nonzero element of the first column as

$$g_{i1}(x) = Q_{i1}(x)g_{11}(x) + r_{i1}(x).$$

Subtract from the j*th column the product of* $Q_{1j}(x)$ *and the first column, or subtract from the* i*th row the product of* $Q_{i1}(x)$ *and the first row. This process gives a new matrix. In the new matrix, element* $g_{1j}(x)$ *or* $g_{i1}(x)$ *now has a degree smaller than the degree of* $g_{11}(x)$*. Replace* $\mathbf{G}(x)$ *by the new matrix. If the new* $\mathbf{G}(x)$ *has any nonzero terms in the first row or the first column, except the corner position, return to Step 1. At each pass through Step 1, the maximum degree of the nonzero polynomials in the first row and the first column is reduced. Eventually, unless the process exits at the start of Step 2, all these polynomials, except the corner position, are zero. When this happens, go to Step 3.*

Step 3. *At this point every nonzero element of the first row and the first column is divisible by* $g_{11}(x)$*. For each such nonzero element, except the corner position, subtract an appropriate multiple of the first column (or first row) from that column (or row) so that the leading element is now zero, which completes the algorithm.*

Algorithm 9.10.6 (Extended Smith Algorithm). *Collect all of the elementary row and column operations on* $\mathbf{G}(x)$ *as premultiplications and postmultiplications by elementary matrices to write* $\mathbf{G}(x) = \mathbf{A}'(x)\mathbf{G}'(x)\mathbf{B}'(x)$.

Corollary 9.10.7 (Invariant Factor Theorem). *Every full-rank k by n matrix over the ring* $F[x]$*, with* $k \leq n$*, can be written as*

$$\mathbf{G}(x) = \mathbf{A}(x)\mathbf{\Gamma}(x)\mathbf{B}(x),$$

where $A(x)$ and $B(x)$ are unimodular matrices and $\Gamma(x)$ is the unique k by n diagonal matrix with the invariant factors $\gamma_i(x)$ of $G(x)$ as diagonal elements.

Proof: This is an immediate consequence of Theorem 9.10.4. □

To see that the extended euclidean algorithm is a special case of the extended Smith algorithm, consider the matrix

$$G(x) = [s(x) \quad t(x)].$$

Recall that the result of the extended euclidean algorithm can be expressed as

$$[s(x) \quad t(x)] = [\gamma(x) \quad 0] \begin{bmatrix} g_{00}(x) & g_{01}(x) \\ g_{10}(x) & g_{11}(x) \end{bmatrix},$$

which is the Smith form. We are now ready to give a formal definition of a convolutional code.

Definition 9.10.8. *An (n, k) convolutional code is a k-dimensional subspace of the vector space $GF(q)(x)^n$.*

Any basis of the k-dimensional subspace consisting of k basis vectors can be used to define a rational generator matrix $G(x)$ by using the basis vectors as rows. We can require that a rational generator matrix for a convolutional code be both delay-free and realizable. The reason that this requirement is imposed is because if all numerators of elements of the generator matrix $G(x)$ had x as a factor, then x could be factored out of $G(x)$ and regarded as a simple frame delay, while if all denominators had x as a factor, then x^{-1} could be factored out of $G(x)$ and regarded as a simple frame advance. To sharpen our conclusions, we exclude these cases by requiring $G(x)$ to be delay-free and realizable.

Theorem 9.10.9. *Every convolutional code has a polynomial generator matrix.*

Proof: Let $G(x)$ be any (realizable and delay-free) generator matrix for a convolutional code, and let $b(x)$ be the least common multiple of all denominators of elements of $G(x)$. Because $b(x)$ does not have x as a factor, the matrix

$$G'(x) = b(x)G(x)$$

is a polynomial, delay-free matrix. It is a generator matrix for the code because if

$$c(x) = a(x)G(x)$$

then

$$c(x) = a'(x)G'(x)$$

where $a'(x) = b(x)^{-1}a(x)$. □

Theorem 9.10.10. *Every convolutional code has a polynomial generator matrix with a polynomial right inverse.*

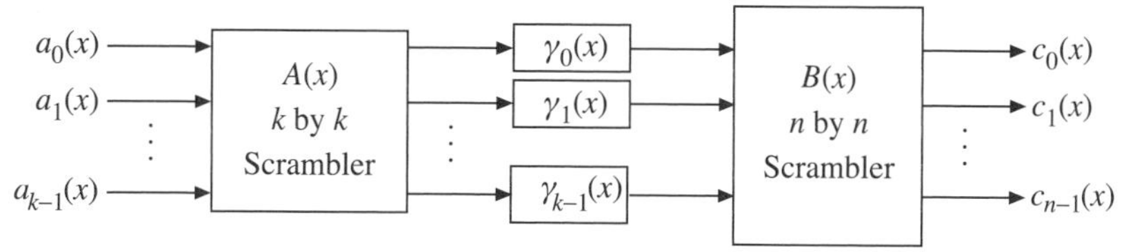

Figure 9.23. An encoder for a convolutional code

Proof: By Theorem 9.10.9, every convolutional code has a polynomial, delay-free generator matrix $\boldsymbol{G}(x)$. Using invariant factor decomposition, this matrix can be written

$$\boldsymbol{G}(x) = \boldsymbol{A}(x)\boldsymbol{\Gamma}(x)\boldsymbol{B}(x)$$

where $\boldsymbol{A}(x)$ and $\boldsymbol{B}(x)$ are unimodular matrices and so have inverses. Let $\boldsymbol{G}'(x)$ be the matrix consisting of the first k rows of $\boldsymbol{B}(x)$. The matrix consisting of the first k columns of $\boldsymbol{B}^{-1}(x)$ is a right inverse for $\boldsymbol{G}(x)$. Let $\boldsymbol{\Gamma}'(x)$ denote the matrix consisting of the first k columns of $\boldsymbol{\Gamma}(x)$. Then

$$\boldsymbol{G}(x) = \boldsymbol{A}(x)\boldsymbol{\Gamma}'(x)\boldsymbol{G}'(x).$$

But $\boldsymbol{A}(x)$ and $\boldsymbol{\Gamma}'(x)$ both have inverses so $\boldsymbol{G}'(x)$ has a right inverse and is a polynomial generator matrix for the convolutional code. □

The Smith form can be used to understand the structure of convolutional codes. Every polynomial generator matrix $\boldsymbol{G}(x)$ has a Smith form, which can be used to specify a standard form of an encoder for the convolutional code generated by $\boldsymbol{G}(x)$. The encoding operation

$$\boldsymbol{c}(x) = \boldsymbol{a}(x)\boldsymbol{G}(x)$$

can be decomposed as

$$\boldsymbol{c}(x) = \boldsymbol{a}(x)\boldsymbol{A}(x)\boldsymbol{\Gamma}(x)\boldsymbol{G}(x).$$

The function of the unimodular matrix $\boldsymbol{A}(x)$, because it is invertible, is to scramble the data symbols. In this role it is called a *scrambler matrix*. Likewise, the role of the unimodular matrix $\boldsymbol{B}(x)$ is to scramble the code bits. It too is called a scrambler matrix. The redundancy is created in the multiplication by $\boldsymbol{\Gamma}(x)$ which, because $\boldsymbol{G}(x)$ has full rank, has exactly k invariant factors.

An encoder based on a generator matrix of rank k factored in the Smith form is shown in Figure 9.23.

Problems

9.1 a. Design an encoder for a systematic, noncatastrophic, binary (4, 1) convolutional code with minimum distance $d_{\min} = 9$ and constraint length $\nu = 3$.
b. What is the free distance of this code?

9.2 A binary convolutional code has polynomial generator matrix

$$\boldsymbol{G}(x) = [x^3 + x^2 + 1 \quad x^3 + x^2 + x + 1].$$

a. Is the generator matrix for the convolutional code catastrophic? Find a way to invert the encoding in the absence of errors.
b. Sketch a trellis for this code.
c. What is the free distance? Which paths define the free distance?
d. Encode the datastream that begins with two ones and is zero forever after.

9.3 The rate $1/2$ convolutional code with polynomial generator matrix

$$\boldsymbol{G}(x) = [x^2 + 1 \quad x^2 + x + 1]$$

is made into a rate $3/4$ trellis code by discarding every second bit of $(x^2 + 1)a(x)$. This is an example of a *punctured convolutional code*.

a. Prove that the free distance of this code is 3.
b. Prove that $a(x)$ can be uniquely recovered from its punctured convolutional codeword, consisting of $(x^2 + x + 1)a(x)$ and the even symbols of $(x^2 + 1)a(x)$.

9.4 Prove that $\boldsymbol{G}(x)$ and $\boldsymbol{G}'(x)$ generate the same convolutional code if and only if $\boldsymbol{G}(x) = \boldsymbol{T}(x)\boldsymbol{G}'(x)$ where $\boldsymbol{T}(x)$ is a rational k by k matrix with nonzero determinant and so is invertible as a rational matrix.

9.5 Design an encoder and a syndrome decoder for the (8, 7) Wyner–Ash code.

9.6 For the example of Figure 9.22, find the syndrome corresponding to the senseword

$$\boldsymbol{v} = 1\,0\,0\,0\,1\,0\,0\,0\,0\,0\,1\,0\,0\,0\,0\,0\,0\,0\,0\,0\ldots.$$

Find the burst error from the syndrome.

9.7 Determine which of the following generator matrices for convolutional codes with rate $1/2$ are catastrophic:

(i) $g_0(x) = x^2, \qquad g_1(x) = x^3 + x + 1.$
(ii) $g_0(x) = x^4 + x^2 + 1, \qquad g_1(x) = x^4 + x^3 + x + 1.$
(iii) $g_0(x) = x^4 + x^2 + x + 1, \qquad g_1(x) = x^4 + x^3 + 1.$
(iv) $g_0(x) = x^6 + x^5 + x^4 + 1, \qquad g_1(x) = x^5 + x^3 + x + 1.$

9.8 Every Hamming code over $GF(q)$ can be used to build a single-error-correcting convolutional code over $GF(q)$.

a. Give a check matrix for a systematic (9, 7) octal Hamming code.
b. Based on this Hamming code, find the check matrix for a Wyner–Ash code over $GF(8)$.
c. Design an encoder and a syndrome decoder.
d. What is the rate of this code? What is the rate of a Wyner–Ash hexadecimal code based on the (17, 15) Hamming code over $GF(16)$?

9.9 Design an encoder for the (4, 3) Wyner–Ash code that uses a single binary shift register of length 3.

9.10 A rate 1/3 convolutional code with constraint length 2 has generator polynomials $g_0(x) = x^2 + x + 1$, $g_1(x) = x^2 + x + 1$, and $g_2(x) = x^2 + 1$; and d_{free} equals 8.

a. What is the minimum distance of the code?

b. What decoding window width should be used if all triple errors are to be corrected?

c. Design a decoder for this code with a decoding window width equal to 9 that will correct all double errors.

9.11 A rate 1/2 convolutional code over $GF(4)$ has generator polynomials $g_0(x) = 2x^3 + 2x^2 + 1$ and $g_1(x) = x^3 + x + 1$.

a. Show that this generator matrix is noncatastrophic.

b. Show that the minimum distance is 6.

c. How many double-error patterns are within the decoding constraint length of 8?

9.12 Find a systematic feedback shift-register encoder for the binary convolutional code with a generator matrix of polynomials

$$\boldsymbol{G}(x) = \begin{bmatrix} x & x^2 & 1 \\ 0 & 1 & x \end{bmatrix}.$$

9.13 Over $GF(2)$, let $p(x) = x^2 + x + 1$, and let $g(x) = x^4 + x + 1$. Find the "Chinese polynomials" $a(x)$ and $b(x)$, satisfying $a(x)p(x) + b(x)g(x) = 1$.

9.14 For the binary (8, 4) convolutional code with generator polynomials

$$g_0(x) = x^3 + x + 1 \quad g_1(x) = x^3 + x^2 + x + 1,$$

find the free distance and the number of ones in the input sequence that corresponds to a minimum-weight codeword.

9.15 Prove that the set $F(x) = \{a(x)/b(x) | a(x), b(x) \in F[x]; b(x) \neq 0\}$ is a field.

9.16 The rational field over $GF(q)$ is the infinite field whose elements are the rational forms $a(x)/b(x)$ where $a(x)$ and $b(x)$ are polynomials over $GF(q)$, and the arithmetic operations are defined in the natural way. Prove that an (n, k) block code over the rational field over $GF(q)$ can be regarded as the rational subcode of a convolutional code.

Notes

The notion of a convolutional code originated with Elias (1954) and was developed by Wozencraft (1957). It was rediscovered by Hagelbarger (1959). Wyner and Ash (1963) gave a general construction, akin to that of the Hamming codes, for a family of single-error-correcting convolutional codes. A class of multiple-error-correcting convolutional codes was found by Massey (1963). Massey's codes are easy to decode, but otherwise are not notable in performance. No general method is yet known for constructing a family of high-performance multiple-error-correcting codes. Costello (1969) has described some ways of finding good convolutional codes of moderate blocklength. The codes now in common use have been found by computer search, principally by Bussgang (1965), Odenwalder (1970), Bahl and Jelinek (1971), Larsen (1973), Paaske (1974), and Johannesson (1975).

The first study of the algebraic structure of convolutional codes was carried out by Massey and Sain (1968). Forney (1970) explored the relationship between convolutional codes and the algebraic structure of the ring of rational forms, and he introduced the use of a trellis to describe convolutional codes. Costello (1969) showed that every convolutional code can be encoded systematically if feedback is allowed in the encoder.

The study of decoders for convolutional codes began with simple syndrome decoders, but then turned to the complex sequential decoders of high performance and slowly evolved toward the simple decoders for simpler codes, such as those using the Viterbi algorithm, that are so popular today.

More extended discussions of convolutional codes can be found in the book by Johannesson and Zigangirov (1999), and the tutorial articles by Forney (1974), Massey (1975), and McEliece (1998).

10 Beyond BCH Codes

A digital communication system may transmit messages consisting of thousands or even millions of bits. While one can always break a long message into short blocks for encoding, in principle, a single, long block code will give better performance because it will protect against both error patterns in which the errors are clustered and error patterns in which the errors are scattered throughout the message. Therefore, there are many occasions where good codes of very long blocklength can be used.

Although short binary cyclic codes can be quite good, the known long binary cyclic codes have a small minimum distance. Codes of large blocklength with a much larger minimum distance do in principle exist, though we know very little about these more powerful codes, nor do we know how to find them. Despite more than fifty years of intense effort, codes of large blocklength and large minimum distance, both binary and nonbinary, still elude us. Even if such codes were found, it may be that their decoding would be too complex. Accordingly, the most successful constructions for codes of large blocklength for many applications combine codes of small blocklength into more elaborate structures. We call such structures *composite codes*. The elementary codes from which composite codes are formed can then be called *basic codes*.

In this chapter we shall study explicit ways of using multidimensional arrays to build composite codes. We will study five distinct kinds of composite code; these are the product codes, the interleaved codes, the concatenated codes, the cross-interleaved codes, and the turbo codes. We shall also discuss the use of multidimensional arrays to construct other codes, called *bicyclic codes*, that need not be composite codes. Then, at the end of the chapter, we will give a construction of a class of codes of very large blocklength with rate and relative distance bounded away from zero.

The most elementary form of a composite code is an interleaved code, often used to protect against burst errors as well as random errors. A slightly stronger class of composite codes is the class of product codes, which can be viewed as a generalization of the class of interleaved codes. We shall begin with a study of product codes.

10.1 Product codes and interleaved codes

The components of a codeword of a linear code with blocklength n can be arranged into an n' by n'' array if $n = n'n''$. This array may be a natural way of treating certain kinds of codes because the rows and columns can be considered individually. In such a case, the code is referred to as a *two-dimensional code*, and the more general case is referred to as a *multidimensional code*. Of course, as a vector space, the dimension of the code is still k – the number of data symbols. It is the array of indices that is referred to as multidimensional. The simplest kind of multidimensional code is a product code.

Definition 10.1.1. *The product of two linear codes, $\mathcal{C}'$ and $\mathcal{C}''$, is the linear code $\mathcal{C} = \mathcal{C}' \times \mathcal{C}''$ whose codewords are all the two-dimensional arrays for which each row is a codeword in $\mathcal{C}'$ and each column is a codeword in $\mathcal{C}''$.*

We can obtain a systematic representation of $\mathcal{C}$ by using systematic representations for $\mathcal{C}'$ and $\mathcal{C}''$ with the first k' and k'' symbols of each row and column, respectively, as data symbols. A codeword for $\mathcal{C}$ then appears as in Figure 10.1. From this figure, it is clear how to encode: First encode each of the first k'' rows with an encoder for $\mathcal{C}'$, then encode each of the n' columns with an encoder for $\mathcal{C}''$. It is simple to verify that one obtains the same codeword if the k' data columns are encoded first, and the n'' rows are encoded second. The codeword array is serialized to form a codeword for passage through the channel, for example, by reading it sequentially by rows or by columns.

If in the array of Figure 10.1, we set k'' equal to n'' and read the codeword to the channel by columns, then all we have done is to interleave the n'' codewords, each from the same (n', k') code, that appear as rows. Thus an interleaved code is a degenerate form of a product code in which $k'' = n''$. A product code is like an interleaved code, but some of the rows contain only check symbols on other rows.

Theorem 10.1.2. *If the minimum distances of the linear codes $\mathcal{C}'$ and $\mathcal{C}''$ are d' and d'', respectively, then the minimum distance of the product code $\mathcal{C}$ is $d'd''$.*

Proof: The code $\mathcal{C}$ is linear. A codeword of minimum weight has at least d' nonzero entries in every nonzero column, and has at least d'' nonzero entries in every nonzero

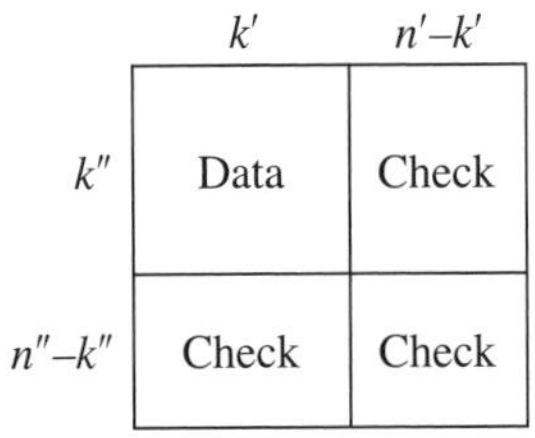

Figure 10.1. Structure of a systematic $(n'n'', k'k'')$ product codeword

row; thus the minimum weight is at least $d'd''$. To show that at least one such codeword exists, take a minimum-weight codeword from C' and a minimum-weight codeword from C''. Then the array $\boldsymbol{c}$ with elements $c_{i'i''} = c'_{i'}c''_{i''}$ is a codeword of $\mathcal{C}$ and has weight $d'd''$.

□

For an example of an interleaved code, take five interleaved codewords of a (255, 223) Reed–Solomon code over $GF(256)$, which can be thought of as a 255 by 5 array serialized by rows. The Reed–Solomon code has $d_{\min} = 33$, so the interleaved code is a (1275, 1115) code also with $d_{\min} = 33$.

For an example of a product code, the product of a (255, 223) Reed–Solomon code over $GF(256)$ and a (5, 4) Reed–Solomon code over $GF(256)$ is a (1275, 892) product code with $d_{\min} = 66$.

Both the rate

$$R = \left(\frac{k'}{n'}\right)\left(\frac{k''}{n''}\right)$$

and fractional minimum distance

$$\left(\frac{d}{n}\right) = \left(\frac{d'}{n'}\right)\left(\frac{d''}{n''}\right)$$

of a product code are smaller than those of the component codes, so the elementary performance of product codes is not attractive. However, the minimum distance is only one measure of performance. A large number of errors well beyond the packing radius of a product code are easily decoded, and a product code can also be used to combine random error correction and burst error correction.

A product code can correct bursts of length $n''b'$ if the row code can correct bursts of length b' and the product codeword is read out to the channel by columns. Because the rows and columns can be interchanged, the product code can be used to correct bursts of length $\max(n''t', n't'')$. The product code can also be used to correct up to $(d'd'' - 1)/2$ random errors. The following theorem says that there is no conflict in these two ways of using the code. Each syndrome points to only one of the two kinds of error pattern, so the code can be used to protect against both kinds of error at the same time.

Theorem 10.1.3. *The product of a linear (n', k', d') code and a linear (n'', k'', d'') code does not have any coset containing both a word of weight t or less where $t = \lfloor (d'd'' - 1)/2 \rfloor$, and when appropriately serialized, a word that is a burst of length $\max(n''t', n't'')$ or less where $t' = \lfloor (d' - 1)/2 \rfloor$ and $t'' = \lfloor (d'' - 1)/2 \rfloor$.*

Proof: To be specific, assume that $n''t' \geq n't''$, and thus the code is to be serialized by columns. If a burst pattern of length $n''t'$ or less and a random pattern of weight t or less are in the same coset, then their difference is a codeword. The weight of each row of such a codeword is either zero or at least d'. Because the burst has a length of at most $n''t'$, it terminates t' columns after it begins, so each nonzero row can contain at most t' nonzeros from the burst, and therefore must contain at least $d' - t'$ of its nonzeros from the random pattern. Thus every nonzero row contains at least $d' - t'$ of the random errors, so the random nonzeros occur in at most $\lfloor t/(d' - t') \rfloor$ rows. Hence the codeword contains at most this many nonzero rows, each with at most t' nonzeros from the burst. The total number of nonzeros is at most $\lfloor t/(d' - t') \rfloor t' + t$. But this is less than $2t$ because $d' - t'$ is greater than t'. Because the code has a minimum weight of at least $2t + 1$, this is a contradiction. This means that such a codeword cannot exist, so the burst pattern and the random pattern cannot be in the same coset. This completes the proof. □

Any product code can be used in the manner allowed by this theorem. One way, (though perhaps not the best way), to build the decoder is to combine a burst-error-correcting decoder and a random-error-correcting decoder side by side in a compound decoder; each decoder corrects up to the required number of errors, and otherwise detects an uncorrectable error pattern. Either both decoders will yield the same error pattern or only one of the two decoders will successfully decode; the other will detect a pattern that it cannot correct.

The codewords in any product code can also be represented as polynomials in two variables. That is, the two-dimensional codeword with components $c_{i'i''}$ for $i' = 0, \ldots, n' - 1$ and $i'' = 0, \ldots, n'' - 1$ has the polynomial representation

$$c(x, y) = \sum_{i'=0}^{n'-1} \sum_{i''=0}^{n''-1} c_{i'i''} x^{i'} y^{i''}.$$

This bivariate polynomial can be parenthesized in either of two ways:

$$c(x, y) = \sum_{i'=0}^{n'-1} \left(\sum_{i''=0}^{n''-1} c_{i'i''} y^{i''} \right) x^{i'},$$

or

$$c(x, y) = \sum_{i''=0}^{n''-1} \left(\sum_{i'=0}^{n'-1} c_{i'i''} x^{i'} \right) y^{i''}.$$

By the definition of a product code, each of the parenthesized terms (one for each value of i') in the first expression is a codeword in $\mathcal{C}''$, and each of the parenthesized terms (one for each value of i'') in the second expression is a codeword in $\mathcal{C}'$.

Suppose that the component codes $\mathcal{C}'$ and $\mathcal{C}''$ are cyclic. The Chinese remainder theorem (Theorem 4.1.5) can be used to determine when this implies that the product code itself is cyclic.

Theorem 10.1.4. *If $\mathcal{C}'$ and $\mathcal{C}''$ are cyclic codes whose blocklengths n' and n'' are coprime, then $\mathcal{C}$ is a cyclic code when appropriately serialized into a one-dimensional representation.*

Proof: By the Chinese remainder theorem, if $0 \le i' \le n' - 1$ and $0 \le i'' \le n'' - 1$, then there is a unique integer i in the interval $0 \le i \le n'n'' - 1$, satisfying

$$\begin{aligned} i &= i' \pmod{n'} \\ i &= i'' \pmod{n''}. \end{aligned}$$

We can replace the codeword $c(x, y)$ by the serialization

$$c(z) = \sum_{i=0}^{n'n''-1} c_{((i))((i))} z^i$$

where the double parentheses in the first index denote modulo n' and in the second index denote modulo n''.

To see that the set of such $c(z)$ forms a cyclic code, notice that $zc(z) \pmod{z^{n'n''} - 1}$ corresponds to $xyc(x, y) \pmod{x^{n'} - 1} \pmod{y^{n''} - 1}$. Because this is a codeword, $zc(z) \pmod{z^{n'n''} - 1}$ is a codeword, as was to be proved. □

The relationship between $c(z)$ and $c(x, y)$ can be understood in terms of the following two-dimensional array:

$$c_{i'i''} = \begin{bmatrix} c_{00} & c_{01} & c_{02} & \cdots & c_{0(n'-1)} \\ c_{10} & c_{11} & c_{12} & \cdots & c_{1(n'-1)} \\ \vdots & \vdots & & & \vdots \\ c_{(n''-1)0} & c_{(n''-1)1} & c_{(n''-1)2} & \cdots & c_{(n''-1)(n'-1)} \end{bmatrix}.$$

The coefficients of $c(z)$ are obtained by reading down the diagonal, starting with c_{00} and folding back at the edges as if the matrix were periodically repeated in both directions. Because n' and n'' are coprime, the extended diagonal does not re-pass through any element until all $n'n''$ elements have been used. Serializing a two-dimensional codeword in this way generates a one-dimensional codeword of a cyclic code.

If n' and n'' are not coprime, the code $\mathcal{C}$ need not be cyclic even though the component codes $\mathcal{C}'$ and $\mathcal{C}''$ are cyclic. However, an interleaved code, as we have already encountered in Section 5.9, is always cyclic if the component codes are cyclic and the array is serialized by columns. The individual codewords

$$c_j(x) = a_j(x)g(x) \quad j = 0, \ldots, n'' - 1$$

are combined into the interleaved codeword

$$\begin{aligned} c(x) &= \left[a_0(x^{n''}) + xa_1(x^{n''}) + \cdots + x^{n''-1}a_{n''-1}(x^{n''})\right]g(x^{n''}) \\ &= a(x)g(x^{n''}). \end{aligned}$$

Replacing $g(x)$ by the generator polynomial $g(x^{n''})$ produces a new cyclic code of blocklength $n'n''$, which is equivalent to interleaving n'' copies of the code generated by $g(x)$.

10.2 Bicyclic codes

Product codes whose two component codes are cyclic codes are members of a class of codes called bicyclic codes. The general definition is as follows. A *bicyclic code* is a linear two-dimensional code with the property that every cyclic shift of a codeword array in the row direction is a codeword and every cyclic shift of a codeword array in the column direction is a codeword.

Just as cyclic codes are intimately related to the one-dimensional Fourier transform, so bicyclic codes are intimately related to the two-dimensional Fourier transform. We shall develop bicyclic codes by exploiting properties of the two-dimensional Fourier transform.

A two-dimensional array can be studied by means of a two-dimensional Fourier transform, provided the array dimensions are compatible with the existence of a Fourier transform of that blocklength. Let $\boldsymbol{v} = \{v_{i'i''}\}$ be an n by n, two-dimensional array over $GF(q)$, which will be called a *two-dimensional signal*. Let ω be an element of $GF(q)$ of order n. The *two-dimensional Fourier transform* of $\boldsymbol{v}$ is the n by n array $\boldsymbol{V}$ with elements

$$V_{j'j''} = \sum_{i'=0}^{n-1} \sum_{i''=0}^{n-1} \omega^{i'j'} \omega^{i''j''} v_{i'i''}.$$

The array $\boldsymbol{V}$ will be called the *two-dimensional spectrum*, the indices j' and j'' will be called the *frequency indices*, or the two-dimensional frequency, and $V_{j'j''}$ will be called the *component* of the spectrum at *bivariate index* (j', j''). It is obvious that

$$v_{i'i''} = \frac{1}{n}\frac{1}{n} \sum_{j'=0}^{n-1} \sum_{j''=0}^{n-1} \omega^{-i'j'} \omega^{-i''j''} V_{j'j''}$$

by inspection of the one-dimensional inverse Fourier transform.

For example, because $GF(8)$ contains an element ω of order 7, there is a seven by seven, two-dimensional Fourier transform in $GF(8)$. Figure 10.2(a) shows a seven by seven, two-dimensional spectrum. Each square in the grid contains an octal symbol.

<table>
<tr><td> </td><td> </td><td> </td><td> </td><td> </td><td> </td><td> </td></tr>
<tr><td> </td><td> </td><td> </td><td> </td><td> </td><td> </td><td> </td></tr>
<tr><td> </td><td> </td><td> </td><td> </td><td> </td><td> </td><td> </td></tr>
<tr><td> </td><td> </td><td> </td><td> </td><td> </td><td> </td><td> </td></tr>
<tr><td> </td><td> </td><td> </td><td> </td><td> </td><td> </td><td> </td></tr>
<tr><td> </td><td> </td><td> </td><td> </td><td> </td><td> </td><td> </td></tr>
<tr><td> </td><td> </td><td> </td><td> </td><td> </td><td> </td><td> </td></tr>
</table>

(a) Unconstrained spectrum

<table>
<tr><td>0</td><td>0</td><td> </td><td> </td><td> </td><td> </td><td> </td></tr>
<tr><td> </td><td> </td><td> </td><td> </td><td>0</td><td> </td><td> </td></tr>
<tr><td> </td><td> </td><td> </td><td> </td><td>0</td><td> </td><td> </td></tr>
<tr><td> </td><td> </td><td> </td><td>0</td><td> </td><td> </td><td> </td></tr>
<tr><td> </td><td> </td><td>0</td><td> </td><td> </td><td> </td><td> </td></tr>
<tr><td> </td><td> </td><td> </td><td> </td><td> </td><td> </td><td> </td></tr>
<tr><td> </td><td>0</td><td> </td><td>0</td><td> </td><td> </td><td> </td></tr>
</table>

(b) Constrained spectrum

Figure 10.2. Two-dimensional spectra over $GF(8)$

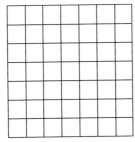

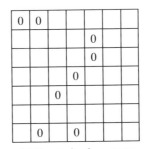

A two-dimensional spectrum need not be square. If the spectrum is an n' by n'' array with $n' \neq n''$, one must deal with the smallest extension field $GF(q^m)$ such that both n' and n'' divide $q^m - 1$. Let $\boldsymbol{v} = \{v_{i'i''}\}$ be an n' by n'' two-dimensional signal. Let β and γ be elements of $GF(q^m)$ of order n' and n'', respectively. Then the two-dimensional Fourier transform of $\boldsymbol{v}$ is

$$V_{j'j''} = \sum_{i'=0}^{n'-1} \sum_{i''=0}^{n''-1} \beta^{i'j'} \gamma^{i''j''} v_{i'i''}.$$

Again, it is obvious that

$$v_{i'i''} = \frac{1}{n'} \frac{1}{n''} \sum_{j'=0}^{n'-1} \sum_{j''=0}^{n''-1} \beta^{-i'j'} \gamma^{-i''j''} V_{j'j''}$$

by inspection of the one-dimensional inverse Fourier transform.

To define an (N, K) bicyclic code over $GF(q)$ with $N = n^2$, choose a set of $N - K$ bivariate indices, denoted $\mathcal{B} = \{j', j''\}$. The elements of $\mathcal{B}$ are called *two-dimensional check frequencies*. Constrain components with these indices to be zero, as in Figure 10.2(b). The remaining K components of the spectrum are filled with K data symbols over $GF(q)$, and the inverse two-dimensional Fourier transform then gives the codeword corresponding to the given set of data symbols. The cyclic code is $\mathcal{C} = \{\{c_{i'i''}\} | \, C_{j'j''} = 0 \text{ for } (j', j'') \in \mathcal{B}\}$. Clearly, $\mathcal{C}$ is a linear code. In general, it is not equivalent to a cyclic code. The *two-dimensional defining set* of the code is the set $\mathcal{A} = \{(\omega^{j'}, \omega^{j''}) | \, (j', j'') \in \mathcal{B}\}$. We would like to choose the two-dimensional defining set $\mathcal{A}$ to make the minimum distance of the code large. However, it is not known whether exceptional codes can be found in this way.

The BCH bound for one-dimensional cyclic codes can be used to give a bound that relates the two-dimensional spectrum of a bicyclic code and the minimum distance of the code. We state the theorem in a form that places the spectral zeros consecutively in one row. The statement is similar if the spectral zeros lie in one column. The theorem is even true with the zeros lying along a "generalized knight's move."

Theorem 10.2.1. *The only n' by n'' two-dimensional array of weight $d - 1$ or less satisfying the spectral condition*

$$C_{j'j''} = 0 \quad j' = 1, \ldots, d-1$$

for any fixed value of j'', also satisfies the stronger condition

$$C_{j'j''} = 0 \quad j' = 0, \ldots, n'-1$$

for that same value of j''.

Proof:

$$C_{j'j''} = \sum_{i'=0}^{n'-1} \beta^{i'j'} \left[\sum_{i''=0}^{n''-1} \gamma^{i''j''} c_{i'i''} \right].$$

With j'' fixed, let

$$b_{i'} = \sum_{i''=0}^{n''-1} \gamma^{i''j''} c_{i'i''}.$$

If $c_{i'i''}$ has weight at most $d-1$, then $\boldsymbol{b}$ also has weight at most $d-1$. Thus

$$C_{j'j''} = \sum_{i'=0}^{n'-1} \beta^{i'j'} b_{i'}.$$

For that fixed value of j'', the condition of the theorem becomes the condition of the BCH bound applied to $\boldsymbol{b}$. The conclusion of the theorem is the conclusion of the BCH bound. □

This theorem can be used to specify codes that are stronger than product codes. Consider the product of the (255, 223) Reed–Solomon code over $GF(256)$ and the (5, 4) Reed–Solomon code over $GF(256)$. This product code is a (1275, 892, 66) code. The two-dimensional spectrum is a 255 by 5 array over $GF(256)$ with thirty-two consecutive columns containing only zeros and one row containing only zeros. But consider what happens if instead that row of spectral zeros is replaced by a row with only sixty-five consecutive zeros. Theorem 10.2.1 says that any codeword in this code either has a weight greater than 65 or has all zeros in that row of the spectrum. That is, every codeword of smaller weight is in the product code, and the only such codeword in the product code is the all-zero codeword. Therefore this alternative defining set gives a (1275, 1082, 66) code over $GF(256)$, which is preferable to the (1275, 892, 66) product code.

The dual of a product code, called a *dual product code*, can be studied with the aid of Theorem 10.2.1. In a two-dimensional array of indices, choose a rectangle b frequencies wide and a frequencies high as the set of check frequencies. The defining set of the code is determined by this rectangle of indices. Any codeword of weight $\min(a, b)$ or less, by Theorem 10.2.1, must have a spectrum that is zero everywhere in any horizontal or vertical stripe passing through the rectangle of check frequencies. This implies, in turn, that the spectrum is zero everywhere, so the codeword is the all-zero codeword. Consequently,

$$d_{\min} \geq 1 + \min(a, b) = \min(a + 1, b + 1).$$

Hence the example gives a (49, 45, 3) code over $GF(8)$. The binary subfield-subcode is a (49, 39) $d \geq 3$ code. A dual product code does not have a large minimum distance, and so it is not suitable for correcting random errors. It may be suitable for correcting multiple low-density burst errors.

10.3 Concatenated codes

One way of constructing very long block codes is to concatenate codes.[1] This technique combines a code on a small alphabet and a code on a larger alphabet to form a code known as a *concatenated code*. A block of q-ary symbols of length kK can be broken into K subblocks, each of k symbols, where each subblock is viewed as an element from a q^k-ary alphabet. A sequence of K such q^k-ary symbols can be first encoded with an (N, K) code over $GF(q^k)$ to form a sequence of N such q^k-ary symbols. Then each of the N q^k-ary symbols can be reinterpreted as k q-ary symbols and coded with an (n, k) q-ary code. In this way, a concatenated code has two distinct levels of coding, an inner code and an outer code.

Let us recapitulate. Choose an (n, k) code for a q-ary channel as the inner code. This code over $GF(q)$ has q^k codewords. The encoder input is a block of k input symbols, which can be viewed as a single element of $GF(q^k)$. This q^k-ary symbol enters the encoder and later leaves the decoder – possibly in error. Hence the combination of the encoder, channel, and decoder can be thought of as a superchannel with the larger symbol alphabet $GF(q^k)$. Now choose a code for this superchannel. As shown in Figure 10.3, the outer code is an (N, K) code over the q^k-ary alphabet.

For example, for the inner code, choose the $(7, 3)$ double-error-correcting Reed–Solomon code over $GF(8)$, and for the outer code, choose the $(511, 505)$ triple-error-correcting Reed–Solomon code over $GF(8^3)$. When these are concatenated, they form a $(3577, 1515)$ code over $GF(8)$, which can correct any pattern of eleven errors and many error patterns with more than eleven errors. A way to visualize a codeword of a concatenated code is shown in Figure 10.4. Each codeword is a vector of 3577 octal characters, shown as a two-dimensional array only to illustrate the method of construction.

Another example of a concatenated code is given by the single codeword shown in Figure 10.5. This is a code for an octal channel obtained by concatenating a $(7, 4, 4)$ Reed–Solomon code over $GF(8)$ with a $(22, 18, 5)$ shortened Reed–Solomon code over $GF(4096)$. The concatenated codeword, which consists of the entire array produced by the inner encoder, contains seventy-two octal data symbols and has a blocklength of 154.

The senseword in this example, also shown in Figure 10.5, has been corrupted by both errors and erasures; the erasures are denoted by dashes and the errors are underlined. Symbols that are not erased may be correct or may be in error. First, the inner decoder

[1] The code is called a *concatenated code* because the encoders and the decoders are concatenated. This usage should be carefully distinguished from the term concatenated codeword – codewords that are strung together side by side.

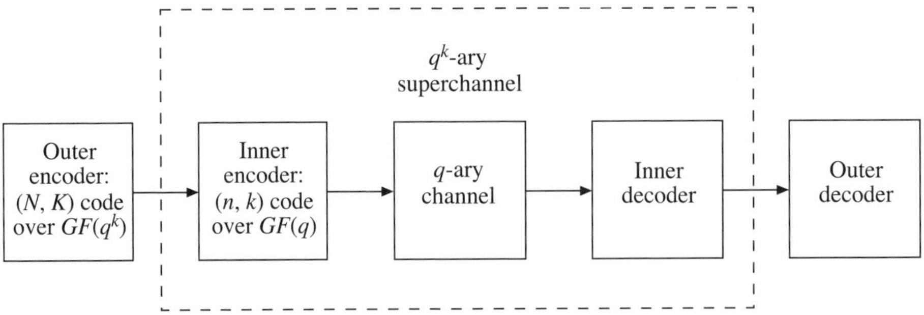

Figure 10.3. Concatenation of codes

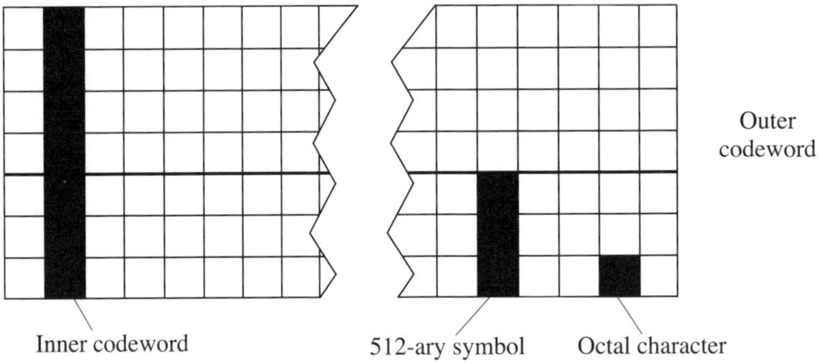

Figure 10.4. A concatenated (511, 505), (7, 3) Reed–Solomon code

corrects each column if possible. Some columns may fail to decode and some may decode incorrectly. Then the three check symbols in each column are discarded. Finally, now regarding each column as a single symbol of $GF(4096)$, the outer decoder decodes the outer codeword.

The example in Figure 10.5 dramatizes the strength of a concatenated code because of the highly broken message that is successfully decoded. Notice that if the symbols are transmitted row by row, the long string of erasures might correspond to a transient malfunction of the receiver or a burst interference. Because the inner encoder generates one column at a time, however, this transmission scheme requires a so-called *corner-turning memory* between the inner encoder and the channel. Here the inner codewords are stored by columns until the entire concatenated codeword is assembled. Then the codeword is read into the channel by rows.

One can also concatenate block codes with convolutional codes. A simple convolutional code will do quite well at correcting occasional errors in a long binary datastream, but too many errors occurring closely together will be decoded into an error burst. An outer Reed–Solomon code can be used to correct these bursts. This technique of concatenating a Reed–Solomon code with a simple convolutional code is a powerful technique for designing decoders for the gaussian-noise channel.

The original data – 72 octal symbols

$$\begin{matrix} 0&1&2&3&4&5&6&7&6&5&4&3&2&1&0&0&1&2\\ 3&4&5&6&7&6&5&4&3&2&1&0&0&1&2&3&4&5\\ 6&7&6&5&4&3&2&1&0&0&1&2&3&4&5&6&7&6\\ 5&4&3&2&1&0&0&1&2&3&4&5&6&7&6&5&4&3 \end{matrix}$$

Outer codeword – (22, 18, 5) Reed–Solomon code over $GF(4096)$
(each column is one symbol, parity block in first four columns)

$$\begin{array}{cccc|cccccccccccccccccc} 0&4&6&5&0&1&2&3&4&5&6&7&6&5&4&3&2&1&0&0&1&2\\ 2&4&6&5&3&4&5&6&7&6&5&4&3&2&1&0&0&1&2&3&4&5\\ 4&0&0&2&6&7&6&5&4&3&2&1&0&0&1&2&3&4&5&6&7&6\\ 2&1&7&2&5&4&3&2&1&0&0&1&2&3&4&5&6&7&6&5&4&3 \end{array}$$

Inner codeword – 22 copies of (7, 4, 4) Reed–Solomon code over $GF(8)$
(each column is one codeword, parity block in first three rows)

$$\begin{array}{cccc|cccccccccccccccccc} 7&7&6&7&4&1&4&5&4&1&0&4&2&5&4&3&6&2&6&4&1&4\\ 4&4&5&6&0&4&7&2&6&5&2&4&2&1&5&4&3&4&6&0&4&7\\ 2&2&5&2&4&6&1&7&1&1&2&5&3&0&7&3&3&2&2&4&6&1\\ \hline 0&4&6&5&0&1&2&3&4&5&6&7&6&5&4&3&2&1&0&0&1&2\\ 2&4&6&5&3&4&5&6&7&6&5&4&3&2&1&0&0&1&2&3&4&5\\ 4&0&0&2&6&7&6&5&4&3&2&1&0&0&1&2&3&4&5&6&7&6\\ 2&1&7&2&5&4&3&2&1&0&0&1&2&3&4&5&6&7&6&5&4&3 \end{array}$$

The received message with errors and erasures

$$\begin{matrix} \underline{4}&7&-&7&4&-&4&5&4&1&-&4&2&\underline{1}&4&-&6&2&6&4&-&4\\ 4&\underline{3}&-&-&0&\underline{2}&7&2&-&-&2&4&2&\underline{4}&5&4&3&4&\underline{5}&0&4&7\\ 2&2&-&2&4&\underline{4}&1&7&1&1&2&5&3&\underline{5}&7&3&3&-&2&4&6&1\\ 0&4&-&-&\underline{1}&\underline{6}&\underline{0}&-&-&-&-&-&-&-&-&-&-&-&-&-&-&-\\ -&-&-&-&-&-&-&-&-&-&-&-&-&-&1&0&0&1&2&\underline{2}&4&5\\ 4&0&-&2&6&\underline{1}&6&5&4&3&2&-&-&-&\underline{4}&-&\underline{0}&-&5&6&-&6\\ 2&1&\underline{5}&2&5&\underline{2}&3&-&1&0&0&1&2&\underline{6}&4&5&6&7&6&5&4&\underline{1} \end{matrix}$$

The message after the inner decoder

$$\begin{matrix} 0&4&-&5&0&-&2&3&4&5&6&7&6&\underline{7}&4&3&2&1&0&0&1&2\\ 2&4&-&5&3&-&5&6&7&6&5&4&3&\underline{3}&1&0&0&1&2&3&4&5\\ 4&0&-&2&6&-&6&5&4&3&2&1&0&\underline{0}&1&2&3&4&5&6&7&6\\ 2&1&-&2&5&-&3&2&1&0&0&1&2&\underline{6}&4&5&6&7&6&5&4&3 \end{matrix}$$

The message after the outer decoder

$$\begin{matrix} 0&1&2&3&4&5&6&7&6&5&4&3&2&1&0&0&1&2\\ 3&4&5&6&7&6&5&4&3&2&1&0&0&1&2&3&4&5\\ 6&7&6&5&4&3&2&1&0&0&1&2&3&4&5&6&7&6\\ 5&4&3&2&1&0&0&1&2&3&4&5&6&7&6&5&4&3 \end{matrix}$$

Figure 10.5. An octal concatenated code

10.4 Cross-interleaved codes

A *cross-interleaved* code is another kind of composite code that achieves the performance of a larger code, but retains the essential simplicity of the basic code. It combines basic codewords on the rows and on the off-diagonals of a codeword array. The ℓth off-diagonal of a two-dimensional array $\{c_{i'i''}\}$ is the vector of elements $\{c_{i,i+\ell}\}$ where the second index is interpreted modulo n''.

A cross-interleaved code can be regarded as a variation of a product code that provides a wider range of code parameters. Like a product code, a cross-interleaved code uses only one alphabet, $GF(q)$, for the basic codes. A two-dimensional cross-interleaved code uses two basic codes: an (n', k') linear code over $GF(q)$, called the inner code, and an (n'', k'') linear code over $GF(q)$ with $n'' > n'$, called the outer code. We shall treat only the case in which both basic codes are Reed–Solomon codes, possibly shortened. This means that $n'' \le q - 1$. The cross-interleaved codeword is an n'' by L array of symbols of $GF(q)$. The code has blocklength $n = Ln''$. Each column of the array is of length $n'' > n'$, of which the first n' symbols form one codeword of the inner code, and every off-diagonal forms one codeword of the outer code.

Definition 10.4.1. *The cross-interleave of two linear codes, $\mathcal{C}'$ and $\mathcal{C}''$, of dimension k' and k'' and blocklength n' and n'', respectively, with $k'' = n'$, is the set of n'' by L two-dimensional arrays $\mathbf{c} = \{c_{i'i''}\ i' = 0, \ldots, n'' - 1, i'' = 0, \ldots, L - 1\}$, for some L, such that every shortened column $c_{i'i''}$, $i'' = 0, \ldots, n' - 1$ is a codeword in code $\mathcal{C}'$ and every off-diagonal is a codeword in code $\mathcal{C}''$.*

The structure of the code, as per the definition, is shown by the simple example given in Figure 10.6. The first five elements of each column are the five components of a $(5, 4)$ shortened Reed–Solomon code over $GF(8)$. This is the inner code. The seven elements of each extended diagonal are the seven components of a $(7, 5)$ Reed–Solomon code over $GF(8)$. The last two elements of each diagonal are check symbols. This is the outer code.

Theorem 10.4.2. *The minimum distance of a cross-interleaved code satisfies*

$$d_{\min} = d'_{\min} d''_{\min}.$$

Proof: Any column that is nonzero in any place must be nonzero in at least $d'_{\min}$ places. The off-diagonal passing through each nonzero place of that column must be nonzero in at least $d''_{\min}$ places. □

$$\begin{bmatrix}
c_{00} & c_{10} & c_{20} & c_{30} & c_{40} & c_{50} & c_{60} & c_{70} & c_{80} & c_{90} \\
c_{01} & c_{11} & c_{21} & c_{31} & c_{41} & c_{51} & c_{61} & c_{71} & c_{81} & c_{91} \\
c_{02} & c_{12} & c_{22} & c_{32} & c_{42} & c_{52} & c_{62} & c_{72} & c_{82} & c_{92} \\
c_{03} & c_{13} & c_{23} & c_{33} & c_{43} & c_{53} & c_{63} & c_{73} & c_{83} & c_{93} \\
c_{04} & c_{14} & c_{24} & c_{34} & c_{44} & c_{54} & c_{64} & c_{74} & c_{84} & c_{94} \\
c_{05} & c_{15} & c_{25} & c_{35} & c_{45} & c_{55} & c_{65} & c_{75} & c_{85} & c_{95} \\
c_{06} & c_{16} & c_{26} & c_{36} & c_{46} & c_{56} & c_{66} & c_{76} & c_{86} & c_{96}
\end{bmatrix}$$

Figure 10.6. A cross-interleaved code

$$\begin{bmatrix} c_{00} & c_{11} & c_{22} & c_{33} & c_{44} & c_{55} & c_{66} & c_{70} & c_{81} & c_{92} \\ c_{01} & c_{12} & c_{23} & c_{34} & c_{45} & c_{56} & c_{60} & c_{71} & c_{82} & c_{93} \\ c_{02} & c_{13} & c_{24} & c_{35} & c_{46} & c_{50} & c_{61} & c_{72} & c_{83} & c_{94} \\ c_{03} & c_{14} & c_{25} & c_{36} & c_{40} & c_{51} & c_{62} & c_{73} & c_{84} & c_{95} \\ c_{04} & c_{15} & c_{26} & c_{30} & c_{41} & c_{52} & c_{63} & c_{74} & c_{85} & c_{96} \\ c_{05} & c_{16} & c_{20} & c_{31} & c_{42} & c_{53} & c_{64} & c_{75} & c_{86} & c_{90} \\ c_{06} & c_{10} & c_{21} & c_{32} & c_{43} & c_{54} & c_{65} & c_{76} & c_{80} & c_{91} \end{bmatrix}$$

Figure 10.7. Alternative structure for a cross-interleaved code

The cross-interleaved code can be regarded as a kind of generalization of a product code, which itself is a kind of generalization of an interleaved code. However, it is not a true generalization because the inner codewords do not occupy a full column of the array.

To describe the cross-interleaved code in an alternative way, cyclically shift the ℓth column of the array vertically by ℓ places. This moves the outer codewords from off-diagonals to rows. The example shown in Figure 10.6 thus changes to the array shown in Figure 10.7, in which the outer codewords, though shorter than rows, are written along rows.

One can think of the array of Figure 10.7 as filled row by row from a long serial sequence of M concatenated codewords. Write this string into an n'' by L array, row by row. The codewords are shorter than the rows of the array, and so the codewords will not be aligned with the rows, one codeword to a row. Rather, the second codeword will begin in the first row and, if necessary, continue in the second row. Moreover, for each ℓ from 0 to $n' - 1$, the ℓth symbols from all of the codewords taken together form a Reed–Solomon codeword of the (n', k') Reed–Solomon code.

Most decoders for cross-interleaved codes are composite decoders. Although a cross-interleaved code can be viewed as a single block code, the composite structure of the code can be used to simplify the design of the decoder. There are many ways of organizing a decoder for a cross-interleaved code, and most reduce complexity by recognizing the composite structure of the code. These decoders may differ considerably in complexity and performance. The simplest decoders treat the outer code and the inner code separately by using, in sequence, two distinct levels of decoder. First the outer code is decoded, then the inner code. If an outer codeword fails to decode, then all symbols of that codeword can be erased and passed to the inner decoder. The inner decoder then corrects both errors and erasures.

For the example of Figure 10.6, both the inner code and the outer code are Reed–Solomon codes. An elementary decoding strategy is first to decode each outer codeword, erasing all symbols of any $(7, 5)$ Reed–Solomon codeword that is uncorrectable. A decoder for the inner code then fills up to one erasure in each $(5, 4)$ Reed–Solomon codeword.

10.5 Turbo codes

A *turbo code* (or a *Berrou code*) is a linear block code that is constructed from a terminated, and possibly punctured, systematically encoded convolutional code in a certain prescribed way using two different codewords of the convolutional code. A turbo code encodes the same data twice in such a way that decoders for the two convolutional codewords can help each other. Turbo codes, when decoded using an iterative soft-input decoder, are among the codes with the best performance known for the additive white gaussian-noise channel.

Let $\mathcal{C}$ be a (2, 1) convolutional code with systematic polynomial generator matrix of the form

$$\boldsymbol{G}(x) = [1 \quad g(x)]$$

where $g(x)$ is a polynomial. Recall that a codeword of this convolutional code is a polynomial of the form

$$\boldsymbol{c}(x) = a(x)\boldsymbol{G}(x),$$

which can be written

$$[c_0(x) \quad c_1(x)] = [a(x) \quad a(x)g(x)]\,.$$

Corresponding to the convolutional code $\mathcal{C}$ is a block code which is obtained by restricting the degree of the datawords $a(x)$. To use the notation of block codes, let $g(x)$ have degree r and let $a(x)$ have degree at most $k-1$. Now a codeword of the terminated convolutional code consists of the pairs $(a(x), a(x)g(x))$ and has blocklength $n = r + 2k$.

In the notation of block codes, the generator matrix for the terminated code is

$$\boldsymbol{G} = [\boldsymbol{I} \quad [g(x)]]$$

where $[g(x)]$ denotes the matrix formed with the coefficients of $g(x)$, cyclically shifted, as rows, as for a cyclic code.

We require one more detail before we are ready to define a turbo code. Note that we could just as well permute the components of $a(x)$ before multiplication by $g(x)$, which would give a different codeword of the block code $\mathcal{C}$. The permutation can be expressed by the k by k permutation matrix $\boldsymbol{\Pi}$. A permutation matrix is a square matrix of zeros and ones with the property that every row has a single one, and every column has a single one. A permutation matrix can be obtained by permuting the columns of an identity matrix.

Definition 10.5.1. *A turbo code is a linear block code with a generator matrix of the form*

$$G = [I \quad [g(x)] \quad \Pi[g(x)]]$$

where Π *is a* k *by* k *permutation matrix.*

The matrix G has k rows and, because $g(x)$ has degree r, has $2r + 3k$ columns. This means that the turbo code is a $(2r + 3k, k)$ block code.

Turbo codes are interesting only for very large dimensions and very large blocklength – often k is equal to one thousand or more. Then, because r is usually small in comparison, the code rate is approximately one-third. However, we shall illustrate the definition by describing a small code. To form a code with $(n, k) = (13, 3)$, let $g(x) = g_2x^2 + g_1x + g_0$, and let

$$\Pi = \begin{bmatrix} 0 & 0 & 1 \\ 1 & 0 & 0 \\ 0 & 1 & 0 \end{bmatrix}.$$

Then the generator matrix of the block code is

$$G = \begin{bmatrix} 1 & 0 & 0 & g_2 & g_1 & g_0 & 0 & 0 & 0 & 0 & g_2 & g_1 & g_0 \\ 0 & 1 & 0 & 0 & g_2 & g_1 & g_0 & 0 & g_2 & g_1 & g_0 & 0 & 0 \\ 0 & 0 & 1 & 0 & 0 & g_2 & g_1 & g_0 & 0 & g_2 & g_1 & g_0 & 0 \end{bmatrix},$$

and the turbo codeword is

$$c = aG$$

where a is a binary dataword of length 3.

The turbo codeword consists of three parts. It can be written as a vector of three polynomials

$$\begin{aligned} c(x) &= (a(x), g(x)a(x), g(x)a'(x)) \\ &= (c_0(x), c_1(x), c_2(x)) \end{aligned}$$

where the coefficients of polynomial $a'(x)$ are a permutation, determined by Π, of the coefficients of $a(x)$.

An example of an encoder for a turbo code is shown in Figure 10.8. The same data sequence a is given both in its natural order and in a scrambled order, described by the permutation matrix Π. The unscrambled data sequence is fed into a systematic convolutional encoder, which produces a first check sequence. The scrambled data sequence is fed into another systematic convolutional encoder, identical to the first, to produce a second check sequence. The data sequence and the two check sequences form the codeword for a turbo code of rate approximately 1/3. The data sequence and the first check sequence form a conventional convolutional codeword. The scrambled data sequence and the second check sequence also form a conventional convolutional

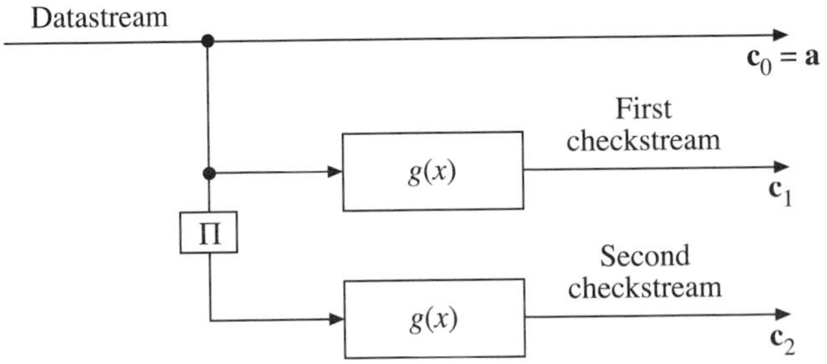

Figure 10.8. Encoder for a turbo code

codeword. The data sequence and both check sequences together form a turbo code. The use of two checkstreams in parallel, as shown in Figure 10.8, is sometimes called "parallel concatenation."

We wish to specify the generator polynomial $g(x)$ and the permutation matrix $\mathbf{\Pi}$ so as to obtain a turbo code with superior properties. However, a formal procedure for doing this is not known. Accordingly, good turbo codes are found by trial and error and are tested by simulation.

A specific example of a simple (15, 3) turbo code can be constructed by using the (2, 1) convolutional code of constraint length 3 with the systematic generator matrix

$$\boldsymbol{G}(x) = [1 \quad x^3 + x + 1].$$

This convolutional code has a free distance equal to 4. The terminated convolutional code has codewords of the form $(a(x), a(x)(x^3 + x + 1))$. This code can be regarded as a block code with generator matrix

$$\boldsymbol{G} = \begin{bmatrix} 1 & 0 & 0 & 1 & 0 & 1 & 1 & 0 & 0 \\ 0 & 1 & 0 & 0 & 1 & 0 & 1 & 1 & 0 \\ 0 & 0 & 1 & 0 & 0 & 1 & 0 & 1 & 1 \end{bmatrix}.$$

The block code formed by the generator matrix has minimum distance equal to 4. To construct the turbo code, choose the permutation matrix

$$\mathbf{\Pi} = \begin{bmatrix} 0 & 0 & 1 \\ 1 & 0 & 0 \\ 0 & 1 & 0 \end{bmatrix}.$$

The (15, 3) turbo code is the block code with generator matrix

$$\boldsymbol{G} = \begin{bmatrix} 1 & 0 & 0 & 1 & 0 & 1 & 1 & 0 & 0 & 0 & 0 & 1 & 0 & 1 & 1 \\ 0 & 1 & 0 & 0 & 1 & 0 & 1 & 1 & 0 & 1 & 0 & 1 & 1 & 0 & 0 \\ 0 & 0 & 1 & 0 & 0 & 1 & 0 & 1 & 1 & 0 & 1 & 0 & 1 & 1 & 0 \end{bmatrix}.$$

The minimum distance of this turbo code is 7.

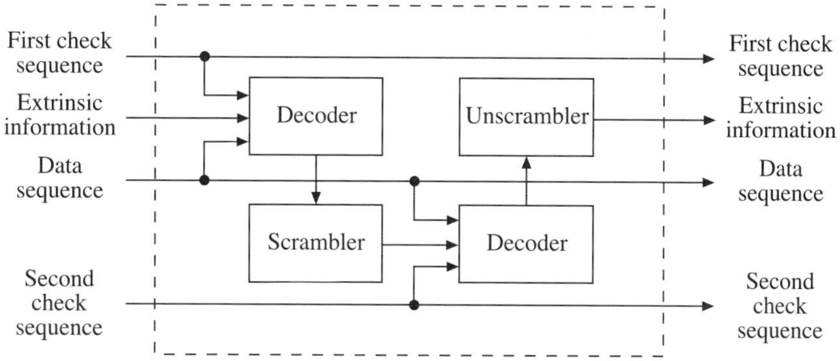

Figure 10.9. A stage of a pipelined decoder

This small example of a turbo code is adequate to illustrate the construction, but the example hides the fact that useful turbo codes are of much larger blocklength. Moreover, the power of turbo codes is not due to their minimum distance, which is not noteworthy. The power of these codes appears to be a consequence of the fact that very few codewords have minimum weight. Indeed, very few codewords have low weight. Other properties of the weight distribution, which is defined in Section 12.1, must play an important role in determining performance.

The usual decoding of a turbo code uses an iterative procedure, viewing the code as two codes. The initial decoding of the first code uses only two of the channel output streams, corresponding to $a(x)$ and $c_1(x)$. The purpose of this first decoder is to produce reliability information about each symbol in datastream $a(x)$. Several algorithms are popular for doing this decoder step. One is the soft-output Viterbi algorithm and the other is the two-way algorithm. Either of these algorithms will update a reliability measure for each data symbol. A vector describing the symbol reliability is then passed from the first decoder to a second decoder which processes the two channel outputs corresponding to $a(x)$ and $c_2(x)$, with the aid of this vector of reliabilities and updates the reliability measure of each symbol of datastream $a(x)$. The new vector of reliabilities is passed back to the first decoder which then repeats the decoding. The decoders alternate in this way until convergence is achieved. It has been found empirically, for large binary turbo codes, that ten to twenty iterations are usually sufficient for convergence. It has also been found empirically that bit error rates of the order of 10^{-6} can be obtained at very small values of E_b/N_0, as defined in Section 12.5.

The structure of the decoding procedure may be easier to understand if it is depicted as one stage of a pipelined architecture, as shown in Figure 10.9. The two codes are each a rate one-half-code whose senseword is partitioned as $\boldsymbol{v} = (\boldsymbol{u}, \boldsymbol{w})$ where $\boldsymbol{u}$ and $\boldsymbol{w}$ correspond to data and check symbols respectively. The per-symbol marginalization in terms of $(\boldsymbol{u}, \boldsymbol{w})$ is

$$p(a_\ell \mid \boldsymbol{u}, \boldsymbol{w}) = \sum_{a_0} \cdots \sum_{a_{\ell-1}} \sum_{a_{\ell+1}} \cdots \sum_{a_{k-1}} p(\boldsymbol{a} \mid \boldsymbol{u}, \boldsymbol{w}).$$

For a binary code, each sum involves only two terms and the log-posterior ratio provides the appropriate reliability.

$$\Lambda_\ell = \log \frac{p(a_\ell = 1 \mid \boldsymbol{u}, \boldsymbol{w})}{p(a_\ell = 1 \mid \boldsymbol{u}, \boldsymbol{w})}.$$

Using the Bayes formula, this is broken into three terms

$$\Lambda_\ell = \log \frac{p(a_\ell = 1)}{p(a_\ell = 1)} + \log \frac{P(u_\ell \mid a_\ell = 1)}{P(u_\ell \mid a_\ell = 1)} + \gamma_\ell$$

where γ_ℓ, known as the *extrinsic information*, is given in terms of marginalizations by

$$\gamma_\ell = \log \frac{\sum_{a:a_\ell=1} Pr[\boldsymbol{w} \mid \boldsymbol{a}] \prod_{\ell=0}^{k-1} Q(u_\ell \mid a_\ell) p(a_\ell)}{\sum_{a:a_\ell=1} Pr[\boldsymbol{w} \mid \boldsymbol{a}] \prod_{\ell=0}^{k-1} Q(u_\ell \mid a_\ell) p(a_\ell)}.$$

The extrinsic information summarizes the relevant information about the ℓth bit that comes from other symbols of the code.

10.6 Justesen codes

The *Justesen codes* form an explicit class of good codes, although still far from the codes promised by the Varshamov–Gilbert bound, which is discussed in Section 12.3. The Justesen codes are discussed, not because of their practical importance, but because they are an explicit class of codes for which the performance can be bounded away from zero.

The codewords of a $(2mN, mK)$ Justesen code over $GF(q)$ are constructed from the codewords of any (N, K) Reed–Solomon code over $GF(q^m)$. From the Reed–Solomon codeword $\boldsymbol{c} = (c_0, c_1, \ldots, c_{N-1})$, first form the $2N$ vector of $GF(q^m)$-ary symbols:

$$(c_0, \alpha^0 c_0, c_1, \alpha^1 c_1, c_2, \alpha^2 c_2, \ldots, c_{N-1}, \alpha^{N-1} c_{N-1}).$$

Each element of the vector is an element of the field $GF(q^m)$. Component c_i of the Reed–Solomon code appears here both as component c_i and modified as component $\alpha^i c_i$. Next, tilt each component into $GF(q)$ by replacing that $GF(q^m)$-ary symbol by m $GF(q)$-ary symbols that are the expansion coefficients with respect to any fixed basis. The resulting vector of $2Nm$ symbols from $GF(q)$ gives one codeword of the Justesen code. Repeating this process for each codeword of the Reed–Solomon code gives the set of codewords of the Justesen code.

The Justesen construction gives a linear code, so to find the minimum distance of the code, it suffices to find the minimum weight. In any codeword, all nonzero pairs $(c_i, \alpha^i c_i)$ are distinct because if $c_i = c_{i'}$, then $\alpha^i c_i \neq \alpha^{i'} c_{i'}$, and $c_i \neq 0$ for at least $N - K + 1$ distinct values of i.

The significance of Justesen codes is a result of Theorem 10.6.3. This theorem says that, asymptotically for large n, for any $\epsilon > 0$, and rate R,

$$\frac{d_{\min}}{n} \geq 0.11(1 - 2R) - \epsilon.$$

In particular, for a fixed rate R less than one-half, the ratio $d_{\min}/n$ is bounded away from zero as n goes to infinity. The Justesen codes are a class of binary codes with an explicitly defined construction for which this property is known to be true.

Theorem 10.6.1. *The minimum distance of the* $(2mN, mK)$ *Justesen code, constructed from an* (N, K) *primitive Reed–Solomon code, is bounded by*

$$d_{\min} \geq \sum_{\ell=1}^{I} \ell (q-1)^{\ell} \binom{2m}{\ell}$$

for any I *that satisfies*

$$\sum_{\ell=1}^{I} (q-1)^{\ell} \binom{2m}{\ell} \leq N - K + 1.$$

Proof: The minimum-weight Reed–Solomon codeword produces $N - K + 1$ distinct nonzero pairs $(c_i, \alpha^i c_i)$ which appear as distinct $2m$-tuples. The weight of the Justesen codeword is at least as large as a word constructed by filling in these $N - K + 1$ $2m$-tuples with the $N - K + 1$ distinct $2m$-tuples of smallest weight. In a $2m$-tuple, there are $\binom{2m}{i}$ ways of picking i nonzero places and $q - 1$ different nonzero values in each place. Hence there is a pair $(c_i, \alpha^i c_i)$ of weight I for every I that satisfies

$$\sum_{i=1}^{I} (q-1)^{i} \binom{2m}{i} \leq N - K + 1.$$

The minimum distance is at least as large as the sum of weights of these pairs. That is

$$d_{\min} \geq \sum_{i=1}^{I} i (q-1)^{i} \binom{2m}{i},$$

which completes the proof of the theorem. □

For large N, the binomial coefficients of Theorem 10.6.1 are impractical to calculate, and the point of the theorem is not evident. The following discussion will describe the asymptotic behavior of binary Justesen codes for large blocklength in terms of the binary entropy function.

Lemma 10.6.2. *The sum* W *of the weights of any* M *distinct binary words of length* $2m$ *satisfies*

$$W \geq 2m\lambda\left(M - 2^{2mH_b(\lambda)}\right)$$

for any λ *between* 0 *and* $\frac{1}{2}$.

Proof: By Theorem 3.5.3, for any $0 < \lambda < \frac{1}{2}$, the number of $2m$-tuples $H_b(\lambda)$ having weight at most $2m\lambda$ satisfies

$$\sum_{\ell=1}^{2m\lambda} \binom{2m}{\ell} < 2^{2mH_b(\lambda)}.$$

Therefore for each λ, there are at least $M - 2^{2mH_b(\lambda)}$ words of weight greater than $2m\lambda$. Hence the sum of the weights of the M binary words in the set satisfies

$$W \geq 2m\lambda\left(M - 2^{2mH_b(\lambda)}\right).$$

This completes the proof of the lemma. □

Theorem 10.6.3. *For fixed positive R smaller than $\frac{1}{2}$, let $K = \lceil 2NR \rceil$. Given $\epsilon > 0$ and sufficiently large n, the (n, k) Justesen code, where $n = 2mN$ and $k = mK$, has a rate at least as large as R and a minimum distance $d_{\min}$ satisfying*

$$\frac{d_{\min}}{n} \geq (1 - 2R)H_b^{-1}\left(\tfrac{1}{2}\right) - \epsilon.$$

Proof: Because $K = \lceil 2NR \rceil$, $K > 2NR$, and the code rate $K/2N$ is greater than R. Now the codeword $\boldsymbol{c}$ contains at least $N - K + 1$ distinct nonzero binary $2m$-tuples. Further,

$$N - K + 1 \geq N(1 - 2R).$$

Next, choose $N(1 - 2R)$ of these $2m$-tuples. This fits the requirements of Lemma 10.6.2. Therefore

$$d_{\min} \geq W \geq 2m\lambda\left[N(1 - 2R) - 2^{2mH_b(\lambda)}\right].$$

Because $N = 2^m$, this becomes

$$\frac{d_{\min}}{n} \geq \lambda(1 - 2R)\left[1 - \frac{2^{[2H_b(\lambda)-1]m}}{1 - 2R}\right].$$

Set

$$\lambda = H_b^{-1}\left(\frac{1}{2} - \frac{1}{\log_2 2m}\right)$$

to give

$$\frac{d_{\min}}{n} \geq H_b^{-1}\left(\frac{1}{2} - \frac{1}{\log_2 2m}\right)(1 - 2R)\left(1 - \frac{2^{-2m/\log_2 m}}{1 - 2R}\right).$$

This completes the proof of the theorem. □

The significance of the Justesen codes is found in the bound of Theorem 10.6.3. The first term of the bound converges to 0.11 as m grows; the second term converges to 1. The theorem then says that, asymptotically for large n,

$$\frac{d_{\min}}{n} \geq 0.11(1 - 2R).$$

For fixed R less than $\frac{1}{2}$, the ratio $d_{\min}/n$ is bounded away from zero. We can conclude that the Justesen codes are an explicit class of codes for which both the rate and the minimum distance can be bounded away from zero as the blocklength tends to infinity.

Problems

10.1 The simplest two-dimensional binary code with N rows and M columns is a single-error-correcting code that uses a single check on each row and a single check on each column. The minimum distance of the code is 4. Suppose this code is used for error detection on a binary channel with crossover probability ϵ. What is the probability of an undetected error? (Only the term with the smallest power of ϵ needs to be included in the calculation.)

10.2 Let $\mathcal{C}$ be an $(n, n-1)$ simple parity-check code, and let $\mathcal{C}^3$ be the $(n^3, (n-1)^3)$ code obtained as the three-dimensional product code with $\mathcal{C}$ as each component code.

a. How many errors can $\mathcal{C}^3$ correct?

b. Give two error patterns of the same weight that are detectable, but which have the same syndrome and so cannot be corrected.

10.3 a. Design a $(3840, 3360)$ interleaved extended Reed–Solomon code over $GF(64)$ to correct bursts of length 240 six-bit symbols. What is the minimum distance of the interleaved code?

b. Upgrade the code to a $(4096, 3360)$ product code that corrects all bursts of length 240 symbols and, in addition, all random error patterns.

c. Describe decoders for the codes in Problems (10.3a) and (10.3b).

10.4 a. Find the generator polynomial for the $(42, 25, 7)$ extended BCH code over $GF(2)$. Draw an encoder.

b. Find the generator polynomial for a $(17, 9, 7)$ extended BCH code over $GF(4)$. Draw an encoder.

10.5 Form a (jmn, jmk) binary code by interleaving j copies of an (n, k) Reed–Solomon code over $GF(2^m)$. What is the longest burst of binary symbols that this code is certain to correct? How does this code compare to the Rieger bound?

10.6 Suppose that $y = f(x)$ is a linear function from $GF(2^m)$ into $GF(2)$. This is a function that satisfies

$$f(ax + bx') = af(x) + bf(x').$$

Prove that the function $f(x)$ can be described by the trace function as

$$f(x) = \text{trace}(\beta x)$$

for some $\beta \in GF(2^m)$.

10.7 Construct a $(75, 40, 12)$ linear binary code as the binary expansion of a Reed–Solomon code with an extra check bit on each symbol.

10.8 Let a and b satisfy $an' + bn'' = 1$, where n' and n'' are coprime. Given an (n', k') cyclic code with the generator polynomial $g_1(x)$ and an (n'', k'') cyclic code with the generator polynomial $g_2(x)$, prove that the cyclic product code with blocklength $n = n'n''$ has the generator polynomial

$$g(x) = \text{GCD}[g_1(x^{bn''}), g_2(x^{an'}), x^n - 1].$$

10.9 Show that if $\mathcal{C}'$ and $\mathcal{C}''$ have generator matrices $\boldsymbol{G}'$ and $\boldsymbol{G}''$ and check matrices $\boldsymbol{H}'$ and $\boldsymbol{H}''$, then the product code $\mathcal{C}' \times \mathcal{C}''$ has generator matrix $\boldsymbol{G}' \times \boldsymbol{G}''$ and check matrix $\boldsymbol{H}' \times \boldsymbol{H}''$ where $\boldsymbol{M}' \times \boldsymbol{M}''$ denotes the Kronecker product of the matrices $\boldsymbol{M}'$ and $\boldsymbol{M}''$.

10.10 How many errors can be corrected by the product of a (7, 4) Hamming code with itself? Using a (7, 4) Hamming decoder as an off-the-shelf building block, design a simple decoder that corrects to the packing radius of the product code.

10.11 Prove that the turbo code constructed with the polynomial $g(x) = x^3 + x + 1$ has a minimum distance equal to 7.

Notes

Product codes were introduced by Elias (1954), who showed that the minimum distance is the product of the minimum distances of the two codes. It was proved by Burton and Weldon (1965) that the product of two cyclic codes is a cyclic code if the dimensions of the two cyclic codes are coprime. The same paper also studied the correction of burst and random errors by product codes. The use of dual product codes for the correction of multiple low-density bursts was discussed by Chien and Ng (1973). Generalizations of cross-interleaved codes were studied by Baggen and Tolhuizen (1997).

The introduction of turbo codes and of iterative decoding by Berrou, Glavieux, and Thitimajshima in 1993 created an instant sensation and eventually a large literature, and opened new research directions into both turbo codes and the maximum-posterior decoding problem, iterative decoding, and the two-way algorithm.

The behavior of the minimum distance and rate of BCH codes of large blocklength has been determined only slowly and by many authors. Studying the asymptotic behavior of the code rate of binary BCH codes by enumerating index sequences was introduced by Mann (1962), with later contributions by Berlekamp (1967). The use of linearized polynomials to relate minimum distance to designed distance for certain BCH codes was developed by Kasami, Lin, and Peterson (1966). Berlekamp (1972) later found an asymptotically precise estimate of the minimum distance of binary BCH codes. The Varshamov–Gilbert bound is the asymptotic limit that can be obtained from either the Gilbert (1952) bound or the Varshamov (1957) bound.

11 Codes and Algorithms Based on Graphs

The theory of algebraic codes draws much of its strength from the structure and properties of finite fields. In earlier chapters these properties were used to develop powerful codes and efficient decoding algorithms. The properties of those codes are tightly coupled to the properties of the fields, and the decoding algorithms use the full structure of the field to realize computational efficiency. Algebraic codes are attractive, not necessarily because of their performance, but because their structure is closely entwined with the rich structure of finite fields, a structure that underlies practical encoder and decoder implementations.

However, not every code is well-suited to the powerful algebraic decoding algorithms that are now known and not every physical channel is well-described by the mathematical structure of a finite field. Moreover, a physical channel is usually a waveform channel that transmits real or complex functions of continuous time. The methods of modulation and demodulation allow the continuous channel to be treated as a discrete channel, transmitting real or complex sequences indexed by discrete time. The sequence of codeword symbols from a finite field must be mapped into the real or complex field to form a sequence of real or complex inputs to the channel. The sequence of real or complex outputs from the channel may be mapped back to the finite field but the map takes many values to one and information is lost.

In this chapter we shall develop an alternative topic, the theory of codes and decoding algorithms based on graphs. The codes and algorithms are stronger than algebraic codes and algebraic algorithms in one sense, and weaker in another sense. On the one hand, these codes and algorithms are stronger in that raw channel output symbols, which may be real-valued or complex-valued symbols, can be processed as such, and do not need to be first mapped into a Galois field, which would lose information. On the other hand, the codes and algorithms are weaker in the sense that the algorithms may be computationally demanding even for rather small codes. The decoding algorithms in the class of algorithms based on graphs do not presuppose much about the structure of the codes, nor about the nature of the channel. This is both a strength and a weakness. Graph-theoretic decoding algorithms, though very general for some codes, are weaker than algebraic methods based on the structure of algebraic fields. One must give up the computational power of algebraic algorithms to obtain the generality of graph-theoretic methods, which obtain their attractive performance by taking full advantage of information in the raw channel symbols.

From an engineering point of view, this chapter makes connections to the subjects of information theory and the theory of modem design. From a mathematical point of view, this chapter makes connections between algebra and graph theory, portraying subspaces of $GF(q)^n$ both as configurations of points in $\boldsymbol{R}^n$ and as graphs.

We shall begin our discussion of codes on graphs with the most useful graph in coding theory, the trellis. The trellis, which was used in Chapter 9 to describe convolutional codes, now will be used to describe block codes. Then we will present an alternative graphical description of block codes known as a *Tanner graph*. Finally, we will describe a more elaborate graph known as a *factor graph*, which combines some features of trellises and of Tanner graphs.

11.1 Distance, probability, and likelihood

Codes described on trellises, or on other graphs, are attractive because their decoders can be based on soft receiver information. Soft information generally consists of real numbers (or complex numbers) at the channel output that are closely related to the method of representation of the code symbols by real numbers at the channel input. This received vector of real numbers is called a *soft senseword*. In contrast, the senseword studied earlier, consisting of elements from the code alphabet, is called a *hard senseword*. Whereas the components of a hard senseword are field elements, the components of a soft senseword can be viewed as confidence indicators.

For a binary code, either a block code or a convolutional code, our standard channel representation will be to map $GF(2)$ into $\boldsymbol{R}$ by $1 \rightarrow +1$, $0 \rightarrow -1$. This map takes the vector space $GF(2)^n$ into the set $\{-1, +1\}^n$. The set $\{-1, +1\} \subset \boldsymbol{R}$ is called the *bipolar alphabet* or the *bipolar constellation*. The image of the code $\mathcal{C}$ is a subset of $\{-1, +1\}^n$. This subset of $\{-1, +1\}^n$ is called the *euclidean image* of the code and is also denoted $\mathcal{C}$. Now the minimum-distance decoding task, when given any senseword $\boldsymbol{v} \in \boldsymbol{R}^n$, is to find the codeword $\hat{\boldsymbol{c}} \in \boldsymbol{R}^n$ at minimum euclidean distance from $\boldsymbol{v}$. The change from Hamming distance to euclidean distance alters the problem considerably. The most useful methods are no longer algebraic, and the theory of graph searching emerges as a central topic.

The squared euclidean distance from codeword $\boldsymbol{c} \in \boldsymbol{R}^n$ to codeword $\boldsymbol{c}' \in \boldsymbol{R}^n$ is defined as

$$d^2(\boldsymbol{c}, \boldsymbol{c}') = \sum_{i=0}^{n-1} (c_i - c_i')^2.$$

The *euclidean distance* between $\boldsymbol{c}$ and $\boldsymbol{c}'$ is the square root of this expression.

Definition 11.1.1. *The minimum euclidean distance of a code in* $\boldsymbol{R}^n$ *is the smallest euclidean distance between any two codewords of the code.*

Two vectors of real numbers may arise by mapping the codebits of a binary block codeword into the bipolar alphabet. Then

$$(d_{\text{min}})_{\text{euclidean}} = 2(d_{\text{min}})_{\text{Hamming}}.$$

Sequences of real numbers may also arise by mapping the codebits of a binary convolutional codeword into the bipolar alphabet. Then

$$(d_{\text{free}})_{\text{euclidean}} = 2(d_{\text{free}})_{\text{Hamming}}$$

is the appropriate statement. Similarly the squared euclidean distance from senseword $\boldsymbol{v} \in \boldsymbol{R}^n$ to codeword $\boldsymbol{c} \in \boldsymbol{R}^n$ is defined as

$$d^2(\boldsymbol{v}, \boldsymbol{c}) = \sum_{i=0}^{n-1} (v_i - c_i)^2.$$

The euclidean distance is the most natural distance measure for codes in $\boldsymbol{R}^n$. The minimum euclidean-distance decoder is defined as

$$\hat{\boldsymbol{c}} = \arg\min_{\boldsymbol{c} \in \mathcal{C}} d^2(\boldsymbol{v}, \boldsymbol{c}).$$

This codeword $\hat{\boldsymbol{c}}$ has the property that it is at minimum euclidean distance from $\boldsymbol{v}$.

Because there are q^k codewords, a minimum euclidean-distance decoder realized by exhaustive search is not computationally tractable. This chapter deals with making this decoder tractable by efficient reorganization of the code to allow a graphical description suitable for graph-searching algorithms. First, we enrich the setting by introducing a general probabilistic formulation that can be specialized to a formulation in terms of euclidean distance.

Let $Q(v|c)$ be the probability that symbol v is the channel output when symbol c is the channel input. Even though c may be a symbol from a small discrete alphabet such as $\{-1, +1\}$, a soft senseword will have a larger alphabet for v – perhaps the real field $\boldsymbol{R}$. Similarly, let $Q(\boldsymbol{v}|\boldsymbol{c})$ be the probability that senseword $\boldsymbol{v}$ is the channel output vector when codeword $\boldsymbol{c}$ is the channel input vector and let $Q(\boldsymbol{v}|\boldsymbol{a})$ be the probability that senseword $\boldsymbol{v}$ is the channel output vector when dataword $\boldsymbol{a}$ is the encoder input. When regarded as a function of $\boldsymbol{c}$ with $\boldsymbol{v}$ fixed, the function $Q(\boldsymbol{v}|\boldsymbol{c})$ is called a *likelihood function*. Similarly, when regarded as a function of $\boldsymbol{a}$ with $\boldsymbol{v}$ fixed, the function $Q(\boldsymbol{v}|\boldsymbol{a})$ also is called a likelihood function. When we discuss minimizing probability of symbol error, it is appropriate to use $Q(\boldsymbol{v}|\boldsymbol{a})$. When we discuss minimizing the probability of block error, it is sufficient to use $Q(\boldsymbol{v}|\boldsymbol{c})$.

The senseword $\boldsymbol{v}$ has n components denoted by v_i for $i = 0, \ldots, n-1$. The codeword $\boldsymbol{c}$ has n components denoted by c_i for $i = 0, \ldots, n-1$, and the dataword has

k components denoted a_ℓ for $\ell = 0, \ldots, k-1$. If the channel is memoryless – transmitting each symbol identically and independently – then the conditional probability distribution on the senseword $\boldsymbol{v}$ is

$$Q(\boldsymbol{v}|\boldsymbol{c}) = \prod_{i=0}^{n-1} Q(v_i|c_i).$$

By the Bayes rule, the *posterior probability*, denoted $P(\boldsymbol{c}|\boldsymbol{v})$, that $\boldsymbol{c}$ is the input to the channel given that $\boldsymbol{v}$ is the output of the channel, is given by

$$P(\boldsymbol{c}|\boldsymbol{v}) = \frac{Q(\boldsymbol{v}|\boldsymbol{c})p(\boldsymbol{c})}{q(\boldsymbol{v})}$$

where $p(\boldsymbol{c})$ is the *prior probability* of codeword $\boldsymbol{c}$, and $q(\boldsymbol{v}) = \sum_{\boldsymbol{c}} p(\boldsymbol{c})Q(\boldsymbol{v}|\boldsymbol{c})$ is the probability that $\boldsymbol{v}$ is the channel output. Even though $Q(\boldsymbol{v}|\boldsymbol{c})$ is a product distribution for a memoryless channel, neither $p(\boldsymbol{c})$ nor $q(\boldsymbol{v})$ is a product distribution, so even if the channel is memoryless, $P(\boldsymbol{c}|\boldsymbol{v})$ need not be a product distribution.

Definition 11.1.2. *The maximum-likelihood block decoder is given by*

$$\hat{\boldsymbol{c}} = \arg\max_{\boldsymbol{c}\in\mathcal{C}} Q(\boldsymbol{v}|\boldsymbol{c}).$$

The maximum-posterior block decoder is given by

$$\hat{\boldsymbol{c}} = \arg\max_{\boldsymbol{c}\in\mathcal{C}} P(\boldsymbol{c}|\boldsymbol{v}).$$

The probability of block decoding error is the probability that the decoded word $\hat{\boldsymbol{c}}$ is not equal to $\boldsymbol{c}$. This probability is

$$p_e = \sum_{\boldsymbol{v}} \Pr[\hat{\boldsymbol{c}} \neq \boldsymbol{c}|\boldsymbol{v}]q(\boldsymbol{v})$$

where $\Pr[\hat{\boldsymbol{c}} \neq \boldsymbol{c}|\boldsymbol{v}]$ is the probability that the decoder output is different from the transmitted codeword when $\boldsymbol{v}$ is the senseword. But this is minimized by independently minimizing each term of the sum, and each term of the sum is minimized by choosing $\hat{\boldsymbol{c}}$ to maximize the posterior probability $P(\boldsymbol{c}|\boldsymbol{v})$. Thus the maximum-posterior block decoder gives the minimum probability of block decoding error.

If $p(\boldsymbol{c})$ is independent of $\boldsymbol{c}$ then, because $q(\boldsymbol{v})$ does not depend on $\boldsymbol{c}$,

$$\arg\max_{\boldsymbol{c}\in\mathcal{C}} P(\boldsymbol{c}|\boldsymbol{v}) = \arg\max_{\boldsymbol{c}\in\mathcal{C}} Q(\boldsymbol{v}|\boldsymbol{c}).$$

In this case, the maximum-posterior decoder is identical to the maximum-likelihood decoder. This means that the maximum-posterior block decoder need not be discussed except when $p(\boldsymbol{c})$ is not the same for all $\boldsymbol{c}$. We shall discuss maximum-posterior decoding in Section 11.8.

A maximum-likelihood decoder realized by exhaustive search is not tractable. Efficient algorithms that provide practical approximations are described in Section 11.2

and Section 11.3. Often, in place of the likelihood function, it is convenient to work with the log-likelihood function, defined as

$$\Lambda(\boldsymbol{c}) = \log Q(\boldsymbol{v}|\boldsymbol{c}).$$

If the channel is memoryless, the log-likelihood function becomes

$$\Lambda(\boldsymbol{c}) = \sum_{i=0}^{n-1} \log Q(v_i|c_i).$$

Because $Q(v_i|c_i) \leq 1$, the summands are negative, or zero, so it is better to work with the negatives of the summands. The term $-\log Q(v_i|c_i)$ is called the *branch metric*.

An important channel is the additive white gaussian-noise channel. This channel takes an input sequence of real numbers as its input and adds white gaussian noise to form another sequence of real numbers as its output. White gaussian noise is a discrete random process for which the components are independent and identically distributed gaussian random variables. The additive white gaussian-noise channel is described by the conditional probability density function:

$$\begin{aligned} Q(\boldsymbol{v}|\boldsymbol{c}) &= \prod_{i=0}^{n-1} \frac{1}{\sqrt{2\pi}\,\sigma} e^{-(c_i - v_i)^2/2\sigma^2} \\ &= \left(\frac{1}{\sqrt{2\pi}\,\sigma}\right)^n e^{-\sum_{i=0}^{n-1}(c_i - v_i)^2/2\sigma^2}. \end{aligned}$$

The negative of the log-likelihood function for this channel is the sum of the branch metrics

$$\begin{aligned} -\log Q(\boldsymbol{v}|\boldsymbol{c}) &= \sum_{i=0}^{n-1}(-\log Q(v_i|c_i)) \\ &= -\sum_{i=0}^{n-1} \log\left(\frac{1}{\sqrt{2\pi}\,\sigma} e^{-(c_i - v_i)^2/2\sigma^2}\right) \\ &= \frac{1}{2\sigma^2}\sum_{i=0}^{n-1}(v_i - c_i)^2 + \frac{1}{2}\log(2\pi\sigma^2). \end{aligned}$$

To simplify the branch metric, we discard the unnecessary constants from the log-likelihood function so that the branch metric is replaced by the squared euclidean distance

$$d_{\mathrm{e}}^2(\boldsymbol{v}, \boldsymbol{c}) = \sum_{i=0}^{n-1}(v_i - c_i)^2.$$

Thus, for a gaussian channel, maximizing the log-likelihood is equivalent to minimizing the euclidean distance. Even if the noise is nongaussian, euclidean distance may still be used as the branch metric. In such cases, the euclidean distance no longer corresponds to the log-likelihood function, and so does not give a maximum-likelihood decoder. It may still be used because of its simple form and because it often gives good performance when the channel likelihood function is not known.

11.2 The Viterbi algorithm

A complete decoder for a convolutional code always makes a decision; an incomplete decoder sometimes declares a decoding failure. Convolutional codewords are infinitely long, but a decoder of finite complexity must begin to make decisions after observing only a finite segment of the senseword.

The following procedure gives a (nearly) complete decoder for a convolutional code. Choose a fixed integer b, called the *decoding window width* or the *decoding delay*, that is larger than the output span N of the code. The decoding window width will be thought of as the width of a window through which the initial segment of the senseword $\boldsymbol{v}$ is viewed. A codeword of $\mathcal{C}$ that is closest in Hamming or euclidean distance to the senseword in the initial segment of length b is found. The first dataframe associated with this codeword is the first dataframe of the decoder output. This dataframe, re-encoded, is subtracted from the senseword, so the modified senseword corresponds to a new dataword that is zero in the first frame. The first frame of the senseword is now discarded, and a new frame is shifted into the decoder. This process is repeated to find the second dataframe. This decoder fails only on a very few rare error patterns including some that could be decoded if b were a little larger.

The *Viterbi algorithm* is a procedure that exploits the trellis structure of a convolutional code to obtain an efficient method of implementing the minimum-distance decoder; it is based on techniques drawn from the subject of dynamic programming.

The Viterbi decoder operates iteratively frame by frame, tracing through the trellis used by the encoder in an attempt to find the path that the encoder followed. Given the senseword, the decoder determines the most likely path to every node of the rth frame as determined by the distance between each path and the senseword. This distance between the most likely path and the senseword is called the *discrepancy* of the path. At the end of the rth iteration, the decoder knows the minimum-distance path to each node in the rth frame, as well as the associated discrepancy. If all of the minimum-distance paths to nodes in the rth frame are the same in the first frame, the decoder deduces how the encoder began and can decode the first frame.

Then, for each node of the $(r + 1)$th frame, the decoder determines the path to that node that is at minimum distance from the senseword. But to get to a given node in frame $r + 1$, the path must pass through one of the nodes in frame r. One can get the candidate paths to a new node by extending to this new node each of the old paths that can be so extended. The minimum-distance path is found by adding the incremental discrepancy of each path extension to the discrepancy of the path to the old node. There are q^k paths to each node in the $(r+1)$th frame. The path with the smallest discrepancy is the minimum-distance path to the node. This process is repeated for each of the q^ν new nodes. At the end of the $(r+1)$th iteration, the decoder knows the minimum-distance path to each of the nodes in the $(r+1)$th frame. If all of these paths begin the same, the second frame can be decoded.

If the decoding window width b is chosen large enough, then a well-defined decision will almost always be made for each dataframe. If the code is properly matched to the channel, usually this decision will be the correct one. However, several things can go wrong. Occasionally the surviving paths might not all go through a common node in the initial frame. This is a decoding failure. A decoder may indicate to the user that a failure has occurred, or may simply guess the ambiguous symbol.

Sometimes, the decoder will reach a well-defined decision – but a wrong one. This is a decoding error. A decoding error should be rarer than a decoding default. When a decoding error occurs, the decoder will necessarily follow this with additional decoding errors. A dependent sequence of decoding errors is called an *error event*. During an error event, the decoder is following the wrong path through the trellis. One wants error events to be infrequent, and wants their duration to be short when they do occur.

Figure 11.1 depicts the Viterbi decoder as a window through which a portion of the trellis of length b may be viewed. On the finite-length section of the trellis within the decoding window, the surviving paths are marked. At each iteration new nodes appear on the right, some paths are extended to them, other paths disappear, and an old column of nodes on the left is shifted out of sight. By the time a column of nodes reaches the left side, there should be a path through only one of its nodes. If there is a path through more than one of its nodes, then there is a decoding failure.

As an example, consider the rate 1/2 convolutional code with generator polynomials $g_0(x) = x^2 + x + 1$, $g_1(x) = x^2 + 1$. We choose a decoder with a decoding-window width b equal to fifteen. Suppose that the binary senseword is

$$v = 0\,1\,1\,1\,1\,0\,1\,0\,0\,1\,0\,1\,0\,0\,0\,0\,0\,0\,0\,0\ldots.$$

The development of the candidate paths through the trellis is shown in Figure 11.2 beginning at the third iteration. At the start of iteration r, the decoder knows the shortest

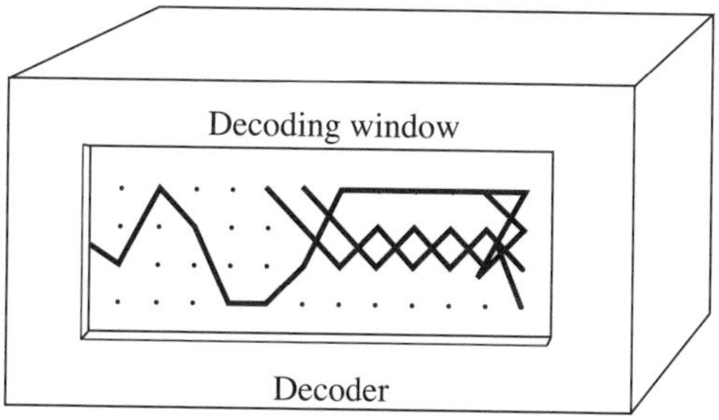

Figure 11.1. Conceptualizing the Viterbi decoder

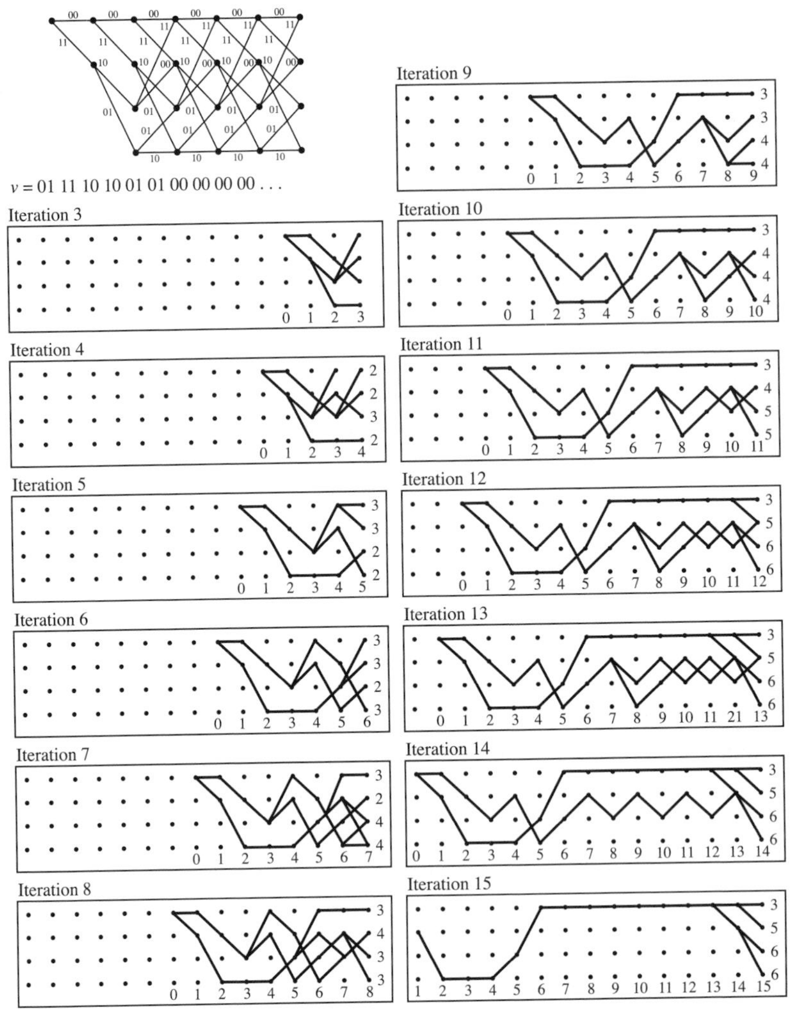

Figure 11.2. Sample of Viterbi algorithm

path to each node of the $(r-1)$th frame. Then during iteration r, the decoder finds the shortest path to nodes of the rth frame by extending the paths for each node of the $(r-1)$th frame and keeping the minimum-distance path to each new node. Whenever a tie occurs, as illustrated by the two ties at iteration 7, the decoder either may break the tie by guessing or may keep both paths in the tie. In the example, ties are retained either until they die or they reach the end of the decoder. As the decoder penetrates into deeper frames, the earlier frames reach the end of the decoder memory. If a path exists to only one node of the oldest frame, the decoding is unambiguous.

In the example, the data sequence has been decoded as

$$\boldsymbol{a} = 1\,1\,1\,1\,0\,0\,0\,0\,\ldots$$

because these are the data bits labeling the demodulated path through the trellis.

The decoder, shown graphically in Figure 11.2, might look much different in its actual implementation. For example, the active paths through the trellis could be represented by a table of four fifteen-bit numbers. At each iteration, each fifteen-bit number is used to form two sixteen-bit numbers by appending a zero and a one. This forms eight sixteen-bit numbers, treated as four pairs of such numbers. Only one of each pair – the one with minimum discrepancy – is saved; the other is discarded. Now the sixteen-bit numbers are shifted left by one bit. This sends one bit to the user and forms a new table of four fifteen-bit numbers, and the iteration is complete.

Technically, the required decoding delay of the Viterbi decoder – or of any minimum-distance trellis decoder – is unbounded because an optimum decision cannot be made until all surviving paths share a common initial subpath and, occasionally, this may take an arbitrarily long time. However, little such degradation occurs if the fixed decoding window width b is sufficiently large.

The Viterbi algorithm, with only a slight adjustment, can decode soft sensewords as easily as hard sensewords. A soft-input Viterbi decoder is the same as a hard-input Viterbi decoder except that, in place of Hamming distance, it uses another branch metric, perhaps euclidean distance. At each iteration, each surviving path is extended in every possible way to a node at the next level, with the branch metrics added to the discrepancy, which is now defined as the total accumulated path metric. The survivor to each new node is kept, and other paths are discarded. Thus the complexity of a Viterbi decoder increases very little when it is revised to decode a soft senseword.

The Viterbi algorithm uses only the trellis structure of a code. Any other structure of the code, such as the algebraic structure, is ignored and is not used to simplify the decoding with the Viterbi algorithm. Any well-behaved branch metric can be used in its computations; it need not have the properties of a distance. When using the Hamming distance, the Viterbi algorithm provides a minimum Hamming-distance decoder for block codes or convolutional codes over $GF(q)$. When using the euclidean distance, it provides a minimum euclidean-distance decoder for the euclidean image of a block code or a convolutional code in the real field. It can also do maximum-likelihood decoding, or use other soft-decision information consisting of a real nonnegative weight attached to each senseword symbol.

As the decoder advances through many frames, the discrepancies continue to grow. To avoid overflow problems, they must be reduced occasionally. One procedure is to subtract the smallest discrepancy from all discrepancies. This does not affect the choice of the maximum discrepancy.

11.3 Sequential algorithms to search a trellis

A convolutional code with a large free distance will have a large constraint length, and a large constraint length implies a formidable decoding task. For a constraint length of

40, for example, several hundred frames may need to be searched before a decision can be made. Even more overwhelming is the number of nodes. There are 2^{40} nodes in each frame, which is approximately 1000 billion nodes. The decoder must find the path through this trellis that most closely matches the senseword in the branch metric.

The Viterbi algorithm becomes unreasonable even at a moderate constraint length because, at a constraint length ν, the decoder must store and search q^{ν} surviving paths. For constraint lengths larger than about 10 or 12, one must devise alternative strategies that explore only part of the trellis. Such search strategies ignore the improbable paths through the trellis – at least for as long as they appear to be improbable – and continue to explore probable paths. Occasionally, the decoder will decide to back up and explore a path that was ignored previously. Such strategies for searching a trellis along only the high-probability paths are known collectively as *sequential decoding*. Sequential decoding is quite general and can be used to decode both hard and soft sensewords. To simplify the discussion, we shall treat a hard senseword, using the Hamming distance as the branch metric.

We shall discuss in detail two algorithms for sequential decoding, the *Fano algorithm* and the *stack algorithm*. The Fano algorithm generally involves more computation and less memory. The stack algorithm involves less computation and more memory because it compiles a list, or stack, of partial computations to avoid repeating them. Because of the way it explores a trellis, a sequential decoder usually works its way along a trellis at an erratic pace. Usually, it moves along smoothly, producing estimated data symbols as output. Occasionally, because of a difficult noise pattern, it has a little trouble and slows down to get past this troublesome spot. More rarely, it has considerable trouble and almost stops while it works itself past this difficult spot.

The Fano algorithm starts at the first frame, chooses the branch closest to the senseword frame as measured by the branch metric, proceeds along that branch to a node of the next senseword frame, and checks the accumulated metric discrepancy against a threshold. Then it repeats this procedure. At each iteration, it looks at a senseword frame, choosing the branch closest in the branch metric to that senseword frame, and proceeds to a single node of the next frame, and checks the discrepancy against a threshold. If there are no errors, then the procedure continues to move forward. If there are errors, however, the decoder will occasionally choose the wrong branch. As it proceeds after a wrong choice, the decoder will soon perceive that it is finding too many errors because the discrepancy rises above a threshold. Then the Fano decoder will back up and explore alternative paths to find another likely path. It proceeds to extend this alternative path as before. The precise rules controling this backward and forward movement will be developed shortly. When the decoder has found a path that penetrates sufficiently far into the trellis, it makes an irrevocable decision that the oldest decoded frame is correct, outputs the corresponding data symbols, and shifts a new frame of senseword symbols into its window, discarding the oldest frame.

The Fano algorithm requires that we know p, the probability of symbol error through the channel, or at least an upper bound on p. As long as the decoder is following the right path, we expect to see about $pn\ell$ errors through the first ℓ frames. The decoder will tolerate an error rate a little higher than this, but if the error rate is too much higher, the decoder concludes that it is probably on the wrong path. Choose a parameter p', perhaps by simulation, that is larger than p and smaller than one-half, and define the tilted distance as[1]

$$t(\ell) = p'n\ell - d(\ell)$$

where $d(\ell)$ is the Hamming distance between the senseword and the current path through the trellis. For the correct path, $d(\ell)$ is approximately $pn\ell$, and $t(\ell)$ is positive and increasing. As long as $t(\ell)$ is increasing, the decoder continues threading its way through the trellis. If $t(\ell)$ starts to decrease, then the decoder concludes that, at some node, it has chosen the wrong branch. Then it backs up through the trellis, testing other paths. It may find a better path and follow it, or it may return to the same node, but now with more confidence, and continue past it. To decide when $t(\ell)$ is beginning to decrease, the decoder uses a running threshold T, which is a multiple of an increment Δ. As long as the decoder is moving forward, it keeps the threshold as large as possible, with the constraints that it is not larger than $t(\ell)$ and that it is a multiple of Δ. The quantization of T allows $t(\ell)$ to decrease a little without falling through the threshold.

The Fano algorithm requires that, at each node, the q^k branches leaving that node be ordered according to some rule that is data-dependent. The ordering rule assigns an index j for $j = 0, \ldots, q^k - 1$ to each branch, but this index need not be stored. It is necessary only that the rule be known so that when the decoder first reaches that node, it knows which branch to take and when it backs up to a node along the branch with known index j, it can order the branches by the rule, find branch j, and thence find branch $j + 1$. The most common rule is the minimum-distance rule. The branches are ordered according to their Hamming distance from the corresponding frame of the senseword, and ties are broken by any convenient subrule. The algorithm will work if the branches are preassigned any fixed order, such as one where the branches are ordered lexicographically according to their data symbols. This latter ordering rule will result in more back-and-forth searching but eliminates the need to compute or recompute an order when reaching each node. For ease of understanding the structure of the algorithm, it is best to leave the ordering rule a little vague.

A shift-register implementation of a decoder based on the Fano algorithm is outlined in Figure 11.3. A flow diagram of the algorithm is shown in simplified form in Figure 11.4. The heart of the decoder is a replica of the encoder together with some auxiliary storage registers. The decoder attempts to insert symbols into the replica encoder in such a way as to generate a codeword that is sufficiently close to the senseword. At each iteration, it has immediate access to the content of the latest frame entered into the replica encoder. It can change the data in this frame, or it can back up to an earlier

[1] We have chosen the sign convention so that a positive sign corresponds to a good number.

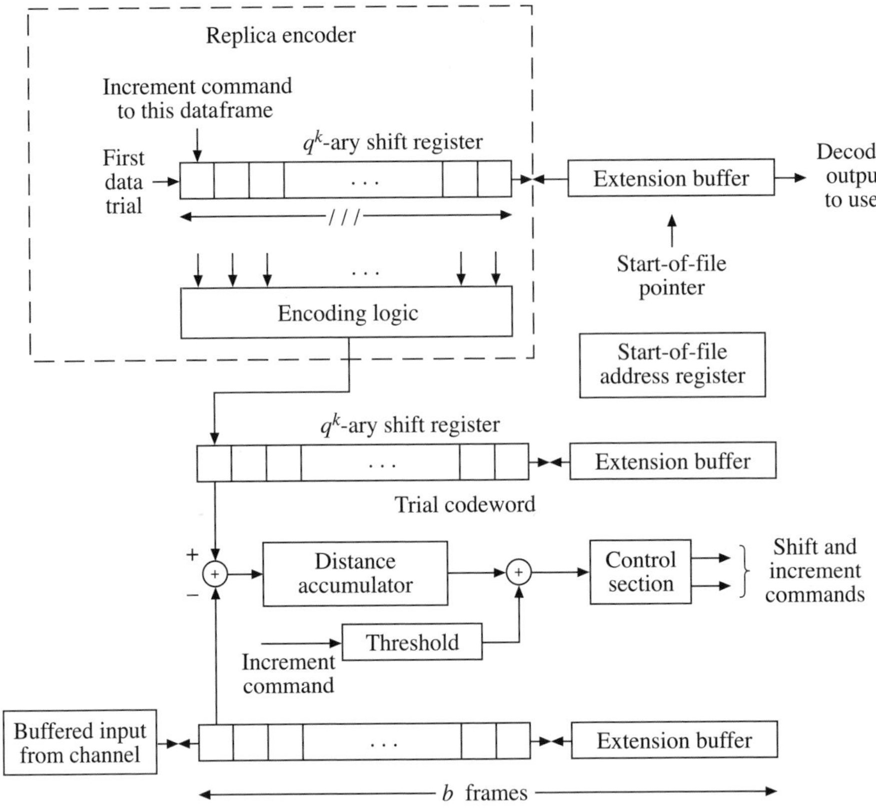

Figure 11.3. Shift-register implementation of the Fano algorithm

frame, or it can move forward to a new frame. It decides what action to take based on the value of the tilted distance $t(\ell)$ as compared to the running threshold T.

When the channel is quiet and errors are rare, the decoder will circle around the rightmost loop of Figure 11.4, each time shifting all the registers of Figure 11.3 one frame to the right. As long as $t(\ell)$ remains above the threshold, the decoder continues to shift right and to raise the threshold to keep it tight. If $t(\ell)$ drops below the threshold, the decoder tests alternative branches in that frame, trying to find one above the threshold. If it cannot find one, it backs up. As we shall see later, once it begins to back up, the logic will force it to continue to back up until it finds an alternative path that stays above the current threshold or until it finds the node at which the current threshold was set. Then the decoder moves forward again with a lowered threshold; but now, as we shall see, the threshold is not raised again until the decoder reaches nodes that were previously unexplored. Each time it visits the same node while moving forward, it has a smaller threshold. The decoder will never advance to the same node twice with the same threshold. Consequently, it can visit any node only a finite number of times. This behavior assures us that the decoder cannot be trapped in a loop. It must continue to work through the data with or without correct decoding.

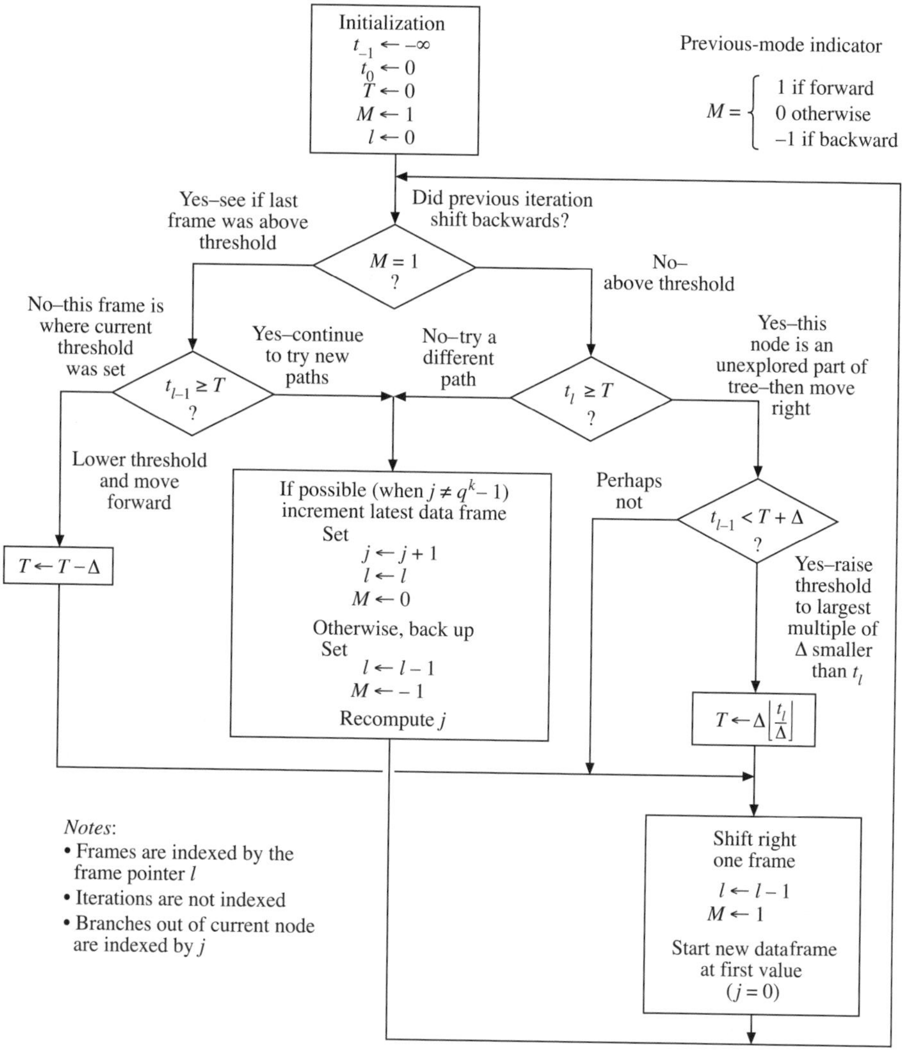

Figure 11.4. An annotated Fano algorithm

Now we will discuss two earlier assertions: That if the decoder cannot find an alternative path, it will move back to the node where the current threshold was set and lower it; and that the decoder will not raise the threshold again until it reaches a previously unexplored node. In the first assertion, it is obvious that if the decoder cannot find a new branch on which to move forward, it must eventually back up to the specified node. But if at any node the tilted distance $t(\ell - 1)$ at the previous frame is smaller than the current threshold T, then the threshold must have been increased at the ℓth frame. This is the test contained in Figure 11.4 to find the node at which the current threshold was set, and at this node, the threshold is now reduced.

To show the second assertion, notice that after the threshold is lowered by Δ, the decoder will search the subsequent branches in exactly the same sequence as before until it finds a place at which the threshold test is now passed where it had failed previously. Until this point, the logic will not allow the threshold T to be changed because once the threshold is lowered by Δ, the tilted distance will never be smaller than $T + \Delta$ at any node where previously it exceeded the original threshold. When it penetrates into a new part of the tree, eventually it will reach the condition that $t(\ell - 1) < T + \Delta$ while $t(\ell) \geq T$. This is the point at which the threshold is raised. Therefore this is the test to determine if a new node is being visited. There is no need to keep an explicit record of the nodes visited previously. This test appears in Figure 11.4.

In practical implementations, one must choose the parameters p' and Δ, usually by computer simulation, and also the decoding-window width b, usually at least several times the constraint length. Figure 11.4 show the important role of b. Whenever the oldest dataframe reaches the end of the buffer, as indicated by the frame pointer, it is passed out of the decoder, and the frame pointer is decremented so that it always points to the oldest available frame. Otherwise, the algorithm may try to back up far enough to look at a frame that has been passed out of the decoder.

The second class of sequential-decoding algorithms is known collectively as the *stack algorithm*. This class was developed to reduce the computational work at the cost of a more complicated structure. The stack algorithm keeps track of the paths it has searched, in contrast to the Fano algorithm, which may search its way to a node and then back up some distance, only to repeat the search to that same node. The stack algorithm keeps more records than the Fano algorithm so that it need not repeat unnecessary work. On the other hand, the stack algorithm needs considerably more memory.

The stack algorithm is easy to understand; a simplified flow diagram is shown in Figure 11.5. The decoder contains a stack of previously searched paths of various lengths stored as a list in memory. Each entry in the stack is a path recorded in three parts: the path length; the variable-length sequence of data symbols defining the path; and the path metric. Initially, the stack contains only the trivial zero-length path.

Any algorithm for sequential decoding has the following properties: first, the decoder performs at least one computation at each node it visits; second, the branches are examined sequentially so that, at any node, the decoder's choice among previously unexplored branches does not depend on senseword branches deeper in the tree. This second condition leads to the peculiar behavior of sequential decoding.

The number of computations made by a sequential decoder to advance one node deeper is a random variable. This computational variability is the main determinant of the complexity required to achieve a given level of performance. When there is little noise, the decoder is usually following the correct path, using only one computation to advance one node deeper into the code tree. When the noise is severe, however, the decoder may proceed along an incorrect path, and a large number of computations may be required before the decoder finds the correct path. The variability in the number of computations means that a large memory is required to buffer the incoming data.

Sometimes the sequential decoder makes so many computations that the input buffer is not big enough. This is called *buffer overflow* and is the major limitation on the

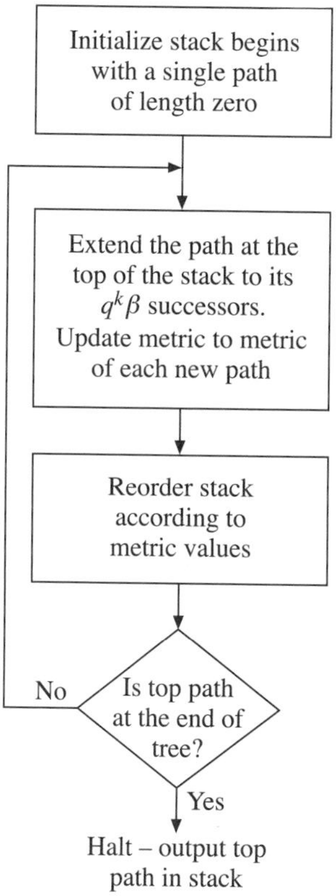

Figure 11.5. Simplified stack algorithm

performance of sequential decoding. Because the decoding delay is a random variable with a "heavy tail," the probability of buffer overflow decreases very slowly with buffer size. Thus no matter how large one makes the buffer, the problem of buffer overflow does not go away completely.

Two ways of handling buffer overflows are available. The most certain method is to insert into the encoder periodically a known sequence of data symbols – typically a sequence of zeros – of a length equal to the constraint length of the convolutional code. Upon buffer overflow, the decoder declares a decoding failure and waits for the start of the next known sequence where it starts decoding again. All intervening data between buffer overflow and the next start-up is lost. This approach reduces the rate of the code slightly and also introduces a time-synchronization problem into the decoder design. Alternatively, if the constraint length is not too large, one can just force the pointer forward. The decoder can recover if it can find a correct node. If it obtains an error-free segment for one constraint length, the decoder has found the correct node and can continue to decode. If the node is wrong, then the buffer will quickly overflow again. This process of forcing the decoder forward is repeated until correct decoding resumes.

11.4 Trellis description of linear block codes

A block code also can be described in terms of a trellis. Recall that a trellis is a directed graph in which every node is at a well-defined depth with respect to a beginning node. A directed graph is a graph in which all branches are one-way branches; a node with a well-defined depth is one for which all paths to it from a starting node have the same number of branches.

A simple example of a trellis for a block code is shown in Figure 11.6. There is no requirement that the number of nodes be the same at each depth. This means that the trellis structure in this chapter is a more general structure than was used in Chapter 9 for convolutional codes. For a block code the branch structure, the branch labeling, and the number of states in the trellis may change from frame to frame. One speaks of a *state-complexity profile* consisting of a sequence of integers giving the number of states at each time. The state-complexity profile for the trellis of Figure 11.6 is $\{1, 2, 4, 2, 4, 2, 2, 1\}$. The *state complexity*, denoted σ, of a trellis is the largest integer in the state complexity profile. In the example above, the state complexity is four.

If each branch of a trellis is labeled with a symbol from the code alphabet, then the labeled trellis specifies a code. To construct the code from a labeled trellis is straightforward. Each codeword of code $\mathcal{C}$ corresponds to a sequence of labels on a distinct path through the trellis. There are sixteen distinct paths in the trellis of Figure 11.6, and so there are sixteen codewords. In contrast, as we shall see later, to construct a labeled trellis that corresponds to a given code is less straightforward.

It is not enough to find just any trellis; we want to find a *minimal trellis*. This is a trellis for which the number of nodes at time ℓ is fewer than or equal to the number of nodes at time ℓ for any other trellis for this code. It is not immediately obvious that a minimal trellis must exist. It may be that minimizing the number of nodes at one time is not compatible with minimizing the number of nodes at another time. Nevertheless, we shall see that for each ordered code, a minimal trellis does always exist.

A linear code $\mathcal{C}$ is a vector subspace of $GF(q)^n$. Until now, except for cyclic codes, the order of the vector components has not been of much importance because the order does not affect the geometrical structure of the code. An exception is the cyclic

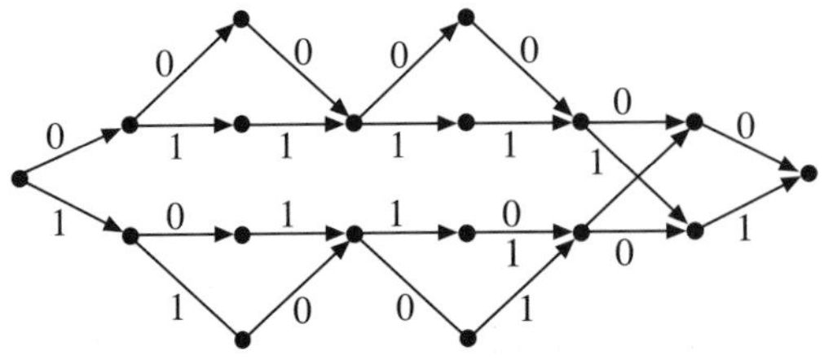

Figure 11.6. Minimal trellis for a (7, 4, 2) code

property of a cyclic code, which depends critically on the order of the codeword components. Likewise, a trellis description of a linear code depends critically on the order of the codeword components. If the components of the code are reordered, the trellis changes.

Producing a trellis for the code $\mathcal{C}$ requires that an order be specified for the components of $\mathcal{C}$. The components of a given linear code $\mathcal{C}$ can be ordered in $n!$ ways. Each of these ordered codes has a minimal trellis description. To introduce a trellis, we require that the components of the codeword be arranged in a fixed order; then the order becomes identified with time, and the terms *past* and *future* are used to refer to this order. Because each state of the trellis corresponds to one possible memory of the relevant past, the notions of past and future are natural for a trellis description.

An ordered linear block code, $\mathcal{C} \subset GF(q)^n$, may be defined by a check matrix $\boldsymbol{H}$ or by a generator matrix $\boldsymbol{G}$. Let $\boldsymbol{H}$ be any r by n matrix over $GF(q)$. The code $\mathcal{C}$ with $\boldsymbol{H}$ as a check matrix is the set of all column n-tuples $\boldsymbol{c}$ over $GF(q)$ such that $\boldsymbol{Hc} = \boldsymbol{0}$. Given a check matrix $\boldsymbol{H}$, a minimal trellis is constructed as follows. The codewords are thought of as generated in time, component by component. At time i, the past i components c_ℓ, for $\ell = 0, \ldots, i-1$, have been generated, while the present component c_i and the future components $c_{i+1}, \ldots, c_{n-1}$, remain to be generated. These components, however, are not arbitrary. The first i components constrain the possible values that the remaining components may take. A state at frame time i consists of the past history insofar as it is relevant in determining the future. At time i, two pasts are equivalent if they have the same possible futures. By using this equivalence principle at each time i, the minimal trellis is easily constructed.

For example, we will construct a minimal trellis for the simple binary block code $\mathcal{C}$, defined by the 3 by 7 check matrix:

$$\boldsymbol{H} = \begin{bmatrix} 1 & 1 & 1 & 0 & 0 & 0 & 0 \\ 1 & 0 & 0 & 1 & 1 & 0 & 0 \\ 1 & 0 & 0 & 0 & 0 & 1 & 1 \end{bmatrix}.$$

The code consists of the sixteen binary codewords of blocklength 7 that satisfy the three equations:

$$c_0 + c_1 + c_2 = 0;$$
$$c_0 + c_3 + c_4 = 0;$$
$$c_0 + c_5 + c_6 = 0.$$

This code has minimum Hamming distance equal to 2 because the vector 0110000 is a codeword of weight 2, and no codeword of $\mathcal{C}$ has weight 1. Hence the code is a (7, 4, 2) code over $GF(2)$. In fact, a trellis for this block code is the one shown in Figure 11.6.

To construct this trellis for the (7, 4, 2) code of our example, note that at time 1, after the first code bit, the set of states is $\mathcal{S}_1 = \{\{0\}, \{1\}\}$ because 0 and 1 are the two possible pasts and they have different futures. Then at time 2, the set of states is

$$\mathcal{S}_2 = \{\{00\}, \{01\}, \{10\}, \{11\}\}.$$

There are four states in $\mathcal{S}_2$ because the past consists of a pair of bits that can take any of four values.

At time 3, the past consists of three bits, but the constraint $c_0 + c_1 + c_2 = 0$ restricts the three bits c_0, c_1, and c_2 to four possible combinations. The set of states is

$$\mathcal{S}_3 = \{\{000, 011\}, \{101, 110\}\}.$$

The reason that there are only two states in $\mathcal{S}_3$ is because c_1 and c_2 do not appear in later constraints, so they do not individually affect the future. Thus it is possible to represent a state of $\mathcal{S}_3$ by a single bit, which corresponds to c_0.

At time 4, the set of states is

$$\mathcal{S}_4 = \{\{0000, 0110\}, \{0001, 0111\}, \{1010, 1100\}, \{1011, 1101\}\}.$$

There are four states in $\mathcal{S}_4$, so it is possible to represent a state of $\mathcal{S}_4$ by two bits, which could be c_0 and c_3.

At time 5, the set of states, $\mathcal{S}_5$, again contains only two states because of the constraint $c_0 + c_3 + c_4 = 0$. At time 6, the set of states, $\mathcal{S}_6$, contains only two states, now because the set of possible futures is limited.

In this way, the trellis of Figure 11.6 is easily constructed for a linear code with ordered components.

An alternative method of constructing the same trellis is to start with the generator matrix

$$\boldsymbol{G} = \begin{bmatrix} 1 & 0 & 1 & 1 & 0 & 0 & 1 \\ 0 & 1 & 1 & 0 & 0 & 0 & 0 \\ 0 & 0 & 0 & 1 & 1 & 0 & 0 \\ 0 & 0 & 0 & 0 & 0 & 1 & 1 \end{bmatrix}$$

and to use the rows of $\boldsymbol{G}$ to define paths through the trellis, as in Figure 11.6.

Both of these constructions, one based on the check matrix and one based on the generator matrix, give a minimal trellis for the given order because no opportunity to use fewer nodes remains unused. Because the seven components of the linear code $\mathcal{C}$ can be ordered in 7! different ways, there are 7! such trellises corresponding to the same linear code. We have constructed only one of these. To find the ordering that gives the best trellis appears to be a difficult task because 7! orderings must be examined.

Another example of a trellis for a larger block code is shown in Figure 11.7, which shows a trellis for an ordering of the (8, 4, 4) binary Hamming code. This trellis is the minimal trellis for the Hamming code. The state-complexity profile of this trellis is $\{1, 2, 4, 8, 4, 8, 4, 2, 1\}$. This trellis has a maximum of eight states, so it is natural to say that the (8, 4, 4) Hamming code has a state complexity of 8 and a constraint length of 3.

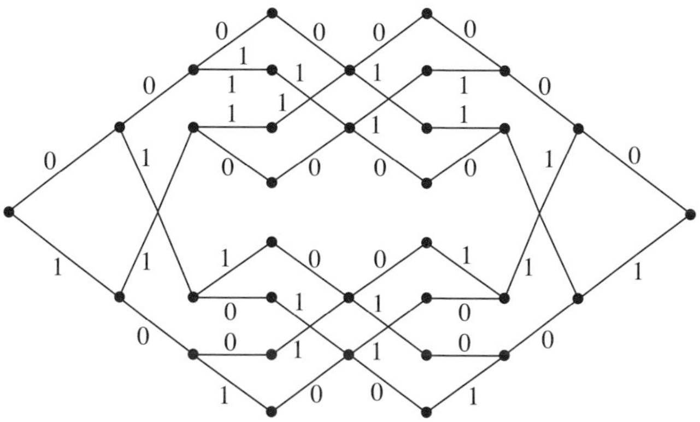

Figure 11.7. Minimal trellis for the (8, 4, 4) Hamming code

Every ordered linear block code can be so described by a trellis. Because there are a great many orders, there are a great many trellis representations of a given linear code. Indeed, there are $n!$ ways to order the n components of the code, and for each order the code has a minimal trellis representation. A simple method of finding the ordering that gives the best minimal trellis is not known. The most important facts concerning the trellis representation of linear block codes are these: for every fixed coordinate ordering, every linear block code has an essentially unique minimal trellis representation, and the minimal state-complexity profile depends on the coordinate ordering. Even with the most efficient ordering, the state complexity of a trellis can depend exponentially on blocklength.

A minimal trellis for a code is uniquely determined by a generator matrix for that code. This means that the state complexity of the trellis must be a consequence of a property of that generator matrix. Consequently, there should be a method of computing the state complexity of a trellis directly from a generator matrix for that trellis.

Define the *span vector* of the jth row of $\boldsymbol{G}$ as the binary vector of length $n+1$ that, for $\ell = 0, \ldots, n$, is equal to zero at position ℓ if $g_{ij} = 0$ for all $i < \ell$ or if $g_{ij} = 0$ for all $i > \ell$, and otherwise is equal to one. It is a sequence of contiguous ones marking where the row is active. The array formed with these span vectors as rows is an $n+1$ by k array of zeros and ones formed from $\boldsymbol{G}$ by placing zeros and ones between the elements of each row, and at the beginning or end of the row. A one is placed if at least one nonzero element occurs somewhere both to the left and to the right; otherwise a zero is placed there. For each ℓ, let ν_ℓ be the number of rows whose span vector is one at that ℓ. Now the state-complexity profile can be described as a property of the matrix $\boldsymbol{G}$.

Definition 11.4.1. *The state-complexity profile of the matrix $\boldsymbol{G}$ is the vector*

$$\sigma = \{q^{\nu_\ell} | \ell = 0, \ldots, n\}.$$

Definition 11.4.2. *The state complexity, denoted σ, of the k by k matrix $\boldsymbol{G}$ is the maximum value of its state-complexity profile. The constraint length of the matrix is the maximum value of ν_ℓ.*

For example, a generator matrix for the extended Hamming code that corresponds to Figure 11.7 is

$$\boldsymbol{G} = \begin{bmatrix} 1 & 1 & 1 & 1 & 0 & 0 & 0 & 0 \\ 0 & 0 & 0 & 0 & 1 & 1 & 1 & 1 \\ 0 & 0 & 1 & 1 & 1 & 1 & 0 & 0 \\ 0 & 1 & 1 & 0 & 0 & 1 & 1 & 0 \end{bmatrix}.$$

The span vectors of this matrix form the array

$$\begin{bmatrix} 0 & 1 & 1 & 1 & 0 & 0 & 0 & 0 & 0 \\ 0 & 0 & 0 & 0 & 0 & 1 & 1 & 1 & 0 \\ 0 & 0 & 0 & 1 & 1 & 1 & 0 & 0 & 0 \\ 0 & 0 & 1 & 1 & 1 & 1 & 1 & 0 & 0 \end{bmatrix}.$$

The vector is always zero in the components at $\ell = 0$ or n because then the condition is vacuously true. Because the states are q^{ν_ℓ} where the ν_ℓ are the number of ones in each column, we conclude that this generator matrix for the Hamming code has a state-complexity profile $\boldsymbol{\sigma} = \{1, 2, 4, 8, 4, 8, 4, 2, 1\}$, and so the constraint length ν is equal to 3.

11.5 Gallager codes

A Gallager code is a binary block code characterized by a special form of the check matrix $\boldsymbol{H}$.

Definition 11.5.1. *An r, s* Gallager code *is a linear code that has a check matrix $\boldsymbol{H}$ satisfying the condition that every column has r ones and every row has s ones.*

As does every linear code, a Gallager code has many check matrices, and most of them will not satisfy the condition in the definition. However, the definition only requires that one check matrix has this form. This check matrix is preferred for describing a Gallager code because it establishes the special structure of a Gallager code.

The Gallager codes of primary interest are those of large blocklength for which r and s are small compared to n because these codes are easy to decode. A Gallager code with small r and s is an example of a *low-density parity-check code*. A low-density parity-check code is a linear code with a check matrix that is sparse. The number of ones in each row is small, and the number of ones in each column is small. If the class of low-density parity-check codes is to be precisely defined, one may take this to mean that r is at most $\log_2 n$.

One may also define an *irregular* low-density parity-check code to be a code whose check matrix $\boldsymbol{H}$ has approximately r ones in each column and s ones in each row.

An example of a check matrix for a Gallager code, with four ones in every row and two ones in every column, is the following

$$\boldsymbol{H} = \begin{bmatrix} 1 & 1 & 1 & 1 & 0 & 0 & 0 & 0 & 0 & 0 \\ 1 & 0 & 0 & 0 & 1 & 1 & 1 & 0 & 0 & 0 \\ 0 & 1 & 0 & 0 & 1 & 0 & 0 & 1 & 1 & 0 \\ 0 & 0 & 1 & 0 & 0 & 1 & 0 & 1 & 0 & 1 \\ 0 & 0 & 0 & 1 & 0 & 0 & 1 & 0 & 1 & 1 \end{bmatrix}.$$

In this example, the matrix includes every possible column that has two ones, so there are $\binom{5}{2}$ columns.

The rows of the check matrix of this example are not independent because they sum to zero. The top row, for example, may be regarded as the sum of the other rows and can be deleted without affecting the row space of $\boldsymbol{H}$. Thus the same code could be defined by the check matrix

$$\boldsymbol{H} = \begin{bmatrix} 1 & 0 & 0 & 0 & 1 & 1 & 1 & 0 & 0 & 0 \\ 0 & 1 & 0 & 0 & 1 & 0 & 0 & 1 & 1 & 0 \\ 0 & 0 & 1 & 0 & 0 & 1 & 0 & 1 & 0 & 1 \\ 0 & 0 & 0 & 1 & 0 & 0 & 1 & 0 & 1 & 1 \end{bmatrix}.$$

This is now seen to be the check matrix for a (10, 6, 3) shortened Hamming code. Hence this Gallager code is also the (10, 6, 3) shortened Hamming code.

In an r, s Gallager code, every code symbol is checked by precisely r equations, and every check equation sums precisely s code symbols. Usually, r and s are small compared to n. A Gallager code, or any low-density parity-check code, can be decoded by a simple iterative *flipping algorithm*. During each iteration, all check sums are first computed. Then each bit is flipped if the majority of its checks so indicate. Then the process is repeated and the procedure stops if and when all of the checks are satisfied, or when a preselected number of trials are completed. The flipping algorithm does not guarantee that all errors up to half the minimum distance are corrected. However, for long blocklengths the flipping algorithm does work remarkably well and, for some channels, gives performance close to that promised by better codes.

11.6 Tanner graphs and factor graphs

Every linear code can also be described by another kind of graph called a *Tanner graph*. A Tanner graph for a linear code $\mathcal{C}$ is a graphical representation of the code based on any set of check equations for that linear code. A Tanner graph is a bipartite graph, which means that it has two kinds of nodes: *symbol nodes* and *check nodes*.

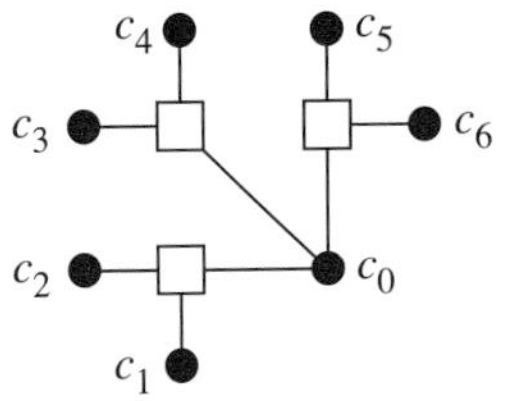

Figure 11.8. Tanner graph for a (7, 4, 2) binary code

Each component of the code is represented in the Tanner graph by a symbol node, so the graph has n symbol nodes. Each check equation of the code is represented in the Tanner graph by a check node, so the graph has $n - k$ check nodes. Each check node is connected by an edge to each symbol node that corresponds to a component that it checks. Thus a Tanner graph is a bipartite graph in which each symbol node is connected only to check nodes, and each check node is connected only to symbol nodes. The important property of a Tanner graph is that it specifies a code in terms of local constraints. These local constraints describe a code by its local behavior. Each constraint specifies the combinations of values that the attached symbol nodes are allowed to have.

For example, the (7, 4, 2) code mentioned in Section 11.4 has check matrix

$$H = \begin{bmatrix} 1 & 1 & 1 & 0 & 0 & 0 & 0 \\ 1 & 0 & 0 & 1 & 1 & 0 & 0 \\ 1 & 0 & 0 & 0 & 0 & 1 & 1 \end{bmatrix}.$$

Figure 11.8 illustrates the Tanner graph corresponding to this code. Each symbol node is shown as a black circle, and each check node is shown as a square. Because the symbol nodes c_0, c_3, and c_4 are connected to the same check node, the values of these symbols must sum to zero. Other check nodes correspond to other check constraints; each specifies a set of symbols that must sum to zero. It is apparent from the Tanner graph of Figure 11.8 that node c_0 has a special property; given c_0, the pairs of variables (c_1, c_2), (c_3, c_4), and (c_5, c_6) become "conditionally independent." This means that if c_0 is known, then the value of the pair (c_1, c_2) is of no help in determining the value of the pair (c_3, c_4). Thus a graph representation can be useful in indicating instances of conditional independence. In this respect, the variable c_0 acts like a sufficient statistic as defined in the subject of statistical inference.

Figure 11.8 is an uncluttered depiction of a Tanner graph for a simple code. For a more general code, to avoid clutter, a more regular arrangement is preferable. Accordingly, Figure 11.9 redraws the Tanner graph in a standard form with all symbol nodes on the left and all check nodes on the right. This form of the Tanner graph is useful when describing certain decoding algorithms as the passing of messages through a graph.

An elaboration of the Tanner graph, known as a *Wiberg–Loeliger graph* or a *factor graph*, provides a representation of a code that is intermediate between the Tanner graph representation and the trellis representation. The Wiberg–Loeliger graph, or

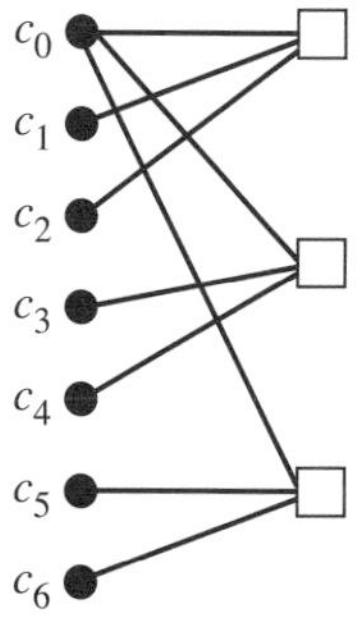

Figure 11.9. Another Tanner graph for a (7, 4, 2) code

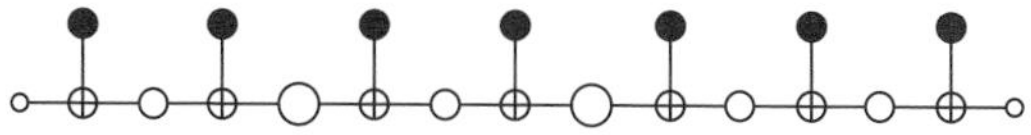

Figure 11.10. Factor graph for a (7, 4, 2) code

factor graph, is defined by the introduction of a third type of node representing an unobserved state. Thus the factor graph has three kinds of nodes: Symbol nodes, check nodes, and state nodes. The introduction of the factor graph greatly increases the kinds of graphs and dependency relationships by which a code can be described.

Figure 11.10 shows an example of a factor graph. This factor graph corresponds to the linear code described by the Tanner graph of Figure 11.8. The factor graph has eight state nodes, and each state node represents one column of states in Figure 11.6. The factor graph has a one-dimensional structure and can be thought of as lying along the time axis. Each state node is represented by an open circle. The size of the open circle depends on the number of states. The checks represent the constraints governing a particular trellis section; only certain pairs of neighboring states are allowed, and each pair that is allowed uniquely determines an associated code symbol.

The factor graph that corresponds to a trellis diagram is a *cycle-free graph.* Every linear block code can be represented by a cycle-free factor graph. This need not be the simplest cycle-free representation. For example, the Tanner graph of Figure 11.8, for most purposes, is a simpler cycle-free graph representation of the (7, 4, 2) code than the trellis representation shown in Figure 11.6. Furthermore, it may be that a graph with cycles is a yet simpler representation of a code.

The factor graph for the tail-biting representation of the Golay code described in Section 11.11 is shown in Figure 11.11. Each symbol node in the figure represents four bits, and each state has sixteen possible values.

11.7 Posterior probabilities

It is often convenient to work with the log-posterior function instead of the log-likelihood function. Because $P(\boldsymbol{c}|\boldsymbol{v}) \leq 1$, the negative log of the posterior, which is

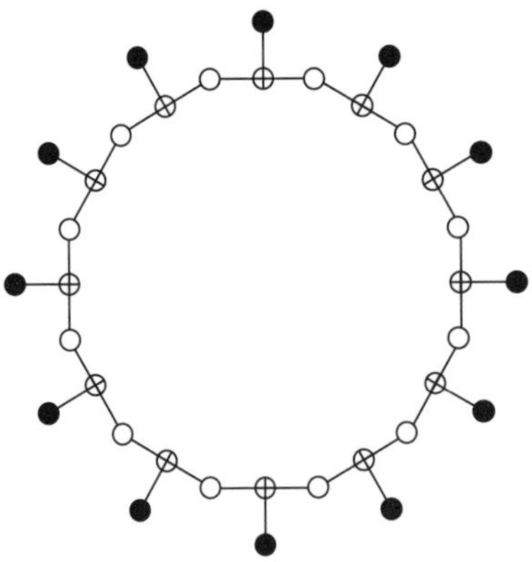

Figure 11.11. Factor graph for a tail-biting Golay code

positive, is preferred. The negative log-posterior function is

$$-\log P(\boldsymbol{c}|\boldsymbol{v}) = \log q(\boldsymbol{v}) - \log p(\boldsymbol{c}) - \log Q(\boldsymbol{v}|\boldsymbol{c}).$$

A maximum-posterior decoder is now written as

$$\widehat{c} = \arg\min_{c \in C} [-\log P(\boldsymbol{c}|\boldsymbol{v})].$$

A maximum-posterior decoder minimizes the message error rate. However, a maximum-posterior block decoder realized by exhaustive search is not tractable.

An alternative approach is to minimize the error probability of a single symbol of the codeword. This gives a smaller symbol error rate, but can give a higher block error rate. The approach requires that the posterior probability $P(c|v)$ be marginalized to the ith symbol of the dataword. This process of marginalization is defined as

$$P(a_i|\boldsymbol{v}) = \sum_{a_0} \cdots \sum_{a_{i-1}} \sum_{a_{i+1}} \cdots \sum_{a_{k-1}} P((a_0, \ldots, a_{k-1}))|\boldsymbol{v})$$

where the componentwise sums are over all possible values of that component. We will abbreviate this as

$$P(a_\ell|v) = \sum_{\boldsymbol{a} \backslash a_\ell} P(\boldsymbol{a}|\boldsymbol{v}).$$

The notion $\boldsymbol{a} \backslash a_\ell$ denotes the vector $\boldsymbol{a}$ punctured by a_ℓ:

$$\boldsymbol{a} \backslash a_\ell = (a_0, \ldots, a_{\ell-1}, a_{\ell+1} \cdots a_{k-1}).$$

Thus,

$$(\boldsymbol{a} \backslash a_\ell, a_\ell) = \boldsymbol{a}$$

The maximum-posterior decoder at the symbol level, which is given by

$$\widehat{a}_\ell = \max_{a_\ell} P(a_\ell|\boldsymbol{v})$$

for $\ell = 0, \ldots, k-1$ then minimizes the symbol probability of error. Because of the marginalization, the maximum-posterior symbol decoder appears to require formidable computations. The two-way algorithm, which is discussed in Section 11.8, organizes the computation in an efficient form, but the work is still substantial.

Definition 11.7.1. *The Fano distance is given by*

$$d_F(\boldsymbol{c}, \boldsymbol{v}) = -\log P(\boldsymbol{c}|\boldsymbol{v}).$$

When the channel is memoryless, the posterior probability is

$$P(\boldsymbol{c}|\boldsymbol{v}) = \frac{\prod_{i=1}^{n} Q(v_i|c_i)}{\prod_{i=1}^{n} q(v_i)} p(\boldsymbol{c}).$$

If codewords are used with equal probability, the prior probability $p(\boldsymbol{c})$ is a constant if $\boldsymbol{c}$ is a codeword, given by

$$p(\boldsymbol{c}) = \frac{1}{M} = e^{nR},$$

and is zero if $\boldsymbol{c}$ is not a codeword. The Fano metric is

$$d_F(\boldsymbol{c}, \boldsymbol{v}) = \left[\sum_{i=1}^{n} \log \frac{Q(v_i|c_i)}{q(v_i)} - R\right].$$

This is an appropriate metric to use for a soft-input decoder. However, the denominator inside the logarithm and the parameter R are constants and can be dropped without affecting which codeword achieves the maximum as long as all codewords in a comparison have the same number of frames N.

11.8 The two-way algorithm

The *two-way algorithm* is an algorithm for computing posterior probabilities on the branches of a finite trellis. Given the soft senseword $\boldsymbol{v}$, the task is to compute, for each branch, the posterior probability that that branch was traversed. The posterior probability of the entire codeword $\boldsymbol{c}$ is $P(\boldsymbol{c}|\boldsymbol{v})$. The posterior probability of a particular branch, denoted b_ℓ, is the marginalization of this posterior by summing out all but branch b_ℓ. This is denoted by the expression

$$\Pr(b_\ell|\boldsymbol{v}) = \sum_{s_0}^{s_{\ell-1}} \sum_{s_\ell}^{s_{n-1}} P(\boldsymbol{b}|\boldsymbol{v})$$

where $s_{\ell-1}$ and s_ℓ are the states at the beginning and end of branch b_ℓ. Recall that the first sum from state s_0 to state $s_{\ell-1}$ specifies a sum over all forward paths beginning at state s_0 and ending at state $s_{\ell-1}$, and the second sum from state s_ℓ to state s_{n-1} specifies a sum over all backward paths beginning at state s_{n-1} and ending at state s_ℓ.

The branch posterior, once computed, can be turned into a hard decision, or can be used as a soft decision, thereby playing the role of a reliability measure on that branch. To form a hard decision, simply choose the symbols on that branch with the largest branch posterior. If desired, this procedure can be modified to declare an erasure whenever no branch posterior is sufficiently large. In other applications, such as a turbo decoder, the branch posterior is computed as an inner step within a larger process. A symbol posterior formed during one step is used as a symbol prior during a subsequent step. The computation of $\Pr(b_\ell|\boldsymbol{v})$ must be done for every branch. Because there are so many branches, an efficient organization of the computations is essential. The two-way algorithm performs two sets of partial computations: one partial computation starts at the beginning of the trellis and works forward until it reaches the end; the other starts at the end of the trellis and works backwards until it reaches the beginning. The two arrays of partial computations are then combined to form the vector of posteriors.

The two-way algorithm is a straightforward computational procedure, in iterative form, for certain decoding problems for finite, cycle-free factor graphs with symbols from a finite alphabet and state nodes from a finite alphabet. The two-way algorithm works directly on the nodes and branches of the trellis representing the code as shown in Figure 11.12. Here, the set of nodes on the left represents the symbols or states; the set of nodes on the right represents the constraints. A symbol or state node is connected to a constraint node in the graph if and only if the corresponding variable participates in the corresponding constraint.

For a systolic implementation of the decoding algorithm, with one processor at each node, information is exchanged along the graph by passing messages from each node to its neighboring nodes. Such a systolic decoding algorithm is called a *message passing algorithm*. For the codeword component that is associated with the branch over which the message is being transmitted, the messages can be interpreted as conveying an estimate of the value of that variable along with some reliability information for that estimate.

Implicit in the symbol posterior probability distribution in such a message is a hard decision. The most likely value for the symbol is the value with the largest posterior probability. The hard decision may perhaps be an erasure if no component of the posterior probability is above a threshold.

A path through a trellis defines a codeword

$$\boldsymbol{c} = (c_0, c_1, \ldots, c_{n-1})$$

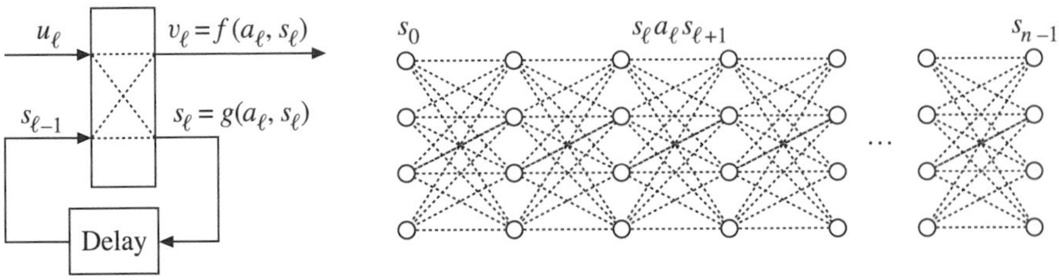

Figure 11.12. State diagram for the two-way algorithm

ℓ	$\Pr(a_\ell \mid v_\ell)$	α_ℓ	β_ℓ
		↓	↑

Figure 11.13. A scratch pad for the two-way algorithm

where c_ℓ is the branch label on the branch taken in the ℓth frame. If the same value can appear on more than one branch in the ℓth frame, then the symbol c_ℓ does not identify the unique branch from which it came. It is only within the context of the entire codeword that the branch can be identified. This means that we must allow the symbol c_ℓ to be augmented with some additional tags that identify the branch from which it came.

For the usual case that we consider, a branch is uniquely determined by the states at time $\ell - 1$ and time ℓ, and by the input data symbol a_ℓ at time ℓ. Therefore it will be convenient to denote a branch as the symbol b_ℓ where $b_\ell = s_\ell a_\ell s_{\ell+1}$, as illustrated in Figure 11.13. We define the precodeword as the string

$$\boldsymbol{b} = (b_0, b_1, \ldots, b_{n-1}).$$

Then c_ℓ is a function of b_ℓ. The reason for this elaboration is that the notation b_ℓ now retains all available information.

The two-way algorithm is a "sum–product" algorithm, meaning that the central computation is in the form of a sum of products. To reduce the computational burden, these calculations are sometimes approximated in the form of "min-sum" by taking the logarithm of each term in the sum, then keeping only the smallest.

The posterior probability of precodeword $\boldsymbol{b}$ is given by $\Pr(\boldsymbol{b} \mid \boldsymbol{v})$. The posterior probability of a single branch b_ℓ is obtained by summation over all paths containing that branch. These are the paths that start at state s_0, pass through states $s_{\ell-1}$ and s_ℓ, and end at state s_{n-1}. The posterior probability of branch b_ℓ is

$$\begin{aligned}\Pr(b_\ell \mid \boldsymbol{v}) &= \sum_{s_0}^{s_{\ell-1}} \sum_{s_\ell}^{s_{n-1}} P(\boldsymbol{b} \mid \boldsymbol{v}) \\ &= \sum_{s_0}^{s_{\ell-1}} \sum_{s_\ell}^{s_{n-1}} \frac{Q(\boldsymbol{v} \mid \boldsymbol{c}) p(\boldsymbol{c})}{q(\boldsymbol{v})}.\end{aligned}$$

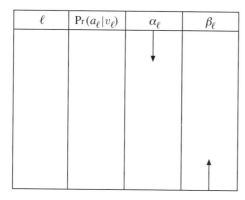

The first summation from state s_0 to state $s_{\ell-1}$ denotes the sum over all forward paths that connect state s_0 to state $s_{\ell-1}$. The second summation from state s_ℓ to state s_{n-1} denotes the sum over all backward paths from state s_{n-1} to state s_ℓ.

For a memoryless channel, $Q(\boldsymbol{v}|\boldsymbol{c})$ is a product distribution; the branch posterior becomes

$$\Pr(b_\ell|\boldsymbol{v}) = \frac{1}{q(\boldsymbol{v})}\sum_{s_0}^{s_{\ell-1}}\prod_{i=1}^{\ell-1} Q(v_i|c_i)p(c_i) + \frac{1}{q(\boldsymbol{v})}\sum_{s_\ell}^{s_{n-1}}\prod_{i=1}^{\ell-1} Q(v_i|c_i)p(c_i).$$

This is a probability distribution on the set of values that b_ℓ can take. This probability must be computed for every branch of a stage and for every stage of the trellis. To organize the computational work into an efficient algorithm, the two-way algorithm computes the two terms in two separate recursions. Thus, let

$$\alpha_\ell = \sum_{s_0}^{s_{\ell-1}}\prod_{i=1}^{\ell-1} Q(v_i|c_i)p(c_i)$$

and

$$\beta_\ell = \sum_{s_{n-1}}^{s_0}\prod_{i=1}^{\ell-1} Q(v_i|c_i)p(c_i).$$

The first term is computed for all ℓ, starting with $\ell = 1$ and increasing ℓ by one at each step. The second term is computed for all ℓ starting with $\ell = n - 1$ and decreasing ℓ by one at each step. Hence the name "two-way algorithm."

A scratch-pad memory for the computations is shown in Figure 11.13. The column labeled α_ℓ is filled in from the top down; the column labeled β_ℓ is filled in from the bottom up. When these two columns are complete, the posterior probability is quickly computed by combining the two columns.

Theorem 11.8.1. *The two-way algorithm converges if the Tanner graph is cycle free.*

Proof: This is straightforward. □

11.9 Iterative decoding of turbo codes

The factor graph of a turbo code, shown in Figure 11.14, gives important insight into the code structure. Each of the component codes is cycle free, but the combination of two component codes introduces other interconnections and this leads to cycles. If the permutation is long, however, then we expect that most of the cycles become very long though there may be a few short cycles. The short cycles only cause a few local errors, and the long cycles have no effect if the number of iterations is not too large. The performance of turbo codes can be attributed to the cycles in the structure of the factor graph.

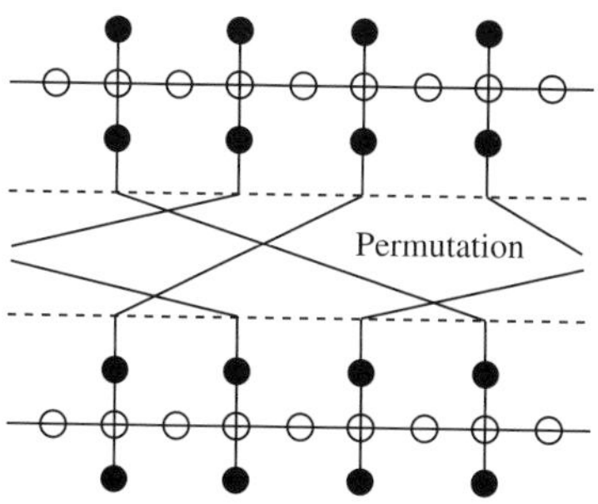

Figure 11.14. Factor graph for a turbo code

Turbo codes are specifically defined in a way that anticipates the use of iterative decoding. A turbo code can be regarded as the combination of two terminated convolutional codes. This combination is referred to as parallel concatenation. Either of the two convolutional codes can be decoded using any posterior algorithm, the two-way algorithm being the favorite – especially when a prior probability is available.

During each pass, the prior probability on one dataword is converted to a posterior probability on that dataword. This posterior becomes the prior probability for decoding the second codeword, again producing a posterior probability. In this way, the two codewords are decoded alternately. Each time the posterior computed at the output of one pass becomes the prior at the input to the other pass. A flow diagram for iterative decoding is shown in Figure 11.15. Convergence of iterated decoding is demonstrated by simulation.

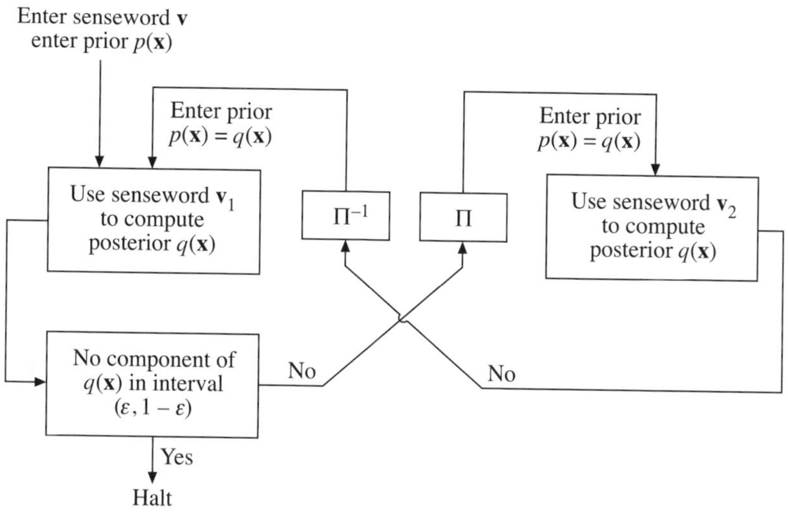

Figure 11.15. Flow diagram for iterative decoding

11.10 Tail-biting representations of block codes

As we have seen in Section 11.4, a linear block code has a constraint length and this constraint length determines the complexity of decoding that code by a graph-searching algorithm. In this section we will look at a technique for embedding a linear block code into a larger block code that has a smaller constraint length. Because it has a smaller constraint length, the larger block code is easier to decode. This suggests that a senseword for the original code can be decoded by somehow using a decoder for the larger code, then making some appropriate adjustments in the answer.

Sometimes a linear block code can be described by a special form of trellis called a *tail-biting trellis*. This is a trellis drawn on a cylinder, as in Figure 11.16, instead of in the plane. The time axis of a tail-biting trellis is a circle rather than an infinite line or a finite line segment.

We need a generator matrix in a special form that is defined in terms of its diagonals and off-diagonals. For this reason, we will require that n/k is a small integer, denoted b. A k by n generator matrix $\boldsymbol{G}$ can be regarded as a square k by k matrix of b-tuples. For example, the eight by four generator matrix for the $(8, 4, 4)$ Hamming code

$$\boldsymbol{G} = \begin{bmatrix} 1 & 0 & 0 & 1 & 1 & 0 & 1 & 0 \\ 0 & 1 & 0 & 1 & 0 & 1 & 1 & 0 \\ 0 & 0 & 1 & 0 & 1 & 1 & 1 & 0 \\ 1 & 1 & 1 & 1 & 1 & 1 & 1 & 1 \end{bmatrix}$$

can be rewritten as the four by four matrix of two-tuples

$$\boldsymbol{G} = \begin{bmatrix} 10 & 01 & 10 & 10 \\ 01 & 01 & 01 & 10 \\ 00 & 10 & 11 & 10 \\ 11 & 11 & 11 & 11 \end{bmatrix}$$

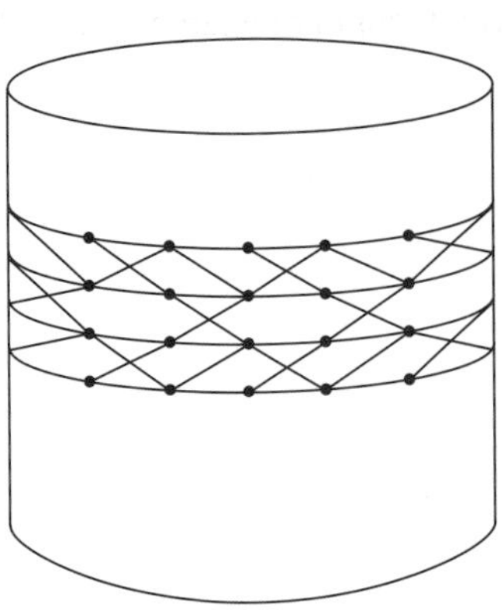

Figure 11.16. A tail-biting trellis

in which each entry is a pair of bits. With this structure, the diagonal of $\boldsymbol{G}$ is the ordered set of pairs of bits $\{10, 01, 11, 11\}$.

In general, the diagonal of a k by bk generator matrix $\boldsymbol{G}$ is the set of b-tuples lying along the diagonal of the matrix $\boldsymbol{G}$ regarded as a k by k matrix of b-tuples. An *off-diagonal* of the array $\{g_{ij}\}$, regarded as a k by k array of b-tuples, is defined in a similar way as the set of elements $\{g_{i,i+\ell} | i = 0, \ldots, k-1\}$ where the index $i + \ell$ is interpreted modulo k. We will be especially interested in the set of elements $\{g_{i,i+\ell} | i = k - \ell, \ldots, k-1\}$, which will be called the *tail* of the off-diagonal.

Definition 11.10.1. *A tail-biting generator matrix of tail-biting constraint length ν is a matrix in which all nonzero terms are confined to the diagonal and the first ν extended upper off-diagonals. The tail-biting constraint length of a linear code $\mathcal{C}$ is the smallest tail-biting constraint length of any generator matrix for that code.*

For example, a tail-biting generator matrix for the (8, 4, 4) Hamming code is

$$\boldsymbol{G} = \begin{bmatrix} 11 & 11 & 00 & 00 \\ 00 & 11 & 01 & 10 \\ 00 & 00 & 11 & 11 \\ 01 & 10 & 00 & 11 \end{bmatrix}.$$

There are two nonzero upper off-diagonals in this matrix. So its tail-biting constraint length is $\nu = 2$. The tail of the generator matrix is the triangular piece of the matrix in the bottom left of $\boldsymbol{G}$. The tail is

$$\begin{bmatrix} 00 & \\ 01 & 10 \end{bmatrix}.$$

The tail in this example happens to have an all-zero element in the first row.

A conventional trellis for matrix $\boldsymbol{G}$ must have constraint length at least equal to 3 because we have already asserted that the (8, 4, 4) Hamming code has a constraint length of 3. The tail-biting constraint length of the (8, 4, 4) Hamming code is 2 because no generator matrix for the code has a smaller tail-biting constraint length. We shall see that the tail-biting constraint length is the constraint length of a larger code into which we can embed this Hamming code.

Definition 11.10.2. *An upper triangular generator matrix of state complexity σ is a matrix in which all nonzero terms are confined to the diagonal and the first σ upper off-diagonals.*

The distinction between this definition and Definition 11.10.1 is that the definition of state complexity does not allow the off-diagonals to be extended. We will, however, allow a nonsquare matrix with the diagonal understood in the natural way.

Now, we are ready to define the two new codes, $\mathcal{C}'$ and $\mathcal{C}''$. The method is to add more rows to $\boldsymbol{G}$ and to move the tail into the new rows, or to add more columns to $\boldsymbol{G}$ and to move the tail into the new columns. In this way, the tail-biting generator matrix

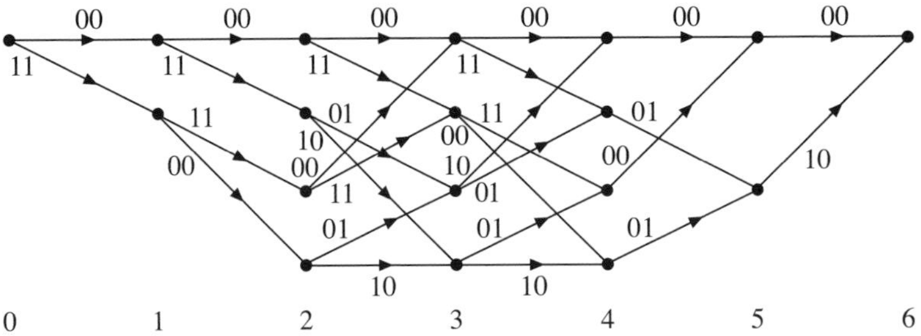

Figure 11.17. Trellis for a (12, 4, 4) binary code

for $\mathcal{C}$ is unfolded to form either a k by $(k+\nu)$ upper-triangular generator matrix $\boldsymbol{G}'$ or a $k+\nu$ by bk generator matrix $\boldsymbol{G}''$ for a linear block code. A similar method can be used to form a generator matrix for a convolutional code, as we shall discuss later.

For the tail-biting generator matrix for the (8, 4, 4) Hamming code, the 4 by 6 upper-triangular generator matrix of two-tuples, which generates a (12, 4, 4) code $\mathcal{C}$, is

$$\boldsymbol{G}' = \begin{bmatrix} 11 & 11 & 00 & 00 & 00 & 00 \\ 00 & 11 & 01 & 10 & 00 & 00 \\ 00 & 00 & 11 & 11 & 00 & 00 \\ 00 & 00 & 00 & 11 & 01 & 10 \end{bmatrix}.$$

This matrix has all elements equal to zero in the third and subsequent off-diagonals. The codewords of the Hamming codeword are easily formed from $\mathcal{C}'$ by adding the last four components of each codeword onto the first four components of that codeword.

A trellis for the (12, 4, 4) code $\mathcal{C}'$, shown in Figure 11.17, is easily constructed. Because the constraint length is 2, the trellis has four states. There are six frames, corresponding to the number of columns in $\boldsymbol{G}'$, and all paths begin and end on the zero node. These conditions determine a trellis and the rows of $\boldsymbol{G}'$ determine the labeling along paths of the trellis. Some states in frame 2 and frame 4 are redundant and can be merged to produce the simpler trellis shown in Figure 11.18. Finally, by regarding the frame index modulo 4, the ends can be overlapped and the overlapped nodes combined to produce the simpler tail-biting trellis for the Hamming code shown in Figure 11.19. The number of states in an overlapped frame is the

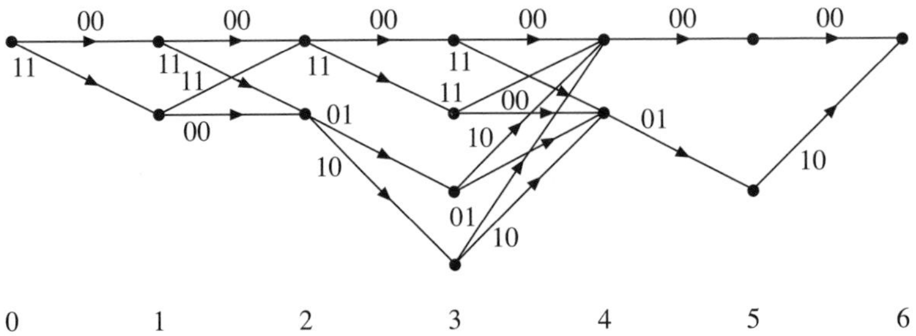

Figure 11.18. A better trellis for the (12, 4, 4) binary code

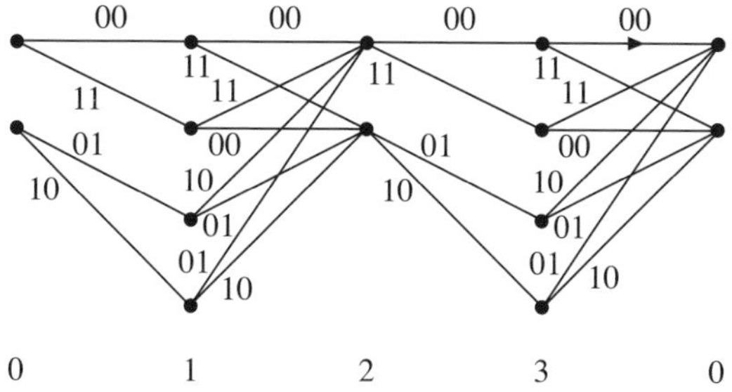

Figure 11.19. Tail-biting trellis for the (8, 4, 4) Hamming code

product of the number of states in each overlapping frame. There are two states at time zero in Figure 11.19, which is the product of the number of states at time zero and time four in Figure 11.18. For a similar reason, there are four states at time one in Figure 11.19.

The nodes at the beginning and the end of the tail-biting trellis are regarded as the same nodes because the trellis is regarded as wrapped around a cylinder. A *proper path* through a tail-biting trellis is a path that returns to itself after one pass around the cylinder. An improper path through a tail-biting trellis returns to itself after two or more passes around the cylinder.

The codewords of the Hamming code can be read from all paths that begin and end on the same node. Although the trellis has constraint length 2, three bits of memory are needed because the structure starting node must be remembered. This bit, however, plays a simple role. It does not directly contribute to the complexity of the Viterbi algorithm.

The set of codewords, proper and improper, forms a linear code $\mathcal{C}''$, which is formed by the generator matrix

$$\boldsymbol{G}'' = \begin{bmatrix} 01 & 10 & 00 & 00 \\ 11 & 11 & 00 & 00 \\ 00 & 11 & 01 & 10 \\ 00 & 00 & 11 & 11 \\ 00 & 00 & 00 & 11 \end{bmatrix}.$$

This matrix is obtained from $\boldsymbol{G}$ by relocating the nonzero portion of the tail to a single new row at the top of the matrix. An improper path corresponds to the concatenation of two or more codewords from $\mathcal{C}''$, but not $\mathcal{C}$. Clearly, the minimum distance of $\mathcal{C}'$ can provide a bound.

Theorem 11.10.3 (Square-Root Bound). *The constraint length of a tail-biting trellis for a linear block code, $\mathcal{C}$, satisfies*

$$k \;\geq \nu \geq \tfrac{1}{2}\log_q \sigma$$

where σ is the state complexity of $\mathcal{C}$.

Proof: The inequality $\nu \leq k$ is trivial, so we only need to prove that $2\nu \geq \log_q \sigma$. Suppose that the constraint length of a tail-biting generator matrix for $\mathcal{C}$ is ν. Then the tail of $\boldsymbol{G}$ occupies at most ν rows. But for any ℓ, at most ν other rows are nonzero, so $\nu_\ell \leq 2\nu$ for all ℓ. Therefore $\nu \geq \log_q \sqrt{\sigma}$. This completes the proof of the theorem. □

The rows of the tail-biting generator matrix for the Hamming code can also be used to form a generator matrix for a convolutional code simply by repeating the pattern of rows indefinitely. This gives a convolutional code with the generator matrix

$$\boldsymbol{G} = \begin{bmatrix} 01 & 10 & 00 & 00 & 00 & 00 & 00 & \dots \\ 11 & 11 & 00 & 00 & 00 & 00 & 00 & \\ 00 & 11 & 01 & 10 & 00 & 00 & 00 & \\ 00 & 00 & 11 & 11 & 00 & 00 & 00 & \\ 00 & 00 & 00 & 11 & 01 & 10 & 00 & \dots \\ \vdots & & & & \dots & & & \end{bmatrix}.$$

The free distance of this convolutional code is four. It has a trellis consisting of continued repetition of the trellis shown in Figure 11.19. This convolutional code is more general than the convolutional codes studied in Chapter 9 because the branch structure is different on alternate frames. This is called a time-varying convolutional code.

11.11 The Golay code as a tail-biting code

Each ordering of the twenty-four coordinates of the (24, 12, 8) binary Golay code gives an ordered code that can be described by a conventional trellis. Choose any set of rows of any generator matrix $\boldsymbol{G}$ of the Golay code. Only two of the twelve generators of the code can both have their support restricted to the same twelve of the coordinates because the largest value of k for which a $(12, k, 8)$ code exists is 2. In particular, for any ordering of the coordinates, at least ten of the generators must be supported by components in the second twelve coordinates. Similarly, not more than two of the generators can be supported only by the second twelve coordinates. Consequently, at least eight of the twelve generators must span the midpoint. This implies that the minimal trellis for the binary Golay code must have at least $2^8 = 256$ states at the midpoint. In contrast, it will be shown that a tail-biting trellis for the binary Golay code with sixteen states does exist.

A binary generator matrix for the Golay code with not more than two of the twelve generators supported by any set of twelve coordinates would be a matrix, if one exists, of the following form:

$$
G = \begin{bmatrix}
\ast\ast & \ast\ast & \ast\ast & \ast\ast & \ast\ast & 00 & 00 & 00 & 00 & 00 & 00 & 00 \\
00 & \ast\ast & \ast\ast & \ast\ast & \ast\ast & \ast\ast & 00 & 00 & 00 & 00 & 00 & 00 \\
00 & 00 & \ast\ast & \ast\ast & \ast\ast & \ast\ast & \ast\ast & 00 & 00 & 00 & 00 & 00 \\
00 & 00 & 00 & \ast\ast & \ast\ast & \ast\ast & \ast\ast & \ast\ast & 00 & 00 & 00 & 00 \\
00 & 00 & 00 & 00 & \ast\ast & \ast\ast & \ast\ast & \ast\ast & \ast\ast & 00 & 00 & 00 \\
00 & 00 & 00 & 00 & 00 & \ast\ast & \ast\ast & \ast\ast & \ast\ast & \ast\ast & 00 & 00 \\
00 & 00 & 00 & 00 & 00 & 00 & \ast\ast & \ast\ast & \ast\ast & \ast\ast & \ast\ast & 00 \\
00 & 00 & 00 & 00 & 00 & 00 & 00 & \ast\ast & \ast\ast & \ast\ast & \ast\ast & \ast\ast \\
\ast\ast & 00 & 00 & 00 & 00 & 00 & 00 & 00 & \ast\ast & \ast\ast & \ast\ast & \ast\ast \\
\ast\ast & \ast\ast & 00 & 00 & 00 & 00 & 00 & 00 & 00 & \ast\ast & \ast\ast & \ast\ast \\
\ast\ast & \ast\ast & \ast\ast & 00 & 00 & 00 & 00 & 00 & 00 & 00 & \ast\ast & \ast\ast \\
\ast\ast & \ast\ast & \ast\ast & \ast\ast & 00 & 00 & 00 & 00 & 00 & 00 & 00 & \ast\ast
\end{bmatrix},
$$

where each matrix element denoted $*$ is either a zero or a one.

To view this matrix as a k by k matrix, regard its entries as pairs of bits. Then the matrix can be described in terms of its diagonal structure. The diagonal is the set $\{g_{ii} \mid i = 0, \ldots, n-1\}$ where the matrix element g_{ii} is now a pair of bits. Furthermore, for each ℓ, we refer to the set of elements $\{g_{i(i+\ell)} \mid i = 0, \ldots, k-1\}$ as the ℓth continued off-diagonal of matrix $\boldsymbol{G}$.

To find a tail-biting generator matrix for the Golay code is not simple. One must, in principle, examine the 24! possible coordinate orderings, and for each ordering, attempt to construct a basis in this form. In fact a generator matrix of this form does exist. The following 12 by 24 tail-biting generator matrix

$$
G = \begin{bmatrix}
11 & 01 & 11 & 01 & 11 & 00 & 00 & 00 & 00 & 00 & 00 & 00 \\
00 & 11 & 11 & 10 & 01 & 11 & 00 & 00 & 00 & 00 & 00 & 00 \\
00 & 00 & 11 & 01 & 10 & 11 & 11 & 00 & 00 & 00 & 00 & 00 \\
00 & 00 & 00 & 11 & 01 & 11 & 01 & 11 & 00 & 00 & 00 & 00 \\
00 & 00 & 00 & 00 & 11 & 01 & 11 & 01 & 11 & 00 & 00 & 00 \\
00 & 00 & 00 & 00 & 00 & 11 & 11 & 10 & 01 & 11 & 00 & 00 \\
00 & 00 & 00 & 00 & 00 & 00 & 11 & 01 & 10 & 11 & 11 & 00 \\
00 & 00 & 00 & 00 & 00 & 00 & 00 & 11 & 01 & 11 & 01 & 11 \\
11 & 00 & 00 & 00 & 00 & 00 & 00 & 00 & 11 & 01 & 11 & 01 \\
01 & 11 & 00 & 00 & 00 & 00 & 00 & 00 & 00 & 11 & 11 & 10 \\
10 & 11 & 11 & 00 & 00 & 00 & 00 & 00 & 00 & 00 & 11 & 01 \\
01 & 11 & 01 & 11 & 00 & 00 & 00 & 00 & 00 & 00 & 00 & 11
\end{bmatrix}
$$

has the right form and does generate a (24, 12) code. Moreover, the minimum distance of this code is 8, so this matrix is indeed the tail-biting generator matrix for a (24, 12, 8) code that is equivalent to the Golay code.

Notice that the rows of $\boldsymbol{G}$ come in groups of four. That is, let

$$\boldsymbol{g} = \begin{bmatrix} 11 & 01 & 11 & 01 & 11 & 00 & 00 & 00 \\ 00 & 11 & 11 & 10 & 01 & 11 & 00 & 00 \\ 00 & 00 & 11 & 01 & 10 & 11 & 11 & 00 \\ 00 & 00 & 00 & 11 & 01 & 11 & 01 & 11 \end{bmatrix} \quad \text{and} \quad \boldsymbol{0} = \begin{bmatrix} 00 & 00 & 00 & 00 \\ 00 & 00 & 00 & 00 \\ 00 & 00 & 00 & 00 \\ 00 & 00 & 00 & 00 \end{bmatrix}.$$

Then, with $\boldsymbol{g}$ partitioned as $\boldsymbol{g} = [\boldsymbol{g}_{\mathrm{L}} \ \ \boldsymbol{g}_{\mathrm{R}}]$, the matrix can be written in the quasi-cyclic form

$$\boldsymbol{G} = \begin{bmatrix} \boldsymbol{g}_{\mathrm{L}} & \boldsymbol{g}_{\mathrm{R}} & \boldsymbol{0} \\ \boldsymbol{0} & \boldsymbol{g}_{\mathrm{L}} & \boldsymbol{g}_{\mathrm{R}} \\ \boldsymbol{g}_{\mathrm{R}} & \boldsymbol{0} & \boldsymbol{g}_{\mathrm{L}} \end{bmatrix}.$$

A cyclic shift of $\boldsymbol{c}$ by eight bits forms another codeword, which is the codeword that would be formed by cyclically shifting the dataword by four bits before encoding.

The tail-biting generator matrix for the Golay code can be unwrapped to form an infinite recurring generator matrix for a convolutional code. This code is called the *Golay convolutional code*. It is a time-varying convolutional code with exceptional properties. The Golay convolutional code has a generator matrix formed by the indefinite repetition of the four rows of $\boldsymbol{G}$ to give

$$\boldsymbol{G} = \begin{bmatrix} \boldsymbol{g}_{\mathrm{R}} & \boldsymbol{0} & \boldsymbol{0} & \boldsymbol{0} & \boldsymbol{0} & \boldsymbol{0} & \boldsymbol{0} & \cdots \\ \boldsymbol{g}_{\mathrm{L}} & \boldsymbol{g}_{\mathrm{R}} & \boldsymbol{0} & \boldsymbol{0} & \boldsymbol{0} & \boldsymbol{0} & \boldsymbol{0} & \cdots \\ \boldsymbol{0} & \boldsymbol{g}_{\mathrm{L}} & \boldsymbol{g}_{\mathrm{R}} & \boldsymbol{0} & \boldsymbol{0} & \boldsymbol{0} & \boldsymbol{0} & \cdots \\ \boldsymbol{0} & \boldsymbol{0} & \boldsymbol{g}_{\mathrm{L}} & \boldsymbol{g}_{\mathrm{R}} & \boldsymbol{0} & \boldsymbol{0} & \boldsymbol{0} & \cdots \\ \boldsymbol{0} & \boldsymbol{0} & \boldsymbol{0} & \boldsymbol{g}_{\mathrm{L}} & \boldsymbol{g}_{\mathrm{R}} & \boldsymbol{0} & \boldsymbol{0} & \cdots \\ & & & & & & \cdots & \end{bmatrix}.$$

This time-varying convolutional code has $d_{\mathrm{min}} = 8$. Moreover, every codeword has a weight divisible by four. The best time-invariant binary convolutional code of constraint length 4 and rate $1/2$ has $d_{\mathrm{min}} = 7$.

One method of decoding a tail-biting code on a trellis is to initialize all values of a starting state to zero, and then to execute the Viterbi algorithm for a few circuits around the loop, at which point a proper path will usually emerge as a stable solution. It can be observed empirically that such an approach works almost as well as minimum euclidean-distance decoding, and it is considerably simpler than passing a single Viterbi algorithm through the larger 256-state trellis. Occasionally, however, the stable solution will be an improper path, which means that the use of a tail-biting trellis has resulted in a decoding failure.

Every k by n tail-biting generator matrix can be made into a $k + \nu$ by n generator matrix by appending ν additional rows at the top, and moving the nonzero entries at the start of the bottom rows to the start of the new rows. For example, one can use the tail-biting generator matrix to specify the generator matrix for a (24, 16, 2) linear block code by "unwrapping" the last four rows. This gives the following larger generator matrix for the augmented code:

$$\boldsymbol{G}' = \begin{bmatrix}
11 & 00 & 00 & 00 & 00 & 00 & 00 & 00 & 00 & 00 & 00 & 00 \\
01 & 11 & 00 & 00 & 00 & 00 & 00 & 00 & 00 & 00 & 00 & 00 \\
10 & 11 & 11 & 00 & 00 & 00 & 00 & 00 & 00 & 00 & 00 & 00 \\
01 & 11 & 01 & 11 & 00 & 00 & 00 & 00 & 00 & 00 & 00 & 00 \\
11 & 01 & 11 & 01 & 11 & 00 & 00 & 00 & 00 & 00 & 00 & 00 \\
00 & 11 & 11 & 10 & 01 & 11 & 00 & 00 & 00 & 00 & 00 & 00 \\
00 & 00 & 11 & 01 & 10 & 11 & 11 & 10 & 00 & 00 & 00 & 00 \\
00 & 00 & 00 & 11 & 01 & 11 & 01 & 11 & 00 & 00 & 00 & 00 \\
00 & 00 & 00 & 00 & 11 & 01 & 11 & 01 & 11 & 00 & 00 & 00 \\
00 & 00 & 00 & 00 & 00 & 11 & 11 & 10 & 01 & 11 & 00 & 00 \\
00 & 00 & 00 & 00 & 00 & 00 & 11 & 01 & 10 & 11 & 11 & 00 \\
00 & 00 & 00 & 00 & 00 & 00 & 00 & 11 & 01 & 11 & 01 & 11 \\
00 & 00 & 00 & 00 & 00 & 00 & 00 & 00 & 11 & 01 & 11 & 01 \\
00 & 00 & 00 & 00 & 00 & 00 & 00 & 00 & 00 & 11 & 11 & 10 \\
00 & 00 & 00 & 00 & 00 & 00 & 00 & 00 & 00 & 00 & 11 & 01 \\
00 & 00 & 00 & 00 & 00 & 00 & 00 & 00 & 00 & 00 & 00 & 11
\end{bmatrix}.$$

When written in block-partitioned form, this matrix is

$$\boldsymbol{G}' = \begin{bmatrix}
g_{\rm R} & \boldsymbol{0} & \boldsymbol{0} \\
g_{\rm L} & g_{\rm R} & g_{\rm R} \\
\boldsymbol{0} & g_{\rm L} & g_{\rm R} \\
\boldsymbol{0} & \boldsymbol{0} & g_{\rm L}
\end{bmatrix}.$$

The matrix $\boldsymbol{G}'$ generates a (24, 16, 2) code $\mathcal{C}'$ that contains the Golay code $\mathcal{C}$. Only four rows of $\boldsymbol{G}'$ span the midpoint, so a trellis for the code $\mathcal{C}'$ has only $2^4 = 16$ states. This code $\mathcal{C}'$ has a much simpler trellis representation than does the Golay code. Nevertheless, this code contains the Golay code. To find the Golay code within this set of 2^{16} codewords, take those 2^{12} values of $\boldsymbol{G}'\boldsymbol{a}$ in which the last four components of $\boldsymbol{a}$ are equal to the first four components of $\boldsymbol{a}$. This expurgated code is the Golay code.

The sixteen-state trellis for $\mathcal{C}'$ can be wrapped around a cylinder to form a tail-biting trellis for the Golay code. The last four frames must be overlapped with the first four frames and the overlapped states merged.

This example of the Golay code shows that a conventional trellis for a block code can have many more states than a tail-biting trellis for that code. Although the trellis for $\mathcal{C}'$ has only sixteen states, the process of expurgation imposes sixteen more states needed to remember the first four data bits. This means that the trellis for $\mathcal{C}$ has 16^2 states.

Problems

11.1 Prove that if $\boldsymbol{G} = [g_{ij}]$ is a k by k tail-biting generator matrix, then

$$G' = \left[g_{((i+1))((j+1))}\right]$$

is a k by k tail-biting generator matrix where the double parentheses denote modulo k on the indices.

11.2 Prove that every nontrivial tail-biting trellis has paths that are not periodic. Prove that a minimum-distance decoder must choose a periodic path.

11.3 Show that for any ℓ less than $n-k$, an $(n-k)$ by n check matrix always exists with $n-k=r$, $n=\binom{r}{\ell}$, and ℓ ones in each column.

11.4 What is the smallest value of n for which there exists a Gallager code with three ones in each column and the number of ones in each row at most $\lceil \log_2 n \rceil$?

11.5 Give an expression for the Fano metric for the case where the difference between the dataword and the senseword is gaussian-noise, independent, and identically distributed from sample to sample.

11.6 Suppose that a turbo code is defined by using a nonsystematic generator matrix of the form

$$\boldsymbol{G}(x) = [[g_2(x)] \;\; [g_1(x)] \;\; \boldsymbol{\Pi}[g_1(x)]].$$

Can iterative decoding be used for this code? Is there any reason in the definition of turbo codes to restrict the generator matrix to be systematic?

11.7 Given that the (2, 1) convolutional code with generator polynomials

$$g_1(x) = x^2 + x + 1$$
$$g_2(x) = x^2 + 1$$

is used on a binary channel, suppose the codeword

1101011011100...

is transmitted. If the channel makes no errors, but loses the first bit because of a synchronization error, the senseword is

1010110111000....

Work through the steps of the Viterbi algorithm for this senseword. Give an ad hoc method of embellishing the Viterbi decoder to both synchronize and correct an incoming datastream.

11.8 An encoder for a (2, 1) convolutional code has a generator matrix

$$\boldsymbol{G}(x) = [1 \quad 1 + x].$$

The channel has a binary output with a probability transition matrix

$$Q_{k|j} = \begin{bmatrix} 0.96 & 0.04 \\ 0.04 & 0.96 \end{bmatrix}.$$

Use the two-way algorithm to decode the senseword 10 01 11 11 11 11.

11.9 A binary convolutional code has a polynomial generator matrix

$$G(x) = [x^3 + x^2 + 1 \quad x^3 + x^2 + x + 1].$$

Trace through the steps of the Viterbi algorithm if the channel makes two errors, one in the first channel bit and one in the third.

11.10 A maximum-posterior decoder involves the repeated computation of the expression

$$z^2 = -\log\left(e^{-x^2} + e^{-y^2}\right),$$

so a fast approximation can be very useful. Prove that

$$-\log\left(e^{-x^2} + e^{-y^2}\right) = \max(x^2, y^2) - \log\left(1 + e^{|x^2 - y^2|}\right).$$

What can be said about the required accuracy of computing the second term?

11.11 Prove that the notions of past and future of a trellis can be interchanged.

11.12 Construct a trellis similar to that of Figure 11.6 by starting with the generator matrix

$$\boldsymbol{G} = \begin{bmatrix} 1 & 0 & 1 & 1 & 0 & 0 & 1 \\ 0 & 1 & 1 & 0 & 0 & 0 & 0 \\ 0 & 0 & 0 & 1 & 1 & 0 & 0 \\ 0 & 0 & 0 & 0 & 0 & 1 & 1 \end{bmatrix}.$$

11.13 Prove the *Wolf bound*: That the minimal trellis for a linear (n, k) code has a state-complexity profile whose components satisfy $\sigma_\ell \leq q^{n-k}$.

11.14 Show that the following simple trellis can be constructed from the minimal trellis for the (8,4,4) Hamming code.

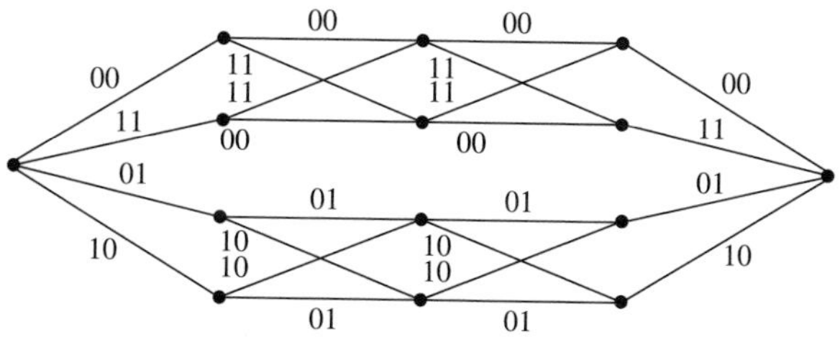

Why does this trellis not contradict the statement that the constraint length of the (8,4,4) Hamming code is 3? Give a trellis for the Hamming code that has only one node in each frame.

Notes

Sequential decoding was introduced by Wozencraft (1957), and described further by Wozencraft and Reiffen (1961). The sequential trellis-searching algorithm that we have described originated with Fano (1963), and was reformulated by Gallager (1968). An alternative sequential algorithm using a stack was developed by Zigangirov (1966), and later by Jelinek (1969). Tutorials by Jelinek (1968) and Forney (1974), and further developments by Chevillat and Costello (1978) as well as by Haccoun and Ferguson (1975), also advanced the subject. Jacobs and Berlekamp (1967) showed that the computational delay of any sequential decoding algorithm is a random variable described by a Pareto probability density function. The two-way algorithm failed for many years to be popularized as a general computational algorithm and so was rediscovered each time that it was needed, both in coding theory and in computer science, and has many names. Notable instances in the coding theory literature were by Gallager (1962) and Bahl, Cocke, Jelinek, and Raviv (1974). An alternative name for the two-way algorithm in the BCJR algorithm.

Viterbi (1967) published his algorithm more as a pedagogical device than as a serious algorithm. Shortly thereafter, Heller (1968) pointed out that the Viterbi algorithm is quite practical if the constraint length is not too large, and Forney (1967, 1973) showed that this algorithm achieves optimality. Hagenauer and Hoeher (1989) overlaid the Viterbi algorithm with a method of annotating each output with a reliability indication. Hence, the Hagenauer–Hoeher decoder takes a soft senseword at its input and converts it to a soft senseword at its output.

The Gallager codes (1962) are now seen as an early example of graph codes. Gallager also introduced the main graph iterative decoding algorithms. Gallager demonstrated reliable decoding of his codes over a noisy binary symmetric channel at rates above a value called the cutoff rate R_0 of the channel. Gallager codes can exceed the performance of turbocodes on the additive white gaussian-noise channel when using a sufficient number of iterations of the two-way algorithm.

Solomon and van Tilborg (1979) recognized the relationship between block and convolutional codes and the special role played by the tail-biting trellis. Ma and Wolf (1988) discussed the tail-biting trellis for another reason, namely to make a convolutional code into a block code. Muder (1988) first observed that every trellis for the Golay codes has at least 256 states. Calderbank, Forney, and Vardy (1999), responding to a challenge put forward by Forney (1996), discovered the tail-biting representation of the Golay code. The Golay convolutional code was listed by Lee (1989) among other results of his exhaustive search, though without then noting the connection to the Golay code.

12 Performance of Error-Control Codes

In any engineering discipline, one looks for methods of solving a given class of problems and then for methods of quantifying the performance and complexity of those solutions. Accordingly, in this chapter, we shall study performance in terms of the composition structure of block codes, and the probabilities of decoding error and of decoding failure.

Later, after satisfactory methods are in hand, one turns to questions of optimality. Are these methods the best methods, and if not, in what ways and by how much are they inferior? To answer such questions, one needs to know how good are the known codes and how good are the best possible codes. Generally, we cannot answer either of these questions satisfactorily. Although for any given rate and blocklength, not too small, the best possible code is not known, a number of bounds are known – bounds beyond which no codes can exist, and bounds within which codes are sure to exist. We shall give some such bounds on the best possible codes later in this chapter. Knowledge of these bounds can deepen our understanding of the subject.

12.1 Weight distributions of block codes

If a linear block code has a minimum distance $d_{\min}$, then we know that at least one codeword of weight $d_{\min}$ exists. Sometimes, we are not content with this single piece of information; we wish to know how many codewords have weight $d_{\min}$, and what are the weights of the other codewords. For example, in Table 5.2, we gave a list of code weights for the (23, 12) binary Golay code. For a small code, it is possible to find such a table of all the weights through an exhaustive search. For a large code, this is not feasible. Instead, one hopes to find analytical techniques. Because even the minimum distance is unknown for many codes, it is clear that, in general, such techniques will be difficult to find.

Let A_ℓ denote the number of codewords of weight ℓ in an (n, k) linear code. The $(n+1)$-dimensional vector, with components A_ℓ for $\ell = 0, \ldots, n$, is called the *weight distribution* of the code. Obviously, if the minimum distance is $d_{\min} = d$, then $A_0 = 1$, $A_1, \ldots, A_{d-1}$ are all zero, and A_d is not zero. To say more than this requires further work.

The weight distribution is unknown for most codes, but for the important case of the Reed–Solomon codes (or any maximum-distance code), an analytical expression is known. This section will give a formula for the number of codewords of each weight in a maximum-distance code. For an arbitrary linear code, we will not be able to give such a formula, but we can give some information. The *MacWilliams identity* relates the weight distribution of a code to the weight distribution of its dual code. This identity is useful if the dual code is small enough for its weight distribution to be found by computer search. First, we will give the exact solution for maximum-distance codes.

Theorem 12.1.1. *In a maximum-distance code, any arbitrary assignment of values to k places uniquely determines a codeword.*

Proof: A code of minimum distance $d_{\min}$ can fill any $d_{\min} - 1$ erasures. Because $d_{\min} = n - k + 1$ for a maximum-distance code, the theorem is proved. □

It is clear from the proof of the theorem that if the code is not a maximum-distance code, then it is not true that any set of k places may be used as data places. This statement applies to all nontrivial binary codes because no binary code is a maximum-distance code except repetition codes and parity-check codes.

For a maximum-distance code, we easily can compute the number of codewords of weight $d_{\min}$. Choose any set of d components of the code, where $d = d_{\min}$. There are $\binom{n}{d}$ ways of choosing these d components. Consider all the codewords that are zero in the $n - d$ other components. Any set of $k = n - d + 1$ components of a maximum-distance codeword uniquely determines that codeword and $n - d$ components have been set to zero. One additional component can be specified. This additional component can take on any of q values, and then the values of the other $d - 1$ components are determined by the structure of the code. Hence there are exactly q codewords for which a given set of $n - d$ places is zero. One of these is the all-zero codeword, and the remaining $q - 1$ are of weight d. Because the $n - d$ zero components can be chosen in $\binom{n}{d}$ ways, we have

$$A_d = \binom{n}{d}(q-1).$$

To find A_ℓ for $\ell > d = d_{\min}$, we use a similar, but considerably more difficult, argument. This is done in proving the following theorem.

Theorem 12.1.2. *The weight distribution of a maximum-distance (n, k) code over $GF(q)$ is given by $A_0 = 1$; $A_\ell = 0$ for $\ell = 1, \ldots, d_{\min} - 1$; and for $\ell \geq d$,*

$$A_\ell = \binom{n}{\ell}(q-1)\sum_{j=0}^{\ell-d}(-1)^j\binom{\ell-1}{j}q^{\ell-d-j}.$$

Proof: The theorem is obvious for $\ell < d$. The proof of the theorem for $\ell \geq d$ is divided into three steps.

Step 1. Partition the set of integers from 0 to $n-1$ into two sets, $\mathcal{T}_\ell$ and $\mathcal{T}_\ell^c$, with $\mathcal{T}_\ell$ having ℓ integers; consider only codewords that are equal to zero in those components indexed by integers in $\mathcal{T}_\ell^c$ and are equal to zero nowhere else. Let M_ℓ be the number of such codewords of weight ℓ. For the total code,

$$A_\ell = \binom{n}{\ell} M_\ell,$$

and thus we only need to prove that

$$M_\ell = (q-1)\sum_{j=0}^{\ell-d}(-1)^j\binom{\ell-1}{j}q^{\ell-d-j}.$$

This will be proved by developing an implicit relationship between M_ℓ and $M_{\ell'}$ for ℓ' less than ℓ and ℓ greater than d.

Any set of $n-d+1$ codeword components uniquely determines the codeword. Choose a set of $n-d+1$ codeword components as follows. Choose all of the $n-\ell$ components indexed by integers in $\mathcal{T}_\ell^c$, and any $\ell-d+1$ of the components indexed by integers in $\mathcal{T}_\ell$. Because the components indexed by $\mathcal{T}_\ell^c$ have been set to zero, the latter $\ell-d+1$ components from $\mathcal{T}_\ell$ determine the codeword. Because one of these is the all-zero codeword, this gives $q^{\ell-d+1}-1$ nonzero codewords, all of weight at most ℓ.

From the set of ℓ places indexed by $\mathcal{T}_\ell$, we can choose any subset of ℓ' places. There will be $M_{\ell'}$ codewords of weight ℓ' whose nonzero components are confined to these ℓ' places. Hence

$$\sum_{\ell'=d}^{\ell}\binom{\ell}{\ell'}M_{\ell'} = q^{\ell-d+1}-1.$$

This recursion implicitly gives M_{d+1} in terms of M_d, gives M_{d+2} in terms of M_d and M_{d+1}, and so forth. Next, we will solve the recursion to give an explicit formula for M_ℓ.

Step 2. In this step we will rearrange the equation to be proved in a form that will be more convenient to prove. We treat q as an indeterminate and manipulate the equations as polynomials in q. To this end, define the notation

$$\left\lceil \sum_{n=-N_1}^{N_2} a_n q^n \right\rceil = \sum_{n=0}^{N_2} a_n q^n$$

as an operator that keeps only coefficients of nonnegative powers of q. This is a linear operator. With this convention, the expression to be proved can be written

$$M_\ell = (q-1)\left\lceil q^{-(d-1)}\sum_{j=0}^{\ell-1}(-1)^j\binom{\ell-1}{j}q^{\ell-1-j}\right\rceil.$$

The extra terms included in the sum correspond to negative powers of q and do not contribute to M_ℓ. This can be collapsed by using the binomial theorem:

$$M_\ell = (q-1)\left\lceil q^{-(d-1)}(q-1)^{\ell-1}\right\rceil.$$

Step 3. Finally, by direct substitution, we will show that the expression for M_ℓ, derived in step 2, solves the recursion derived in step 1. Thus:

$$\begin{aligned}
\sum_{\ell'=d}^{\ell}\binom{\ell}{\ell'}M_{\ell'} &= \sum_{\ell'=0}^{\ell}\binom{\ell}{\ell'}M_{\ell'} \\
&= (q-1)\sum_{\ell'=0}^{\ell}\binom{\ell}{\ell'}\left[q^{-(d-1)}(q-1)^{\ell'-1}\right] \\
&= (q-1)\left[q^{-(d-1)}(q-1)^{-1}\sum_{\ell'=0}^{\ell}\binom{\ell}{\ell'}(q-1)^{\ell'}\right] \\
&= (q-1)\left[q^{-d}\left(1-\frac{1}{q}\right)^{-1}q^{\ell}\right] \\
&= (q-1)\left[\sum_{i=0}^{\infty}q^{\ell-d-i}\right] \\
&= (q-1)\sum_{i=0}^{\ell-d}q^{\ell-d-i} \\
&= q^{\ell-d+1}-1,
\end{aligned}$$

as was to be proved. □

Corollary 12.1.3. *The weight distribution of a maximum-distance (n, k) code over $GF(q)$ is given by $A_0 = 1$; $A_\ell = 0$ for $\ell = 1, \ldots, d-1$; and for $\ell \geq d$,*

$$A_\ell = \binom{n}{\ell}\sum_{j=0}^{\ell-d}(-1)^j\binom{\ell}{j}\left(q^{\ell-d+1-j}-1\right).$$

Proof: Use the identity

$$\binom{\ell}{j} = \binom{\ell-1}{j} + \binom{\ell-1}{j-1}$$

to rewrite the equation to be proved as

$$A_\ell = \binom{n}{\ell}\sum_{j=0}^{\ell-d}(-1)^j\left[\binom{\ell-1}{j}+\binom{\ell-1}{j-1}\right]\left(q^{\ell-d+1-j}-1\right).$$

The second term is zero for $j = 0$ and is unaffected by including an extra term equal to zero. Thus

$$A_\ell = \binom{n}{\ell}\left[\sum_{j=0}^{\ell-d}(-1)^j\binom{\ell-1}{j}\left(qq^{\ell-d-j}-1\right) - \sum_{j=1}^{\ell-d+1}(-1)^{j-1}\binom{\ell-1}{j-1}\left(q^{\ell-d+1-j}-1\right)\right]$$

ℓ	A_ℓ
0	1
1–16	0
17	8.22×10^{9}
18	9.59×10^{10}
19	2.62×10^{12}
20	4.67×10^{13}
21	7.64×10^{14}
22	1.07×10^{16}
23	1.30×10^{17}
24	1.34×10^{18}
25	1.17×10^{19}
26	8.37×10^{19}
27	4.81×10^{20}
28	2.13×10^{21}
29	6.83×10^{21}
30	1.41×10^{22}
31	1.41×10^{22}

Figure 12.1. Approximate weight distribution for the (31, 15) Reed–Solomon code

Replace j by i in the first term and replace $j-1$ by i in the second term to get

$$A_\ell = \binom{n}{\ell} \sum_{i=0}^{\ell-d} (-1)^i \binom{\ell-1}{i} (q-1) q^{\ell-d-i},$$

which is known to be true from Theorem 12.1.2. □

Corollary 12.1.3 is useful for calculating the weight distribution of a Reed–Solomon code. For example, the weight distribution of the (31, 15) Reed–Solomon code is shown in Figure 12.1. Even for small Reed–Solomon codes, such as this one, the number of codewords of each weight can be very large. This is why, generally, it is not practical to find the weight distribution of a code by simple enumeration of the codewords.

We have nothing like Theorem 12.1.2 for the case of codes that are not maximum-distance codes. For small n, the weight distribution can be found by a computer search, but for large n, this quickly becomes impractical.

The strongest tool we have is an expression of the relationship between the weight distribution of a linear code and the weight distribution of its dual code – the so-called MacWilliams identity. The MacWilliams identity holds for any linear code and is based on the vector-space structure of linear codes, and on the fact that the dual code of a code $\mathcal{C}$ is the orthogonal complement of $\mathcal{C}$. Before we can derive the MacWilliams identity, we must return to the study of abstract finite-dimensional vector spaces, which was started in Section 2.5. We need to introduce the ideas of the intersection and direct sum of two subspaces and prove some properties.

Let U and V be subspaces of F^n, with dimensions denoted $\dim[U]$ and $\dim[V]$. Then $U \cap V$, called the *intersection* of U and V, denotes the set of vectors that are in both U and V; and $U \oplus V$, called the *direct sum* of U and V, denotes the set of all linear combinations $a\boldsymbol{u} + b\boldsymbol{v}$ where $\boldsymbol{u}$ and $\boldsymbol{v}$ are in U and V, respectively, and a and b are scalars. Both $U \cap V$ and $U \oplus V$ are subspaces of F^n, with dimension denoted $\dim[U \cap V]$ and $\dim[U \oplus V]$.

Theorem 12.1.4.

$$\dim[U \cap V] + \dim[U \oplus V] = \dim[U] + \dim[V].$$

Proof: A basis for $U \cap V$ has $\dim[U \cap V]$ vectors. This basis can be extended to a basis for U by adding $\dim[U] - \dim[U \cap V]$ more basis vectors, and can be extended to a basis for V by adding $\dim[V] - \dim[U \cap V]$ more basis vectors. All of these basis vectors, taken together, form a basis for $U \oplus V$. That is,

$$\dim[U \oplus V] = \dim[U \cap V] + (\dim[U] - \dim[U \cap V]) + (\dim[V] - \dim[U \cap V]),$$

from which the theorem follows. □

Theorem 12.1.5.

$$U^\perp \cap V^\perp = (U \oplus V)^\perp.$$

Proof: The subspace U is contained in $U \oplus V$, so the subspace $(U \oplus V)^\perp$ is contained in $U^\perp$. Similarly, $(U \oplus V)^\perp$ is contained in $V^\perp$. Therefore $(U \oplus V)^\perp$ is contained in $U^\perp \cap V^\perp$. On the other hand, write an element of $U \oplus V$ as $a\boldsymbol{u} + b\boldsymbol{v}$, and let $\boldsymbol{w}$ be any element of $U^\perp \cap V^\perp$. Then $\boldsymbol{w} \cdot (a\boldsymbol{u} + b\boldsymbol{v}) = 0$, and thus $U^\perp \cap V^\perp$ is contained in $(U \oplus V)^\perp$. Hence the two are equal. □

Let A_ℓ for $\ell = 0, \ldots, n$ and B_ℓ for $\ell = 0, \ldots, n$ be the weight distributions of a linear code and its dual code, respectively. Define the weight-distribution polynomials

$$A(x) = \sum_{\ell=0}^{n} A_\ell x^\ell \quad \text{and} \quad B(x) = \sum_{\ell=0}^{n} B_\ell x^\ell.$$

The following theorem relates these two polynomials and allows one to be computed from the other.

Theorem 12.1.6 (MacWilliams Theorem). *The weight-distribution polynomial $A(x)$ of an (n, k) linear code over $GF(q)$ and the weight-distribution polynomial of its dual code are related by*

$$q^k B(x) = [1 + (q-1)x]^n A\left(\frac{1-x}{1+(q-1)x}\right).$$

Proof: Let $\mathcal{C}$ be the code, and let $\mathcal{C}^\perp$ be the dual code. The proof will be in two parts. In part 1, we shall prove that

$$\sum_{i=0}^{n} B_i \binom{n-i}{m} = q^{n-k-m} \sum_{j=0}^{n} A_j \binom{n-j}{n-m}$$

for $m = 0, \ldots, n$. In part 2, we shall prove that this is equivalent to the condition of the theorem.

Part 1. For a given m, partition the integers from 0 to $n-1$ into two subsets, T_m and T_m^{c}, with set T_m having m elements. In the vector space $GF(q)^n$, let V be the m-dimensional subspace consisting of all vectors that have zeros in components indexed by the elements of T_m. Then $V^\perp$ is the $(n-m)$-dimensional subspace consisting of all vectors that have zeros in components indexed by the elements of T_m.

By Theorem 12.1.5,

$$(\mathcal{C} \cap V)^\perp = \mathcal{C}^\perp \oplus V^\perp.$$

Therefore

$$\dim[\mathcal{C}^\perp \oplus V^\perp] = n - \dim[\mathcal{C} \cap V].$$

On the other hand, by Theorem 12.1.4,

$$\dim[\mathcal{C}^\perp \oplus V^\perp] = (n-k) + (n-m) - \dim[\mathcal{C}^\perp \cap V^\perp].$$

Equating these gives

$$\dim[\mathcal{C}^\perp \cap V^\perp] = \dim[\mathcal{C} \cap V] + n - k - m.$$

Now for each choice of T_m, there are $q^{\dim[\mathcal{C} \cap V]}$ vectors in $\mathcal{C} \cap V$, and $q^{\dim[\mathcal{C}^\perp \cap V^\perp]}$ vectors in $\mathcal{C}^\perp \cap V^\perp$. Consider $\{T_m\}$, the collection of all such T_m. Enumerate the vectors in each of the $\mathcal{C} \cap V$ that can be produced from some subset T_m in the collection $\{T_m\}$. There will be $\sum_{\{T_m\}} q^{\dim[\mathcal{C} \cap V]}$ vectors in the enumeration, many of them repeated appearances. Similarly, an enumeration of all vectors in each $\mathcal{C}^\perp \cap V^\perp$, produced from T_m in $\{T_m\}$, is given by

$$\sum_{\{T_m\}} q^{\dim[\mathcal{C}^\perp \cap V^\perp]} = q^{n-k-m} \sum_{\{T_m\}} q^{\dim[\mathcal{C} \cap V]}.$$

To complete part 1 of the proof, we must evaluate the two sums in the equation. We do this by counting how many times a vector $\boldsymbol{v}$ of weight j in $\mathcal{C}$ shows up in a set $\mathcal{C} \cap V$. The vector $\boldsymbol{v}$ is in $\mathcal{C} \cap V$ whenever the j nonzero positions of $\boldsymbol{v}$ lie in the m positions in which vectors in V are nonzero, or equivalently, whenever the $n-m$ positions where vectors in V must be zero lie in the $n-j$ zero positions of the codeword. There are $\binom{n-j}{n-m}$ choices for the $n-m$ zero components, and thus the given codeword of weight

j shows up in $\binom{n-j}{n-m}$ sets. There are A_j codewords of weight j. Therefore

$$\sum_{\{T_m\}} q^{\dim[C\cap V]} = \sum_{j=0}^{n} A_j \binom{n-j}{n-m}.$$

Similarly, we can count the vectors in $C^{\perp} \cap V^{\perp}$. The earlier equation then becomes

$$\sum_{i=0}^{n} B_i \binom{n-i}{m} = q^{n-k-m} \sum_{j=0}^{n} A_j \binom{n-j}{n-m}.$$

Because m is arbitrary, the first part of the proof is complete.

Part 2. Starting with the conclusion of part 1, write the polynomial identity

$$\sum_{m=0}^{n} y^m \sum_{i=0}^{n} B_i \binom{n-i}{m} = \sum_{m=0}^{n} y^m q^{n-k-m} \sum_{j=0}^{n} A_j \binom{n-j}{n-m}.$$

Interchange the order of the summations

$$\sum_{i=0}^{n} B_i \sum_{m=0}^{n} \binom{n-i}{m} y^m = q^{n-k} \sum_{j=0}^{n} A_j \sum_{m=0}^{n} \binom{n-j}{n-m} \left(\frac{y}{q}\right)^n \left(\frac{q}{y}\right)^{n-m},$$

recalling that $\binom{n-i}{m} = 0$ if $m > n-i$. Using the binomial theorem, this becomes

$$\sum_{i=0}^{n} B_i (1+y)^{n-i} = q^{n-k} \sum_{j=0}^{n} A_j \left(\frac{y}{q}\right)^n \left(1+\frac{q}{y}\right)^{n-j}.$$

Finally, make the substitution $y = (1/x) - 1$, to get

$$q^k x^{-n} \sum_{i=0}^{n} B_i x^i = q^n \sum_{j=0}^{n} A_j \left(\frac{1-x}{xq}\right)^n \left(\frac{1+(q-1)x}{1-x}\right)^{n-j},$$

or

$$q^k \sum_{i=0}^{n} B_i x^i = (1+(q-1)x)^n \sum_{j=0}^{n} A_j \left(\frac{1-x}{1+(q-1)x}\right)^j,$$

which completes the proof of the theorem. □

We close this section with a simple application of this theorem. From Table 1.1, we see that the weight distribution of the (7, 4) Hamming code is given by

$$(A_0, A_1, \ldots, A_7) = (1, 0, 0, 7, 7, 0, 0, 1),$$

and thus its weight-distribution polynomial is

$$A(x) = x^7 + 7x^4 + 7x^3 + 1.$$

The dual code is the binary cyclic code known as the *simplex code*. Its generator polynomial

$$g(x) = x^4 + x^3 + x^2 + 1$$

has zeros at α^0 and α^1. By Theorem 12.1.6, the weight-distribution polynomial $B(x)$ of the simplex code is given by

$$\begin{aligned} 2^4 B(x) &= (1+x)^7 A\left(\frac{1-x}{1-x}\right) \\ &= (1-x)^7 + 7(1+x)^3(1-x)^4 + 7(1+x)^4(1-x)^3 + (1+x)^7. \end{aligned}$$

This reduces to

$$B(x) = 7x^4 + 1.$$

The (7, 3) simplex code has one codeword of weight 0 and seven codewords of weight 4.

12.2 Performance of block codes

An incomplete decoder for a t-error-correcting code that corrects to the packing radius corrects all error patterns of weight t or less, and corrects no error patterns of weight larger than t. When more than t errors occur, the decoder sometimes declares that it has an uncorrectable message and sometimes makes a decoding error. Hence the decoder output can be the correct message, an incorrect message, or a decoding failure (an erased message). Figure 12.2 illustrates the three regions into which the senseword can fall. Each codeword has a sphere of radius t drawn around it. This sphere encompasses all sensewords that will be decoded into that codeword. Between the decoding spheres lie many other sensewords that do not lie within distance t of any codeword, and so are not decoded. The probability of correct decoding is the probability that the senseword is in the central circle. The probability of incorrect decoding is the probability that the senseword is in the shaded region. The probability of decoding default is the probability that the senseword lies in the white region. The sum of these three probabilities equals

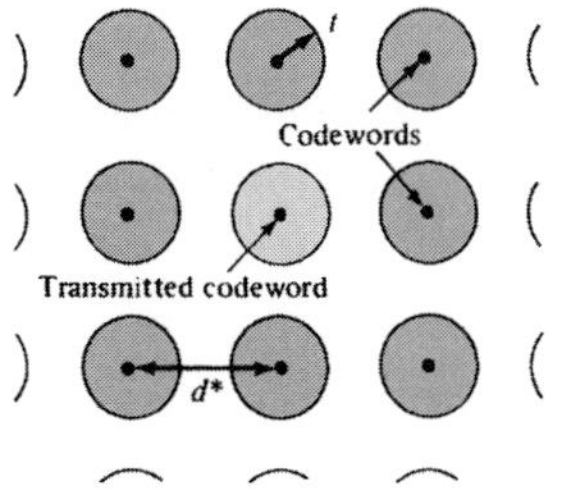

Figure 12.2. Decoding regions

one, and thus only formulas for two of them are needed. In general, it is not known how to compute the probabilities of these events, but for some special cases of practical interest, satisfactory formulas are known.

We shall study the case of linear codes used on channels that make symbol errors independently and symmetrically. The probability expressions are in terms of the weight distribution $\{A_\ell\}$ of the code, and thus are useful only when the weight distribution is known. In the previous section we saw that the weight distribution is known for all Reed–Solomon codes; thus for these codes, we can compute the probability of decoding error and the probability of decoding failure.

The channels we will consider are q-ary channels that make independent errors with probability P in each component and transmit correctly with probability $1 - P$. Each of the $q - 1$ wrong symbols occurs with probability $P/(q - 1)$. Each pattern of k errors has a probability

$$p(k) = \left(\frac{P}{q-1}\right)(1 - P)^{n-k}$$

of occurring. A bounded-distance decoder will decode every senseword to the closest codeword, provided that it is within distance t of the codeword, where t is a fixed number that satisfies

$$2t + 1 \leq d.$$

For a linear code, we can analyze the decoder performance when the all-zero word is transmitted. Every other codeword will have the same conditional probabilities, and thus the conditional probabilities are also the unconditional probabilities. First, let us dispense with the easy part of the work.

Theorem 12.2.1. *A decoder that corrects to the packing radius of a code has a probability of correct decoding given by*

$$p_c = \sum_{\nu=0}^{t} \binom{n}{\nu} P^\nu (1 - P)^{n-\nu}.$$

Proof: There are $\binom{n}{\nu}$ ways that the ν places with errors can be selected: Each occurs with probability $P^\nu(1 - P)^{n-\nu}$. The theorem follows. □

Although the theorem holds for any channel alphabet size, only the probability of making an error enters the formula. It does not matter how this probability of symbol error is broken down among the individual symbols. In the next theorem, dealing with decoding failure, it is necessary to count the ways in which errors can be made. Figure 12.3 illustrates some of the error patterns whose probability must be summed.

Let $N(\ell, h; s)$ be the number of error patterns of weight h that are at distance s from a codeword of weight ℓ. Clearly, $N(\ell, h; s)$ is the same for every codeword of weight ℓ.

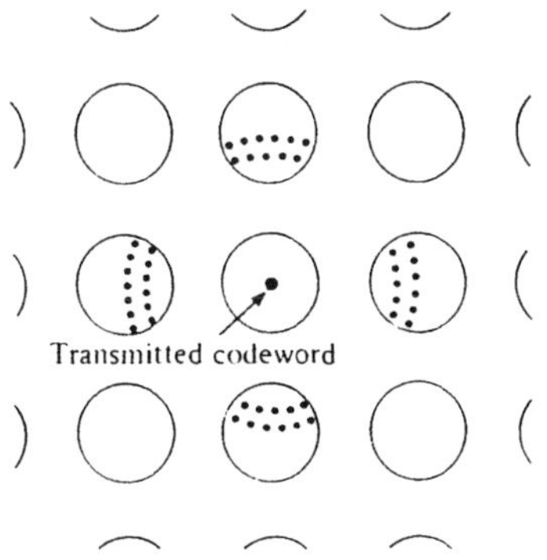

Figure 12.3. Some words causing a decoding error

Notice that if $N(\ell, h; s) \neq 0$, then $\ell - s \leq h \leq \ell + s$. In the following theorem, this is taken care of automatically by using the convention $\binom{n}{m} = 0$ if m is negative or larger than n.

Theorem 12.2.2. *The number of error patterns of weight h that are at distance s from a particular codeword of weight ℓ is*

$$N(\ell, h; s) = \sum_{\substack{0 \leq i \leq n \\ 0 \leq j \leq n \\ i+2j+h=s+\ell}} \binom{n-\ell}{j+h-\ell} \binom{\ell}{i} \binom{\ell-i}{j} (q-1)^{j+h-\ell} (q-2)^i.$$

Proof: An equivalent expression is

$$N(\ell, h; s) = \sum_{\substack{i,j,k \\ i+j+k=s \\ \ell+k-j=h}} \left[\binom{n-\ell}{k}(q-1)^k\right] \left[\binom{\ell}{i}(q-2)^i\right] \left[\binom{\ell-i}{j}\right],$$

which can be verified with the substitution $k = j + h - \ell$. This form has three summation indices and two constraints. The summand is the number of words that can be obtained by changing any k of the $n - \ell$ zero components of the codewords to any of the $q - 1$ nonzero elements, any i of the nonzero elements to any of the other $q - 2$ nonzero elements, and any j of the remaining nonzero elements to zeros. The constraint $i + j + k = s$ ensures that the resulting word is at distance s from the codeword. The constraint $\ell + k - j = h$ ensures that the resulting word has a weight of h. □

Theorem 12.2.3. *If $2t + 1 \leq d_{\min}$, and the decoding is limited to t apparent errors, the probability of decoding error is*

$$p_e = \sum_{h=0}^{n} \left(\frac{P}{q-1}\right)^h (1-P)^{n-h} \sum_{s=0}^{t} \sum_{\ell=1}^{n} A_\ell N(\ell, h; s).$$

Proof: The number of error words of weight h that will be decoded is $\sum_{s=0}^{t} \sum_{\ell=1}^{n} A_\ell N(\ell, h; s)$ where no point has been counted twice because $2t + 1 \leq d_{\min}$.

Each such word occurs with probability $[P/(q-1)]^h(1-P)^{n-h}$, and the theorem follows. □

Theorems 12.2.2, 12.2.3, and 12.1.2 (or Corollary 12.1.3) contain all of the equations that are necessary to compute the error probability for a Reed–Solomon code. Evaluating the error probability by using a computer program is straightforward. These equations are easily modified for an errors-and-erasures decoder. For ρ erasures, simply replace n with $n-\rho$, and replace p_{e} with the conditional error probability $p_{\mathrm{e}|\rho}$. Then if the probability distribution on ρ, $Q(\rho)$, is known, p_{e} can be obtained as $p_{\mathrm{e}} = \sum_{\rho} p_{\mathrm{e}|\rho} Q(\rho)$.

12.3 Bounds on minimum distance of block codes

An error-control code of blocklength n and rate R is judged by its minimum distance $d_{\min}$, which often is the most meaningful and most accessible property of the weight distribution. If we have a code $\mathcal{C}$, we would like to know if $d_{\min}$ is as large as it can be for any code of this n and R. Generally, unless n is quite small, we have no satisfactory answer to this question. Except when the blocklength is very small, only coarse bounds are known. We have encountered the Singleton bound previously, which is a very coarse bound when n is much greater than q, as well as the Hamming bound, which also is a coarse bound except for some special cases of small blocklength.

In this section, we shall develop the *Varshamov–Gilbert bound*, which asserts that long codes of good performance do exist. For any $\delta \in [0, 0.5]$ and sufficiently large blocklength n, there is a binary code of blocklength n and minimum distance δn whose rate R is arbitrarily close to $1 - H_{\mathrm{b}}(\delta)$.

Thus, rate and fractional minimum distance are both bounded away from zero as n goes to infinity. This is quite different from the behavior of binary BCH codes, but consistent with the behavior of the Justesen codes. The derivation of the Varshamov–Gilbert bound, however, is nonconstructive; it gives little insight into how to find such codes.

Recall Theorem 3.5.3, which states that the volume of a Hamming sphere of radius $d-1$ satisfies the bound

$$V_{d-1} = \sum_{\ell=0}^{d-1} \binom{n}{\ell} \leq 2^{nH_{\mathrm{b}}\left(\frac{d-1}{n}\right)}$$

where $H_{\mathrm{b}}(x)$ is the binary entropy function.

Theorem 12.3.1 (Binary Gilbert Bound). *For any integers n and d not larger than $n/2$, there exists a binary code of blocklength n and minimum distance at least d with* *rate R, satisfying*

$$R \geq 1 - H_b\left(\frac{d-1}{n}\right).$$

Proof: For any binary code not necessarily linear, of blocklength n, with a minimum distance at least d and M codewords, consider a Hamming sphere of radius $d-1$ centered at each codeword. If the union of these spheres does not contain 2^n points, then some point in the space is at a distance at least d from every codeword and can be appended to the code, thereby enlarging the code. Hence, for a specified d and n, there exists a code with M codewords and $d_{\min} \geq d$ such that $MV \geq 2^n$. But $M = 2^{nR}$, so by Theorem 3.5.3, this becomes

$$2^{nR} 2^{nH_b\left(\frac{d-1}{n}\right)} \geq 2^n,$$

from which the theorem follows. □

The bound in Theorem 12.3.1 can be strengthened in several ways. A similar bound exists for nonbinary codes and holds even for the restricted class of linear codes. We shall give two lower bounds on the minimum distance of nonbinary, linear codes of finite blocklength; these are the Gilbert bound and the Varshamov bound. Asymptotically, each of the two bounds leads to the Varshamov–Gilbert bound.

Recall that the number of points in a Hamming sphere $\mathcal{S}$ of radius $d-1$ in $GF(q)^n$ is

$$V_{d-1} = \sum_{\ell=0}^{d-1} \binom{n}{\ell} (q-1)^\ell.$$

Theorem 12.3.2 (Gilbert Bound for Linear Codes). *For any integers n and d, such that $2 \leq d \leq n/2$, a q-ary (n, k) linear code exists with $d_{\min} \geq d$ whose dimension k satisfies*

$$\sum_{\ell=0}^{d-1} \binom{n}{\ell} (q-1)^\ell \geq q^{n-k}$$

Proof: There are q^k codewords and q^k Hamming spheres of radius $d-1$ around codewords, each with V_{d-1} points. Given any linear code such that

$$\sum_{\ell=0}^{d-1} \binom{n}{\ell} (q-1)^\ell < q^{n-k},$$

then the number of points in the union of all Hamming spheres of radius $d-1$ centered on codewords satisfies

$$q^k \sum_{\ell=0}^{d-1} \binom{n}{\ell} (q-1)^\ell < q^n$$

so there must be a point $\boldsymbol{v}$ that is not in any Hamming sphere. But then, for any codeword $\boldsymbol{c}$ and for any scalar β, the vector $\boldsymbol{c} + \beta\boldsymbol{v}$ is in the same coset as $\boldsymbol{v}$ and also cannot be in any Hamming sphere because if it were, then $\boldsymbol{v}$ would be in such a Hamming sphere around another codeword. Then the vector subspace spanned by $\mathcal{C}$ and $\boldsymbol{v}$ is a linear code larger than $\mathcal{C}$ and with minimum distance at least d. Thus any linear code that does not satisfy the condition of the theorem can be made into a larger linear code by appending another basis vector. □

For example, let $q = 2$ and $n = 10$. Then $2^5 < V_2 = 56 < 2^6$. Upon taking $d = 3$, Theorem 12.3.2 guarantees the existence of a binary $(10, k)$ code with $k \geq 5$ and $d_{\min} \geq 3$. This linear code has at least $2^5 = 32$ codewords.

The Varshamov bound is an alternative bound. The bound is suggested by Corollary 3.2.3, which says that $d_{\min}$ is equal to the size of the smallest set of columns of a check matrix $\boldsymbol{H}$ that forms a linearly dependent set. The Varshamov construction fixes the redundancy $r = n - k$ of the dual code rather than fixing the blocklength n.

Theorem 12.3.3 (Varshamov Bound for Linear Codes). *For any integers r and d, such that $3 \leq d \leq r + 1$, a q-ary $(k + r, k)$ linear code exists with $d_{\min} \geq d$ whose dimension k satisfies*

$$\sum_{\ell=0}^{d-2} \binom{k+r}{\ell} (q-1)^{\ell} \geq q^r.$$

Proof: There are q^{r-1} nonzero column vectors of length r. A check matrix $\boldsymbol{H}$ consists of $k + r$ columns from this set. Let $\boldsymbol{H}$ be a $k + r$ by k check matrix with every set of $d - 1$ columns linearly independent. Then $\boldsymbol{H}$ is the check matrix for a $(k + r, k, d)$ linear code over $GF(q)$. Suppose that

$$\sum_{\ell=0}^{d-2} \binom{k+r}{\ell} (q-1)^{\ell} < q^r.$$

Because there are $\binom{k+r}{\ell}(q-1)^{\ell}$ linear combinations of exactly ℓ columns of $\boldsymbol{H}$, the inequality states that at least one column vector exists that is not a linear combination of at most $d - 2$ columns of $\boldsymbol{H}$. Hence this column can be adjoined to $\boldsymbol{H}$, thereby forming a check matrix for a $(k + 1 + r, k + 1, d)$ linear code. Hence whenever the inequality of the theorem is not satisfied, k can be increased. Thus the theorem is proved. □

It appears at a casual glance that one could substitute $n = k + r$ into the inequality of Theorem 12.3.2 to obtain

$$\sum_{\ell=0}^{d-2} \binom{n}{\ell} (q-1)^{\ell} \geq q^{n-k},$$

so, superficially, it appears that the Varshamov bound is stronger than the Gilbert bound. In fact, the bounds are not directly comparable because the Gilbert bound treats the set of linear codes with n fixed, and the Varshamov bound treats the set of linear codes with $n - k$ fixed.

The performance of codes of large blocklength can be studied in terms of asymptotic bounds on the fractional minimum distance in the limit of infinite blocklength. A great deal of detail is discarded in presenting the asymptotics but, even so, the asymptotic behavior is not known; only lower and upper bounds are known.

For each $GF(q)$, define

$$d(n, R) = \max_{\mathcal{C}} d_{\min}(\mathcal{C})$$

where the maximum is over all codes over $GF(q)$ of blocklength n and rate R, and $d_{\min}(\mathcal{C})$ denotes the minimum distance of code $\mathcal{C}$. The function $d(n, R)$ is the largest minimum distance of any code over $GF(q)$ of rate R and blocklength n. Except for very small n, we do not know $d(n, R)$. Next, define

$$d(R) = \lim_{n\to\infty} \frac{1}{n} d(n, R),$$

provided the limit exists. For large enough n, the best block code of rate R has a minimum distance approximately equal to $nd(R)$. Thus if it were known, the function $d(R)$ would give us a way by which to judge error-control codes. Also, the derivation of $d(R)$ might give some clues about the structure of good codes and how to find them.

Although we would like to know the functions $d(n, R)$ and $d(R)$, we do not know them. All that we know are lower bounds and upper bounds. We will study only those bounds that can be derived with moderate effort; in particular, we will derive bounds known as the *Varshamov–Gilbert lower bound* and the *Elias upper bound.*

The Varshamov–Gilbert bound will be stated in terms of the q-ary entropy function

$$H_q(x) = x \log_q(q-1) - x \log_q x - (1-x)\log_q(1-x).$$

Theorem 12.3.4 (Varshamov–Gilbert Bound). *For each R, $d(R) \geq \delta$ for all δ that satisfy*

$$R \geq 1 - H_q(\delta).$$

Proof: Theorem 12.3.2 states that

$$q^{n-k} \leq \sum_{\ell=0}^{d-1} \binom{n}{\ell} (q-1)^{\ell}.$$

Theorem 3.5.3 can be generalized to the inequality

$$\sum_{\ell=0}^{d-1} \binom{n}{\ell} (q-1)^{\ell} \leq q^{nH_q\left(\frac{d-1}{n}\right)}.$$

Combining these gives

$$q^{n-k} \leq q^{nH_q\left(\frac{d-1}{n}\right)},$$

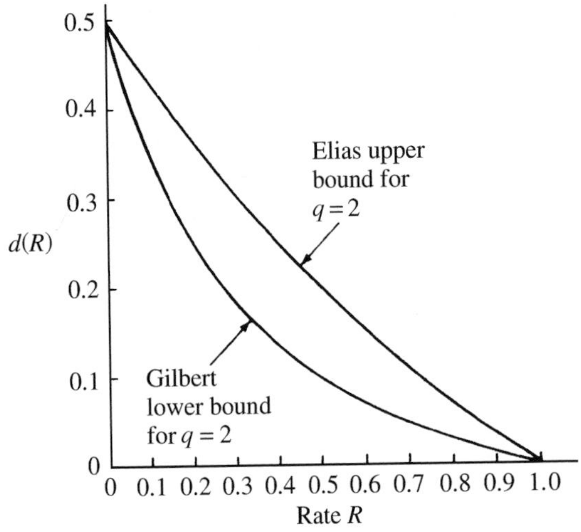

Figure 12.4. Some known bounds on the function $d(R)$

which becomes

$$1 - \frac{k}{n} \leq H_q\left(\frac{d-1}{n}\right),$$

and the theorem follows. □

The Varshamov–Gilbert bound on $d(R)$ for $q = 2$ is shown graphically in Figure 12.4. Also shown is the Elias upper bound, which is derived next. The unknown function $d(R)$, if it exists, must lie between these two bounds.

The Elias bound is derived by studying only those codewords that lie on the surface of some chosen Hamming sphere of radius t in the vector space $GF(q)^n$. Let A_t be the number of vectors on the surface of a Hamming sphere of radius t around any point. Each vector of $GF(q)^n$ has a total of

$$A_t = (q-1)^t \binom{n}{t}$$

vectors at a distance t from it, so there are this many words on the surface of a sphere of radius t.

Proposition 12.3.5. *For any (n, k) linear code C, the average number of codewords on the surface of a sphere around an arbitrary point in the space is*

$$\overline{T}_s = q^k \frac{A_t}{q^n}.$$

Proof: Because a codeword has A_t points at distance t, it lies on the surface of A_t Hamming spheres of radius t. Therefore the number of codewords on a sphere's surface summed over all q^n spheres is $q^k A_t$, and the average number of codewords per sphere surface is $(q^k A_t)/q^n$. □

To derive the Elias bound, we will choose any such sphere with at least the average number of codewords on its surface. It is convenient to visualize the all-zero vector as the center of this sphere, so we will translate the entire space, including the codewords, to place the center of the chosen sphere at the all-zero vector. This has no effect on the minimum distance of the code.

We now have a sphere of radius t centered at the origin, with T codewords on its surface, where T is at least as large as $A_t q^{k-n}$. Later, we can choose the radius t so that T is large enough for our needs. The average distance between two codewords on the sphere's surface is

$$d_{\text{av}} = \frac{1}{T(T-1)} d_{\text{tot}}$$

where d_{tot} is the total distance obtained as the sum of the distances between all unordered pairs of codewords on the surface, not necessarily distinct:

$$d_{\text{tot}} = \sum_{\boldsymbol{c},\boldsymbol{c}'} d(\boldsymbol{c}, \boldsymbol{c}').$$

Define the *joint composition* of codewords $\boldsymbol{c}$ and $\boldsymbol{c}'$ as the q by q array $\{n_{\ell\ell'}^{\boldsymbol{cc}'}\}$ wherein the element $n_{\ell\ell'}^{\boldsymbol{cc}'}$ gives the number of components in which codeword $\boldsymbol{c}$ contains an ℓ and codeword $\boldsymbol{c}'$ contains an ℓ'. Hence

$$d(\boldsymbol{c}, \boldsymbol{c}') = \sum_{\substack{\ell,\ell' \\ \ell \neq \ell'}} n_{\ell\ell'}^{\boldsymbol{cc}'}.$$

Then

$$d_{\text{tot}} = \sum_{\boldsymbol{c},\boldsymbol{c}'} \sum_{\substack{\ell,\ell' \\ \ell \neq \ell'}} n_{\ell\ell'}^{\boldsymbol{cc}'}.$$

Now we can simplify the sum on $\boldsymbol{c}$ and $\boldsymbol{c}'$. Let $n_{\ell\ell'|i}^{\boldsymbol{cc}'}$ equal one if $\boldsymbol{c}$ has an ℓ in the ith place and $\boldsymbol{c}'$ has an ℓ' in the ith place, and otherwise let $n_{\ell\ell'|i}^{\boldsymbol{cc}'}$ equal zero. Then

$$n_{\ell\ell'}^{\boldsymbol{cc}'} = \sum_{i=0}^{n-1} n_{\ell\ell'|i}^{\boldsymbol{cc}'}$$

and

$$\begin{aligned} d_{\text{tot}} &= \sum_{\boldsymbol{c},\boldsymbol{c}'} \sum_{\substack{\ell,\ell' \\ \ell \neq \ell'}} \sum_{i=0}^{n-1} n_{\ell\ell'|i}^{\boldsymbol{cc}'} \\ &= \sum_{i=0}^{n-1} \sum_{\substack{\ell,\ell' \\ \ell \neq \ell'}} \sum_{\boldsymbol{c},\boldsymbol{c}'} n_{\ell\ell'|i}^{\boldsymbol{cc}'}. \end{aligned}$$

Let $T_{\ell|i}$ be the number of codewords on the sphere's surface having an ℓ in the ith place. Then

$$d_{\text{tot}} = \sum_{i=0}^{n-1} \sum_{\substack{\ell,\ell' \\ \ell \neq \ell'}} T_{\ell|i} T_{\ell'|i}.$$

Next, define

$$\lambda_{\ell|i} = \frac{T_{\ell|i}}{T}.$$

This is the fraction of codewords on the sphere's surface with an ℓ in the ith place. The total distance between these codewords is

$$d_{\text{tot}} = T^2 \sum_{i=0}^{n-1} \sum_{\ell,\ell'} \lambda_{\ell|i} \lambda_{\ell'|i} D_{\ell\ell'}$$

where $D_{\ell\ell'}$ equals one if $\ell \neq \ell'$, and otherwise equals zero. We do not know $\lambda_{\ell|i}$, but we do know that it satisfies the following two constraints:

$$\sum_{\ell=0}^{q-1} \lambda_{\ell|i} = 1 \quad \text{and} \quad \sum_{\ell=1}^{q-1} \sum_{i=0}^{n-1} \lambda_{\ell|i} = t.$$

Because we do not know $\lambda_{\ell|i}$, we will choose the worst case for d_{tot} by making the right side as large as possible, consistent with the known constraints.

Lemma 12.3.6. *Let $D_{\ell\ell'}$ equal one if $\ell \neq \ell'$ and otherwise equal zero. If $\boldsymbol{\lambda}$ satisfies the constraints*

$$\sum_{\ell=0}^{q-1} \lambda_{\ell|i} = 1 \quad \textit{and} \quad \sum_{\ell=1}^{q-1} \sum_{i=0}^{n-1} \lambda_{\ell|i} = t,$$

then

$$\sum_{i} \sum_{\ell,\ell'} \lambda_{\ell|i} \lambda_{\ell'|i} D_{\ell\ell'} \leq t\left[2 - \left(\frac{q}{q-1}\right)\frac{t}{n}\right].$$

Proof: First, notice that the constraints imply that

$$\sum_{i=0}^{n-1} \lambda_{0|i} = n - t.$$

Define

$$\lambda^*_{\ell|i} = \begin{cases} 1 - (t/n) & \ell = 0 \\ (t/n)(q-1) & \ell = 1, \ldots, q-1 \end{cases}$$

for all i. Notice that

$$\sum_{\ell=0}^{q-1} \lambda^*_{\ell|i} D_{\ell\ell'} = \begin{cases} t/n & \ell' = 0 \\ 1 - [(t/n)/(q-1)] & \ell' \neq 0. \end{cases}$$

Therefore for all λ that satisfy the constraints

$$\sum_{i=0}^{n-1}\sum_{\ell'=0}^{q-1}\lambda_{\ell'|i}\sum_{\ell=0}^{q-1}\lambda^*_{\ell|i}D_{\ell\ell'} = \sum_{\ell'=1}^{q-1}\sum_{i=0}^{n-1}\lambda_{\ell'|i}\left(1-\frac{t/n}{q-1}\right)+\sum_{i=0}^{n-1}\lambda_{0|i}\frac{t}{n}$$
$$= t\left(1-\frac{t/n}{q-1}\right)+(n-t)\frac{t}{n},$$

including $\boldsymbol{\lambda} = \boldsymbol{\lambda}^*$. Hence by expanding the left side below and cancelling terms, we have

$$\sum_i\sum_{\ell,\ell'}(\lambda_{\ell|i}-\lambda^*_{\ell|i})(\lambda_{\ell'|i}-\lambda^*_{\ell'|i})D_{\ell\ell'} = \sum_i\sum_{\ell,\ell'}(\lambda_{\ell|i}\lambda_{\ell'|i}-\lambda^*_{\ell|i}\lambda^*_{\ell'|i})D_{\ell\ell'}.$$

It suffices to show that the left side of this equation is negative. Define

$$c_{\ell|i} = \lambda_{\ell|i} - \lambda^*_{\ell|i},$$

and notice that $\sum_\ell c_{\ell|i} = 0$. We need to prove that for each i,

$$\sum_{\ell,\ell'} c_{\ell|i}c_{\ell'|i}D_{\ell\ell'} \leq 0.$$

But for each i,

$$\sum_{\ell,\ell'} c_{\ell|i}c_{\ell'|i}D_{\ell\ell'} = \sum_{\ell,\ell'} c_{\ell|i}c_{\ell'|i} - \sum_\ell c^2_{\ell|i} = 0 - \sum_\ell c^2_{\ell|i} \leq 0.$$

Evaluating $F(\boldsymbol{\lambda}^*)$ completes the proof of the lemma. □

Theorem 12.3.7 (Elias Bound). *For any t, let*

$$T = q^{-(n-k)}(q-1)^t\binom{n}{t}.$$

Every code over $GF(q)$ of rate R and blocklength n has a pair of codewords at distance d satisfying

$$d \leq \frac{T}{T-1}t\left[2-\frac{q}{(q-1)}\frac{t}{n}\right],$$

provided $T > 1$.

Proof: Choose t in Proposition 12.3.5 such that $\overline{T}_s$ is greater than one. Then for some sphere, the number of codewords, denoted T, on the surface of the sphere is greater than one. The average distance between these codewords satisfies

$$d_{av} \leq \frac{T}{T-1}\sum_i\sum_{\ell,\ell'}\lambda_{\ell|i}\lambda_{\ell'|i}D_{\ell\ell'}.$$

Finally, to complete the proof, apply Lemma 12.3.6 and notice that some pair of codewords must be at least as close as the average. □

Corollary 12.3.8 (Elias Bound). *For binary codes of rate R, the function $d(R)$ asymptotically satisfies*

$$d(R) \leq 2\rho(1-\rho)$$

for any ρ that satisfies $1 - H_b(\rho) < R$.

Proof: Set $q = 2$ and $t/n = \rho$ in Theorem 12.3.7, and write Theorem 12.3.7 as

$$\frac{d^*}{n} \leq \frac{T}{T-1} 2\rho(1-\rho)$$

and

$$T = 2^{n\delta + o(1)}$$

where

$$\delta = R - (1 - H_b(\rho)).$$

Because $T/(T-1)$ goes to one as n goes to infinity, δ is positive. The corollary follows. □

12.4 Binary expansions of Reed–Solomon codes

The purpose of this section is to show that binary codes that meet the Varshamov–Gilbert bound can be obtained directly from appropriately chosen generalized Reed–Solomon codes. Unfortunately it is not known how to recognize the appropriate generalized Reed–Solomon codes, so the method is not practical.

Any linear (N, K) code over $GF(q^m)$ can be converted into a linear (n, k) code over $GF(q)$ simply by "tilting" each q^m-ary symbol into a sequence of m q-ary symbols, where $n = mN$ and $k = mK$. The $d_{\min}$ nonzero symbols of a minimum-weight codeword will tilt into $md_{\min}$ symbols, but not all of these will be nonzero. The rate of the new code is the same as that of the original code, but the minimum distance is a smaller fraction of blocklength.

This tilting construction gives a simple form of a code that can correct random errors and multiple burst errors. If the original code has a minimum weight of $2t + 1$, then any single burst of q-ary symbols of length $m(t - 1) + 1$ can be corrected, as well as many patterns consisting of multiple shorter bursts. For example, an (N, K) Reed–Solomon code over $GF(2^m)$ can be tilted into an (mN, mK) binary code for correction of multiple burst errors.

Each symbol of an (N, K) Reed–Solomon code over the field $GF(2^m)$ can be represented by m bits, so the Reed–Solomon code can be converted into an (mN, mK) code over $GF(2)$. The minimum distance of the binary (n, k) code is at least as large as the minimum distance $N - K + 1$ of the Reed–Solomon code, so for the binary expansion, we have the bound

$$d_{\min} \geq \frac{n-k}{m} + 1,$$

or

$$\frac{d_{\min}}{n} \geq \frac{1-R}{\log_2 n} + \frac{1}{n}.$$

This is clearly a rather weak bound because the right side goes to zero with n. The reason the bound is weak is because it assumes that each symbol of the minimum-weight Reed–Solomon codeword is replaced by an m-bit binary word with a single one. If, instead, we knew that an appreciable fraction of the bits in most of those binary words were ones, then a much better bound might be stated. However, even this would not be enough because there is no assurance that a minimum-weight Reed–Solomon codeword corresponds to a minimum-weight binary codeword.

To form the binary expansion of a symbol of $GF(2^m)$, one may choose any pair of complementary bases

$$\{\beta_0, \beta_1, \ldots, \beta_{m-1}\}$$
$$\{\gamma_0, \gamma_1, \ldots, \gamma_{m-1}\}$$

for the field $GF(2^m)$. An arbitrary m-bit binary number $\boldsymbol{b} = (b_0, b_1, \ldots, b_{m-1})$ can be mapped into $GF(2^m)$ by

$$\beta = \sum_{i=0}^{m-1} b_i \beta_i.$$

The binary number can be recovered from β by using the trace operator and Theorem 8.8.1 to write

$$b_\ell = \text{trace}(\beta \lambda_\ell).$$

Every such choice of a complementary basis gives a different binary expansion of a Reed–Solomon code, and so a different linear binary code. Even more binary codes can be obtained by using a different basis for each component of the code.

Another method is to multiply the symbol of the codeword by a given constant in $GF(2^m)$ and then expand the product by using any fixed basis such as a polynomial basis. Were it known, one would choose the basis for which the minimum weight of any pair of complementary bases is as large as possible.

If the symbol multiplying each component of the Reed–Solomon code is different, the code is a generalized Reed–Solomon code. In this section we shall use a binary expansion of a generalized Reed–Solomon code as a way of representing a Reed–Solomon code over $GF(2^m)$ as a binary codeword.

The binary expansion of a Reed–Solomon code (using the polynomial basis) need not have a minimum distance very much larger than the minimum distance of the Reed–Solomon code, even though the blocklength is m times as large. Our goal is to show that the binary expansions of some of the generalized Reed–Solomon codes have large minimum distances. Because we do not know how to specify which of the generalized Reed–Solomon codes are the good ones, we will use a nonconstructive argument based on averaging over an ensemble of codes.

To begin, consider the set of binary expansions, using the polynomial basis, of all vectors over $GF(2^m)$ in which a fixed set of components is nonzero (and the remaining $N - \ell$ components are always zero).

Lemma 12.4.1. *The weight-distribution polynomial for the set of vectors of ℓ nonzero binary n-tuples is*

$$G^{(\ell)}(x) = ((x+1)^m - 1)^\ell.$$

Proof: For a single m-bit symbol, there are $\binom{m}{j}$ patterns with j ones. Hence, because the weight-distribution polynomial of a single symbol is $(x+1)^m$ and the all-zero symbol must be excluded, we have

$$\begin{aligned} G^{(1)}(x) &= \sum_{j=1}^{m} \binom{m}{j} x^j \\ &= (x+1)^m - 1. \end{aligned}$$

This implies the recursion

$$\begin{aligned} G^{(\ell)}(x) &= (x+1)^m G^{(\ell-1)}(x) - G^{(\ell-1)}(x) \\ &= G^{(1)}(x) G^{(\ell-1)}(x), \end{aligned}$$

which completes the proof of the lemma. □

Now we can consider the binary weight distribution of a binary code that is the expansion of a generalized Reed–Solomon code. The weight distribution of an (N, K, D) Reed–Solomon code as given by Corollary 12.1.3 is

$$A_\ell = \binom{N}{\ell} \sum_{j=0}^{\ell-D} (-1)^j \left(q^{\ell-D+1-j} - 1\right),$$

for $\ell \geq D$. The weight distributions of the binary expansions are much more difficult to find, and no general solution is known. It is easy, however, to find the weight distribution for the set of all codewords from the binary expansions of all generalized Reed–Solomon codes that can be obtained from a given Reed–Solomon code. Each component of the generalized Reed–Solomon codes will take all possible values an equal number of times, and the contribution of each column to the overall weight distribution is decoupled from the contribution of other columns.

Theorem 12.4.2. *The binary weight-distribution polynomial for the set of all binary expansions of all codewords of all generalized Reed–Solomon codes obtained from a given (N, K, D) Reed–Solomon code over $GF(2^m)$ of blocklength $N = 2^m - 1$ and weight distribution A_ℓ, is*

$$G(x) = \sum_{\ell=0}^{N} A_\ell N^{N-\ell}((x+1)^m - 1)^\ell.$$

Proof: There are $2^m - 1$ nonzero elements of $GF(2^m)$, and the Reed–Solomon code has blocklength $N = 2^m - 1$. Hence there are N^N distinct templates, and so N^N distinct, generalized Reed–Solomon codes are derived from a given Reed–Solomon code. Any codeword of weight ℓ from the Reed–Solomon code will generate N^N codewords in the set of all generalized Reed–Solomon codes. However, because the codeword is zero in $N - \ell$ places, there will be $N^{N-\ell}$ identical copies of each distinct binary pattern in the set of binary patterns formed from the image of this codeword in the set of generalized Reed–Solomon codes.

There are A_ℓ Reed–Solomon codewords of weight ℓ; each produces $N^{N-\ell}$ identical copies of each of N patterns. By Lemma 12.4.1 each pattern has a weight distribution

$$G^{(\ell)}(x) = ((x+1)^m - 1)^\ell.$$

Consequently, the weight distribution for the list of N^Ncodes, each with $(2^m)^K$ codewords, is

$$G(x) = \sum_{\ell=0}^{q-1} A_\ell N^{N-\ell}((x+1)^m - 1)^\ell,$$

as asserted by the theorem. □

The next theorem states that binary codes obtained by such expansions asymptotically achieve the Varshamov–Gilbert bound. The proof uses the well-known inequality

$$\left(\frac{n+1}{n}\right)^n < e,$$

and also the inequality

$$\sum_{\ell=0}^{t} \binom{n}{\ell} \leq 2^{nH_b(t/n)}$$

given in Theorem 3.5.3.

Theorem 12.4.3. *Choose $\epsilon > 0$. For a sufficiently large blocklength n, generalized Reed–Solomon codes exist whose binary expansions have a minimum distance d and* *rate k/n satisfying*

$$H_b\left(\frac{d}{n}\right) > 1 - \frac{k}{n} - \epsilon.$$

Proof: For a given rate k/n, choose $\delta < 1/2$ satisfying $H_b(\delta) = 1 - (k/n) - \epsilon$. By Theorem 12.4.2, the weight-distribution polynomial for the set of binary expansions of all generalized Reed–Solomon codes is

$$G(x) = \sum_{\ell=0}^{N} A_\ell N^{N-\ell}((x+1)^m - 1)^\ell$$

where

$$A_\ell = \binom{N}{\ell} \sum_{j=0}^{\ell-D} (-1)^j \binom{\ell}{j} \left(q^{\ell-D+1-j} - 1\right).$$

We will show that

$$\sum_{i=D}^{\delta mn} G_i < \frac{N^N}{2}.$$

Because there are N^N generalized Reed–Solomon codes, half of them do not contribute to this sum, and so they must have a minimum weight larger than $\delta m N$. First, bound $G(x)$ as follows

$$G(x) \le N^N \sum_{\ell=0}^{N} \frac{A_\ell}{N^\ell}(x+1)^{m\ell},$$

which gives a bound on G_i

$$G_i \le N^N \sum_{\ell=0}^{N} \frac{A_\ell}{N^\ell} \binom{m\ell}{i}.$$

In the expression for A_ℓ, the term with $j = 0$ has the largest magnitude, and $N < q$, so we have the bound

$$A_\ell \le \binom{N}{\ell} N q^{\ell-D} < \binom{N}{\ell} q^{\ell+1-D}.$$

Combining these gives

$$\begin{aligned}
\frac{1}{N^N} G_i &< \sum_{\ell=0}^{N} \binom{N}{\ell} \left(\frac{q}{N}\right)^\ell q^{1-D} \binom{m\ell}{i} \\
&< e q^{1-D} \sum_{\ell=0}^{N} \binom{N}{\ell} \binom{m\ell}{i}.
\end{aligned}$$

The largest term in the sum is when $\ell = N$, and $q = 2^m$, so

$$\frac{1}{N^N} G_i < e q^{2-D} \binom{mN}{i} = e(2^m)^{K-N+1} \binom{mN}{i}$$
$$< 2^2 2^{k-n+m} \binom{n}{i}.$$

Now sum over the low-weight codewords

$$\frac{1}{N^N} \sum_{i=D}^{\delta n} G_i \le 2^{k-n+2+m} \sum_{i=D}^{\delta n} \binom{n}{i}$$
$$< 2^{k-n+2+\log_2 n + n H_b(\delta)} = n 2^{2-n\epsilon}.$$

For large enough n, the right side is smaller than one-half. Hence

$$\sum_{i=D}^{\delta n} G_i < \frac{N^N}{2}$$

for large enough N. Thus some binary code has

$$d_{\min} > \delta n,$$

as asserted by the theorem. □

12.5 Symbol error rates on a gaussian-noise channel

In the remainder of this chapter we shall turn away from the algebraic properties of codes, such as the minimum distance, to briefly study the performance of signaling on a communication channel. The performance of a signaling waveform is given by the probability of bit error at the demodulator output. In this section we will study the probability of error of a single symbol. In the next section we will study the probability of error for sequence demodulation, that is, for decoding codewords.

The simplest example of mapping code symbols into real or complex numbers is the mapping of the symbols $\{0, 1\}$ of the Galois field $GF(2)$ into the bipolar alphabet (or the bipolar constellation) $\{-1, +1\}$. The standard map is $1 \to +1, 0 \to -1$. It will be convenient to choose instead $\{-A, +A\}$ as the bipolar alphabet.

A real, additive noise channel is a channel that takes a real-valued input c and produces a real-valued output v given by

$$v = c + e.$$

The channel input and output are real numbers differing by the error signal e, which is a zero-mean random variable of variance σ^2. If e is a gaussian random variable, then the channel is called a gaussian-noise channel.

It is conventional to regard the constellation points $\pm A$ as having an energy, which we define as the bit energy $E_b = A^2 T$ where T is a constant, called the symbol duration, that will not play a significant role. It is also conventional to express the noise variance in terms of a parameter N_0 by the definition $\sigma^2 = N_0 T/2$. It is reasonable to expect that if the signal power and noise power are increased in the same proportion, then the error rate will not change. Hence, the ratio $(A/\sigma)^2 = 2E_b/N_0$ should play an important role, and it does.

Let $p_0(x)$ denote the probability density function on the channel output, denoted x, when a zero is transmitted, as denoted by $-A$; let $p_1(x)$ denote the probability density function on the channel output when a one is transmitted, as denoted by $+A$. If e is a gaussian random variable, which is the case we consider, these probability density functions are

$$p_0(x) = \frac{1}{\sqrt{2\pi}\,\sigma} e^{-(x+A)^2/2}$$

$$p_1(x) = \frac{1}{\sqrt{2\pi}\,\sigma} e^{-(x-A)^2/2}.$$

By symmetry, the best decision rule decides in favor of zero if x is negative, and in favor of one if x is positive. Therefore, when a zero is transmitted, the probability of error is

$$p_{e|0} = \int_0^\infty \frac{1}{\sqrt{2\pi}\,\sigma} e^{-(x+A)^2/2\sigma^2} dx.$$

When a one is transmitted, the probability of error is

$$p_{e|1} = \int_{-\infty}^0 \frac{1}{\sqrt{2\pi}\,\sigma} e^{-(x-A)^2/2\sigma^2} dx.$$

These two integrals are illustrated in Figure 12.5. With a simple change of variables, either $y = (x + A)/\sigma$ or $y = (x - A)/\sigma$, both integrals reduce to

$$p_{e|0} = p_{e|1} = \int_{A/\sigma}^\infty \frac{1}{\sqrt{2\pi}} e^{-y^2/2} dy.$$

The lower limit will be rewritten using the relationship $A/\sigma = \sqrt{2E_b/N_0}$.

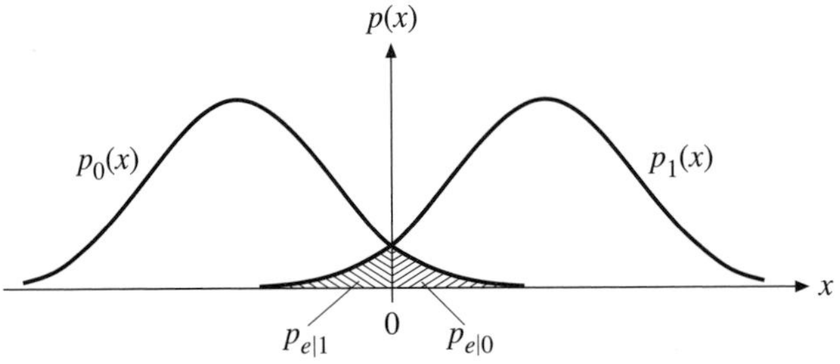

Figure 12.5. Error probabilities for bipolar signaling

The probability of error is commonly expressed in terms of a function $Q(x)$, defined as

$$Q(x) = \int_x^\infty \frac{1}{\sqrt{2\pi}} e^{-y^2/2} dy.$$

The function $Q(x)$ is closely related to a function known as the *complementary error function*,

$$\text{erfc}(x) = \frac{2}{\sqrt{\pi}} \int_x^\infty e^{-y^2} dy.$$

The error function $1 - \text{erfc}(x)$ is a widely tabulated function.

Theorem 12.5.1. *The average symbol error probability for the bipolar alphabet in gaussian noise is*

$$p_\text{e} = Q(\sqrt{2E_\text{b}/N_0}).$$

Proof: The average error probability is

$$p_\text{e} = \tfrac{1}{2} p_{\text{e}|0} + \tfrac{1}{2} p_{\text{e}|1}.$$

Because the two integrals for $p_{\text{e}|0}$ and $p_{\text{e}|1}$ are equal, we need to evaluate only one of them. Then

$$p_\text{e} = Q\left(\frac{A}{\sigma}\right).$$

The theorem follows, by setting $(A/\sigma)^2 = 2E_\text{b}/N_0$. □

The error probability for bipolar signaling in gaussian noise is shown in Figure 12.6.

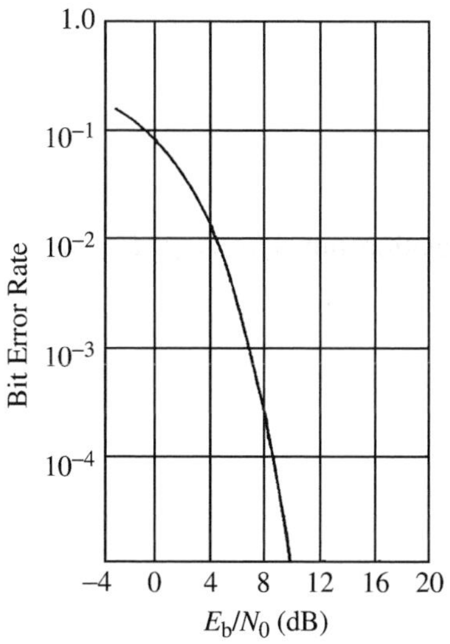

Figure 12.6. Error probability for bipolar signaling

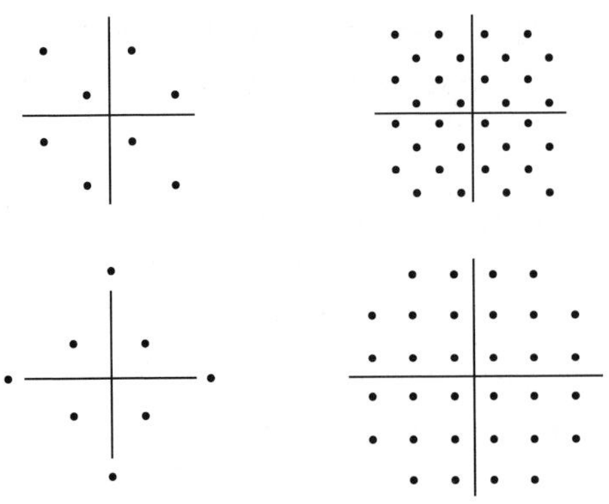

Figure 12.7. Signal constellations

The bipolar alphabet is a simple example of a signal constellation. The bipolar alphabet has only two points: A general signal constellation is a set $\mathcal{S}$ of q points in the complex plane. An example with $q = 8$ is shown in Figure 12.7. We use these q points to represent elements of $GF(q)$ by defining any convenient map between the q elements of $GF(q)$ and the q elements of the signal constellation.

The euclidean distance between two points, c and c', of the signal constellation is

$$d(c, c') = |c - c'|^2.$$

Definition 12.5.2. *The minimum euclidean distance of signal constellation $\mathcal{S}$ is*

$$d_{\min} = \min_{\substack{c,c' \in \mathcal{S} \\ c \neq c'}} d(c, c').$$

The minimum distance of a signal constellation is the smallest euclidean distance in the complex plane between two points of the signal constellation. The minimum distance is an important predictor of error performance.

The channel input is any element of the signal constellation. The channel output is

$$v = c + e$$

where e is zero-mean random variable of variance $\sigma^2 = N_0/2$ which we will take to be a gaussian random variable. The probability density function, given that point c is transmitted, is

$$\begin{aligned} p(v|c) &= \frac{1}{2\pi\sigma^2} e^{-|v-c|^2/2\sigma^2} \\ &= \left(\frac{1}{\sqrt{2\pi}\sigma} e^{-(v_R - c_R)^2/2\sigma^2}\right)\left(\frac{1}{\sqrt{2\pi}\sigma} e^{-(v_I - c_I)^2/2\sigma^2}\right). \end{aligned}$$

When given v, a demodulator chooses the point c that maximizes $p(v|c)$. Equivalently, the demodulator chooses the c that minimizes $|v - c|^2$. This is written

$$\hat{c} = \arg\min |v - c|^2.$$

The *decision region* for c, denoted $\mathcal{S}_c$, is the set of points of the complex plane that are closer to c than to any other point of the signal constellation. In general, regions defined in this way for a fixed set of points in the plane are called *Voronoi regions*. The decision region for a point c is a polygon, often an irregular polygon, whose sides are the perpendicular bisectors of the lines between c and the neighbors of c. A neighbor of c is another point of the signal constellation whose decision region shares a common border with the decision region of c.

The calculation of the error probability in principle is straightforward. The probability of error when c is the correct point is given by

$$p_{e|c} = \int\int_{S_c^c} \frac{1}{2\pi\sigma^2} e^{-|v-c|^2/2\sigma^2} dv_R dv_I$$

where the integral is over the complement of the decision region for c. This exact expression is not informative as it stands. For many purposes, it is useful to have a simple approximation. This is given using the union bound of probability theory

$$\Pr[\cup_i \mathcal{E}_i] \leq \Sigma_i \Pr[\mathcal{E}_i]$$

where the $\mathcal{E}_i$ are the events of any collection. Then, because a decision region is the intersection of half planes, and the complement of a Decision region is the union of half planes, the probability of error is bounded by

$$p_{e|c} \leq \sum_{c'} \int\int \frac{1}{2\pi\sigma^2} e^{-|v-c|/2\sigma^2}.$$

where the sum is over all c' that are neighbors of c and each double integral is over a half plane. Each half plane is defined by the perpendicular bisector of the line joining c to a neighbor.

12.6 Sequence error rates on a gaussian-noise channel

In this section, we shall study the probability of error of sequence demodulation by relating this topic to the probability of error of symbol demodulation. A real, discrete-time, additive noise channel is a channel that takes a real-valued input c_ℓ at time ℓT and produces an output at time ℓT given by

$$v_\ell = c_\ell + n_\ell.$$

The ℓth channel input and output are real numbers differing by the noise signal n_ℓ, which is a zero-mean random variable of variance σ^2. The noise process is *white* if n_ℓ and $n_{\ell'}$ are independent whenever $\ell \neq \ell'$. The channel is a gaussian-noise channel if, for each ℓ, n_ℓ is a gaussian random variable. The channel input c_ℓ at time ℓ is any element of the signal constellation.

A sequence demodulator finds the symbol sequence closest to the received sequence in the sense of minimum euclidean sequence distance. We will analyze this sequence demodulator to determine the probability that an incorrect sequence is closer to the received sequence than is the correct sequence. Of course, for an infinitely long sequence, we may always expect that occasionally there will be some errors, so even though the error rate may be quite small, the probability of at least one error in an infinitely long sequence is one. To discuss error rates in a meaningful way, we must carefully define our notions of errors in a sequence.

In general, we distinguish between the probability that a symbol is in error, the probability that a symbol is contained within an error event, and the probability that a symbol is the start of an error event. However, each of these probabilities is difficult to compute, and we settle for those bounds that can be derived conveniently.

Definition 12.6.1. *The minimum sequence euclidean distance, $d_{\min}$, of a set of real-valued or complex-valued sequences is the minimum of the sequence euclidean distance $d(\boldsymbol{c}, \boldsymbol{c}')$ over all pairs $(\boldsymbol{c}, \boldsymbol{c}')$ of distinct sequences in the set.*

Let $\boldsymbol{c}$ and $\boldsymbol{c}'$ be two sequences that establish $d_{\min}$, that is, such that $d(\boldsymbol{c}, \boldsymbol{c}')$ equals the minimum distance and that these two sequences have different branches in the first frame of the trellis. Suppose that whenever either of these two sequences is transmitted, a genie tells the demodulator that the sequence was chosen from this pair. The demodulator needs only to decide which one. Such side information from the genie cannot make the probability of demodulation error worse than it would be without the genie. Hence the probability of demodulation error for the two-sequence problem can be used to form a lower bound on the probability of demodulation error whenever $\boldsymbol{c}$ is the transmitted sequence.

Theorem 12.6.2. *The average error probability for separating two sequences in white gaussian noise of variance σ^2 is*

$$p_{\mathrm{e}} = Q\left(\frac{d(\boldsymbol{c}, \boldsymbol{c}')}{2\sigma}\right).$$

Proof: For white gaussian noise, the maximum-likelihood sequence estimator is equivalent to the minimum-distance demodulator. The minimum-distance demodulator chooses $\boldsymbol{c}$ or $\boldsymbol{c}'$ according to the distance test: Choose $\boldsymbol{c}$ if

$$\sum_{\ell} |c_\ell - v_\ell|^2 < \sum_{\ell} |c'_\ell - v_\ell|^2,$$

and otherwise choose $\boldsymbol{c}'$. Suppose that $\boldsymbol{c}$ is the correct codeword. Then $v_\ell = c_\ell + n_\ell$, and the decision rule for choosing $\boldsymbol{c}$ is

$$\sum_{\ell} n_\ell^2 < \sum_{\ell} |(c'_\ell - c_\ell) - n_\ell|^2,$$

or

$$\sum_\ell |c'_\ell - c_\ell|^2 - 2\,\text{Re}\left[\sum_\ell n_\ell (c'_\ell - c_\ell)^*\right] > 0.$$

There will be an error if

$$\sum_\ell [n_{\ell R}(c'_{\ell R} - c_{\ell R}) + n_{\ell I}(c'_{\ell I} - c_{\ell I})] \leq \frac{1}{2}\sum_\ell |c'_\ell - c_\ell|^2 = \frac{1}{2}d(\boldsymbol{c}, \boldsymbol{c}')^2.$$

The left side is a gaussian random variable, denoted n, which has a variance

$$\begin{aligned}\text{var}(n) &= \left[\sum_\ell \sigma^2(c'_{\ell R} - c_{\ell R})^2 + \sigma^2(c'_{\ell I} - c_{\ell I})^2\right], \\ &= \sigma^2 d^2(\boldsymbol{c}, \boldsymbol{c}').\end{aligned}$$

We see now that if we rewrite the condition for an error as

$$\frac{n}{d(\boldsymbol{c}, \boldsymbol{c}')} \leq \frac{1}{2}d(\boldsymbol{c}, \boldsymbol{c}'),$$

then the left side has a variance σ^2, and this is the same form as the problem studied in the previous section, and the probability of error has the same form. Because the conclusion is symmetric in $\boldsymbol{c}$ and $\boldsymbol{c}'$, the same conclusion holds if $\boldsymbol{c}'$ is the correct sequence. □

Next, we will use a union-bound technique to obtain an upper bound for the case where there are many sequences. The union bound uses the pairwise probability of error in order to bound the probability of error in the general problem. Then the probability of demodulation error given that $\boldsymbol{c}$ is the correct sequence satisfies

$$p_{\text{e}|\boldsymbol{c}} \leq \sum_{\boldsymbol{c}' \neq \boldsymbol{c}} Q\left(\frac{d(\boldsymbol{c}, \boldsymbol{c}')}{2\sigma}\right).$$

We can group together all terms of the sum for which $d(\boldsymbol{c}, \boldsymbol{c}')$ has the same value. Let d_i for $i = 0, 1 \ldots$ denote the distinct values that the distance can assume. Then

$$p_{\text{e}|\boldsymbol{c}} \leq \sum_{i=0}^{\infty} N_{d_i}(\boldsymbol{c}) Q\left(\frac{d_i}{2\sigma}\right)$$

where $N_{d_i}(\boldsymbol{c})$ is the number of sequences $\boldsymbol{c}'$ at distance d_i from $\boldsymbol{c}$. For values of d smaller than the minimum distance $d_{\min}$, the number of sequences at distance d is zero, and $d_0 = d_{\min}$. Consequently, averaging over all $\boldsymbol{c}$ gives

$$\begin{aligned}p_{\text{e}} &\leq \sum_{i=0}^{\infty} N_{d_i} Q\left(\frac{d_i}{2\sigma}\right) \\ &= N_{d_{\min}} Q\left(\frac{d_{\min}}{2\sigma}\right) + \sum_{i=1}^{\infty} N_{d_i} Q\left(\frac{d_i}{2\sigma}\right).\end{aligned}$$

There is a difficulty, however. Because there are so many codewords in a code the union bound can diverge. Therefore the union bound analysis is deeply flawed, yet usually gives a good approximation. This is because the function $Q(x)$ is a steep function of its argument – at least on its tail, which is where our interest lies. Consequently, the sum on the right is dominated by the first term. Then we have the approximation to the upper bound

$$p_{\mathrm{e}} \lesssim N_{d_{i_{\min}}} Q\left(\frac{d_{\min}}{2\sigma}\right).$$

The approximation to the upper bound is usually quite good and is asymptotically tight, as can be verified by simulation. However, the derivation is no more than a plausibility argument. There are so many terms in the union bound, the right side of the union bound may very well be infinite. It is not correct to discard these small terms in the last step under the justification that they are individually small.

This is the probability of error in the first symbol, and the probability of error for the original problem can only be larger. Consequently, as the lower bound on $p_{\mathrm{e}|c}$ we have the probability of demodulator error in the first symbol, given that sequence $\boldsymbol{c}$ is the correct sequence,

$$p_{\mathrm{e}|c} \geq Q\left(\frac{d_{\min}}{2\sigma}\right).$$

This inequality is mathematically precise, but rather loose. If we discard precision and average over all $\boldsymbol{c}$, we may reason that a sequence $\boldsymbol{c}$ has on average $N_{d_{\min}}$ neighbors at distance $d_{\min}$, and the decision regions for these neighbors have a negligible intersection with regard to the probability distribution centered at $\boldsymbol{c}$. Therefore we may expect that the expression

$$p_{\mathrm{e}} > N_{d_{\min}} Q\left(\frac{d_{\min}}{2\sigma}\right)$$

gives a rough approximation to a lower bound, where p_{e} denotes the probability of error in the first symbol of a sequence starting from an arbitrary trellis state. The bound is not a true lower bound because properly accounting for the intersection of the pairwise decision regions would reduce the right side of the inequality.

12.7 Coding gain

The performance of a data-transmission code used on an additive white gaussian-noise channel is expressed in terms of the probability of channel symbol error as a function of E_{b}/N_0 of the channel waveform. Convention requires that E_{b} is the average energy per data bit. This energy is n/k larger than the energy per channel bit because the energy in the check symbols is reallocated to the data bits for this calculation. It is common to judge the code, not by the reduction in the bit error rate, but by the reduction in

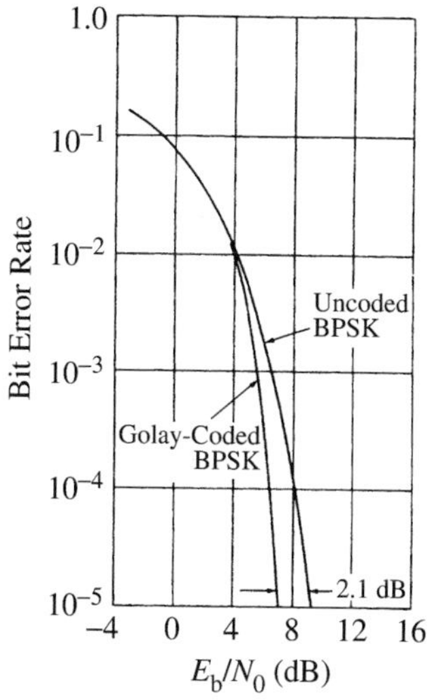

Figure 12.8. Illustration of the notion of coding gain

the E_b/N_0 needed to ensure the specified bit error rate. The reduction in the required E_b/N_0 at the same bit error rate is called the *coding gain*. For example, a simple binary communication system on an additive white gaussian-noise channel using the bipolar signal constellation operates at a bit error rate of 10^{-5} at an E_b/N_0 of 9.6 dB. By adding a sufficiently strong code to the communication system, the ratio E_b/N_0 could be reduced. If the code requires only 6.6 dB for a bit error rate of 10^{-5}, then we say that the code has a coding gain of 3 dB at a bit error rate of 10^{-5}. This coding gain is just as meaningful as doubling the area of an antenna, or increasing the transmitted power by 3 dB. Because it is usually far easier to obtain coding gain than to increase power, the use of a code is preferred.

In most applications, the average transmitted power is fixed. Check symbols can be inserted into the channel codestream only by taking power from the data symbols. Therefore coding gain is defined in terms of the required E_b/N_0. The bit energy E_b is defined as the total energy in the message divided by the number of data bits. This is equal to the energy received per channel bit divided by the code rate.

Figure 12.8 shows the probability of error versus E_b/N_0 for a simple binary code, the Golay (23, 12) code, used on an additive, white gaussian-noise channel. The coding gain at a bit error rate of 10^{-5} is 2.1 dB. For a larger code, the coding gain will be larger.

Arbitrarily large coding gains are not possible. Figure 12.9 shows the region where the coding gain must lie. There is one region for a hard-decision demodulator, given by $E_b/N_0 > 0.4$ dB, and another for a soft-decision demodulator, given by $E_b/N_0 > -1.6$ dB.

When a decoder is used with a demodulator for a gaussian-noise channel, the decoder is called a *hard-input decoder* if the output of the demodulator is simply a demodulated symbol. Sometimes the demodulator passes other data to the decoder such as the digitized output of the matched filters. Then the decoder is called a *soft-input decoder*.

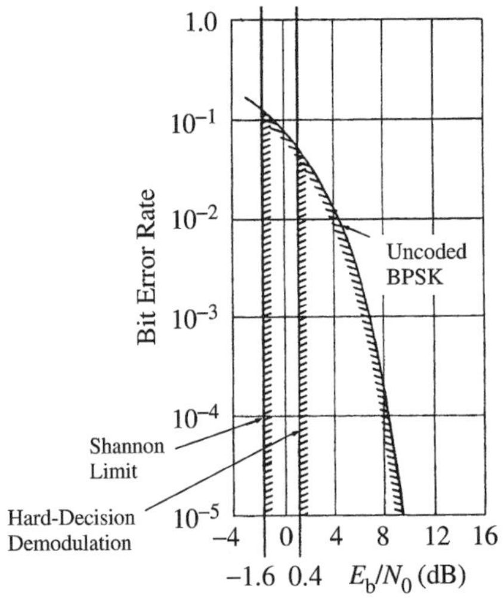

Figure 12.9. Regions of possible coding gain

When used in conjunction with a demodulator for an additive gaussian-noise channel, the performance of the code is described by a quantity known as the coding gain. The coding gain is defined as the difference between the E_b/N_0 required to achieve a performance specification, usually probability of error, when a code is used and when a code is not used. Figure 12.8 illustrates the coding gain with BPSK as the uncoded reference. The binary code used for this illustration is one known as a Golay (23, 12) triple-error-correcting code. The figure shows that at a bit error rate of 10^{-5}, the Golay code has a coding gain of 2.1 dB.

It is natural to ask if the coding gain can be increased indefinitely by devising better and better codes. Because we cannot signal at any E_b/N_0 smaller than -1.6 dB, there is a very definite negative answer to this question. At a bit error rate of 10^{-5}, the maximum coding gain, relative to BPSK, of any coding system whatsoever is 11.2 dB. Moreover, if a hard-decision BPSK demodulator is used, then a soft-decision decoder is inappropriate, and a coding gain of 2 dB is not available. Then the maximum coding gain is 9.2 dB. Figure 12.9 shows the regions in which all coding gains must lie.

The coding gain must be evaluated numerically for each code of interest. It cannot be expressed by a simple equation, but it has a clearly defined meaning. In contrast, the asymptotic coding gain, given in the next definition, has a simple definition but its meaning is not as clear-cut. Rather, it is an approximation to the coding gain in the limit as E_b/N_0 goes to infinity, and t is small compared to n. The asymptotic coding gain sharpens our understanding but should be used with care because of the approximate analysis – which is sometimes meaningless – and because we are usually interested in moderate values of E_b/N_0.

Definition 12.7.1. *A t-error-correcting block code of rate R has an asymptotic coding gain of*

$$G = R(t + 1)$$

when used with a hard-input decoder, and

$$G = R(2t + 1)$$

when used with a soft-input decoder.

The motivation for the definitions of asymptotic coding gain is a pair of approximate analyses that are valid asymptotically for large signal-to-noise ratios, and if t is small compared to n. Let p_e denote the probability of bit error at the output of the hard-decision BPSK demodulator. Let p_{em} denote the probability of not decoding the block correctly, either because of block decoding error or because of block decoding default. Then

$$\begin{aligned} p_{em} &= \sum_{\ell=i+1}^{n} \binom{n}{\ell} p_e^{\ell}(1 - p_e)^{n-\ell} \\ &\approx \binom{n}{t+1} p_e^{t+1} \\ &\approx \binom{n}{t+1} \left[Q\left(\sqrt{\frac{2RE_b}{N_0}} \right) \right]^{t+1}, \end{aligned}$$

using the expression for the probability of error of BPSK. For large x, $Q(x)$ behaves like $e^{-x^2/2}$, so for large x, $[Q(x)]^s \approx Q(x\sqrt{s})$. Therefore

$$p_{em} \approx \binom{n}{t+1} Q\left(\sqrt{R(t+1)\frac{2E_b}{N_0}} \right).$$

Asymptotically, as $E_b/N_0 \to \infty$, the argument of the Q function is as if E_b were amplified by $R(t + 1)$ – hence the term "asymptotic coding gain." However, the approximate analysis is meaningful only if the binomial coefficient appearing as a coefficient is of second-order importance. This means that t must be small compared to n. In this case, we may also write the asymptotic approximation

$$p_{em} \sim e^{-1/2Rd_{\min}E_b/N_0}$$

to express the asymptotic behavior in E_b/N_0 of hard-input decoding.

For a soft-input decoder, the motivation for defining asymptotic coding gain is an argument based on minimum euclidean distance between codewords. Suppose that there are two codewords at Hamming distance $2t + 1$. Then in euclidean distance, they are separated by $(2t + 1)RE_b$. The probability of decoding error, conditional on the premise that one of these two words was transmitted, is

$$p_{\text{em}} \approx Q\left(\sqrt{R(2t+1)\frac{2E_{\text{b}}}{N_0}}\right).$$

On average, if a codeword has N_n nearest neighbors at distance $2t+1$, then the union bound dictates that the probability of error can be at most N_n times larger,

$$p_{\text{em}} \approx N_n Q\left(\sqrt{R(2t+1)\frac{2E_{\text{b}}}{N_0}}\right).$$

Again, for large E_{b}/N_0, the argument of the Q function is as if E_{b} were amplified by $R(2t+1)$, and again this is described by the term "asymptotic coding gain." The significance of this term, however, requires that the number of nearest neighbors N_m is not so large that it offsets the exponentially small behavior of $Q(x)$. In this case, we may again write an asymptotic approximation

$$p_{\text{em}} \sim e^{-Rd_{\min}E_{\text{b}}/N_0}$$

to express the asymptotic behavior in E_{b}/N_0 of soft-decision decoding.

Just as the minimum distance between points of a signal constellation plays a major role in determining the probability of demodulation error in gaussian noise of a multilevel signaling waveform, so the minimum distance between sequences plays a major role in determining the probability of error in sequence demodulation. By analogy with block codes, we have the following definition of asymptotic coding gain for a convolutional code used with a soft-decision decoder.

Definition 12.7.2. *The asymptotic coding gain of a binary convolutional code $\mathcal{C}$ of rate R and free distance $d_{\min}$, modulated into the real (or complex) number system, is*

$$G = Rd_{\min}.$$

By analogy with block codes, we may expect that an approximate formula for the probability of decoding error is

$$p_{\text{e}} \approx N_{d_{\min}} Q\left(\sqrt{G\frac{2E_{\text{b}}}{N_0}}\right).$$

This formula can be verified by simulation. Alternatively, we could show that this approximation follows roughly from a union bound. Thus the asymptotic coding gain $G = Rd_{\min}$ measures the amount that E_{b} can be reduced by using the code.

12.8 Capacity of a gaussian-noise channel

The *energy* in a sequence of blocklength N is given by

$$E_{\mathrm{m}} = \sum_{\ell=0}^{N-1} c_\ell^2.$$

The average power of an infinitely long message is defined as

$$S = \lim_{T\to\infty} \frac{1}{T} E_{\mathrm{m}}(T)$$

where $E_{\mathrm{m}}(T)$ is the energy in a message of duration T. The average power must be finite for all infinitely long messages that are physically meaningful. For a finite-length message containing K information bits and message energy E_{m}, the bit energy E_{b} is defined by

$$E_{\mathrm{b}} = \frac{E_{\mathrm{m}}}{K}.$$

For an infinite-length, constant-rate message of rate R information bits/second, E_{b} is defined by

$$E_{\mathrm{b}} = \frac{S}{R}$$

where S is the average power of the message.

The bit energy E_{b} is not the energy found in a single symbol of the message unless $R = 1$. It is calculated from the message energy. The definition of E_{b} is referenced to the number of information bits at the input to the encoder. At the channel input, one will find a message structure with a larger number of channel symbols because some are check symbols for error control. These symbols do not represent transmitted information. Only information bits are used in calculating E_{b}.

A digital communication system comprises an encoder and a modulator at the transmitter, which map the datastream into a sequence of inputs to the real-valued channel, and a demodulator and a decoder at the receiver, which map the channel output sequence into the reconstructed datastream. The quality of this kind of system is judged, in part, by the probability of bit error, or the *bit error rate* (BER).

The reception of a signal cannot be affected if both the energy per bit E_{b} and the noise power spectral density N_0 are both doubled. It is only the ratio E_{m}/N_0 or E_{b}/N_0 that affects the bit error rate. Two different signaling schemes are compared by comparing their respective graphs of BER versus required E_{b}/N_0.

For a fixed digital communication system that transmits digital data through an additive gaussian-noise channel, the bit error rate can always be reduced by increasing E_{b}/N_0 at the receiver, but it is by the performance at low values of E_{b}/N_0 that one judges the quality of a digital communication system. Therefore the better of two digital communication systems is the one that can achieve a desired bit error rate with the lower value of E_b/N_0.

Surprisingly, it is possible to make precise statements about values of E_b/N_0 for which good waveforms exist. Our starting point for developing bounds on E_b/N_0 is the channel capacity of Shannon. The capacity formula we need is the one for the additive gaussian-noise channel. This formula says that any signaling waveform $s(t)$ of power $S = E_b R$, whose spectrum $S(f)$ is constrained to be zero for $|f| > W/2$, can transmit no more than

$$C = W \log_2 \left(1 + \frac{S}{N_0 W}\right)$$

bits per second through an additive white gaussian-noise channel at an arbitrarily low bit error rate. Conversely, there always exists a signaling waveform $s(t)$ that comes arbitrarily close to this rate and bandwidth constraint at any specified and arbitrarily small bit error rate. This celebrated formula, derived in any information-theory textbook, provides a point of reference against which a digital communication system can be judged.

Theorem 12.8.1. *Any digital communication system for the additive white gaussian-noise channel requires an energy per bit satisfying*

$$E_b/N_0 > \log_e 2 = 0.69$$

where E_b is in joules and N_0 is the noise spectral density in watts/hertz.

Proof: The bandwidth W is not constrained in the conditions of the theorem. Replace S by RE_b in the Shannon capacity formula, and let W go to infinity. Then

$$R \leq \frac{RE_b}{N_0} \log_2 e,$$

and the theorem follows immediately. □

The expression of the theorem holds if the channel output is a soft senseword. If the channel output is a hard senseword, then the corresponding expression is

$$E_b/N_0 > \frac{\pi}{2} \log_e 2$$

which is left to prove as an exercise.

If the ratio E_b/N_0 is measured in decibels, then the inequality of the theorem becomes

$$E_b/N_0 \geq -1.6 \text{ dB}.$$

The value of $E_b/N_0 = -1.6$ dB, and therefore provides a lower limit. Experience shows that it is easy to design waveforms that operate with a small bit error rate if E_b/N_0 is about 12 dB or greater. Hence the province of the waveform designer is the interval between these two numbers – a spread of about 14 dB. This is the regime within which design comparisons are conducted for the gaussian-noise channel.

When only a limited bandwidth is available, the required energy is greater. Define the spectral bit rate r (measured in bits per second per hertz) by

$$r = \frac{R}{W}.$$

The spectral bit rate r and E_b/N_0 are the two most important figures of merit in a digital communication system.

Corollary 12.8.2. *The spectral bit rate of any digital communication system for an additive white gaussian-noise channel satisfies*

$$\frac{E_b}{N_0} \geq \frac{2^r - 1}{r}.$$

Furthermore, the bound cannot be made tighter.

Proof: The Shannon capacity formula says that for any positive ϵ, an information rate R can be achieved, satisfying

$$R \leq W \log_2 \left(1 + \frac{RE_b}{WN_0}\right) \leq R(1+\epsilon).$$

The corollary follows immediately. □

This last corollary tells us that increasing the bit rate per unit bandwidth increases the required energy per bit. This is the basis of the energy/bandwidth trade of digital communication theory where increasing the bandwidth at a fixed information rate can reduce power requirements.

The content of Corollary 12.8.2 is shown in Figure 12.10. Any communication system can be described by a point lying below this curve and, for any point below the curve,

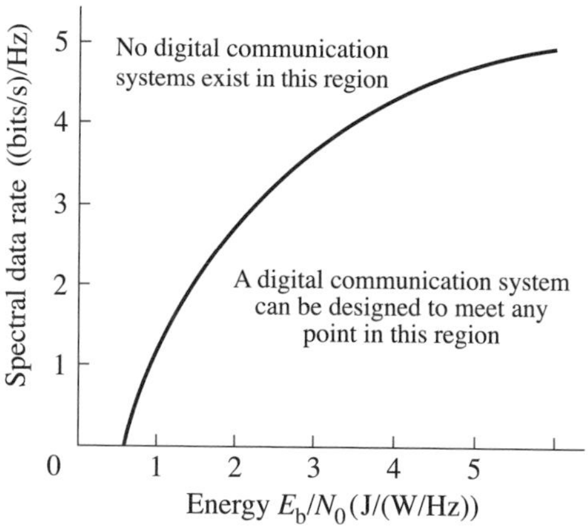

Figure 12.10. Capacity of an additive white gaussian-noise channel

one can design a communication system that has as small a bit error rate as one desires. The history of digital communications can be described, in part, as a series of attempts to move ever closer to this limiting curve with systems that have a very low bit error rate. Such systems employ both modem techniques and error-control techniques.

Problems

12.1 Consider a $(31, 27, 15)$ Reed–Solomon code over $GF(32)$.

a. How many codewords have weight 5?

b. How many error patterns have weight 3?

c. What fraction of error patterns of weight 3 are undetected?

d. Choose a word at random (equiprobable). What is the probability that it will be decoded?

12.2 Use the MacWilliams identity and Table 6.4 to compute the weight-distribution vector of the dual of the (24, 12) extended Golay code. (In fact, the (24, 12) extended Golay code is its own dual; it is called a self-dual code.)

12.3 A subset of k coordinate positions of an (n, k) block code over $GF(q)$ is called an *information set* of the code if every q-ary k-tuple appears, ordered, in those positions in exactly one codeword. Prove that a block code over $GF(q)$ is a maximum-distance code if and only if every subset of k coordinate positions is an information set.

12.4 a. A (7, 4) Hamming code is used to correct errors on a binary symmetric channel with error probability ϵ. What is the probability of decoding error?

b. If the code is used only to detect errors, but not to correct them, what is the probability of undetected error?

c. A (7, 5) Reed–Solomon code is used to correct errors on an octal symmetric channel with error probability ϵ. What is the probability of decoding error?

12.5 Imitate the proof of Theorem 3.5.3 using the free parameter $\lambda = \log_q [(1-p)/p] + \log_q(q-1)$ to prove that

$$\sum_{\ell=0}^{t} \binom{n}{\ell} (q-1)^{\ell} \leq q^{nH_q(t/n)}$$

where the q-ary *entropy function* is

$$H_q(x) = x \log_q(q-1) - x \log_q x - (1-x)\log_q(1-x).$$

12.6 Evaluate the sum of the weight distribution components $\sum_{\ell=0}^{n} A_\ell$.

12.7 By differentiating the MacWilliams identity

$$q^k B(x) = [1 + (q-1)x]^n A\left(\frac{1-x}{1+(q-1)x}\right)$$

prove the first of the *Pless power moments*

$$\sum_{j=0}^{n} jA_j = n(q-1)q^{n-k-1}$$

for the weight distribution of any linear block code that has a generator matrix with no all-zero column. The second of the Pless power moments can be obtained in the same way by a second derivative.

12.8 A t-error-correcting (n, k) Reed–Solomon code has a minimum distance of $d = n - k + 1 = 2t + 1$ and a weight distribution component $A_d = \binom{n}{d}(q-1)$.

a. How many burst-error patterns of weight $t+1$ and length $t+1$ exist?
b. Find the number of burst-error patterns of weight $t+1$ and length $t+1$ that are decoded incorrectly by a decoder that decodes to the packing radius.
c. What fraction of burst-error patterns of weight $t+1$ and length $t+1$ is decoded incorrectly?
d. Evaluate the expression derived for part 12.8c for $q = 32$, $t = 5$, and for $q = 64$, $t = 6$.

12.9 A Reed–Solomon code over $GF(2^m)$ is used as a multiple-burst-correcting code over $GF(2)$. Discuss the decoding of sensewords that are found outside the packing radius of the code.

12.10 a. Suppose that a Fourier transform of blocklength n exists. Show that A_ℓ, the number of codewords of weight ℓ in an (n, k) Reed–Solomon code, is equal to the number of vectors $\boldsymbol{C}$ of length n that have linear complexity ℓ and

$$C_{j_0}, C_{j_0+1}, \ldots, C_{j_0+d-2} = 0.$$

b. Use part 12.10a to verify Corollary 12.1.3 for the case $\ell = d$, and the case $\ell = d + 1$.
c. Use part 12.10a to verify Corollary 12.1.3 for the case $\ell = d + i$.

12.11 Prove that for large values of x, the function $Q(x)$ can be approximated as

$$Q(x) \approx \frac{e^{-x^2/2}}{x\sqrt{2\pi}}.$$

12.12 Use Stirling's approximation:

$$\sqrt{2n\pi}\left(\frac{n}{e}\right)^n < n! < \sqrt{2n\pi}\left(\frac{n}{e}\right)^n\left(1 + \frac{1}{12n-1}\right)$$

to show that

$$e^{n[H(\boldsymbol{p})-o_2(1)]} < \frac{n!}{\prod_\ell (n_\ell !)} < e^{n[H(\boldsymbol{p})-o_1(1)]}$$

where $p_\ell = n_\ell / n$, and

$$o_1(1) = \frac{1}{2n}(q-1)\log(2n\pi) + \frac{1}{2n}\sum_\ell \log p_\ell - \frac{1}{n}\log\left(1 + \frac{1}{12n-1}\right)$$

$$o_2(1) = \frac{1}{2n}(q-1)\log(2n\pi) + \frac{1}{2n}\sum_\ell \log p_\ell + \frac{1}{n}\sum_\ell \log\left(1 + \frac{1}{12np_\ell - 1}\right).$$

12.13 Given that the additive white gaussian-noise channel with bipolar input and hard senseword output has capacity

$$C = 2W(1 - H_b(\rho)),$$

where

$$\rho = Q\left(\sqrt{r\frac{E_b}{N_0}}\right),$$

prove that

$$\frac{E_b}{N_0} > \frac{\pi}{2}\log_e 2.$$

Hint: A Taylor series expansion and the inequality

$$Q(x) \geq \frac{1}{2} - \frac{x}{\sqrt{2\pi}}$$

will be helpful.

Notes

The weight distribution of a maximum-distance code was derived independently by Assmus, Mattson, and Turyn (1965), by Forney (1966), and by Kasami, Lin, and Peterson (1966). The original proof of the MacWilliams identities in 1963 was based on combinatorial methods, as we have used. A simple proof of these identities using probabilistic methods was given by Chang and Wolf (1980). The same paper also treated the relationship between the weight distribution and the probability of decoding error. Other treatments may be found in works by Forney (1966), Huntoon and Michelson (1977), and Schaub (1988).

The lower bounds on the minimum distance of block codes were reported by Gilbert (1952) and by Varshamov (1957). Massey (1985) gave a careful comparison of the Gilbert bound and the Varshamov bounds for finite blocklengths. Elias never published his upper bound, which first appeared in his classroom lectures in 1960. The bound was also discovered by Bassalygo (1965). Better upper bounds are known, but they are far more difficult to prove. The best upper bound known today is due to McEliece *et al.* (1977). Tsfasman, Vladut, and Zink (1982) proved, in the field $GF(49)$, the existence of codes that are better than the Varshamov–Gilbert bound for that field.

The weight distribution of a binary expansion of a Reed–Solomon code as a function of the basis was studied by Imamura, Yoshida, and Nakamura (1986), Kasami and Lin (1988), and Retter (1991).

13 Codes and Algorithms for Majority Decoding

Majority decoding is a method of decoding by voting that is simple to implement and is extremely fast. However, only a small class of codes can be majority decoded, and usually these codes are not as good as other codes. Because code performance is usually more important than decoder simplicity, majority decoding is not important in most applications. Nevertheless, the theory of majority-decodable codes provides another well-developed view of the subject of error-control codes. The topic of majority decoding has connections with combinatorics and with the study of finite geometries.

Most known codes that are suitable for majority decoding are cyclic codes or extended cyclic codes. For these codes, the majority decoders can always be implemented as Meggitt decoders and characterized by an especially simple logic tree for examining the syndromes. Thus one can take the pragmatic view and define majority-decodable codes as those cyclic codes for which the Meggitt decoder can be put in a standard simple form. But in order to find these codes, we must travel a winding road.

13.1 Reed–Muller codes

Reed–Muller codes are a class of linear codes over $GF(2)$ that are easy to describe and have an elegant structure. They are an example of a code that can be decoded by majority decoding. For these reasons, the Reed–Muller codes are important even though, with some exceptions, their minimum distances are not noteworthy. Each Reed–Muller code can be punctured to give one of the cyclic Reed–Muller codes that were studied in Section 6.7.

For each integer m, and for each integer r less than m, there is a Reed–Muller code of blocklength 2^m called the rth-order Reed–Muller code of blocklength 2^m. The Reed–Muller codes will be defined here by a procedure for constructing their generator matrices in a nonsystematic form that suits majority decoding.

The *componentwise product* of the vectors $\boldsymbol{a}$ and $\boldsymbol{b}$ over any field F, given by

$$\boldsymbol{a} = (a_0, a_1, \ldots, a_{n-1}),$$
$$\boldsymbol{b} = (b_0, b_1, \ldots, b_{n-1})$$

is defined as the vector obtained by multiplying the components as follows:

$$\boldsymbol{ab} = (a_0 b_0, a_1 b_1, \ldots, a_{n-1} b_{n-1}).$$

We will use the componentwise product to define the rows of a generator matrix for the rth-order Reed–Muller code of blocklength $n = 2^m$ follows. Let $\boldsymbol{G}_0$ be the 1 by 2^m matrix equal to the all-ones vector of length n. Let $\boldsymbol{G}_1$ be an m by 2^m matrix that has each binary m-tuple appearing once as a column. For definiteness, we may take the columns of $\boldsymbol{G}_1$ to be the binary m-tuples in increasing order from the left, with the low-order bit in the bottom row. Let $\boldsymbol{G}_2$ be an $\binom{m}{2}$ by 2^m matrix constructed by taking its rows to be the componentwise multiplication of pairs of rows of $\boldsymbol{G}_1$. Let $\boldsymbol{G}_\ell$ be constructed by taking its rows to be the componentwise multiplication of each choice of ℓ rows of $\boldsymbol{G}_1$. There are $\binom{m}{\ell}$ ways to choose the ℓ rows of $\boldsymbol{G}_1$, so $\boldsymbol{G}_\ell$ has $\binom{m}{\ell}$ rows.

A Reed–Muller generator matrix $\boldsymbol{G}$ is an array of submatrices:

$$\boldsymbol{G} = \begin{bmatrix} \boldsymbol{G}_0 \\ \boldsymbol{G}_1 \\ \vdots \\ \boldsymbol{G}_r \end{bmatrix}.$$

The code defined by $\boldsymbol{G}$ is the rth-order Reed–Muller code of blocklength n.

Because $\boldsymbol{G}_\ell$ is an $\binom{m}{\ell}$ by 2^m matrix, the number of rows in $\boldsymbol{G}$ is $\sum_{\ell=0}^{r} \binom{m}{\ell}$. This means that

$$\begin{aligned} k &= 1 + \binom{m}{1} + \cdots + \binom{m}{r} \\ n - k &= 1 + \binom{m}{1} + \cdots + \binom{m}{m-r-1}, \end{aligned}$$

provided the rows of $\boldsymbol{G}$ are linearly independent. The zeroth-order Reed–Muller code is an $(n, 1)$ code. It is a simple repetition code and can be decoded by a majority vote. It has a minimum distance of 2^m.

As an example, let $n = 16$. Then

$$\boldsymbol{G}_0 = [1\ 1\ 1\ 1\ 1\ 1\ 1\ 1\ 1\ 1\ 1\ 1\ 1\ 1\ 1\ 1] = [\boldsymbol{g}_0],$$

$$\boldsymbol{G}_1 = \begin{bmatrix} 0&0&0&0&0&0&0&0&1&1&1&1&1&1&1&1 \\ 0&0&0&0&1&1&1&1&0&0&0&0&1&1&1&1 \\ 0&0&1&1&0&0&1&1&0&0&1&1&0&0&1&1 \\ 0&1&0&1&0&1&0&1&0&1&0&1&0&1&0&1 \end{bmatrix} = \begin{bmatrix} \boldsymbol{g}_1 \\ \boldsymbol{g}_2 \\ \boldsymbol{g}_3 \\ \boldsymbol{g}_4 \end{bmatrix}.$$

Because $\boldsymbol{G}_1$ has four rows, $\boldsymbol{G}_2$ has $\binom{4}{2}$ rows,

$$\boldsymbol{G}_2 = \begin{bmatrix} 0&0&0&0&0&0&0&0&0&0&0&0&1&1&1&1 \\ 0&0&0&0&0&0&0&0&0&0&1&1&0&0&1&1 \\ 0&0&0&0&0&0&0&0&0&1&0&1&0&1&0&1 \\ 0&0&0&0&0&0&1&1&0&0&0&0&0&0&1&1 \\ 0&0&0&0&0&1&0&1&0&0&0&0&0&1&0&1 \\ 0&0&0&1&0&0&0&1&0&0&0&1&0&0&0&1 \end{bmatrix} = \begin{bmatrix} \boldsymbol{g}_1\boldsymbol{g}_2 \\ \boldsymbol{g}_1\boldsymbol{g}_3 \\ \boldsymbol{g}_1\boldsymbol{g}_4 \\ \boldsymbol{g}_2\boldsymbol{g}_3 \\ \boldsymbol{g}_2\boldsymbol{g}_4 \\ \boldsymbol{g}_3\boldsymbol{g}_4 \end{bmatrix},$$

and $\boldsymbol{G}_3$ has $\binom{4}{3}$ rows,

$$\boldsymbol{G}_3 = \begin{bmatrix} 0&0&0&0&0&0&0&0&0&0&0&0&0&0&1&1 \\ 0&0&0&0&0&0&0&0&0&0&0&0&0&1&0&1 \\ 0&0&0&0&0&0&0&0&0&0&0&1&0&0&0&1 \\ 0&0&0&0&0&0&0&1&0&0&0&0&0&0&0&1 \end{bmatrix} = \begin{bmatrix} \boldsymbol{g}_1\boldsymbol{g}_2\boldsymbol{g}_3 \\ \boldsymbol{g}_1\boldsymbol{g}_2\boldsymbol{g}_4 \\ \boldsymbol{g}_1\boldsymbol{g}_3\boldsymbol{g}_4 \\ \boldsymbol{g}_2\boldsymbol{g}_3\boldsymbol{g}_4 \end{bmatrix}.$$

Then the generator matrix for the third-order Reed–Muller code of blocklength 16 is the 15 by 16 matrix:

$$\boldsymbol{G} = \begin{bmatrix} \boldsymbol{G}_0 \\ \boldsymbol{G}_1 \\ \boldsymbol{G}_2 \\ \boldsymbol{G}_3 \end{bmatrix}.$$

This generator matrix gives a (16, 15) code over $GF(2)$. (In fact, it is a simple parity-check code.)

The second-order Reed–Muller code from these same matrices is obtained by choosing r equal to two. This generator matrix,

$$\boldsymbol{G} = \begin{bmatrix} \boldsymbol{G}_0 \\ \boldsymbol{G}_1 \\ \boldsymbol{G}_2 \end{bmatrix},$$

gives a (16, 11, 3) code over $GF(2)$. The code is equivalent to the (15, 11) Hamming code expanded by an extra check bit.

From the definition of the generator matrices, it is clear that the Reed–Muller codes of blocklength n are nested: $\mathcal{C}_r \subset \mathcal{C}_{r-1}$. An rth-order Reed–Muller code can be obtained by augmenting an $(r-1)$th-order Reed–Muller code, and an $(r-1)$th-order Reed–Muller code can be obtained by expurgating an rth-order Reed–Muller code. Clearly, because the rth-order Reed–Muller code contains the $(r-1)$th-order Reed–Muller code, its minimum distance cannot be larger.

Every row of $\boldsymbol{G}_\ell$ has weight $2^{m-\ell}$, which is even. Because the sum of two binary vectors of even weight must have an even weight, all linear combinations of rows of $\boldsymbol{G}$ have even weight. Thus all codewords have even weight. The matrix $\boldsymbol{G}_r$ has rows of weight 2^{m-r}, and thus the minimum weight of an rth-order Reed–Muller code is not larger than 2^{m-r}. In fact it is equal to 2^{m-r}.

Theorem 13.1.1. *The rth order Reed–Muller code of blocklength n has dimension*

$$k = \sum_{\ell=0}^{r} \binom{m}{\ell}$$

and minimum distance 2^{m-r}.

Proof: We must show that the rows of $\boldsymbol{G}$ are linearly independent and we must show that the minimum weight is 2^{m-r}. Rather than give a direct proof, we shall show both of these indirectly by developing a decoding algorithm that recovers the k data bits in the presence of $(\frac{1}{2} \cdot 2^{m-r} - 1)$ errors. This will imply that the minimum distance is at least $2^{m-r} - 1$, and because it is even, that it is at least 2^{m-r}. □

We shall describe two algorithms that are designed specifically to decode Reed–Muller codes, the *Reed algorithm* and the *Dumer algorithm*. Both the Reed algorithm and the Dumer algorithm recover the true data bits directly from the senseword and do not explicitly compute the error. Intermediate variables, such as syndromes, are not used.

The Reed algorithm is an iterative algorithm that decodes an rth-order Reed–Muller codeword with $(\frac{1}{2} \cdot 2^{m-r} - 1)$ errors by recovering $\binom{m}{r}$ data bits, thereby reducing the senseword to a senseword of an $(r-1)$th-order Reed–Muller code that can correct $(\frac{1}{2} \cdot 2^{m-(r-1)} - 1)$ errors. Because the zeroth-order Reed–Muller code can be decoded by majority vote, the successive iterations will complete the decoding.

Break the data vector into $r+1$ segments, written $\boldsymbol{a} = [\boldsymbol{a}_0, \boldsymbol{a}_1, \ldots, \boldsymbol{a}_r]$, where segment $\boldsymbol{a}_\ell$ contains $\binom{m}{\ell}$ data bits. Each segment multiplies one block of $\boldsymbol{G}$. The encoding can be represented as a block vector-matrix product

$$\boldsymbol{c} = [\boldsymbol{a}_0, \boldsymbol{a}_1, \ldots, \boldsymbol{a}_r] \begin{bmatrix} \boldsymbol{G}_0 \\ \boldsymbol{G}_1 \\ \vdots \\ \boldsymbol{G}_r \end{bmatrix} = \boldsymbol{a}\boldsymbol{G}.$$

Consider the data sequence broken into such sections. Each section corresponds to one of the r blocks of the generator matrix and is multiplied by this section of $\boldsymbol{G}$ during encoding. If we can recover the data bits in the rth section, then we can compute their contribution to the senseword and subtract this contribution. This reduces the problem to that of decoding a smaller code, an $(r-1)$th-order Reed–Muller code. The decoding procedure for the bits of $\boldsymbol{a}_r$ is a succession of majority votes.

The senseword is

$$v = [\boldsymbol{a}_0, \boldsymbol{a}_1, \ldots, \boldsymbol{a}_r] \begin{bmatrix} \boldsymbol{G}_0 \\ \boldsymbol{G}_1 \\ \vdots \\ \boldsymbol{G}_r \end{bmatrix} + \boldsymbol{e}.$$

The decoding algorithm will first recover $\boldsymbol{a}_r$ from $\boldsymbol{v}$. Then it computes

$$\begin{aligned} v' &= v - \boldsymbol{a}_r \boldsymbol{G}_r \\ &= [\boldsymbol{a}_0, \boldsymbol{a}_1, \ldots, \boldsymbol{a}_{r-1}] \begin{bmatrix} \boldsymbol{G}_0 \\ \boldsymbol{G}_1 \\ \vdots \\ \boldsymbol{G}_{r-1} \end{bmatrix} + \boldsymbol{e}, \end{aligned}$$

which is a noisy codeword of an $(r-1)$th-order Reed–Muller code.

First, consider decoding the data bit that multiplies the last row of $\boldsymbol{G}_r$. This bit is decoded by setting up 2^{m-r} linear check sums in the 2^m received bits; each such check sum involves 2^r bits of the received word, and each received bit is used in only one check sum. The check sums will be formed so that a_{k-1} contributes to only one bit of each check sum, and every other data bit contributes to an even number of bits in each check sum. Hence each check sum is equal to a_{k-1} in the absence of errors. But if there are at most $(\frac{1}{2} \cdot 2^{m-r} - 1)$ errors, the majority of the check sums will still equal a_{k-1}.

The first check sum is the modulo-2 sum of the first 2^r bits of the senseword; the second is the modulo-2 sum of the second 2^r bits, and so forth. There are 2^{m-r} such check sums, and by assumption, $(\frac{1}{2} \cdot 2^{m-r} - 1)$ errors, and hence a majority vote of the check sums gives a_{k-1}.

For example, the (16, 11, 4) Reed–Muller code with $r = 2$ constructed earlier has the generator matrix

$$\boldsymbol{G} = \begin{bmatrix} 11 & 11 & 11 & 11 & 11 & 11 & 11 & 11 \\ 00 & 00 & 00 & 00 & 11 & 11 & 11 & 11 \\ 00 & 00 & 11 & 11 & 00 & 00 & 11 & 11 \\ 00 & 11 & 00 & 11 & 00 & 11 & 00 & 11 \\ 01 & 01 & 01 & 01 & 01 & 01 & 01 & 01 \\ 00 & 00 & 00 & 00 & 00 & 00 & 11 & 11 \\ 00 & 00 & 00 & 00 & 00 & 11 & 00 & 11 \\ 00 & 00 & 00 & 00 & 01 & 01 & 01 & 01 \\ 00 & 00 & 00 & 11 & 00 & 00 & 00 & 11 \\ 00 & 00 & 01 & 01 & 00 & 00 & 01 & 01 \\ 00 & 01 & 00 & 01 & 00 & 01 & 00 & 01 \end{bmatrix}$$

and data vector

$$\boldsymbol{a} = ((a_{00}), (a_{10}, a_{11}, a_{12}, a_{13}), (a_{20}, a_{21}, a_{22}, a_{23}, a_{24}, a_{25})).$$

If the columns of $\boldsymbol{G}$ are summed in groups of four, an eleven by four array is formed with ten rows of zeros and a row of ones in the last row. This means that if codeword bits are summed in groups of four, each sum is equal to a_{25}.

The four estimates of a_{10} are:

$$\begin{aligned}
\hat{a}_{25}^{(1)} &= v_0 + v_1 + v_2 + v_3 \\
\hat{a}_{25}^{(2)} &= v_4 + v_5 + v_6 + v_7 \\
\hat{a}_{25}^{(3)} &= v_8 + v_9 + v_{10} + v_{11} \\
\hat{a}_{25}^{(4)} &= v_{12} + v_{13} + v_{14} + v_{15}.
\end{aligned}$$

If only one error occurs, only one of these estimates is wrong; a majority vote of the four estimates gives a_{10}. If there are two errors, there will be no majority, and a double-error pattern is detected.

In the same way, each data bit of $\boldsymbol{a}_2$ can be decoded. Because each row of $\boldsymbol{G}_r$ has an equivalent role, no bit of $\boldsymbol{a}_2$ is preferred. By a permutation of columns, each row of $\boldsymbol{G}_r$ can be made to look like the last row. Hence the same check sums can be used if the indices are permuted. Each bit is decoded by setting up 2^{m-r} linear check sums in the 2^m received bits, followed by a majority vote.

After these data bits are known, their contribution to the codeword is subtracted from the senseword. This results in the equivalent of a codeword from an $(r-1)$th-order Reed–Muller code. It, in turn, can have its last section of data bits recovered by the same procedure.

The Reed algorithm repeats this step until all data bits are recovered.

There is another way to represent a Reed–Muller code which is sometimes convenient. The *Plotkin representation* of a Reed–Muller code expresses the Reed–Muller code $\mathcal{C}$ of blocklength $n = 2^m$ in terms of two Reed–Muller codes $\mathcal{C}'$ and $\mathcal{C}''$, each of blocklength 2^{m-1}.

Theorem 13.1.2. *The rth-order Reed–Muller code of blocklength 2^m is the set of concatenated codewords $\boldsymbol{c} = |\boldsymbol{b}'|\boldsymbol{b}' + \boldsymbol{b}''|$ where $\boldsymbol{b}'$ and $\boldsymbol{b}''$ are, respectively, codewords from an rth-order and an $(r-1)$th-order Reed–Muller code of blocklength 2^{m-1}.*

Proof: We can write the Plotkin representation as

$$\begin{vmatrix} \boldsymbol{b}' \\ \boldsymbol{b}' + \boldsymbol{b}'' \end{vmatrix} = \begin{vmatrix} 1 & 0 \\ 1 & 1 \end{vmatrix} \begin{vmatrix} \boldsymbol{b}' \\ \boldsymbol{b}'' \end{vmatrix}.$$

Let $\boldsymbol{G}'$ be the generator matrix of an rth-order Reed–Muller code of blocklength 2^{m-1}. Let $\boldsymbol{G}''$ be the generator matrix of an $(r-1)$th-order Reed–Muller code of blocklength

2^{m-1}. We will show that the matrix

$$G = \begin{bmatrix} G' & G' \\ \mathbf{0} & G'' \end{bmatrix}$$

is the generator matrix for an rth-order Reed–Muller code of blocklength n.

However, in the construction of the generator matrix G of the Reed–Muller code, the matrix G_1 starts with a row whose first 2^{m-1} elements are zeros and whose second 2^{m-1} elements are ones. Thus

$$G_1 = \begin{bmatrix} \mathbf{0} & \mathbf{1} \\ G_1' & G_1' \end{bmatrix}.$$

Rearrange the rows of G so that this first row $[\mathbf{0}, \mathbf{1}]$ and all rows computed from it come last.

The generator matrix for the (16, 11, 4) Reed–Muller code constructed using the Plotkin representation is

$$G = \begin{bmatrix}
11 & 11 & 11 & 11 & 11 & 11 & 11 & 11 \\
00 & 00 & 11 & 11 & 00 & 00 & 11 & 11 \\
00 & 11 & 00 & 11 & 00 & 11 & 00 & 11 \\
01 & 01 & 01 & 01 & 01 & 01 & 01 & 01 \\
00 & 00 & 00 & 11 & 00 & 00 & 00 & 11 \\
00 & 00 & 01 & 01 & 00 & 00 & 01 & 01 \\
00 & 01 & 00 & 01 & 00 & 01 & 00 & 01 \\
00 & 00 & 00 & 00 & 11 & 11 & 11 & 11 \\
00 & 00 & 00 & 00 & 00 & 00 & 11 & 11 \\
00 & 00 & 00 & 00 & 00 & 11 & 00 & 11 \\
00 & 00 & 00 & 00 & 01 & 01 & 01 & 01
\end{bmatrix}.$$

The Dumer algorithm is an alternative method of decoding Reed–Muller codes that is suitable for decoding a soft senseword consisting of a vector of probabilities on the codeword. The senseword corresponding to codeword c is the vector p of real numbers with components $p_i \in (0, 1)$, which we regard as the probability that the ith symbol is a zero. We may also write this senseword more fully by setting $q_i = 1 - p_i$ and forming the array of probability distributions

$$p = [(p_0, q_0), (p_1, q_1), \ldots, (p_{n-1}, q_{n-1})]$$

where p_i is the probability that the ith bit is a zero and $q_i = 1 - p_i$ is the probability that the ith bit is a one.

The Dumer algorithm has a recursive structure, based on the Plotkin representation, of the kind known as *decimation*. At each step, the task of decoding a Reed–Muller code is broken into two tasks, each of which is the task of decoding a Reed–Muller code with half the blocklength. The Dumer algorithm is not a majority decoding algorithm.

It is a soft-input decoding algorithm in which the ith senseword symbol is a real number $p_i \in (0, 1)$, which can be thought of as the probability that c_i is a zero.

A codeword can be written in the partitioned form

$$c = |c'|c''| = |b'|b' + b''|,$$

and the array of probability vectors can be partitioned in a similar way, as

$$\{(p, q)\} = \{|(p', q')|(p'', q'')|\}.$$

The recursion starts with a probability vector on c and, in general, computes probability vectors on b' and b''. The exceptional case is when c has blocklength one, in which case the single bit of c is estimated from its probability distribution. Because $b'' = c' + c''$, component b''_i is a zero if c'_i and c''_i are both zero or if c'_i and c''_i are both one. Therefore the probability that b''_i is a zero is $p'_i p''_i + q'_i q''_i$, and the probability that b''_i is a one is $p'_i q''_i + q'_i p''_i$. This gives the rule for computing the probability vector on b'', namely

$$(\widetilde{p}, \widetilde{q}) = \{(p'_i p''_i + q'_i q''_i, \; p'_i q''_i + q'_i p''_i)\}.$$

□

Step 1. Use this formula to compute $(\widetilde{p}, \widetilde{q})$ from (p, q) and then call any soft decoder for an $(r-1)$th-order Reed–Muller code of blocklength 2^{m-1} to find b''. The Dumer algorithm would be an appropriate algorithm for this purpose.

Step 2. After b'' is computed, there are two independent soft sensewords for c'. Because $b' = c'$, one soft senseword for b' can be immediately written as $(\widetilde{p}', \widetilde{q}') = (p', q')$. A second soft senseword for b' is formed by writing $b' = c'' + b''$. This implies that the probability distribution on b' is

$$(p', q') = \begin{cases} p'_i, q'_i & \text{if } b''_i = 0 \\ q'_i, p'_i & \text{if } b''_i = 1. \end{cases}$$

Now merge $(\widetilde{p}', \widetilde{q}')$ and $(\widetilde{p}'', \widetilde{q}'')$ into (p, q). The formulas

$$p_i = \frac{\widetilde{p}'_i \widetilde{p}''_i}{\widetilde{p}'_i \widetilde{p}''_i + \widetilde{q}'_i \widetilde{q}''_i} \qquad q_i = \frac{\widetilde{q}'_i \widetilde{q}''_i}{\widetilde{p}'_i \widetilde{p}''_i + \widetilde{q}'_i \widetilde{q}''_i}$$

are suitable. Now call any decoder for an rth-order Reed–Muller code of blocklength 2^{m-1} to find b'. The Dumer algorithm would be an appropriate algorithm for this purpose.

When b' is returned, $c = |b'|b' + b''|$ is computed and the correction of errors is complete.

Of course, it is very natural for step 1 and step 2 to call the same procedure, and this is how the recursive structure of the Dumer algorithm arises.

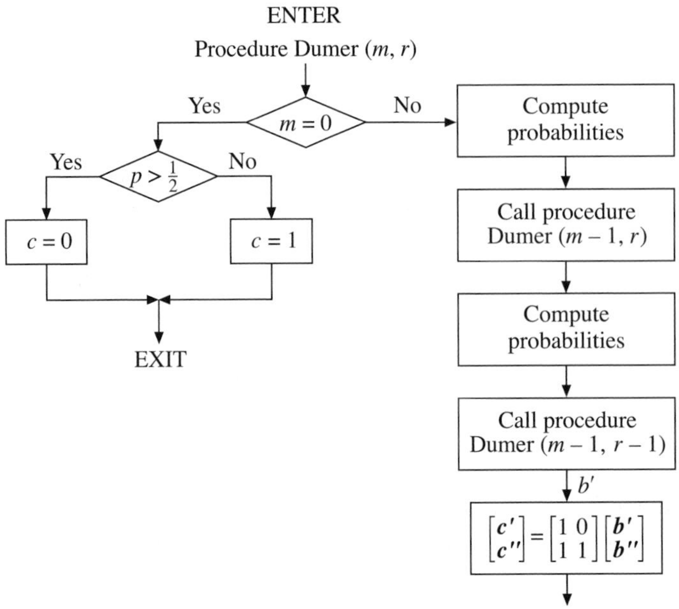

Figure 13.1. The Dumer algorithm

A flow diagram of the Dumer algorithm is shown in Figure 13.1. It has a recursive structure that calls itself twice. This means that it must be executed by using a push-down stack in which intermediate variables are pushed down each time the procedure is recalled and popped up each time the procedure is finished. If the blocklength is one, then the decoder simply sets the single bit to zero if $p_i > \frac{1}{2}$, and otherwise sets the bit to one. If the blocklength is not one, the decoder splits the codeword in half and calls decoders for each of the two smaller codes. It may call any decoders, but it is natural to use the same decoder.

For example, to decode the (16, 11, 4) Reed–Muller code with $r = 2$, the Dumer algorithm calls a decoder for the (8, 7, 2) Reed–Muller code with $r = 2$ and the (8, 4, 4) Reed–Muller code with $r = 1$. In turn these call decoders for the (4, 4, 1) Reed–Muller code with $r = m = 2$, the (4, 3, 2) Reed–Muller code with $r = 1$, again the (4, 3, 2) Reed–Muller code with $r = 1$, and the (4, 1, 4) Reed–Muller code with $r = 0$. There are two terminal cases: The first and the last.

13.2 Decoding by majority vote

The Reed–Muller codes can be decoded by taking a majority vote of a set of the check equations for each data symbol. The Reed–Muller codes are not the only codes that can be majority decoded.

Recall that any linear (n, k) code over $GF(q)$ has a check matrix $\boldsymbol{H}$, and the codewords satisfy $\boldsymbol{c}\boldsymbol{H}^{\mathrm{T}} = \boldsymbol{0}$. If we restrict attention to the jth row of $\boldsymbol{H}$, we have the check equation

$$\sum_{i=0}^{n-1} H_{ij} c_i = 0.$$

Any linear combination of rows of $\boldsymbol{H}$ forms another check equation. A total of q^{n-k} check equations can be formed in this way. The art of majority decoding lies in choosing an appropriate subset of these check equations.

In this and the next section, we will construct majority decoders. In later sections, we shall study the construction of codes for which majority decoding is possible.

Definition 13.2.1. *A set of check equations is concurring on coordinate k if c_k appears in every check equation of the set, and for every $j \neq k$, c_j appears in at most one check equation of the set.*

As shown in the following theorem, for each k the majority decoder estimates e_k by a majority vote of a set of J check equations. The majority vote will always be correct if the number of errors is less than $J/2$. If the number of check equations, J, is even, then ties are broken in the direction of no error.

Theorem 13.2.2. *If a set of J check equations is concurring on coordinate k, then c_k can be correctly estimated, provided that no more than $J/2$ errors occurred in the received word.*

Proof: The proof is given by exhibiting the correction procedure. Because c_k appears in every check equation in this set, H_{kj} is not zero. Set

$$s_j = H_{kj}^{-1} \sum_{i=0}^{n-1} H_{ij} v_i = H_{kj}^{-1} \sum_{i=0}^{n-1} H_{ij} e_i.$$

The division by H_{kj} ensures that the term multiplying e_k equals one. But now $s_j = e_k$ for at least half of the values of j because there are J equations, at most $J/2$ of the other e_i are nonzero, and each of the other e_i appears at most once in the set of check equations. Hence at least $J/2$ of the s_j are equal to e_k if e_k equals zero, and more than $J/2$ of the s_j are equal to e_k if e_k is not equal to zero. We can find e_k by a majority vote over the s_j, as was to be proved. □

Corollary 13.2.3. *If each coordinate has a set of J check equations concurring on it, the code can correct $[J/2]$ errors.*

Proof: The proof is immediate. □

If J is even, $J/2$ is an integer, and the code can correct $J/2$ errors. This implies that the code has a minimum distance of at least $J+1$. The next theorem gives a direct proof of this and includes the case where J is odd. Hence $J+1$ is called the *majority-decodable distance*, denoted d_{MD}. The true minimum distance may be larger.

Theorem 13.2.4. *If each coordinate of a linear code has a set of J check equations concurring on it, the code has a minimum distance of at least $J + 1$.*

Proof: Choose any nonzero codeword, and pick any place k where it is nonzero. There are J check equations concurring on k. Each equation involves a nonzero component c_k, and thus it must involve at least one other nonzero component because the check equation is equal to zero. Hence there are at least J other nonzero components, and the theorem is proved. □

Corollary 13.2.5. *If any coordinate of a cyclic code has a set of J check equations concurring on it, then the minimum distance of the code is at least $J + 1$.*

Proof: Every cyclic shift of a codeword in a cyclic code is another codeword in the same code. Therefore a set of J check equations on one coordinate can be readily used to write down a set of J check equations on any other coordinate. □

Theorem 13.2.6. *Let $\overline{d}$ be the minimum distance of the dual code of $\mathcal{C}$, a code over $GF(q)$. Not more than $\frac{1}{2}(n-1)/(\overline{d}-1)$ errors can be corrected by a majority decoder for $\mathcal{C}$.*

Proof: Linear combinations of rows of $\boldsymbol{H}$ are codewords in the dual code $\mathcal{C}^{\perp}$. Consider a set of J check equations concurring on the first coordinate. Each equation has at least $\overline{d} - 1$ nonzero components in the remaining $n - 1$ coordinates, and each of the remaining $n - 1$ coordinates is nonzero at most once in the J equations. Hence

$$J(\overline{d}-1) \leq n-1,$$

and $J/2$ errors can be corrected by a majority decoder. □

The theorem states that majority decoding cannot decode to the packing radius of a code unless

$$d_{\min} \leq \frac{n-1}{\overline{d}-1} + 1.$$

This is a necessary, but not sufficient, condition. Most interesting codes fail to meet this condition.

The majority decoder is quite simple, but generally it can be used only for codes of low performance. Therefore we introduce something a little more complex – an L-step majority decoder. A two-step majority decoder uses a majority decision to localize an error to a set of components rather than to a specific component. Then another majority decision is made within this set to find the error. An L-step majority decoder uses L levels of majority voting. At each level, the logic begins with an error already localized to a set of components, and by a majority decision, further localizes the error to a subset of these components.

Definition 13.2.7. *A set of check equations is concurring on the set of components* $i_1, i_2, \ldots, i_r$ *if, for some coefficients* $A_1, A_2, \ldots, A_r$*, the sum* $A_1c_{i_1} + A_2c_{i_2} + \cdots + A_rc_{i_r}$ *appears in every check equation of the set, and every* c_i *for* $i \neq i_1, i_2, \ldots, i_r$ *appears in at most one check equation of the set.*

An L-step majority decoder may correct more errors than a single-step majority decoder, but in general, it will not reach the packing radius of the code, as shown by the following theorem.

Theorem 13.2.8. *Let* $\overline{d}$ *be the minimum distance of the dual code of* $\mathcal{C}$*, a code over* $GF(q)$*. Not more than* $(n/\overline{d}) - \frac{1}{2}$ *errors can be corrected by an* L*-step majority decoder for* $\mathcal{C}$*.*

Proof: To correct t errors, it is first necessary to construct J check equations, with $J = 2t$, concurring on some set $\mathcal{B}$ that contains b components. Each such equation corresponds to a linear combination of rows of $\boldsymbol{H}$, and these linear combinations are codewords in the dual code $\mathcal{C}^{\perp}$. Excluding components in the set $\mathcal{B}$, let the number of nonzero components in the ℓth check equation be denoted by a_ℓ for $\ell = 1, \ldots, J$. These equations correspond to codewords in the dual code, and thus

$$b + a_\ell \geq \overline{d}.$$

Summing these J equations gives

$$Jb + \sum_{\ell=1}^{J} a_\ell \geq J\overline{d}.$$

Because each component other than those in set $\mathcal{B}$ can be nonzero in at most one of the check equations,

$$\sum_{\ell=1}^{J} a_\ell \leq n - b.$$

Eliminating b gives

$$Jn + \sum_{\ell=1}^{J} a_\ell \geq J\left(\overline{d} + \sum_{\ell=1}^{J} a_\ell\right).$$

Now we need to derive a second condition. Subtracting two codewords in the dual code produces another codeword in which the b places have zeros. That is, for $\ell \neq \ell'$,

$$a_\ell + a_{\ell'} \geq \overline{d}.$$

There are $J(J-1)$ such equations, and each a_ℓ appears in $2(J-1)$ of them. Adding all such equations together gives

$$2(J-1)\sum_{\ell=1}^{J} a_\ell \geq J(J-1)\overline{d}.$$

Finally, eliminate $\sum_{\ell=1}^{J} a_\ell$, using this equation and the earlier equation. This gives

$$2J(n-\overline{d}) \geq J(J-1)\overline{d}.$$

Because a majority decoder can correct $J/2$ errors, the theorem follows. □

As an application of this theorem, we observe that the (23, 12) Golay code cannot be decoded by an L-step majority decoder.

13.3 Circuits for majority decoding

Majority-decodable codes are studied because the decoders are simple and fast. Thus an important part of their story is a description of the decoders. We shall study the decoders by way of examples.

As a first example, we take the (7, 3) code that is dual to the (7, 4) Hamming code. This code, known as a *simplex code*, has minimum distance 4 and can correct one error. Because $1 \leq \frac{1}{2}(7-1)/(3-1) = 3/2$, Theorem 13.2.6 does not rule out majority decoding. One way to get a check matrix for this code is to note that the cyclic code has α^0 and α^3 as zeros in $GF(8)$. Hence

$$\boldsymbol{H} = \begin{bmatrix} \alpha^0 & \alpha^0 & \alpha^0 & \alpha^0 & \alpha^0 & \alpha^0 & \alpha^0 \\ \alpha^4 & \alpha & \alpha^5 & \alpha^2 & \alpha^6 & \alpha^3 & \alpha^0 \end{bmatrix}.$$

Over $GF(2)$, this becomes

$$\boldsymbol{H} = \begin{bmatrix} 1 & 1 & 1 & 1 & 1 & 1 & 1 \\ 1 & 0 & 1 & 1 & 1 & 0 & 0 \\ 1 & 1 & 1 & 0 & 0 & 1 & 0 \\ 0 & 0 & 1 & 0 & 1 & 1 & 1 \end{bmatrix}.$$

There are sixteen different linear combinations of rows of $\boldsymbol{H}$, each of which is a check equation. Of these, the equations that check c_0 have a one in the rightmost component. There are eight such equations, given by

$$\begin{matrix} 1 & 1 & 1 & 1 & 1 & 1 & 1 \\ 0 & 0 & 1 & 0 & 1 & 1 & 1 \\ 0 & 1 & 0 & 0 & 0 & 1 & 1 \\ 0 & 0 & 0 & 1 & 1 & 0 & 1 \\ 1 & 0 & 0 & 1 & 0 & 1 & 1 \\ 1 & 1 & 0 & 0 & 1 & 0 & 1 \\ 1 & 0 & 1 & 0 & 0 & 0 & 1 \\ 0 & 1 & 1 & 1 & 0 & 0 & 1. \end{matrix}$$

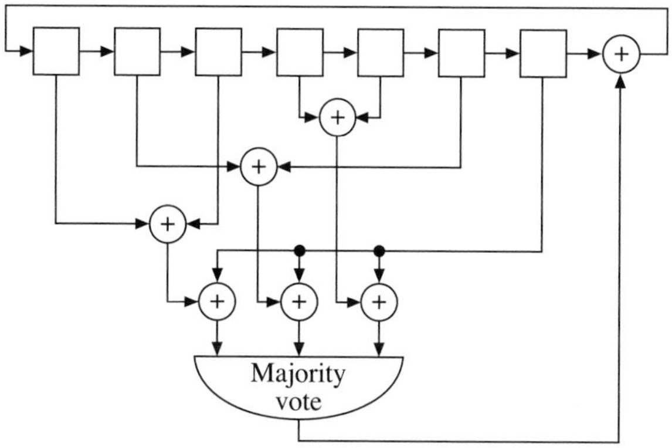

Figure 13.2. A majority decoder for the (7, 3) simplex code

From these we choose the third, fourth, and seventh to obtain the set of check equations

$$c_0 + c_1 + c_5 = 0$$
$$c_0 + c_2 + c_3 = 0$$
$$c_0 + c_4 + c_6 = 0$$

as the desired set of three equations concurring on c_0. Based on these equations, we have the decoder shown in Figure 13.2. If the shift register initially contains the received vector, after seven shifts it will contain the corrected codeword, provided a single error took place.

Of course, to correct one error, we only need to use any two of the above check equations. Adding the third equation gives a decoder that is based on a two-out-of-three decision rather than a two-out-of-two decision. As long as only one error occurs, the two decoders behave in the same way. But if two errors occur, the decoders behave differently. The two-out-of-three decoder will correct a slightly larger fraction of the double-error patterns, although both decoders will fail to correct or will miscorrect most double-error patterns.

If desired, the decoder could require all three checks to be satisfied. Then the decoder would correct all single-error patterns and detect all double-error patterns.

It is possible to turn the structure of a majority decoder into that of a Meggitt decoder. Recall that the syndrome polynomial is defined either as

$$s(x) = R_{g(x)}v(x),$$

or as

$$s(x) = R_{g(x)}[x^{n-k}v(x)].$$

The syndrome polynomial is related linearly to the senseword polynomial $v(x)$ by the check matrix. We can write each coefficient as

$$s_j = \sum_{i=0}^{n-1} H_{ij} v_i \quad j = 0, \ldots, n-k-1.$$

Any other check relationship is defined by some linear combination of the rows of $\boldsymbol{H}$, that is, as a linear combination of syndromes. Therefore all of the checks used by a majority decoder can be obtained as a linear combination of the syndrome coefficients of a Meggitt decoder. In this way, any majority decoder for a cyclic code can be implemented as a Meggitt decoder.

The simplex (7, 3) code is a cyclic code and has a generator polynomial $g(x) = x^4 + x^2 + x + 1$. Let

$$\begin{aligned} s(x) &= R_{g(x)}[x^4 v(x)] \\ &= v_6 x^3 + v_5 x^2 + v_4 x + v_3 + v_2(x^3 + x + 1) + \\ &\quad v_1(x^3 + x^2 + x) + v_0(x^2 + x + 1). \end{aligned}$$

We can write this in the form

$$[s_0\ s_1\ s_2\ s_3] = [v_0\ v_1\ v_2\ v_3\ v_4\ v_5\ v_6] \begin{bmatrix} 1 & 1 & 1 & 0 \\ 0 & 1 & 1 & 1 \\ 1 & 1 & 0 & 1 \\ 1 & 0 & 0 & 0 \\ 0 & 1 & 0 & 0 \\ 0 & 0 & 1 & 0 \\ 0 & 0 & 0 & 1 \end{bmatrix}.$$

For majority decoding, we used the check equations

$$\begin{aligned} c_0 + c_1 + c_5 &= 0 \\ c_0 + c_2 + c_3 &= 0 \\ c_0 + c_4 + c_6 &= 0. \end{aligned}$$

The first of these is the same as s_2, the second is the same as s_0, and the third is the same as $s_1 + s_3$. Figure 13.3 shows the majority decoder implemented as a Meggitt decoder.

Majority decoders, without any additional circuitry, will decode many error patterns beyond the packing radius of the code. This might seem like an attractive advantage for majority decoders. To employ majority decoding, however, usually one must use a code whose packing radius is small compared to other codes of comparable rate and blocklength. The ability of the decoder to decode beyond the packing radius can only be viewed as a partial compensation for using an inferior code.

When using a majority decoder, implemented as a Meggitt decoder, it is necessary to add feedback if the decoder is to correct many patterns beyond the packing radius. This is the function of the feedback path in Figure 13.3.

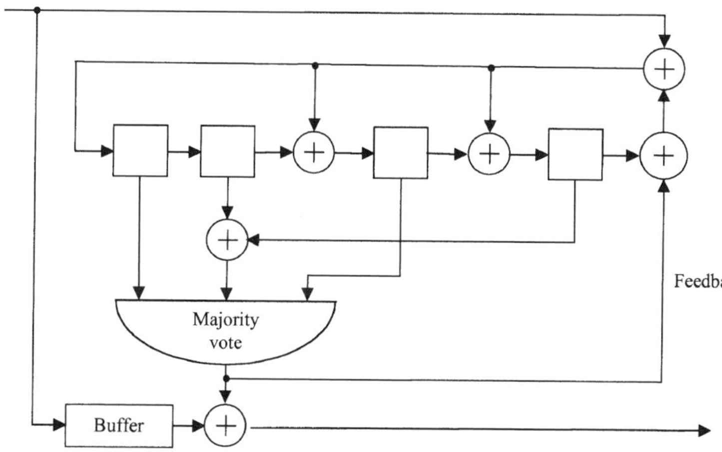

Figure 13.3. A Meggitt decoder for the (7, 3) simplex code

13.4 Affine permutations for cyclic codes

A cyclic shift of any codeword in a cyclic code gives another codeword. Hence the code is invariant under a cyclic shift. A cyclic shift is a simple example of automorphism of a code. A cyclic code may also be invariant under other permutations. In this section we shall show that many cyclic codes, when extended, are invariant under a large group of permutations called *affine permutations*. The affine permutations can be used to design codes, to analyze codes, or to design decoders – particularly majority decoders.

Let $\mathcal{C}$ be a cyclic code of blocklength $n = q^m - 1$ generated by the polynomial $g(x)$. Let $\overline{\mathcal{C}}$ be the extended code of blocklength q^m obtained from $\mathcal{C}$ by appending an overall check symbol to every codeword in $\mathcal{C}$. That is, if $(c_{n-1}, \ldots, c_1, c_0)$ is a codeword in $\mathcal{C}$, then $(c_{n-1}, \ldots, c_1, c_0, c_\infty)$ is a codeword in $\overline{\mathcal{C}}$ where c_∞ is the overall check symbol, given by $c_\infty = -(c_{n-1} + \cdots + c_1 + c_0)$.

To index the components of a codeword, we will use $GF(q^m)$ as a set of location numbers. The nonzero elements in $GF(q^m)$ are given by α^i where α is primitive. The zero element 0 in $GF(q^m)$ will be represented by α^∞. We number the components of a vector $(c_{n-1}, \ldots, c_1, c_0, c_\infty)$ in $\overline{\mathcal{C}}$ by the elements of $GF(q^m)$ as follows: Component c_∞ is numbered α^∞; for $0 \leq i < q^m - 1$, component c_i is numbered α^i.

A group of permutations is called *transitive* if, for every pair of locations (X, Y) in a codeword, there is a permutation in the group that interchanges them, possibly rearranging other locations as well. A group of permutations is called *doubly transitive* if, for every two pairs of locations $((X_1, Y_1), (X_2, Y_2))$ with $X_1 \neq X_2$ and $Y_1 \neq Y_2$, there is a permutation in the group that interchanges the locations in the first pair and also interchanges the locations in the second pair, possibly rearranging other locations as well.

An affine permutation is one that carries the component with location number X to the component with location number $aX + b$ where a and b are any fixed elements in $GF(q^m)$ and $a \neq 0$. The set of all affine permutations is a group under composition because: (1) if location X goes to location $Y = aX + b$, and location Y goes to location $Z = a'Y + b'$, then X goes to $Z = a'aX + a'b + b'$; (2) the permutation $a^{-1}X - a^{-1}b$ is the inverse permutation to $aX + b$. The group of affine permutations is doubly transitive because, given the pairs (X_1, Y_1) and (X_2, Y_2), the pair of equations

$$Y_1 = aX_1 + b$$
$$Y_2 = aX_2 + b$$

has a unique solution for a and b.

Theorem 13.4.1. *Any code of blocklength $n = q^m$ that is invariant under the group of affine permutations can be made into a cyclic code by dropping the component with location number α^∞.*

Proof: Let α be the primitive element used to define the location numbers. The permutation $Y = \alpha X$ is an affine permutation. But this permutation is a cyclic shift of all locations except α^∞. Hence the code is cyclic. □

It is more difficult to give a condition that works in the other direction, that is, stating when a cyclic code can be extended to obtain a code that is invariant under the group of affine permutations. Such cyclic codes, identified in Theorem 13.4.4, are called cyclic codes with the doubly-transitive-invariant property. Before we can study this theorem, we need some background.

The radix-q representation of an integer j is the unique expression

$$j = j_0 + j_1 q + j_2 q^2 + \cdots + j_{m-1} q^{m-1} \quad 0 \leq j_\ell < q.$$

Likewise, the radix-q representation of an integer k is

$$k = k_0 + k_1 q + k_2 q^2 + \cdots + k_{m-1} q^{m-1} \quad 0 \leq k_\ell < q.$$

Definition 13.4.2. *An integer k is called a radix-q descendent of j if the radix-q coefficients satisfy*

$$k_\ell \leq j_\ell \quad \ell = 0, \ldots, m-1.$$

It may not be immediately apparent whether a given integer k is a radix-q descendent of an integer j. For the case where q is a prime p, the next theorem will be used to give a simple, yet equivalent, condition in terms of $\binom{j}{k}$. In reading this theorem, recall the convention that if $m > n$, $\binom{n}{m} = 0$.

Theorem 13.4.3 (Lucas Theorem). *The radix-p representation, where p is a prime, of two integers j and k satisfies the following congruence:*

$$\binom{j}{k} = \prod_{\ell=0}^{m-1} \binom{j_\ell}{k_\ell} \pmod{p}.$$

Further, $\binom{j}{k}$ *equals zero modulo p if and only if k is not a radix-p descendent of j.*

Proof: The last statement follows from the first because $\binom{j_\ell}{k_\ell}$ is not equal to zero modulo p for all ℓ if and only if k is a radix-p descendent of j. We only need to prove the first statement of the theorem. The proof will consist of expanding the polynomial $(1+x)^j$ in two different ways and then equating coefficients of like powers of x. Using Theorem 4.6.10, we can write the polynomial identity over $GF(p)$:

$$(1+x)^{p^i} = 1 + x^{p^i}.$$

Then for arbitrary j, we have

$$\begin{aligned}(1+x)^j &= (1+x)^{j_0 + j_1 p + j_2 p^2 + \cdots + j_{m-1} p^{m-1}} \\ &= (1+x)^{j_0}(1+x^p)^{j_1}(1+x^{p^2})^{j_2} \ldots (1+x^{p^{m-1}})^{j_{m-1}}.\end{aligned}$$

Next, use the binomial expansion on both sides of the equation above to write it in the following form:

$$\sum_{k=0}^{j} \binom{j}{k} x^k = \left[\sum_{k_0=0}^{j_0} \binom{j_0}{k_0} x^{k_0}\right]\left[\sum_{k_1=0}^{j_1} \binom{j_1}{k_1} x^{k_1 p}\right] \cdots \left[\sum_{k_{m-1}=0}^{j_{m-1}} \binom{j_{m-1}}{k_{m-1}} x^{k_{m-1} p^{m-1}}\right]$$

where each k_ℓ is less than p. Now multiply out the polynomials on the right. The radix-p expansion of k

$$k = k_0 + k_1 p + k_2 p^2 + \cdots + k_{m-1} p^{m-1}$$

is unique so there can be only one term in the sum because equating coefficients of x^k on both sides gives

$$\binom{j}{k} = \binom{j_0}{k_0}\binom{j_1}{k_1} \cdots \binom{j_{m-1}}{k_{m-1}} \pmod{p}.$$

This completes the proof of the theorem. □

Now we are ready to characterize those extended cyclic codes that are invariant under the group of affine permutations. The characterization is in terms of the zeros of the generator polynomial. We exclude from consideration those generator polynomials $g(x)$ with a zero at α^0 because, otherwise, indeterminate expressions would occur.

Theorem 13.4.4. *Let* α *be a primitive element of* $GF(q^m)$*, a field of characteristic p. Let C be a cyclic code of blocklength* $q^m - 1$ *generated by* $g(x)$*, a polynomial for* *which α^0 is not a zero. The extended code $\overline{\mathcal{C}}$, obtained by appending an overall check symbol, is invariant under the group of affine permutations if and only if $\alpha^{k'}$ is a zero of $g(x)$ whenever α^k is a zero of $g(x)$ and k' is a nonzero radix-p descendent of k.*

Proof: Let $X' = aX + b$ denote an affine permutation. Let $X_1, X_2, \ldots, X_\nu$ be the location numbers of the nonzero components of the codeword $\boldsymbol{c}$, and let $Y_1, Y_2, \ldots, Y_\nu$ denote the values of these components. Further, let $X'_1, X'_2, \ldots, X'_\nu$ denote the location numbers of the nonzero components of the codeword under the affine permutation. That is, $X'_i = aX_i + b$.

First, suppose that whenever α^k is a zero of $g(x)$, then so is $\alpha^{k'}$ for every k' that is a nonzero radix-p descendent of k. We must show that the permutation of a codeword produces another codeword.

The codeword polynomial satisfies $c(x) = g(x)d(x)$, and the extension symbol is $c_\infty = -(c_{n-1} + \cdots + c_0)$. Let $C_j = c(\alpha^j) = \sum_{i=0}^{n-1} \alpha^{ij} c_i$. Then

$$C_j = \sum_{i=0}^{n-1} c_i \alpha^{ij} = \sum_{\ell=1}^{\nu} Y_\ell X_\ell^j = 0$$

for every j for which α^j is a zero of $g(x)$. Notice that there may be a nonzero symbol at location number α^∞. But α^∞ is a representation for the zero element, thus this X_ℓ is zero. Even though the term $Y_\ell X_\ell^j$ when X_ℓ is zero was not included in the definition of C_j, we may consider it to be included on the right because it is zero anyway.

The permuted word $\boldsymbol{c}'$ is a codeword if $c'_\infty + c'_{n-1} + \cdots + c'_0 = 0$, which, clearly, is still true after the permutation, and if

$$C'_j = \sum_{\ell=1}^{\nu} Y_\ell X_\ell'^{j} = 0$$

for every j for which α^j is a zero of $g(x)$. Consider such j:

$$\begin{aligned} C'_j &= \sum_{\ell=1}^{\nu} Y_\ell (aX_\ell + b)^j \\ &= \sum_{\ell=1}^{\nu} Y_\ell \sum_{k=0}^{j} \binom{j}{k} a^k X_\ell^k b^{j-k} \\ &= \sum_{k=0}^{j} \binom{j}{k} b^{j-k} a^k C_k. \end{aligned}$$

By Theorem 13.4.3, $\binom{j}{k}$ is equal to zero unless k is a radix-p descendent of j, and by assumption, C_k is zero for such k. Hence in every term of the sum, either $\binom{j}{k}$ equals zero or C_k equals zero. Thus C'_j equals zero, and the permuted codeword is again a codeword.

Now prove the converse. Assume that the extended code is invariant under the affine group of permutations. Then every codeword satisfies

$$C'_j = \sum_{\ell=1}^{\nu} Y_\ell (aX_\ell + b)^j = 0$$

for every a and b, and for every j for which α^j is a zero of $g(x)$. As before, this becomes

$$C'_j = \sum_{k=0}^{j} \binom{j}{k} b^{j-k} a^k C_k = 0.$$

Let K be the number of radix-p descendents of j, and let k for $\ell = 1, \ldots, K$ index them. The sum can be written as

$$C'_j = \sum_{\ell=0}^{K} \binom{j}{k_\ell} b^{j-k_\ell} a^{k_\ell} C_{k_\ell} = 0.$$

Now a and b are arbitrary. We choose, in turn, b equal to one and a equal to the first K successive powers of α to get

$$\begin{bmatrix} (\alpha^{k_1})^0 & (\alpha^{k_2})^0 & \ldots & (\alpha^{k_K})^0 \\ (\alpha^{k_1})^1 & (\alpha^{k_2})^1 & \ldots & (\alpha^{k_K})^1 \\ \vdots & & & \vdots \\ (\alpha^{k_1})^{K-1} & (\alpha^{k_2})^{K-1} & \ldots & (\alpha^{k_K})^{K-1} \end{bmatrix} \begin{bmatrix} \binom{j}{k_1} C_{k_1} \\ \binom{j}{k_2} C_{k_2} \\ \binom{j}{k_K} C_{k_K} \end{bmatrix} = \begin{bmatrix} 0 \\ 0 \\ \vdots \\ 0 \end{bmatrix}.$$

This matrix is a Vandermonde matrix. It is invertible because all columns are distinct. Therefore $C_{k_\ell} = 0$ for $\ell = 1, \ldots, K$. Hence α^{k_ℓ} is a zero of $g(x)$ whenever k_ℓ is a radix-p descendent of j. This completes the proof of the theorem. □

13.5 Cyclic codes based on permutations

Now we will describe a method for constructing some one-step, majority-decodable codes. The technique makes use of the affine permutations described in Section 13.4. This technique can be used to obtain cyclic codes over $GF(q)$ of blocklength n, provided that n is composite. In addition, the field characteristic p and the factors of n are coprime because n is a divisor of $q^m - 1$.

Choose a primitive and composite blocklength n equal to $q^m - 1$, and let $n = L \cdot J$. Then

$$\begin{aligned} x^n - 1 &= (x^J)^L - 1 \\ &= (x^J - 1)\big(x^{J(L-1)} + x^{J(L-2)} + \cdots + x^J + 1\big). \end{aligned}$$

Denote the second term in this factorization by $a(x)$ so that

$$x^n - 1 = (x^J - 1)a(x).$$

The nonzero elements of $GF(q^m)$ are zeros either of $x^J - 1$ or of $a(x)$. If α is primitive in $GF(q^m)$, then $\alpha^{LJ} = 1$, and $\alpha^{L(J-1)}, \alpha^{L(J-2)}, \ldots, \alpha^L, 1$ are the J zeros of $x^J - 1$. Therefore for each j, α^j is a zero of $a(x)$ unless j is a multiple of L.

Define as follows a polynomial $h(x)$ whose reciprocal will be the check polynomial for the desired code. For each j, α^j is a zero of $\widetilde{h}(x)$ unless j or a radix-q descendent of j is a multiple of L. Because every zero of $\widetilde{h}(x)$ is then a zero of $a(x)$, we see that $a(x)$ is a multiple of $h(x)$.

Let $h(x)$ be the reciprocal polynomial of $\widetilde{h}(x)$, and let $g(x)$ be defined by

$$g(x)h(x) = x^n - 1.$$

We now have two cyclic codes that are duals. Let $\mathcal{C}$ be the cyclic code with generator polynomial $g(x)$, and let $\mathcal{C}^\perp$ be the cyclic code with generator polynomial $\widetilde{h}(x)$. By the definition of $\widetilde{h}(x)$ and Theorem 13.4.4, the code $\mathcal{C}^\perp$ is a cyclic code having the doubly-transitive-invariant property. That is, we can extend $\mathcal{C}^\perp$ to a code $\mathcal{C}^\perp$ that is invariant under the group of affine permutations.

We will form J check equations for the cyclic code $\mathcal{C}$ that are concurring on a single component. Because the code is cyclic, therefore it is majority decodable, and the minimum distance is at least $J + 1$. We will find these check equations for the code $\mathcal{C}$ by working with the dual code $\mathcal{C}^\perp$.

As we have seen previously, the polynomial

$$a(x) = x^{J(L-1)} + x^{J(L-2)} + \cdots + x^J + 1$$

has α^j as a zero if and only if j is not a multiple of L. Hence $a(x)$ is a multiple of $\widetilde{h}(x)$ and is a codeword in the dual code. Therefore the elements of the set

$$\{a(x), xa(x), x^2a(x), \ldots, x^{J-1}a(x)\}$$

of polynomials are all codewords in $\mathcal{C}^\perp$. Each of these designated codewords has Hamming weight L and, by inspection of $a(x)$, it is clear that no two of them have a common nonzero component.

Now we will find another set of codewords in $\mathcal{C}^\perp$ that can be used to define check equations for $\mathcal{C}$. We will do this by adding temporarily an extended symbol to $\mathcal{C}^\perp$. This symbol will later serve as a kind of pivot on which to develop the J concurring check equations. In order to move the extended symbol into the interior of the cyclic code, it will be permuted to another component, and then the new extended symbol will be dropped.

Thus append an overall check symbol to each of the designated codewords. This gives J codewords in the extended code $\overline{\mathcal{C}^\perp}$ of blocklength $\overline{n} = n + 1$. If $L \neq 0$ (modulo p), the extended symbol is nonzero and is the same for every one of the J codewords that is constructed. Divide the codewords by this extended symbol to get a new set of codewords. We now have established the existence of a set of J codewords in $\mathcal{C}^{\perp}$ with the following properties.

1. Each codeword in the set of J codewords has a one in location α^{∞}.
2. One and only one codeword in the set of J codewords has a nonzero component at location α^j for $j = 0, 1, \ldots, n-1$.

Now we are ready to make use of Theorem 13.4.4. In fact, the preceding definitions were designed so that the theorem applies. Because the code $\mathcal{C}^{\perp}$ has the doubly-transitive-invariant property, it is invariant under any affine permutation. In particular, choose

$$Y = \alpha X + \alpha^{n-1}.$$

This permutation carries the set of codewords in $\overline{\mathcal{C}^{\perp}}$ into another set of codewords in $\overline{\mathcal{C}^{\perp}}$ with the following properties.

1. Each codeword in the new designated set of J codewords has a one in location α^{n-1}.
2. One and only one codeword in the new set of J codewords has a nonzero component at location α^j for $j = \infty, 0, 1, \ldots, n-2$.

We now can drop location α^{∞} to get a set of codewords in $\mathcal{C}^{\perp}$ that are concurring on location $n-1$. In fact, these J codewords are orthogonal to the codewords of code $\mathcal{C}$ and thus form check equations. There are J check equations concurring on location $n-1$, and the code is a cyclic code. We now are ready for the following theorem.

Theorem 13.5.1. *Let $n = q^m - 1$ be factorable as $n = J \cdot L$. Suppose that a cyclic code $\mathcal{C}$ has a generator polynomial, $g(x)$, constructed by the following rule. For each j, if j is a multiple of L, or any radix-q descendent of j is a multiple of L, then α^{-j} is a zero of $g(x)$, and otherwise α^{-j} is not such a zero. Then $\mathcal{C}$ is a majority-decodable code over $GF(q)$ with a minimum distance of at least J.*

Proof: By the definition of $\widetilde{h}(x)$, α^{-j} is not a zero of $g(x)$ if and only if α^j is a zero of $\widetilde{h}(x)$. Then $\widetilde{h}(x)$ has α^j as a zero if and only if j is not a nonzero multiple of L and if every nonzero radix-q descendent of j is not a multiple of L. Hence as we have seen by reference to the dual code, there are J concurring check equations on each component.

The code is over $GF(q)$ if each conjugate of a zero of $g(x)$ is also a zero. If α^{-j} is a zero because j is a multiple of L, then qj is a multiple of $qL(\bmod LJ)$, and thus α^{-jq} is a zero also. Otherwise, if α^{-j} is a zero because j', a radix-q descendent of j, is a multiple of L, then $qj'(\bmod LJ)$, a radix-q descendent of $qj(\bmod LJ)$, is a multiple of $qL(\bmod LJ)$. □

We will illustrate the theorem with a simple example over $GF(2)$. Because

$$\begin{aligned} x^{15} - 1 &= (x^3 - 1)(x^{12} + x^9 + x^6 + x^3 + 1) \\ &= (x^5 - 1)(x^{10} + x^5 + 1), \end{aligned}$$

we can take $J = 3$ and $L = 5$, or $J = 5$ and $L = 3$, to obtain either a single-error-correcting code or a double-error-correcting code. We will construct the single-error-correcting code.

The nonzero multiples of 5 are 5 and 10; 7 and 13 have 5 as a binary descendent; and 11 and 14 have 10 as a binary descendent. Hence the zeros of $\widetilde{h}(x)$ are $\alpha, \alpha^2, \alpha^3, \alpha^4, \alpha^6, \alpha^8, \alpha^9$, and α^{12}, where α is a primitive element in $GF(16)$, and the zeros of $g(x)$ are $\alpha, \alpha^2, \alpha^4, \alpha^5, \alpha^8, \alpha^{10}$, and α^0. Hence

$$\begin{aligned} g(x) &= (x-1)(x-\alpha)(x-\alpha^2)(x-\alpha^4)(x-\alpha^5)(x-\alpha^8)(x-\alpha^{10}) \\ &= x^7 + x^3 + x + 1. \end{aligned}$$

Thus the code encodes eight data bits.

Table 13.1 tabulates some of these codes and also gives the parameters of some corresponding BCH codes. In each case, the minimum distance listed is that promised by the respective bounds, but the actual minimum distance may be larger.

Notice from the table that a few cases occur where the choice of a majority-decodable code does not involve a serious penalty in k, although in most cases the penalty is substantial. It is important to realize, however, that the majority decoder is quick and simple, and usually will readily correct many patterns with more than j errors. Simulations are worth doing before a final judgment can be made for a given application.

Table 13.1. *Some binary one-step-decodable codes compared with BCH codes*

	Majority-decodable		BCH	
n	k	d_{MD}	k	d_{BCH}
15	8	3	11	3
	6	5	7	5
63	48	3	57	3
	36	9	39	9
	12	21	18	21
255	224	3	247	3
	206	5	239	5
	174	17	191	17
	36	51	91	51
	20	85	47	85
511	342	7	484	7
	138	73	241	73
1023	960	3	1013	3
	832	11	973	11
	780	33	863	33
	150	93	598	93
	30	341	123	341

13.6 Convolutional codes for majority decoding

Some convolutional codes can be decoded by a majority decoder. Just as for block codes, these decoders are simple and extremely fast, but the codes are not as strong as other codes. A code with a higher rate is usually preferable.

Majority-decodable codes are decoded by taking a majority vote of certain of the syndromes, or a majority vote of certain linear combinations of the syndromes. A number of majority-decodable convolutional codes have been found by computer search (for example, see Figure 13.7).

The convolutional code with generator matrix

$$G(x) = [x^2 + x + 1 \quad x^2 + 1]$$

can be decoded by majority logic. We shall give two different majority decoders for this code to illustrate variations in the nature of such decoders. Figure 13.4 shows the syndromes of all single-error patterns. Modified syndromes computed according to two different rules are shown. Notice that for either rule, the modified syndromes are concurring on the first bit position. In either case, because there are four modified syndromes, an error in the first bit position can be corrected even in the presence of a second error. Figure 13.5 shows a majority decoder based on the first set of modified

Single-error patterns	Syndromes						Modified syndromes Rule 1				Modified syndromes Rule 2			
	s_5	s_4	s_3	s_2	s_1	s_0	s_3'	s_2'	s_1'	s_0'	s_3''	s_2''	s_1''	s_0''
0 0 0 0 0 0 0 0 0 0 0 1	1	1	1	0	0	1	1	1	1	1	1	1	1	1
0 0 0 0 0 0 0 0 0 0 1 0	0	0	0	0	0	1	0	0	0	1	0	0	0	0
0 0 0 0 0 0 0 0 0 1 0 0	1	1	0	0	1	0	0	1	0	0	1	0	0	0
0 0 0 0 0 0 0 0 1 0 0 0	0	0	0	0	1	0	0	0	1	0	0	1	0	0
0 0 0 0 0 0 0 1 0 0 0 0	1	0	0	1	0	0	0	0	0	0	1	0	0	0
0 0 0 0 0 0 1 0 0 0 0 0	0	0	0	1	0	0	0	1	0	0	0	0	0	0
0 0 0 0 0 1 0 0 0 0 0 0	0	0	1	0	0	0	1	0	0	0	0	0	1	0
0 0 0 0 1 0 0 0 0 0 0 0	0	0	1	0	0	0	1	0	0	0	0	0	1	0
0 0 0 1 0 0 0 0 0 0 0 0	0	1	0	0	0	0	0	0	1	0	0	1	0	0
0 0 1 0 0 0 0 0 0 0 0 0	0	1	0	0	0	0	0	0	1	0	0	1	0	0
0 1 0 0 0 0 0 0 0 0 0 0	1	0	0	0	0	0	0	1	0	0	1	0	0	0
1 0 0 0 0 0 0 0 0 0 0 0	1	0	0	0	0	0	0	1	0	0	1	0	0	0

Rule 1	Rule 2
$s_0' = s_0$	$s_0'' = s_0$
$s_1' = s_1 + s_4$	$s_1'' = s_3$
$s_2' = s_2 + s_5$	$s_2'' = s_1 + s_4$
$s_3' = s_3$	$s_3'' = s_5$

Figure 13.4. Modified syndromes for a (12, 6) convolutional code

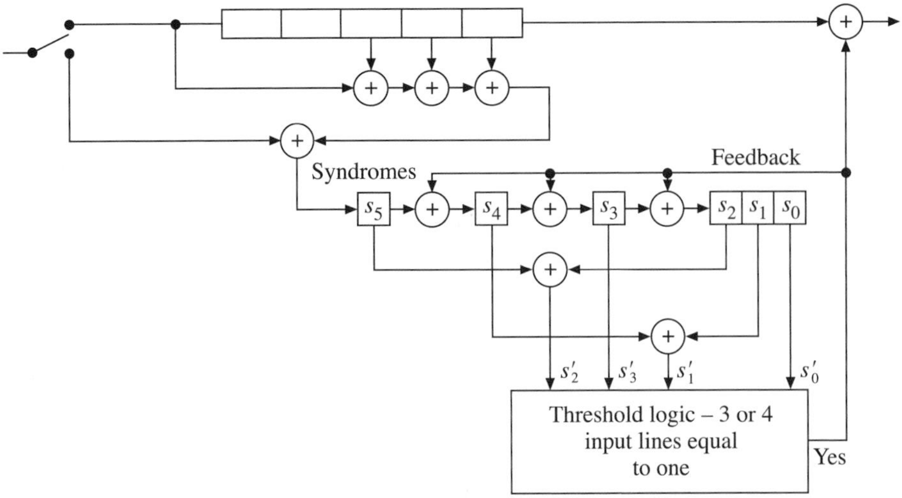

Figure 13.5. A majority decoder for a (12, 6) convolutional code

syndromes. This decoder will correct any double-error pattern and is simpler than a syndrome decoder, discussed in Section 8.6. The majority decoder has a fatal flaw, however, which is typical of many majority decoders for convolutional codes. Some patterns of more than two errors will pass the majority test and cause a false correction. For some of these, the feedback signal will modify the syndrome so that the same syndrome is reproduced even in the absence of further errors. For example, in the absence of further errors, the syndrome pattern 011010 leads to no error corrected, followed by the syndrome 001101 with one error corrected, followed by the syndrome 011010, which renews the cycle. This kind of an error is called *ordinary error propagation*. It is a property of a majority decoder, and it can occur even with a noncatastrophic code. Ordinary error propagation cannot occur in a normal syndrome decoder.

Figure 13.6, shows another majority decoder for the same code, but now based on the second set of modified syndromes from Figure 13.4. This decoder is not subject to ordinary error propagation. From this example, we see that some majority decoders for convolutional codes are subject to ordinary error propagation, but that it may be possible to eliminate the error propagation by careful selection of the modified syndromes. If this is not possible for the desired code, then a normal syndrome decoder should be used.

A short list of convolutional codes that can be corrected by majority logic is given in Figure 13.7. These codes were obtained by computer search.

13.7 Generalized Reed–Muller codes

The original Reed–Muller codes were introduced as binary codes. We now will study the Reed–Muller codes over a general Galois field $GF(q)$. The class of generalized

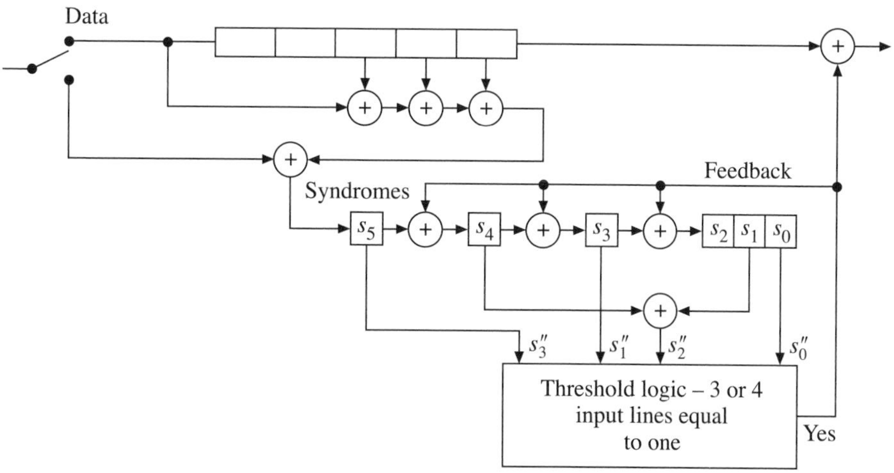

Figure 13.6. Another majority decoder

(n_0, k_0)	(n, t)	t_{MD}	$G(x)$
(2, 1)	(6, 3)	1	$[1 \quad x+1]$
(2, 1)	(14, 7)	2	$[1 \quad x^6+x^5+x^2+1]$
(2, 1)	(36, 18)	3	$[1 \quad x^{17}+x^{16}+x^{13}+x^7+x^2+1]$
(2, 1)	(72, 36)	4	$[1 \quad x^{35}+x^{31}+x^{30}+x^{18}+x^{16}+x^{10}+x^7+1]$
(2, 1)	(112, 56)	5	$[1 \quad x^{55}+x^{54}+x^{49}+x^{45}+x^{32}+x^{29}+x^{21}+x^{14}+x^2+1]$
(3, 2)	(9, 6)	1	$[1 \quad x+1 \quad x^2+1]$
(3, 2)	(42, 28)	2	$[1 \quad x^{12}+x^9+x^8+1 \quad x^{13}+x^{11}+x^6+1]$
(3, 2)	(123, 82)	3	$[1 \quad x^{30}+x^{29}+x^{24}+x^6+x^2+1 \quad x^{36}+x^{35}+x^{28}+x^{25}+x^3+1]$
(4, 3)	(16, 12)	1	$[1 \quad x+1 \quad x^2+1 \quad x^3+1]$
(4, 3)	(80, 60)	2	$[1 \quad x^{19}+x^{15}+x^3+1 \quad x^{18}+x^{17}+x^8+1 \quad x^{13}+x^{11}+x^6+1]$

Figure 13.7. A short list of majority-decodable convolutional codes

Reed–Muller (GRM) codes contains subclasses that are majority decodable, but it also contains many codes that are of no practical interest. All of these are carried along in the same general theory.

The GRM codes, including the binary codes, will be defined by extending a cyclic code called a *cyclic GRM code*. We will limit the discussion in this section to cyclic GRM codes of primitive blocklength $q^m - 1$, and to GRM codes obtained from these by appending an overall check symbol.

The zeros of the codewords are specified in a seemingly roundabout and obscure way. We must define the radix-q weight of an integer.

Definition 13.7.1. *Let j be an integer with radix-q representation*

$$j = j_0 + j_1 q + j_2 q^2 + \cdots + j_{m-1} q^{m-1}.$$

The radix-q weight of j is

$$w_q(j) = j_0 + j_1 + j_2 + \cdots + j_{m-1}$$

with addition as integers.

Definition 13.7.2. *The cyclic GRM code*[1] *of order r and blocklength $n = q^m - 1$ over the field $GF(q)$ is the cyclic code whose generator polynomial $g(x)$ has zeros at all α^j for $j = 1, \ldots, q^m - 1$, such that*

$$0 < w_q(j) \leq (q-1)m - r - 1.$$

The expansion of this cyclic code by a single check sum to a code of blocklength $n = q^m$ is called a GRM code of order r.

In Definition 13.7.1, notice that j and $jq(\bmod\ q^m - 1)$ have the same radix-q weight. Therefore if β is a zero of $g(x)$, then so are all conjugates of β, and Definition 13.7.2 does, indeed, define a code over $GF(q)$.

When $q = 2$, the GRM code reduces to a code that is equivalent to the Reed–Muller code; it is the same except for a permutation of the components – hence the name generalized Reed–Muller code. This code can correct $2^{m-r-1} - 1$ errors by majority logic.

We can restate the definition of the cyclic GRM code in terms of the spectrum of the codewords. The cyclic GRM code of order r and blocklength $n = q^m - 1$ over the field $GF(q)$ is the set of words whose spectral component C_j equals zero for all j, satisfying

$$0 < w_q(j) \leq (q-1)m - r - 1.$$

We can encode such a code in the frequency domain by setting the spectrum at all these frequencies equal to zero, filling the remaining frequencies with data insofar as the conjugacy constraints allow, and taking the inverse Fourier transform. A time-domain encoder usually will be simpler, however.

The minimum distance of cyclic GRM codes must satisfy the BCH bound. The following theorem evaluates this bound.

Theorem 13.7.3. *The cyclic GRM code over $GF(2)$ of order r and blocklength $n = 2^m - 1$ is a subcode of the BCH code of designed distance $d = 2^{m-r} - 1$, and thus has a minimum distance at least this large.*

Proof: In radix 2, $2^{m-r} - 1$ is represented by an $(m-r)$-bit binary number consisting of all ones. Numbers smaller than $2^{m-r} - 1$ have less than $m - r$ ones. Therefore

[1] An alternative definition is to choose $g(x)$ with zeros for all j, satisfying $0 \leq w_q(j) \leq (q-1)m - r - 1$ where the first inequality now includes equality. The GRM code then will be obtained by lengthening the cyclic GRM code, adding the all-one vector to $\boldsymbol{G}$ rather than to $\boldsymbol{H}$. With our definition, the Hamming codes are cyclic GRM codes.

$w_2(j) \le m - r - 1$ for $j = 1, 2, \ldots, 2^{m-r} - 2$, and $g(x)$ has α^j as a zero for $j = 1, 2, \ldots, 2^{m-r} - 2$. Consequently, the code is a subcode of the BCH code of designed distance $d = 2^{m-r} - 1$. □

For nonbinary codes, the cyclic GRM code of order r is again a subcode of the BCH code of designed distance $q^{m-r} - 1$, provided that r is less than $q - 1$. Otherwise, the condition is somewhat more complicated, as given by the following theorem.

Theorem 13.7.4. *Let $r = (q-1)Q + R$ with $0 \le R < q - 1$. The cyclic GRM code over $GF(q)$ of order r and blocklength $n = q^m - 1$ is a subcode of the BCH code over $GF(q)$ of designed distance*

$$d = (q - R)q^{(m-Q-1)} - 1,$$

and thus has a minimum distance at least this large.

Proof: Consider the q-ary representation of $(q - R)q^{(m-Q-1)} - 1$. It is equal to $q - R - 1$ in the most significant digit, and is equal to $q - 1$ in each of the other $m - Q - 1$ digits. Hence

$$\begin{aligned} w_q\big((q - R)q^{(m-Q-1)} - 1\big) &= (q-1)(m - Q - 1) + q - R - 1 \\ &= (q-1)m - r. \end{aligned}$$

But then for all j less than $(q - R)q^{(m-Q-1)} - 1$, the q-ary weight is smaller than $(q-1)m - r$. Hence for all j satisfying $0 < j \le (q - R)q^{(m-Q-1)} - 2$, α^j is a zero of the generator polynomial, and the cyclic GRM code is a subcode of the BCH code with these zeros. This completes the proof. □

The performance of some generalized Reed–Muller codes is shown in Figure 13.8. When q equals two, the codes are equivalent to Reed–Muller codes. The minimum distance listed is actually the lower bound on minimum distance given by Theorem 13.7.4. The number of data symbols is obtained by counting. Because the definition of the code specifies all zeros of $g(x)$, one can easily find the number of check symbols by counting the positive integers less than n for which $w_q(j) \le (q-1)m - r - 1$.

Theorem 13.7.5. *The dual of a GRM code over $GF(q)$ of order r and blocklength $n = q^m$ is equivalent to a GRM code of order $(q-1)m - r - 1$ and blocklength $n = q^m$.*

Proof: Let $\overline{\mathcal{C}}$ be the GRM code, and let $\mathcal{C}$ be the cyclic GRM code obtained by shortening $\overline{\mathcal{C}}$. The proof will be in three steps. In step 1, a generator polynomial, $\widetilde{h}(x)$, for the dual $\mathcal{C}^\perp$ of the cyclic GRM code $\mathcal{C}$ is found. In step 2, the check polynomials of both $\mathcal{C}$ and $\mathcal{C}^\perp$ are observed to have $x - 1$ as a factor. In step 3, $\mathcal{C}$ and $\mathcal{C}^\perp$ are extended to GRM codes $\overline{\mathcal{C}}$ and $\overline{\mathcal{C}}^\perp$, which are shown to be duals.

Step 1. $\mathcal{C}$ is the set of words with check frequencies at all j satisfying

$$0 < j \le q^m - 1,$$

$q=2$				$q=4$				$q=8$			
m	r	(n,k)	d_{BCH}	m	r	(n,k)	d_{BCH}	m	r	(n,k)	d_{BCH}
4	1	(15, 5)	7	2	1	(15, 3)	11	2	1	(63, 3)	55
4	2	(15, 11)	3	2	2	(15, 6)	7	2	2	(63, 6)	47
5	1	(31, 6)	15	2	3	(15, 10)	3	2	3	(63, 10)	39
5	2	(31, 16)	7	2	4	(15, 13)	2	2	4	(63, 15)	31
5	3	(31, 26)	3	3	1	(63, 4)	47	2	5	(63, 21)	23
6	1	(63, 7)	31	3	2	(63, 10)	31	2	6	(63, 28)	15
6	2	(63, 22)	15	3	3	(63, 20)	15	2	7	(63, 36)	7
6	3	(63, 42)	7	3	4	(63, 32)	11	2	8	(63, 43)	6
6	4	(63, 57)	3	3	5	(63, 44)	7	2	9	(63, 49)	5
7	1	(127, 8)	63	3	6	(63, 54)	3	2	10	(63, 54)	4
7	2	(127, 29)	31	3	7	(63, 60)	2	2	11	(63, 58)	3
7	3	(127, 64)	15	4	1	(255, 5)	191	2	12	(63, 61)	2
7	4	(127, 99)	7	4	2	(255, 15)	127	3	1	(511, 4)	447
7	5	(127, 120)	3	4	3	(255, 35)	63	3	2	(511, 10)	383
8	1	(225, 9)	127	4	4	(255, 66)	47	3	3	(511, 20)	319
8	2	(255, 37)	63	4	5	(255, 106)	31	3	4	(511, 35)	255
8	3	(255, 93)	31	4	6	(255, 150)	15	3	5	(511, 56)	191
8	4	(255, 163)	15	4	7	(255, 190)	11	3	6	(511, 84)	127
8	5	(255, 219)	7	4	8	(255, 221)	7	3	7	(511, 120)	63
8	6	(255, 247)	3	4	9	(255, 241)	3	3	8	(511, 162)	55
9	1	(511, 10)	255	4	10	(255, 251)	2	3	9	(511, 208)	47
9	2	(511, 46)	127					3	10	(511, 256)	39
9	3	(511, 130)	63					3	11	(511, 304)	31
9	4	(511, 256)	31					3	12	(511, 350)	23
9	5	(511, 382)	15					3	13	(511, 392)	15
9	6	(511, 466)	7					3	14	(511, 428)	7
9	7	(511, 502)	3					3	15	(511, 456)	6
								3	16	(511, 477)	5
								3	17	(511, 492)	4
								3	18	(511, 502)	3
								3	19	(511, 508)	2

Figure 13.8. Parameters of some generalized Reed–Muller codes

and

$$w_q(j) \le (q-1)m - r - 1.$$

The generator polynomial $g(x)$ has as zeros all α^j for such j. For convenience, we have replaced zero with $q^m - 1$ in the range of j so that $w_q(j) > 0$ for all j. By the theory of cyclic codes, the check frequencies of the dual code are described if j is replaced by $q^m - 1 - j$ and the inequality is reversed. That is, the check frequencies are indexed by those j for which

$$w_q(q^m - 1 - j) > (q-1)m - r - 1.$$

But if

$$j = j_0 + j_1 q + j_2 q^2 + \cdots + j_{m-1} q^{m-1},$$

then

$$q^m - 1 - j = (q - 1 - j_0) + (q - 1 - j_1)q + \cdots + (q - 1 - j_{m-1})q^{m-1},$$

which can be verified by adding both sides of the equations. Hence $w_q(q^m - 1 - j) = (q - 1)m - w_q(j)$. Thus the dual code has check frequencies at j, satisfying $0 < j \leq q^{m-1}$ and

$$\begin{aligned} w_q(j) &= (q-1)m - w_q(q^m - 1 - j) \\ &< r + 1. \end{aligned}$$

The generator polynomial $\widetilde{h}(x)$ for the dual code $\mathcal{C}^\perp$ has such α^j as zeros.

Step 2. α^0 is not a zero of either $g(x)$ or of $\widetilde{h}(x)$. Hence the check polynomials of both $\mathcal{C}$ and $\mathcal{C}^\perp$ have $x - 1$ as a factor.

Step 3. Now extend $\mathcal{C}$ and $\mathcal{C}^\perp$ to obtain two GRM codes. The check matrices for the extended codes are

$$\begin{bmatrix} 1 & 1 \ldots 1 \\ 0 & \\ \vdots & \boldsymbol{H} \\ 0 & \end{bmatrix} \quad \text{and} \quad \begin{bmatrix} 1 & 1 \ldots 1 \\ 0 & \\ \vdots & \boldsymbol{H}' \\ 0 & \end{bmatrix}$$

where $\boldsymbol{H}$ and $\boldsymbol{H}'$ are check matrices for codes $\mathcal{C}$ and $\mathcal{C}^\perp$. Excluding the first row of each matrix for the moment, all rows of the matrix on the left are orthogonal to all rows of the matrix on the right because this is true of $\boldsymbol{H}$ and $\boldsymbol{H}'$. Next, the all-one row is orthogonal to itself because the blocklength is a multiple of q. Finally, the all-one row is orthogonal to every other row of both matrices because $x - 1$ is a factor of both check polynomials. Therefore the matrices are orthogonal, and the dimensions of the lengthened cyclic codes have sum q^m. The proof is complete. □

Nonprimitive GRM codes can also be defined in an obvious way. The following definition will end this section.

Definition 13.7.6. *Let b divide $q^m - 1$. The cyclic nonprimitive GRM code of order r and blocklength $n = (q^m - 1)/b$ over the field $GF(q)$ is the cyclic code whose generator polynomial $g(x)$ has zeros at all α^{bj} for $j = 1, \ldots, (q^m - 1)/b$, such that $0 < w_q(bj) \leq (q - 1)m - r - 1$.*

13.8 Euclidean-geometry codes

A finite geometry is a finite set within which one specifies certain subsets called lines, planes, or flats that satisfy a collection of axioms. A finite geometry takes its terminology from conventional elementary geometry. This latter case consists of a set, containing an infinite number of points, together with subsets known as lines and planes. Each finite geometry is defined by its own set of axioms, and within each, one can develop a list of theorems. The most meaningful finite geometries are the euclidean geometries and the projective geometries. These finite geometries can be used to invent codes.

Several classes of codes that are defined in terms of finite geometries are majority decodable: The euclidean-geometry codes and the projective-geometry codes. We will introduce these classes of codes in terms of the generalized Reed–Muller codes. Originally, both of these classes were introduced in a different way, using the theory of finite geometries, and this is how they got their names. In this section we will treat euclidean-geometry codes, and in Section 13.9 we will treat projective-geometry codes. We will only study codes over the prime field $GF(p)$. Three Galois fields play a role: $GF(p)$, the symbol field of the euclidean-geometry code; $GF(q)$, the symbol field of the GRM code, which is an extension of $GF(p)$ with $q = p^s$; and the locator field $GF(q^m)$.

Definition 13.8.1. *Let r, s, and m be any positive integers, and let $q = p^s$ with p as a prime. The euclidean-geometry code over $GF(p)$ of blocklength $n = q^m$ and order r is the dual of the subfield-subcode of the GRM code over $GF(q)$ of blocklength q^m and of order $(q-1)(m-r-1)$.*

Equivalently, a euclidean-geometry code is the extended form of a cyclic code over $GF(p)$, with blocklength $q^m - 1$, which is also called a *euclidean-geometry code*, given by the following theorem.

Theorem 13.8.2. *Let α be a primitive element of $GF(p^{sm})$. A euclidean-geometry code over $GF(p)$ with parameters r and s and blocklength q^m is an extended cyclic code generated by a polynomial having zeros at α^j for $0 < j \leq q^m - 1$, if j satisfies*

$$0 < \max_{0 \leq i \leq s} w_q(jp^i) \leq (q-1)(m-r-1).$$

Proof: The $GF(p)$ subfield-subcode of the cyclic GRM code of order $(q-1)(m-r-1)$ has a generator polynomial with zero at α^j for $0 < j \leq q^m - 1$, if j satisfies

$$\begin{aligned} w_q(j) &\leq (q-1)m - [(q-1)(m-r-1)] - 1 \\ &= (q-1)(r+1) - 1, \end{aligned}$$

or if any p-ary conjugate of j satisfies this inequality. Conversely, α^j is a zero of the check polynomial $h(x)$ if

$$w_q(j') > (q-1)(r+1) - 1$$

for every j' that is a p-ary conjugate of j. But now, by the theory of cyclic codes, the generator polynomial of the dual code is the reciprocal polynomial of $h(x)$. Then j' is a zero of this reciprocal polynomial if $n - j'$ is a zero of $h(x)$, that is, if

$$w_q(n - j') > (q-1)(r+1) - 1$$

t	EG code (n, k)	BCH code (n, k)
4	(63, 37)	(63, 39)
10	(63, 13)	(63, 18)
2	(255, 231)	(255, 239)
8	(255, 175)	(255, 191)
10	(255, 127)	(255, 179)
42	(255, 19)	(255, 47)

Figure 13.9. Parameters of some euclidean-geometry codes and some BCH codes

for every j' that is a p-ary conjugate of j. But as shown in the proof of Theorem 13.7.5, $w_q(n - j') = (q - 1)m - w_q(j')$. Hence

$$w_q(j') \leq (q - 1)(m - r - 1)$$

for every j' that is a p-ary conjugate of j. The theorem follows. □

The simplest example of a euclidean-geometry code is when $s = 1$ and $p = 2$. Then

$$w_2(j) \leq m - r - 1$$

if α^j is a zero of $g(x)$. This is just a Reed–Muller code of order r. The euclidean-geometry (EG) codes are a generalization of the Reed–Muller codes. The parameters of the other euclidean-geometry codes are given in Figure 13.9.

Our goal is to show that the euclidean-geometry codes can be majority decoded in $r + 1$ steps, and that $d_{\text{MD}} = (q^{m-r} - 1)/(q - 1)$. We will develop the decoder in terms of a suggestive geometric language, the language of a euclidean geometry, which we will now introduce.

A euclidean geometry, denoted by $EG(m, q)$, of dimension m over the field $GF(q)$ consists of q^m points (the vector space $GF(q)^m$) together with certain subsets called *flats* (or *affine subspaces*[2]), which are defined recursively as follows. The 0-flats (zero-dimensional affine subspaces) are the points of $EG(m, q)$. They are the q^m m-tuples of the vector space. The $(t + 1)$-flats are obtained from the t-flats. Each translation of the smallest vector subspace containing a t-flat is a $(t + 1)$-flat. That is, if E_t is a t-flat, then the set $\{\gamma_t \boldsymbol{u} | \gamma_t \in GF(q), \boldsymbol{u} \in E_t\}$ is the smallest vector subspace containing E_t. If $\boldsymbol{v}_{t+1}$ is any specified point in $GF(q)^m$ not belonging to E_t, then a $(t + 1)$-flat E_{t+1} is obtained as the set

$$E_{t+1} = \{\boldsymbol{v}_{t+1} + \gamma_t \boldsymbol{u} \mid \gamma_t \in GF(q), \boldsymbol{u} \in E_t\}.$$

A more formal definition follows.

[2] A vector subspace must contain the origin of the vector space. An affine subspace is a translation of a vector subspace (i.e., a coset).

Definition 13.8.3. *The euclidean geometry $EG(m, q)$ is the vector space $GF(q)^m$ together with all subspaces and translations of subspaces of $GF(q)^m$.*

The next theorem follows immediately from this definition.

Theorem 13.8.4. *A t-flat contains exactly q^t points and itself has the structure of a euclidean geometry $EG(t, q)$. A t-flat can be expressed as the set of q^t points*

$$v_t + \gamma_{i_{t-1}} v_{t-1} + \gamma_{i_{t-2}} v_{t-2} + \cdots + \gamma_{i_0} v_0$$

where γ_i ranges over all elements of $GF(q)$, including the zero element, and the $GF(q)^m$-vectors $v_0, v_1, \ldots, v_t$ are a fixed set of linearly independent elements of the t-flat.

One also might call 1-flats and 2-flats *lines* and *planes* because the definition of euclidean geometry corresponds to the idea of conventional euclidean space, only it is built on a finite field instead of on the real field. We could think of $EG(m, q)$, $GF(q^m)$, and $GF(q)^m$ as containing the same elements, but when we call it $GF(q^m)$, we are thinking of its field structure, rules of multiplication, and so forth. When we call it $EG(m, q)$, we are interested in the geometric structure created by Definition 13.8.3. When we call it $GF(q)^m$, we are interested in its vector-space structure.

The euclidean geometry $EG(3, 2)$ is given in Figure 13.10. This example suggests that the number of t-flats in a euclidean geometry can be quite large. We will count the number of t-flats in $EG(m, q)$. This will make use of the quantities known as q-ary gaussian coefficients, defined by

$$\begin{bmatrix} m \\ i \end{bmatrix} = \prod_{j=0}^{i-1} \frac{q^m - q^j}{q^i - q^j}$$

for $i = 1, 2, \ldots, m$, and $\begin{bmatrix} m \\ 0 \end{bmatrix} = 1$.

Theorem 13.8.5.

(i) $EG(m, q)$ contains $q^{m-t} \begin{bmatrix} m \\ t \end{bmatrix}$ distinct t-flats for $t = 0, 1, \ldots, m$.
(ii) For any s and t, with $0 \le s \le t \le m$, each s-flat is contained properly in exactly $\begin{bmatrix} m-s \\ t-s \end{bmatrix}$ distinct t-flats in $EG(m, q)$.

Proof:
(i) We can construct a t-dimensional subspace by choosing an ordered set of t linearly independent points in $EG(m, q)$. This can be done in

$$(q^m - 1)(q^m - q) \ldots (q^m - q^{t-1})$$

different ways. Many of these sets of independent points, however, will lead to the same

0-Flats	1-Flats	2-Flats
000	000,001	000,001,010,011
001	000,010	000,001,100,101
010	000,011	000,001,110,111
011	000,100	000,010,100,110
100	000,101	000,010,101,111
101	000,110	000,011,100,111
110	000,111	000,011,101,110
111	001,010	001,011,101,111
	001,011	001,011,100,110
	001,100	001,010,101,110
	001,101	001,010,100,111
	001,110	010,011,110,111
	001,111	010,011,100,101
	010,011	100,101,110,111
	010,100	
	010,101	
	010,110	
	010,111	
	011,100	
	011,101	
	011,110	
	011,111	
	100,101	
	100,110	
	100,111	
	101,110	
	101,111	
	110,111	

Figure 13.10. The euclidean geometry $EG(3, 2)$

t-dimensional subspace; in fact,

$$(q^t - 1)(q^t - q) \dots (q^t - q^{t-1})$$

sets will lead to the same t-dimensional subspace because this is the number of ways of picking an ordered sequence of independent points from the t-dimensional subspace. Hence there are

$$\frac{(q^m - 1)(q^m - q) \dots (q^m - q^{t-1})}{(q^t - 1)(q^t - q) \dots (q^t - q^{t-1})} = \begin{bmatrix} m \\ t \end{bmatrix}$$

distinct t-dimensional subspaces. Each subspace has q^{m-t} cosets, and thus there are $q^{m-t} \begin{bmatrix} m \\ t \end{bmatrix}$ t-flats.

(ii) Given an s-flat, it can be extended to a t-flat by choosing a sequence of $t - s$ independent points not yet included in the s-flat. This can be done in

$$(q^m - q^s)(q^m - q^{s+1})\ldots(q^m - q^{t-1})$$

ways. Many of these sequences of independent points, however, will extend the s-flat to the same t-flat; in fact,

$$(q^t - q^s)(q^t - q^{s+1})\ldots(q^t - q^{t-1})$$

sequences will lead to the same t-flat because this is the number of ways of picking an ordered sequence of independent points from the t-flat without using points from the s-flat. The ratio of these two products is the number of distinct t-flats in which the s-flat is contained. Because this is equivalent to the statement being proved, the proof is complete. □

Definition 13.8.6. *The incidence vector of a subset of a set of q^m elements indexed by i is a vector of length q^m that has a one at component i if the element indexed by i is in the subset, and otherwise has a zero at component i.*

In our description of a GRM code, the elements of $GF(q^m)$ are used to number the components of the vector space of dimension $n = q^m$ (which itself contains q^{q^m} vectors). In other words, the elements of the vector space $GF(q)^m$ are being used to index the n components of the vector space $GF(q)^{q^m}$. The incidence vector of a subset of $GF(q)^m$ is a vector in $GF(q)^{q^m}$. When $GF(q)^m$ is given the structure of a euclidean geometry, $EG(m, q)$, then the incidence vector of a flat in $EG(m, q)$ is a vector in $GF(q)^{q^m}$. As we shall see, the incidence vector of a flat in $EG(m, q)$ is a codeword in a GRM code contained in $GF(q)^{q^m}$. We now are ready to find an alternative way in which to define the euclidean-geometry codes.

Theorem 13.8.7. *An rth-order euclidean-geometry code of blocklength $n = q^m$ over $GF(p)$ is the largest linear code over $GF(p)$ having in its null space the incidence vectors of all of the $(r + 1)$-flats in $EG(m, q)$.*

Proof: It suffices to prove that the GRM code that contains the dual of the euclidean-geometry code is the smallest linear code over $GF(q)$ that contains all of the incidence vectors. This is because the components of an incidence vector can only be zero or one, and thus it always has all components in the subfield $GF(p)$. Hence an incidence vector is in the dual of the euclidean-geometry code if it is in the GRM code that contains the dual.

The incidence vector is in the GRM code if it is in the cyclic GRM code and also has the correct extension symbol. But the incidence vector of an $(r + 1)$-flat has q^{r+1} nonzero components that add to zero modulo p, and thus the extension symbol is always correct. It is only necessary to show that the incidence vector, with the last component deleted, is in the cyclic GRM code. That is, we must show that the Fourier transform of each incidence vector $\boldsymbol{f}$ has component F_j equal to zero if

$$w_q(j) \leq (q-1)m - [(q-1)(m-r-1)] - 1,$$

which reduces to

$$w_q(j) < (q-1)(r+1).$$

The proof consists of evaluating F_j and showing that it is zero for all such j, but may be nonzero for other j.

Step 1. Use the representation of Theorem 13.8.4 to express the $(r+1)$-flat as the set $\{\boldsymbol{v}_{r+1} + \gamma_{i_r}\boldsymbol{v}_r + \gamma_{i_{r-1}}\boldsymbol{v}_{r-1} + \cdots + \gamma_{i_0}\boldsymbol{v}_0\}$ where

$$\begin{aligned} i_0 &= 0, 1, \ldots, q-1 \\ i_1 &= 0, 1, \ldots, q-1 \\ &\vdots \qquad \vdots \\ i_r &= 0, 1, \ldots, q-1 \end{aligned}$$

index the q elements of $GF(q)$, and $\boldsymbol{v}_0, \ldots, \boldsymbol{v}_{r+1}$ are a fixed set of independent points in the $(r+1)$-flat. The incidence vector $\boldsymbol{f}$ has a one in component i when the field element α^i is in the preceding set, and otherwise has a zero. The spectral component

$$F_j = \sum_{i=0}^{n-1} \alpha^{ij} f_i$$

now can be written as a sum of those terms where f_i equals one, dropping the terms where f_i equals zero:

$$F_j = \sum_{i_0=0}^{q-1}\sum_{i_1=0}^{q-1}\cdots\sum_{i_r=0}^{q-1}\left(v_{r+1} + \gamma_{i_r}v_r + \gamma_{i_{r-1}}v_{r-1} + \cdots + \gamma_{i_0}v_0\right)^j$$

where now $v_0, \ldots, v_{r+1}$ are thought of as elements of $GF(q^{r+1})$. We need to determine those values of j for which F_j equals zero. A multinomial expansion gives

$$F_j = \sum_{i_0=0}^{q-1}\cdots\sum_{i_r=0}^{q-1}\sum_{\boldsymbol{h}} \frac{j!}{h_0!h_1!\cdots h_{r+1}!} v_{r+1}^{h_{r+1}}\left(\gamma_{i_r}v_r\right)^{h_r}\cdots\left(\gamma_{i_0}v_0\right)^{h_0}$$

where the sum on $\boldsymbol{h}$ is over all $(r+1)$-tuples $(h_0, h_1, \ldots, h_{r+1})$ such that $h_0 + h_1 + \cdots + h_{r+1} = j$. Now interchange the summations and work with the terms of the form $\sum_{i=0}^{q-1}(\gamma_i v)^h$ with h fixed.

Step 2. The sum $\sum_{i=0}^{q-1}\gamma_i^h$ is over all field elements of $GF(q)$, and thus can be rewritten in terms of the primitive element α. For h not equal to zero,

$$\sum_{i=0}^{q-1}\gamma_i^h = 0 + \sum_{k=0}^{q-2}\alpha^{kh}.$$

This is the hth component of the Fourier transform of the all-one vector. It is zero except when h is a multiple of $q-1$. Then it is equal to $q-1$ modulo p, which is equal to -1. For h equal to zero,

$$\sum_{i=0}^{q-1} \gamma_i^0 = \sum_{i=0}^{q-1} 1 = 0 \pmod{p},$$

using the convention $\gamma^0 = 1$ for all γ in $GF(q)$.

Every term in the sum for F_j is zero and can be dropped except when h_ℓ is a nonzero multiple of $(q-1)$ for $\ell = 0, \ldots, r$. The sum becomes

$$F_j = (-1)^r \sum_{h_0} \sum_{h_1} \cdots \sum_{h_{r+1}} \frac{j!}{h_0! h_1! \ldots h_{r+1}!} v_0^{h_0} v_1^{h_1} \ldots v_{r+1}^{h_{r+1}}$$

where the sum is over $(h_0, \ldots, h_{r+1})$ such that

$$\sum_{\ell=0}^{r+1} h_\ell = j,$$

h_ℓ is a nonzero multiple of $q-1$ for $\ell = 0, \ldots, r$, and $h_{r+1} \geq 0$.

Step 3. By the Lucas theorem (Theorem 13.4.3), in the equation for F_j, the multinomial coefficient is zero in every term for which h_ℓ is not a radix-p descendent of j for $\ell = 0, \ldots, r+1$. Therefore if h_ℓ contributes to the sum, h_ℓ is a radix-p descendent of j for $\ell = 0, \ldots, r+1$; hence h_ℓ is a radix-q descendent of j for $\ell = 0, \ldots, r+1$. This follows from the Lucas theorem by writing

$$\begin{aligned}\frac{j!}{h_0! h_1! \cdots h_{r+1}!} &= \frac{j!}{h_0!(j-h_0)!} \frac{(j-h_0)!}{h_1! \cdots h_{r+1}!} \\ &= \frac{j!}{h_0!(j-h_0)!} \frac{(j-h_0)!}{h_1!(j-h_0-h_1)!} \frac{(j-h_0-h_1)!}{h_2! \cdots h_{r+1}!}.\end{aligned}$$

Further, any sum of the h_ℓ contributing to the sum is a radix-p descendent of j, and therefore is a radix-q descendent of j.

We now can summarize the conditions on the terms that contribute to F_j as follows.

(i) $\sum_{\ell=0}^{t+1} h_\ell = j$.
(ii) h_ℓ is a nonzero multiple of $q-1$ for $\ell = 0, \ldots, r$, and $h_{r+1} \geq 0$.
(iii) Each place in the radix-q representation of j is the sum of the corresponding places of the radix-q representations of the h_ℓ.

To complete the proof, we must show that there are no such terms if j satisfies

$$w_q(j) < (q-1)(r+1).$$

Step 4. Consider a radix-q representation of some integer k:

$$k = k_0 + k_1 q + k_2 q^2 + \cdots + k_{m-1} q^{m-1}.$$

Then $k(\bmod q-1)$ can be evaluated as follows:

$$\begin{aligned} k &= k_0 + k_1 + k_2 + \cdots + k_{m-1} \pmod{q-1} \\ &= w_q(k) \end{aligned}$$

because q can be replaced by one without changing values modulo $q-1$. Hence

$$k = w_q(k) \pmod{q-1}.$$

If k is a nonzero multiple of $q-1$, then $w_q(k)$ is a nonzero multiple of $q-1$.

Now consider a radix-q representation of j and h_ℓ. Each place in the radix-q representation of j is the sum of the corresponding places of the radix-q representations of the h_ℓ. Hence

$$w_q(j) = \sum_{\ell=0}^{r+1} w_q(h_\ell),$$

and for $\ell = 0, \ldots, r$, $w_q(h_\ell)$ is a nonzero multiple of $q-1$. Hence

$$w_q(j) \geq (q-1)(r+1)$$

if F_j can be nonzero. The theorem is proved. □

The proof of the next theorem contains the algorithm for majority decoding.

Theorem 13.8.8. *Let $q = p^s$. An rth-order euclidean-geometry code of blocklength $n = q^m$ over $GF(p)$ can be majority decoded in $r+1$ steps in the presence of up to $\frac{1}{2}(q^{m-r}-1)/(q-1)$ errors.*

Proof: The proof uses a recursive argument, showing that from a set of check equations based on the incidence vectors of t-flats, one can obtain check equations based on the incidence vectors of the $(t-1)$-flats. This shows that the check equations based on the incidence vectors of the $(r+1)$-flats, as established by Theorem 13.8.7, can be pushed down in $r+1$ steps to check equations on the individual symbols. The argument is as follows.

It follows from the definition of a t-flat in a finite geometry that the incidence vectors of all t-flats that contain a given $(t-1)$-flat define a set of $r = \left[\begin{smallmatrix} m-(t-1) \\ t-(t-1) \end{smallmatrix}\right]$ checks that are concurring on the sum of the error symbols associated with the points of that $(t-1)$-flat. Given a $(t-1)$-flat E, any point not belonging to E is contained in precisely one t-flat that contains E. Hence by a majority decision, a new check equation that corresponds to the incidence vector of the $(t-1)$-flat E can be obtained. This can be done for all of the $(t-1)$-flats that contain a given $(t-2)$-flat, which, in turn, define a set of checks concurring on that $(t-2)$-flat. Hence by induction, after t steps we obtain a set of checks concurring on a 0-flat, that is, on a single error symbol. At the ith step, the number of concurring checks that can be used is $\left[\begin{smallmatrix} m-t+1 \\ 1 \end{smallmatrix}\right]$. Hence there are at least $\left[\begin{smallmatrix} m-t+1 \\ 1 \end{smallmatrix}\right]$ checks at each step, and the error-correction capability of the algorithm is given by

$$\frac{1}{2}\begin{bmatrix} m-t+1 \\ 1 \end{bmatrix} = \frac{\frac{1}{2}(q^{m-t+1}-1)}{(q-1)}.$$

This completes the proof of the theorem. □

13.9 Projective-geometry codes

The *projective-geometry codes* constitute a class that is similar to the class of euclidean-geometry codes. The codes differ in that they are constructed from nonprimitive cyclic GRM codes with blocklength $(q^m - 1)/(q - 1)$, rather than from GRM codes with blocklength q^m. The development will proceed as before. Three fields play a role: $GF(p)$, the symbol field of the code, with p as a prime; $GF(q)$, the symbol field of a nonprimitive GRM code, where $q = p^s$; and $GF(q^m)$, the locator field.

Definition 13.9.1. *Let r, s, and m be any positive integers, and let $q = p^s$ with p as a prime. The projective-geometry code over $GF(p)$ of blocklength $n = (q^m - 1)/(q - 1)$ and order r is the dual of the subfield-subcode of the nonprimitive cyclic GRM code over $GF(q)$ of the same blocklength and of order $(q - 1)(m - r - 1)$.*

Equivalently, a projective-geometry code is a nonprimitive cyclic code with a generator polynomial given by the following theorem.

Theorem 13.9.2. *A projective-geometry code over $GF(p)$ with parameters r and s and blocklength $(q^m - 1)/(q - 1)$ is a cyclic code generated by the polynomial that has zeros at β^j for $0 < j \leq (q^m - 1)/(q - 1)$, if j satisfies*

$$0 < \max_{0 \leq i < s} w_q(j(q-1)p^i) \leq (q-1)(m-r-1)$$

where $q = p^s$, $\beta = \alpha^{q-1}$, and α is primitive in $GF(q)$.

Proof: The proof is the same as that of Theorem 13.8.2. □

The parameters of some projective-geometry codes are given in Figure 13.11. A projective-geometry code is r-step majority decodable. Our remaining tasks are to develop the majority decoder and to show that $d_{\text{MD}} = 1 + (q^{m-r+1} - 1)/(q - 1)$.

The decoding procedure can be developed with the geometric language of a projective geometry. A projective geometry is closely related to a euclidean geometry. In fact, for a quick description, one may say that a projective geometry is a euclidean geometry augmented by some extra points, which may be called points at infinity, and with some new flats that involve these extra points defined. A formal definition is quite technical.

	Projective-geometry code	Shortened primitive BCH code
t	(n, k)	(n, k)
2	(21, 11)	(21, 11)
4	(73, 45)	(73, 45)
2	(85, 68)	(85, 71)
10	(85, 24)	(85, 22)
8	(273, 191)	(273, 201)
2	(341, 315)	(341, 323)
10	(341, 195)	(341, 251)
42	(341, 45)	(341, 32)
4	(585, 520)	(585, 545)
36	(585, 184)	(585, 250)
16	(1057, 813)	
2	(1365, 1038)	
10	(1365, 1063)	
42	(1365, 483)	
170	(1365, 78)	

Figure 13.11. Parameters of some projective-geometry codes and some BCH codes

The projective geometry $PG(m, q)$ has $(q^{m+1} - 1)/(q - 1)$ points and is defined by working with the nonzero points of $GF(q)^{m+1}$. There are $q^{m+1} - 1$ such nonzero points, and they are divided into $(q^{m+1} - 1)/(q - 1)$ sets, each set constituting one point of $PG(m, q)$. Each point of $PG(m, q)$ has $q - 1$ points of $GF(q)^{m+1}$ mapped onto it, and this projection suggests the name projective geometry.

The rule for forming these sets is that whenever $\boldsymbol{v}$ is a nonzero vector in $GF(q)^{m+1}$ and λ is a nonzero field element from $GF(q)$, $\boldsymbol{v}$ and $\lambda\boldsymbol{v}$ are in the same set V, that is, they are mapped onto the same point of $PG(m, q)$. There are $q - 1$ such nonzero λ, and so $q - 1$ vectors in the set.

Hence the points of the projective geometry can be identified with the $(q^{m+1} - 1)/(q - 1)$ distinct one-dimensional subspaces of $GF(q)^{m+1}$.

The projective geometry $PG(m, q)$ of dimension m over the field $GF(q)$ is the set of these $(q^{m+1} - 1)/(q - 1)$ points together with collections of subsets, called t-flats, for $t = 0, 1, \ldots, m$.

The 1-flat containing V_0 and V_1 is defined as follows. Let $\boldsymbol{v}_0 \in V_0$ and $\boldsymbol{v}_1 \in V_1$; it does not matter which elements are chosen (see Problem 13.11). Then the 1-flat (or line) containing V_0 and V_1 is the set of points in $PG(m, q)$, consisting of the images of points $\beta_0\boldsymbol{v}_0 + \beta_1\boldsymbol{v}_1$ in $GF(q)^{m+1}$, where β_0 and β_1 are arbitrary field elements not both zero. There are $q^2 - 1$ choices for β_0 and β_1, and these map into $(q^2 - 1)/(q - 1)$ points in $PG(m, q)$. Thus a 1-flat in $PG(m, q)$ has $q + 1$ points.

Similarly, the t-flat containing V_i for $i = 0, \ldots, t$ is defined as follows. Let $\boldsymbol{v}_i \in V_i$ for $i = 0, \ldots, t$. The set of points $\boldsymbol{v}_i$ must be linearly independent in $GF(q)^{m+1}$ because the set of V_i consists of distinct points in $PG(m, q)$. Then the t-flat containing $V_0, \ldots, V_t$ is the set of points in $PG(m, q)$ that consists of the images of the points

$$\beta_0 \boldsymbol{v}_0 + \beta_1 \boldsymbol{v}_1 + \cdots + \beta_t \boldsymbol{v}_t$$

in $GF(q)^{m+1}$ where $\beta_0, \ldots, \beta_t$ are arbitrary field elements not all zero. There are $q^{t+1} - 1$ choices for $\beta_0, \ldots, \beta_t$, and these map into $(q^{t+1} - 1)/(q - 1)$ points in $PG(m, q)$. Thus a $(t + 1)$-flat has $q^t + q^{t-1} + \cdots + q + 1$ points.

Theorem 13.9.3. *A t-flat in $PG(m, q)$ contains $\left[{t-1 \atop 1}\right]$ points and itself has the structure of a projective geometry $PG(t, q)$.*

Proof: The number of points is immediate because $\left[{t-1 \atop 1}\right] = (q^{t+1} - 1)/(q - 1)$. The structure of the t-flat is inherited from the structure of $PG(t, q)$. □

Theorem 13.9.4.

(i) *$PG(m, q)$ contains $\left[{m+1 \atop t+1}\right]$ distinct t-flats for $t = 0, 1, \ldots, m$.*
(ii) *For any s and t with $0 \le s \le t \le m$, each s-flat is properly contained in exactly $\left[{m-s \atop t-s}\right]$ distinct t-flats in $PG(m, q)$.*

Proof: The proof is essentially the same as that of Theorem 13.8.5. □

An alternative way in which to define the projective-geometry codes is given by the following theorem. This theorem parallels Theorem 13.8.7, and the proof is the same. It will be used to show that projective-geometry codes are majority decodable, just as Theorem 13.8.7 was used to show that euclidean-geometry codes are majority decodable.

Theorem 13.9.5. *An rth-order projective-geometry code of blocklength $n = (q^m - 1)/(q - 1)$ over $GF(p)$ is the largest linear code over $GF(p)$ having in its null space the incidence vectors of all of the r-flats in $PG(m, q)$.*

Proof: It suffices to prove that the nonprimitive cyclic GRM code that contains the dual of the projective-geometry code is the smallest linear code over $GF(q)$ that contains all of the incidence vectors of the r-flats. This is because the components of an incidence vector can only be zero or one, and thus it always has all components in the subfield $GF(p)$. Hence an incidence vector is in the dual of the projective-geometry code if it is in the GRM code that contains the dual.

The incidence vector $\boldsymbol{f}$ is the nonprimitive cyclic GRM code if the Fourier transform has component F_j equal to zero for

$$w_q((q - 1)j) \le (q - 1)m - [(q - 1)(m - r - 1)] - 1,$$

or

$$w_q((q-1)j) < (q-1)(r+1).$$

The proof consists of evaluating F_j and showing that it is zero for such j.

Step 1. Express the r-flat as the image of the set

$$\{\gamma_{i_r}\boldsymbol{v}_r + \gamma_{i_{r-1}}\boldsymbol{v}_{r-1} + \cdots + \gamma_{i_0}\boldsymbol{v}_0\}$$

where $i_0, i_1, \ldots, i_r$ each indexes the q elements of $GF(q)$, and $\boldsymbol{v}_0, \ldots, \boldsymbol{v}_r$ represent a fixed set of independent points in the r-flat. It is not necessary here to exclude the case where all the γ coefficients are zero. This point will not contribute to F_j. The incidence vector $\boldsymbol{f}$ has a one in component i when the field element α^i is in this set, and otherwise has a zero. Let $j' = (q-1)j$ for $j = 0, \ldots, (q^m-1)/(q-1)-1$, and compute the spectral component $F_{j'}$ by

$$\begin{aligned} F_{j'} &= \sum_{i=0}^{n-1} \alpha^{ij'} f_i \\ &= \sum_{i_0=0}^{q-1}\sum_{i_i=0}^{q-1}\cdots\sum_{i_r=0}^{n-1}\left(\gamma_{i_r}v_r + \gamma_{i_{r-1}}v_{r-1} + \cdots + \gamma_{i_0}v_0\right)^{j'} \end{aligned}$$

where now $v_0, \ldots, v_{r-1}$ are thought of as elements of $GF(q^r)$. We need to determine those values of j' for which this equals zero. A multinomial expansion gives

$$F_{j'} = \sum_{i_0=0}^{q-1}\cdots\sum_{i_r=0}^{q-1}\sum_{\boldsymbol{h}} \frac{j'!}{h_0!h_1!\ldots h_r!}\left(\gamma_{i_r}v_r\right)^{h_r}\ldots\left(\gamma_{i_0}v_0\right)^{h_0}$$

where the sum on $\boldsymbol{h}$ is over all solutions of $h_0 + h_1 + \cdots + h_r = j'$. Now interchange the summations and work with the terms $\sum_{i=0}^{q-1}(\gamma_i v)^h$.

Step 2. The sum $\sum_{i=0}^{q-1}(\gamma_i v)^h$ is over all field elements, and thus can be rewritten in terms of the primitive element α. As in the proof of Theorem 13.8.7, this will imply that only terms with h_ℓ as a nonzero multiple of $q-1$ contribute to $F_{j'}$. Then

$$F_{j'} = (-1)^r \sum_{h_0}\sum_{h_1}\cdots\sum_{h_{r+1}} \frac{j'!}{h_0!h_1!\cdots h_r!} v_0^{h_0} v_1^{h_1}\ldots v_r^{h_r}$$

where the sum is over $(h_0, \ldots, h_r)$, such that

$$\sum_{\ell=0}^{r+1} h_\ell = j',$$

and h_ℓ is a nonzero multiple of $q-1$ for $\ell = 0, \ldots, r$.

Steps 3 and 4. As in the proof of Theorem 13.8.7, the Lucas theorem implies that for each ℓ, h_ℓ is a radix-q descendent of j'. Hence

$$w_q(j') = \sum_{\ell=0}^{r} w_q(h_\ell).$$

But as shown in the proof of Theorem 13.8.7, $w_q(h)$ is a nonzero multiple of $q-1$ whenever h_ℓ is. Hence

$$w_q(j') \geq (q-1)(r+1)$$

if $F_{j'}$ can be nonzero. The theorem is proved. □

The proof of the following theorem contains the majority-decoding algorithm.

Theorem 13.9.6. *Let $q = p^s$. An rth-order projective-geometry code of blocklength $n = (q^m - 1)/(q-1)$ over $GF(p)$ can be majority decoded in r steps in the presence of up to $\frac{1}{2}(q^{m-r+1} - 1)/(q-1)$ errors.*

Proof: The proof uses a recursive argument showing that, from a set of check equations based on the incidence vectors of t-flats, one can obtain check equations based on the incidence vectors of the $(t-1)$-flats. This shows that the check equations based on the incidence vectors of the r-flats can be pushed down in r steps to checks on the individual symbols. The argument is the same as that given in the proof of Theorem 13.8.8. □

Problems

13.1 Prove that

$$\binom{m-1}{r} + \binom{m-1}{r-1} = \binom{m}{r}.$$

13.2 Show that the binary (15, 7) double-error-correcting BCH code is majority decodable by implementing the Meggitt decoder as a majority decoder.

13.3 Design a two-step majority decoder for the (7, 4) binary Hamming code.

13.4 Construct a majority decoder for the binary (21, 12) double-error-correcting, projective-geometry code.

13.5 Find the generator polynomial for a (15, 6) triple-error-correcting code over $GF(4)$ that is majority decodable. Design the decoder. Find the generator polynomial for a triple-error-correcting BCH code over $GF(4)$.

13.6 Develop an algorithm for decoding a Reed–Muller codeword in the presence of $2^{m-r} - 1$ erasures.

13.7 If $GF(q)$ is a field of characteristic 2, some majority-decodable codes over $GF(q)$ of modest performance can be obtained by taking $J = (q^m - 1)/q - 1$ and using a permutation technique. Verify the existence of the following codes:

$$(255, 156), t = 8, \text{ over } GF(16),$$
$$(4095, 2800), t = 32, \text{ over } GF(64),$$
$$(65535, 47040), t = 128, \text{ over } GF(256).$$

If this is done manually, some kind of four-dimensional array describing the radix-q representation of the exponents should be used to reduce the work.

13.8 Prove that no Reed–Solomon code can be decoded to its packing radius by a majority decoder.

13.9 Construct a majority decoder for the (42, 28) convolutional code given in Figure 13.7.

13.10 Show that if an r-flat of $EG(m, q)$ contains the origin, then it is a linear subspace of $EG(m, q)$ regarded as a vector space. Given an r-flat in $EG(m, q)$, how many other r-flats are disjoint from it?

13.11 In the definition of a 1-flat in a projective geometry, show that it does not matter which points representing V_0 and V_1 in $GF(q)^m$ are chosen.

13.12 Prove that no two lines are parallel in a projective geometry, that is, prove that every pair of 1-flats has a nonzero intersection.

13.13 Construct a high-speed decoder for the (63, 24, 15) binary BCH code by using in parallel four majority-logic decoders designed for the (63, 22, 15) GRM code.

Notes

The subject of majority-decodable codes started with the discovery of the Reed–Muller codes. The Reed–Muller codes were discovered by Muller in 1954. In the same year, Reed discovered the decoding algorithm for them. The decoding algorithm of Reed is unusual in that it makes decisions by majority vote. The Plotkin representation appeared in 1960, and Dumer presented his algorithm in 1999, influenced by several earlier contributions in this direction.

The notion that majority-decodable codes constitute a special subclass of error-control codes was crystallized by Massey (1963) who established the general framework for treating such codes. Convolutional codes were of special interest to Massey, and his work included defining a class of majority-decodable convolutional codes. These convolutional codes were developed further by Robinson and Bernstein (1967), and by Wu (1976).

Finding constructive classes of majority-decodable block codes has proved to be a difficult task and this work moves forward slowly. Key contributions have been the application of finite geometries to code construction, first introduced by Rudolph (1964, 1967); the realization by Kolesnik and Mironchikov (1965, 1968) and by Kasami, Lin, and Peterson (1968a,b) that Reed–Muller codes are cyclic (but for an overall check bit); and the generalization of the observation of Zierler (1958) that first-order Reed–Muller codes are cyclic. This latter work showed how the Reed–Muller codes can be generalized to an arbitrary alphabet. The generalized Reed–Muller codes are from Kasami, Lin, and Peterson (1968a). These leads were quickly developed by Weldon (1967, 1968); by Kasami, Lin, and Peterson (1968b) in a second paper; by Goethals and Delsarte (1968); and by Delsarte, Goethals, and MacWilliams (1970). The simple method of construction of Section 13.4 is taken from Lin and Markowsky (1980). Finite-geometry codes were discussed in detail in a tutorial by Goethals (1975).

Bibliography

1. N. A. Abramson, Class of Systematic Codes for Non-Independent Errors, *IRE Transactions on Information Theory*, vol. IT-5, pp. 150–157, 1959.
2. N. A. Abramson, A Note on Single Error-Correcting Binary Codes, *IRE Transactions on Information Theory*, vol. IT-6, pp. 502–503, 1960.
3. V. I. Andryanov and V. N. Saskovets, Decoding Codes, *Akad. Nauk. Ukr. SSR Kibernetika*, Part 1, 1966.
4. S. Arimoto, Encoding and Decoding of p-ary Group Codes and the Correction System (in Japanese), *Information Processing in Japan*, vol. 2, pp. 321–325, 1961.
5. E. F. Assmus, Jr. and H. F. Mattson, Jr., Coding and Combinatorics, *SIAM Review*, vol. 16, pp. 349–388, 1974.
6. E. F. Assmus, Jr., H. F. Mattson, Jr., and R. J. Turyn, *Cyclic Codes*, AFCRL-65-332, Air Force Cambridge Research Labs, Bedford, MA, 1965.
7. C. P. M. J. Baggen and L. M. G. M. Tolhuizen, On Diamond Codes, *IEEE Transactions on Information Theory*, vol. IT-43, pp. 1400–1411, 1997.
8. L. R. Bahl and F. Jelinek, Rate 1/2 Convolutional Codes with Complementary Generators, *IEEE Transactions on Information Theory*, vol. IT-17, pp. 718–727, 1971.
9. L. R. Bahl, J. Cocke, F. Jelinek, and J. Raviv, Optimal Decoding of Linear Codes for Minimizing Symbol Error Rate, *IEEE Transactions on Information Theory*, vol. IT-20, pp. 284–287, 1974.
10. T. C. Bartee and D. I. Schneider, Computation with Finite Fields, *Information and Control*, vol. 6, pp. 79–98, 1963.
11. L. A. Bassalygo, New Upper Bounds for Error Correcting Codes, *Problemy Peredachi Informatsii*, vol. 1, pp. 41–45, 1965.
12. B. Baumslag and B. Chandler, *Theory and Problems of Group Theory*, Schaum's Outline Series, McGraw-Hill, New York, NY, 1968.
13. E. R. Berlekamp, The Enumeration of Information Symbols in BCH Codes, *Bell System Technical Journal*, vol. 46, pp. 1861–1880, 1967.
14. E. R. Berlekamp, *Algebraic Coding Theory*, McGraw-Hill, New York, NY, 1968.
15. E. R. Berlekamp, Long Primitive Binary BCH Codes Have Distance $d \sim 2n \ln R^{-1}/\log n \ldots$, *IEEE Transactions on Information Theory*, vol. IT-18, pp. 415–416, 1972.
16. E. R. Berlekamp, Bit Serial Reed–Solomon Encoders, *IEEE Transactions on Information Theory*, vol. IT-28, pp. 869–874, 1982.
17. C. Berrou and A. Glavieux, Near Optimum Error Correcting Coding and Decoding: Turbo-Codes, *IEEE Transactions on Communications*, vol. COM-44, pp. 1261–1271, 1996.
18. C. Berrou, A. Glavieux, and P. Thitimajshima, Near Shannon Limit Error-Correcting Coding and Decoding: Turbo Codes, *Proceedings of the IEEE International Conference on Communications*, Geneva, Switzerland, IEEE, Piscataway, NJ, 1993.

19. G. Birkhoff and S. MacLane, *A Survey of Modern Algebra*, Revised Edition, Macmillan, New York, NY, 1953.
20. R. E. Blahut, Algebraic decoding in the frequency domain, in *Algebraic Coding Theory and Practice*, G. Longo, editor, Springer-Verlag, New York, NY, 1979.
21. R. E. Blahut, Transform Techniques for Error-Control Codes, *IBM Journal of Research and Development*, vol. 23, pp. 299–315, 1979.
22. R. E. Blahut, On Extended BCH Codes, *Proceedings of the Eighteenth Annual Allerton Conference on Communication, Control, and Computers*, pp. 50–59, University of Illinois, Monticello, IL, 1980.
23. R. E. Blahut, A Universal Reed–Solomon Decoder, *IBM Journal of Research and Development*, vol. 28, pp. 150–158, 1984.
24. R. A. Blum and A. D. Weiss, *Further Results in Error-Correcting Codes*, S.M. Thesis, Massachusetts Institute of Technology, Cambridge, MA, 1960.
25. R. C. Bose and D. K. Ray-Chaudhuri, On a Class of Error-Correcting Binary Group Codes, *Information and Control*, vol. 3, pp. 68–79, 1960.
26. H. O. Burton, Inversionless Decoding of Binary BCH Codes, *IEEE Transactions on Information Theory*, vol. IT-17, pp. 464–466, 1971.
27. H. O. Burton and E. J. Weldon, Jr., Cyclic Product Codes, *IEEE Transactions on Information Theory*, vol. IT-11, pp. 433–439, 1965.
28. J. J. Bussgang, Some Properties of Binary Convolutional Code Generators, *IEEE Transactions on Information Theory*, vol. IT-11, pp. 90–100, 1965.
29. A. R. Calderbank, G. D. Forney, Jr., and A. Vardy, Minimal Tail-Biting Trellis: The Golay Code and More, *IEEE Transactions on Information Theory*, vol. IT-45, pp. 1435–1455, 1999.
30. S. C. Chang and J. K. Wolf, A Simple Derivation of the MacWilliams Identity for Linear Codes, *IEEE Transactions on Information Theory*, vol. IT-26, pp. 476–477, 1980.
31. P. Charpin, Open Problems on Cyclic Codes, *Handbook of Coding Theory*, pp. 963–1063, V. S. Pless and W. C. Huffman, editors, Elsevier, Amsterdam, 1998.
32. C. L. Chen, Computer Results on the Minimum Distance of Some Binary Cyclic Codes, *IEEE Transactions on Information Theory*, vol. IT-16, pp. 359–360, 1970.
33. P. R. Chevillat and D. J. Costello, Jr., An Analysis of Sequential Decoding for Specific Time-Invariant Convolutional Codes, *IEEE Transactions on Information Theory*, vol. IT-24, pp. 443–451, 1978.
34. R. T. Chien, Cyclic Decoding Procedures for Bose–Chaudhuri–Hocquenghem Codes, *IEEE Transactions on Information Theory*, vol. IT-10, pp. 357–363, 1964.
35. R. T. Chien, Burst-Correcting Code with High-Speed Decoding, *IEEE Transactions on Information Theory*, vol. IT-15, pp. 109–113, 1969.
36. R. T. Chien, A New Proof of the BCH Bound, *IEEE Transactions on Information Theory*, vol. IT-18, p. 541, 1972.
37. R. T. Chien and D. M. Choy, Algebraic Generalization of BCH–Goppa–Helgert Codes, *IEEE Transactions on Information Theory*, vol. IT-21, pp. 70–79, 1975.
38. R. T. Chien and S. W. Ng, Dual Product Codes for Correction of Multiple Low-Density Burst Errors, *IEEE Transactions on Information Theory*, vol. IT-19, pp. 672–678, 1973.
39. T. K. Citron, *Algorithms and Architectures for Error-Correcting Codes*, Ph.D. Dissertation, Stanford University, Stanford, CA, 1986.
40. J. Cocke, Lossless Symbol Coding with Nonprimes, *IRE Transactions on Information Theory*, vol. IT-5, pp. 33–34, 1959.

41. D. J. Costello, Jr., A Construction Technique for Random-Error-Correcting Codes, *IEEE Transactions on Information Theory*, vol. IT-15, pp. 631–636, 1969.
42. D. J. Costello, Jr., Free Distance Bounds for Convolutional Codes, *IEEE Transactions on Information Theory*, vol. IT-20, pp. 356–365, 1974.
43. P. Delsarte, On Subfield Subcodes of Modified Reed–Solomon Codes, *IEEE Transactions on Information Theory*, vol. IT-21, pp. 575–576, 1975.
44. P. Delsarte, J. M. Goethals, and F. J. MacWilliams, On Generalized Reed–Muller Codes and Their Relatives, *Information and Control*, vol. 16, pp. 402–442, 1970.
45. I. Dumer, Recursive Decoding of Reed–Muller Codes, *Proceedings of the Thirty-Seventh Annual Allerton Conference on Communication, Control, and Computing*, pp. 61–69, University of Illinois, Monticello, IL, 1999.
46. P. Elias, Error-Free Coding, *IRE Transactions on Information Theory*, vol. IT-4, pp. 29–37, 1954.
47. B. Elspas, A Note on P-nary Adjacent-Error-Correcting Binary Codes, *IRE Transactions on Information Theory*, vol. IT-6, pp. 13–15, 1960.
48. R. M. Fano, A Heuristic Discussion of Probabilistic Decoding, *IEEE Transactions on Information Theory*, vol. IT-9, pp. 64–74, 1963.
49. G. L. Feng, A VLSI Architecture for Fast Inversion in $GF(2^m)$, *IEEE Transactions on Computers*, vol. 38, pp. 1383–1386, 1989.
50. P. Fire, *A Class of Multiple-Error Correcting Binary Codes for Non-Independent Errors*, Sylvania Report RSL-E-2, Sylvania Reconnaissance Systems Lab., Mountain View, CA, 1959.
51. G. D. Forney, Jr., On Decoding BCH Codes, *IEEE Transactions on Information Theory*, vol. IT-11, pp. 549–557, 1965.
52. G. D. Forney, Jr., *Concatenated Codes*, MIT Press, Cambridge, MA, 1966.
53. G. D. Forney, Jr., *Final Report on a Coding System Design for Advanced Solar Mission*, Contract NASA-3637, NASA Ames Research Center, Moffet Field, CA, 1967.
54. G. D. Forney, Jr., Convolutional Codes I: Algebraic Structure, *IEEE Transactions on Information Theory*, vol. IT-16, pp. 720–738, 1970.
55. G. D. Forney, Jr., Burst-Correcting Codes for the Classic Bursty Channel, *IEEE Transactions on Communication Technology*, vol. COM-19, pp. 772–781, 1971.
56. G. D. Forney, Jr., The Viterbi Algorithm, *Proceedings of the IEEE*, vol. 61, pp. 268–276, 1973.
57. G. D. Forney, Jr., Minimal Bases of Rational Vector Spaces with Applications to Multiple Variable Linear Systems, *SIAM Journal of Control*, vol. 13, pp. 493–502, 1973.
58. G. D. Forney, Jr., Structural Analysis of Convolutional Codes via Dual Codes, *IEEE Transactions on Information Theory*, vol. IT-19, pp. 512–518, 1973.
59. G. D. Forney, Jr., Convolutional Codes II: Maximum-Likelihood Decoding and Convolutional Codes III: Sequential Decoding, *Information and Control*, vol. 25, pp. 222–297, 1974.
60. G. D. Forney, Jr., The Forward–Backward Algorithm, *Proceedings of the Thirty-Fourth Annual Allerton Conference on Communication, Control, and Computing*, pp. 432–446, University of Illinois, Monticello, IL, 1996.
61. G. D. Forney, Jr., On Iterative Decoding and the Two-Way Algorithm, *Proceedings of the International Symposium on Turbo Codes and Related Topics*, Brest, France, 1997.
62. J. B. Fraleigh, *A First Course in Abstract Algebra*, Second Edition, Addison-Wesley, Reading, MA, 1976.
63. R. G. Gallager, Low-Density Parity-Check Codes, *IRE Transactions on Information Theory*, vol. IT-8, pp. 21–28, 1962.
64. R. G. Gallager, *Low-Density Parity-Check Codes*, The MIT Press, Cambridge, MA, 1963.

65. R. G. Gallager, *Information Theory and Reliable Communications*, Wiley, New York, NY, 1968.
66. F. R. Gantmacher, *The Theory of Matrices*, Volume 1 (Second Edition), Chelsea, New York, 1959.
67. E. N. Gilbert, A Comparison of Signaling Alphabets, *Bell System Technical Journal*, vol. 31, pp. 504–522, 1952.
68. J. M. Goethals, Threshold Decoding – A Tentative Survey, in *Coding and Complexity*, G. Longo, editor, Springer-Verlag, New York, NY, 1975.
69. J. M. Goethals and P. Delsarte, On a Class of Majority-Logic Decodable Cyclic Codes, *IEEE Transactions on Information Theory*, vol. IT-14, pp. 182–189, 1968.
70. M. J. E. Golay, Notes on Digital Coding, *Proceedings of the IRE*, vol. 37, p. 657, 1949.
71. M. J. E. Golay, Notes on the Penny-Weighing Problem, Lossless Symbol Coding with Nonprimes, etc., *IRE Transactions on Information Theory*, vol. IT-4, pp. 103–109, 1958.
72. V. C. Goppa, A New Class of Linear Error-Correcting Codes, *Problemy Peredachi Informatsii*, vol. 6, pp. 24–30, 1970.
73. W. C. Gore, Transmitting Binary Symbols with Reed–Solomon Codes, *Proceedings of the Princeton Conference on Information Science Systems*, pp. 495–497, Princeton, NJ, 1973.
74. D. C. Gorenstein and N. Zierler, A Class of Error-Correcting Codes in p^m Symbols, *Journal of the Society of Industrial and Applied Mathematics*, vol. 9, pp. 207–214, 1961.
75. M. W. Green, Two Heuristic Techniques for Block Code Construction, *IEEE Transactions on Information Theory*, vol. IT-12, p. 273, 1966.
76. D. Haccoun and M. J. Ferguson, Generalized Stack Algorithms for Decoding Convolutional Codes, *IEEE Transactions on Information Theory*, vol. IT-21, pp. 638–651, 1975.
77. D. W. Hagelbarger, Recurrent Codes: Easily Mechanized Burst Correcting Binary Codes, *Bell System Technical Journal*, vol. 38, pp. 969–984, 1959.
78. J. Hagenauer and P. Hoeher, A Viterbi Algorithm with Soft-Decision Outputs and Its Applications, *Proceedings of the IEEE GLOBECOM '89*, pp. 47.1.1–47.1.7, 1989.
79. R. W. Hamming, Error Detecting and Error Correcting Codes, *Bell System Technical Journal*, vol. 29, pp. 147–160, 1950.
80. C. R. P. Hartmann, Decoding Beyond the BCH Bound, *IEEE Transactions on Information Theory*, vol. IT-18, pp. 441–444, 1972.
81. J. A. Heller, *Short Constraint Length Convolutional Codes*, Jet Propulsion Laboratory Space Program Summary, 37-54 III, pp. 171–177, 1968.
82. A. Hocquenghem, Codes Correcteurs D'erreurs, *Chiffres*, vol. 2, pp. 147–156, 1959.
83. K. Huber, Some Comments on Zech's Logarithms, *IEEE Transactions on Information Theory*, vol. IT-36, pp. 946–950, 1990.
84. Z. McC. Huntoon and A. M. Michelson, On the Computation of the Probability of Post-Decoding Error Events for Block Codes, *IEEE Transactions on Information Theory*, vol. IT-23, pp. 399–403, 1977.
85. K. Imamura, Self-Complementary Bases of Finite Fields, *Book of Abstracts – 1983 IEEE International Symposium on Information Theory*, St. Jovite, Canada, 1983.
86. K. Imamura, W. Yoshida, and N. Nakamura, Some Notes on the Binary Weight Distribution of Reed–Solomon Codes, *Book of Abstracts – 1986 IEEE International Symposium on Information Theory,* Ann Arbor, MI, 1986.
87. I. M. Jacobs and E. R. Berlekamp, A Lower Bound to the Distribution of Computations for Sequential Decoding, *IEEE Transactions on Information Theory*, vol. IT-13, pp. 167–174, 1967.
88. F. Jelinek, *Probabilistic Information Theory*, McGraw-Hill, New York, 1968.
89. F. Jelinek, An Upper Bound on Moments of Sequential Decoding Effort, *IEEE Transactions on Information Theory*, vol. IT-15, pp. 140–149, 1969.

90. J. M. Jensen, On Decoding Doubly Extended Reed–Solomon Codes, *Proceedings – 1995 IEEE International Symposium on Information*, Whistler, BC, Canada, 1995.
91. R. Johannesson, Robustly Optimal Rate One-Half Binary Convolutional Codes, *IEEE Transactions on Information Theory*, vol. IT-21, pp. 464–468, 1975.
92. R. Johannesson and K. Sh. Zigangirov, *Fundamentals of Convolutional Coding*, IEEE Press, Piscataway, NJ, 1999.
93. J. Justesen, A Class of Constructive Asymptotically Good Algebraic Codes, *IEEE Transactions on Information Theory*, vol. IT-18, pp. 652–656, 1972.
94. M. Kasahara, Y. Sugiyama, S. Hirasawa, and T. Namekawa, A New Class of Binary Codes Constructed on the Basis of BCH Codes, *IEEE Transactions on Information Theory*, vol. IT-21, pp. 582–585, 1975.
95. T. Kasami, Systematic Codes Using Binary Shift Register Sequences, *Journal of the Information Processing Society of Japan*, vol. 1, pp. 198–200, 1960.
96. T. Kasami, Optimum Shortened Cyclic Codes for Burst-Error Correction, *IEEE Transactions on Information Theory*, vol. IT-9, pp. 105–109, 1963.
97. T. Kasami, A Decoding Procedure for Multiple-Error-Correcting Cyclic Codes, *IEEE Transactions on Information Theory*, vol. IT-10, pp. 134–139, 1964.
98. T. Kasami and S. Lin, The Binary Weight Distribution of the Extended $(2^m, 2^m - 4)$ Code of the Reed–Solomon Code over $GF(2^m)$ with Generator Polynomial $(x - \alpha)(x - \alpha^2)(x - \alpha^3)$, *Linear Algebra and Its Applications*, vol. 98, pp. 291–307, 1988.
99. T. Kasami and N. Tokura, Some Remarks on BCH Bounds and Minimum Weights of Binary Primitive BCH Codes, *IEEE Transactions on Information Theory*, vol. IT-15, pp. 408–412, 1969.
100. T. Kasami, S. Lin, and W. W. Peterson, *Some Results on Cyclic Codes Which are Invariant Under the Affine Group*, Air Force Cambridge Research Labs Report, Cambridge, MA, 1966.
101. T. Kasami, S. Lin, and W. W. Peterson, Some Results on Weight Distributions of BCH Codes, *IEEE Transactions on Information Theory*, vol. IT-12, p. 274, 1966.
102. T. Kasami, S. Lin, and W. W. Peterson, New Generalizations of the Reed–Muller Codes – Part I: Primitive Codes, *IEEE Transactions on Information Theory*, vol. IT-14, pp. 189–199, 1968.
103. T. Kasami, S. Lin, and W. W. Peterson, Polynomial Codes, *IEEE Transactions on Information Theory*, vol. IT-14, pp. 807–814, 1968.
104. Z. Kiyasu, Research and Development Data No. 4, Electrical Communications Laboratory, Nippon Telegraph Corporation, Tokyo, Japan, 1953.
105. V. D. Kolesnik and E. T. Mironchikov, Some Cyclic Codes and a Scheme for Decoding by a Majority of Tests, *Problemy Peredachi Informatsii*, vol. 1, pp. 1–12, 1965.
106. V. D. Kolesnik and E. T. Mironchikov, Cyclic Reed–Muller Codes and Their Decoding, *Problemy Peredachi Informatsii*, vol. 4, pp. 15–19, 1968.
107. K. J. Larsen, Short Convolutional Codes with Maximal Free Distance for Rate 1/2, 1/3, and 1/4, *IEEE Transactions on Information Theory*, vol. IT-19, pp. 371–372, 1973.
108. B. A. Laws, Jr. and C. K. Rushforth, A Cellular-Array Multiplier for $GF(2^m)$, *IEEE Transactions on Computers*, vol. C-20, pp. 1573–1578, 1971.
109. P. J. Lee, There are Many Good Periodically Time-Varying Convolutional Codes, *IEEE Transactions on Information Theory*, vol. IT-35, pp. 460–463, 1989.
110. A. Lempel, Characterization and Synthesis of Self-Complementary Normal Bases in Finite Fields, *Linear Algebra and Its Applications*, vol. 98, pp. 331–346, 1988.
111. A. Lempel and S. Winograd, A New Approach to Error-Correcting Codes, *IEEE Transactions on Information Theory*, vol. IT-23, pp. 503–508, 1977.

112. R. Lidl and H. Niederreiter, *Finite Fields: Encyclopedia of Mathematics and its Applications*, vol. 20, Cambridge University Press, Cambridge, UK, 1984.
113. S. Lin and G. Markowsky, On a Class of One-Step Majority-Logic Decodable Cyclic Codes, *IBM Journal of Research and Development*, vol. 24, pp. 56–63, 1980.
114. J. H. Ma and J. K. Wolf, On Tail-Biting Convolutional Codes, *IEEE Transactions on Communications*, vol. COM-34, pp. 1049–1053, 1988.
115. F. J. MacWilliams, A Theorem on the Distribution of Weights in a Systematic Code, *Bell System Technical Journal*, vol. 42, pp. 79–94, 1963.
116. F. J. MacWilliams, Permutation Decoding of Systematic Codes, *Bell System Technical Journal*, vol. 43, pp. 485–505, 1964.
117. D. M. Mandelbaum, On Decoding of Reed–Solomon Codes, *IEEE Transactions on Information Theory*, vol. IT-17, pp. 707–712, 1971.
118. D. M. Mandelbaum, Decoding Beyond the Designed Distance for Certain Algebraic Codes, *Information and Control*, vol. 35, pp. 209–228, 1977.
119. D. M. Mandelbaum, Construction of Error-Correcting Codes by Interpolation, *IEEE Transactions on Information Theory*, vol. IT-25, pp. 27–35, 1979.
120. H. B. Mann, On the Number of Information Symbols in Bose–Chaudhuri Codes, *Information and Control*, vol. 5, pp. 153–162, 1962.
121. J. L. Massey, *Threshold Decoding*, MIT Press, Cambridge, MA, 1963.
122. J. L. Massey, Shift-Register Synthesis and BCH Decoding, *IEEE Transactions on Information Theory*, vol. IT-15, pp. 122–127, 1969.
123. J. L. Massey, Error bounds for tree codes, trellis codes and convolutional codes with encoding and decoding procedures, in *Coding and Complexity*, G. Longo, editor, Springer-Verlag, New York, NY, 1975.
124. J. L. Massey, Coding theory, Chapter 16 of *Handbook of Applicable Mathematics*, Volume V, Part B of *Combinatorics and Geometry*, W. Ledermann and S. Vajda, editors, Wiley, Chichester and New York, NY, 1985.
125. J. L. Massey and M. K. Sain, Inverses of Linear Sequential Circuits, *IEEE Transactions on Computers*, vol. C-17, pp. 330–337, 1968.
126. H. F. Mattson and G. Solomon, A New Treatment of Bose–Chaudhuri Codes, *Journal of the Society of Industrial and Applied Mathematics*, vol. 9, pp. 654–669, 1961.
127. U. Maurer and R. Viscardi, *Running-Key Generators with Memory in the Nonlinear Combining Function*, Diploma Thesis, Swiss Federal Institute of Technology, Zurich, 1984.
128. R. J. McEliece, *Finite Fields for Computer Scientists and Engineers*, Kluwer, Boston, MA, 1987.
129. R. J. McEliece, The algebraic theory of convolutional codes, in *Handbook of Coding Theory*, pp. 1065–1138, V. S. Pless and W. C. Huffman, editors, Elsevier, Amsterdam, 1998.
130. R. J. McEliece, D. J. C. MacKay, and J. F. Cheng, Turbo Decoding as an Instance of Pearl's 'Belief Propagation' Algorithm, *IEEE Journal on Selected Areas in Communication*, vol. SAC-16, pp. 140–152, 1998.
131. R. J. McEliece and I. Onyszchuk, *The Analysis of Convolutional Codes via the Extended Smith Algorithm*, Jet Propulsion Laboratory TDA Progress Report, vol. 42-112, pp. 22–30, 1993.
132. R. J. McEliece, E. R. Rodemich, H. Rumsey, Jr., and L. R. Welch, New Upper Bounds on the Rate of a Code Via the Delsarte–MacWilliams Inequalities, *IEEE Transactions on Information Theory*, vol. IT-23, pp. 157–166, 1977.
133. J. E. Meggitt, Error-Correcting Codes for Correcting Bursts of Errors, *IBM Journal of Research and Development*, vol. 4, pp. 329–334, 1960.
134. J. E. Meggitt, Error-Correcting Codes and Their Implementation, *IRE Transactions on Information Theory*, vol. IT-7, pp. 232–244, 1961.

135. A. Michelson, A Fast Transform in Some Galois Fields and an Application to Decoding Reed-Solomon Codes, *IEEE Abstracts of Papers – IEEE International Symposium on Information Theory*, Ronneby, Sweden, 1976.
136. M. E. Mitchell, *Error-Trap Decoding of Cyclic Codes*, G.E. Report No. 62MCD3, General Electric Military Communications Department, Oklahoma City, OK, 1962.
137. M. Morii, M. Kasahara, and D. L. Whiting, Efficient Bit-Serial Multiplication and the Discrete-Time Wiener–Hopf Equation over Finite Fields, *IEEE Transactions on Information Theory*, vol. IT-35, pp. 1177–1183, 1989.
138. D. J. Muder, Minimal Trellises for Block Codes, *IEEE Transactions on Information Theory*, vol. IT-34, pp. 1049–1053, 1988.
139. D. E. Muller, Application of Boolean Algebra to Switching Circuit Design and to Error Detection, *IRE Transactions on Electronic Computers*, vol. EC-3, pp. 6–12, 1954.
140. J. P. Odenwalder, *Optimal Decoding of Convolutional Codes*, Ph.D. Dissertation, University of California at Los Angeles, Los Angeles, CA, 1970.
141. J. K. Omura and J. L. Massey, Computational Method and Apparatus for Finite Field Arithmetic, U.S. Patent 4 587 627, May 6, 1986 (Filed September 14, 1982).
142. E. Paaske, Short Binary Convolutional Codes with Maximal Free Distance for Rates 2/3, and 3/4, *IEEE Transactions on Information Theory*, vol. IT-20, pp. 683–689, 1974.
143. R. H. Paschburg, *Software Implementation of Error-Correcting Codes*, M.S. Thesis, University of Illinois, Urbana, IL, 1974.
144. W. W. Peterson, Encoding and Error-Correction Procedures for the Bose–Chaudhuri Codes, *IRE Transactions on Information Theory*, vol. IT-6, pp. 459–470, 1960.
145. W. W. Peterson, *Error Correcting Codes*, MIT Press, Cambridge, MA, and Wiley, New York, NY, 1961.
146. W. W. Peterson and E. J. Weldon, Jr., *Error Correcting Codes*, Second Edition, MIT Press, Cambridge, MA, 1972.
147. V. Pless, Power Moment Identities on Weight Distributions in Error Correcting Codes, *Information and Control*, vol. 6, pp. 147–152, 1963.
148. V. S. Pless, On the Uniqueness of the Golay Codes, *Journal of Combinatoric Theory*, vol. 5, pp. 215–228, 1968.
149. M. Plotkin, Binary Codes with Specified Minimum Distances, *IEEE Transactions on Information Theory*, vol. IT-6, pp. 445–450, 1960.
150. J. M. Pollard, The Fast Fourier Transform in a Finite Field, *Mathematics of Computation*, vol. 25, pp. 365–374, 1971.
151. E. Prange, *Cyclic Error-Correcting Codes in Two Symbols*, AFCRC-TN-57-103, Air Force Cambridge Research Center, Cambridge, MA, 1957.
152. E. Prange, *Some Cyclic Error-Correcting Codes with Simple Decoding Algorithms*, AFCRC-TN-58-156, Air Force Cambridge Research Center, Bedford, MA, 1958.
153. I. S. Reed, A Class of Multiple-Error-Correcting Codes and the Decoding Scheme, *IRE Transactions on Information Theory*, vol. IT-4, pp. 38–49, 1954.
154. I. S. Reed and G. Solomon, Polynomial Codes over Certain Finite Fields, *Journal of the Society of Industrial and Applied Mathematics*, vol. 8, pp. 300–304, 1960.
155. C. T. Retter, The Average Binary Weight-Enumerator for a Class of Generalized Reed–Solomon Codes, *IEEE Transactions on Information Theory*, vol. IT-37, pp. 346–349, 1991.
156. J. P. Robinson and A. J. Bernstein, A Class of Recurrent Codes with Limited Error Propagation, *IEEE Transactions on Information Theory*, vol. IT-13, pp. 106–113, 1967.
157. L. D. Rudolph, *Geometric Configuration and Majority-Logic Decodable Codes*, M.E.E. Thesis, University of Oklahoma, Norman, OK, 1964.

158. L. D. Rudolph, A Class of Majority-Logic Decodable Codes, *IEEE Transactions on Information Theory*, vol. IT-14, pp. 305–307, 1967.
159. L. Rudolph and M. E. Mitchell, Implementation of Decoders for Cyclic Codes, *IEEE Transactions on Information Theory*, vol. IT-10, pp. 259–260, 1964.
160. D. V. Sarwate and R. D. Morrison, Decoder Malfunction in BCH Decoders, *IEEE Transactions on Information Theory*, vol. IT-36, pp. 884–889, 1990.
161. D. V. Sarwate and N. R. Shanbhag, High-Speed Architecture for Reed–Solomon Decoders, *IEEE Transactions on VLSI Systems*, vol. VLSI-9, pp. 641–655, 2001.
162. J. E. Savage, Minimum Distance Estimates of the Performance of Sequential Decoding, *IEEE Transactions on Information Theory*, vol. IT-15, pp. 128–140, 1969.
163. T. Schaub, *A Linear Complexity Approach to Cyclic Codes*, Doctor of Technical Sciences Dissertation, Swiss Federal Institute of Technology, Zurich, 1988.
164. J. Schönheim, On Linear and Nonlinear Single-Error-Correcting q-nary Perfect Codes, *Information and Control*, vol. 12, pp. 23–26, 1968.
165. C. E. Shannon, A Mathematical Theory of Communication, *Bell System Technical Journal*, vol. 27, pp. 379–423, 1948 (Part I), pp. 623–656 (Part II) reprinted in book form with postscript by W. Weaver, University of Illinois Press, Urbana, IL, 1949; Anniversary Edition, 1998.
166. H. M. Shao, T. K. Truong, L. J. Deutsch, J. H. Yuen, and I. S. Reed, A VLSI Design of a Pipeline Reed–Solomon Decoder, *IEEE Transactions on Computers*, vol. C-34, pp. 393–403, 1985.
167. R. C. Singleton, Maximum Distance q-nary Codes, *IEEE Transactions on Information Theory*, vol. IT-10, pp. 116–118, 1964.
168. D. Slepian, A Class of Binary Signaling Alphabets, *Bell System Technical Journal*, vol. 35, pp. 203–234, 1956.
169. D. Slepian, Some Further Study of Group Codes, *Bell System Technical Journal*, vol. 39, pp. 1219–1252, 1960.
170. N. J. A. Sloane, S. M. Reddy, and C. L. Chen, New Binary Codes, *IEEE Transactions on Information Theory*, vol. IT-18, pp. 503–510, 1972.
171. H. J. S. Smith, On Systems of Linear Indeterminant Equations and Congruences, *Philosophical Transactions of the Royal Society*, vol. 151, pp. 293–326, 1861.
172. G. Solomon and H. C. A. van Tilborg, A Connection Between Block and Convolutional Codes, *SIAM Journal of Applied Mathematics*, vol. 37, pp. 358–369, 1979.
173. G. Strang, *Linear Algebra and Its Applications*, Second Edition, Academic Press, New York, NY, 1980.
174. M. Sudan, Decoding of Reed–Solomon Codes Beyond the Error-Correction Bound, *Journal of Complexity*, vol. 13, pp. 180–183, 1997.
175. M. Sudan, Decoding of Reed–Solomon Codes Beyond the Error-Correction Diameter, *Proceedings of the Thirty-Fifth Annual Allerton Conference on Communication, Control, and Computing*, University of Illinois, IL, 1997.
176. Y. Sugiyama, M. Kasahara, S. Hirasawa, and T. Namekawa, A Method for Solving Key Equation for Decoding Goppa Codes, *Information and Control*, vol. 27, pp. 87–99, 1975.
177. R. M. Tanner, A Recursive Approach to Low-Complexity Codes, *IEEE Transactions on Information Theory*, vol. IT-27, pp. 535–547, 1981.
178. R. M. Thrall and L. Tornheim, *Vector Spaces and Matrices*, Wiley, New York, NY, 1957.
179. A. Tietäväinen, A Short Proof for the Nonexistence of Unknown Perfect Codes over $GF(q)$, $q > 2$, *Annales Acad. Scient. Fennicae*, Ser. A., no. 580, pp. 1–6, 1974.
180. M. A. Tsfasman, S. G. Vladut, and Th. Zink, Modular Curves, Shimura Curves, and Goppa Codes Better Than the Varshamov–Gilbert Bound, *Math. Nachr.*, vol. 104, pp. 13–28, 1982.

181. G. Ungerboeck, Channel Coding with Multilevel Phase Signals, *IEEE Transactions on Information Theory*, vol. IT-28, pp. 55–67, 1982.
182. B. L. Van der Waerden, *Modern Algebra* (two volumes), translated by F. Blum and T. J. Benac, Frederick Ungar, New York, NY, 1950 and 1953.
183. J. H. Van Lint, A Survey of Perfect Codes, *Rocky Mountain Journal of Mathematics*, vol. 5, pp. 199–224, 1975.
184. J. Vanderhorst and T. Berger, Complete Decoding of Triple-Error-Correcting Binary BCH Codes, *IEEE Transactions on Information Theory*, vol. IT-22, pp. 138–147, 1976.
185. A. Vardy, Trellis Structure of Codes, in *Handbook of Coding Theory*, editors, V. S. Pless and W. C. Huffman, Elsevier Science, Amsterdam, 1999.
186. R. R. Varshamov, Estimate of the Number of Signals in Error Correcting Codes, *Doklady Akad. Nauk SSSR*, vol. 117, pp. 739–741, 1957.
187. Y. L. Vasilýev, On Nongroup Close-Packed Codes, *Problemi Cybernetica*, vol. 8, pp. 337–339, 1962.
188. A. J. Viterbi, Error Bounds for Convolutional Codes and an Asymptotically Optimum Decoding Algorithm, *IEEE Transactions on Information Theory*, vol. IT-13, pp. 260–269, 1967.
189. A. J. Viterbi and J. K. Omura, *Principles of Digital Communication and Coding*, McGraw-Hill, 1979.
190. L. R. Welch and E. R. Berlekamp, Error Correction for Algebraic Block Codes, US Patent 4 633 470, 1983.
191. L. R. Welch and R. A. Scholtz, Continued Fractions and Berlekamp's Algorithm, *IEEE Transactions on Information Theory*, vol. IT-25, pp. 19–27, 1979.
192. E. J. Weldon, Jr., euclidean Geometry Cyclic Codes, *Proceedings of the Symposium on Combinatorial Mathematics*, University of North Carolina, Chapel Hill, NC, 1967.
193. E. J. Weldon, Jr., New Generalizations of the Reed–Muller Codes – Part II: Non-Primitive Codes, *IEEE Transactions on Information Theory*, vol. IT-14, pp. 199–206, 1968.
194. N. Wiberg, *Codes Depending on General Graphs*, Doctor of Philosophy Dissertation, Department of Electrical Engineering, University of Linkoping, Sweden, 1996.
195. N. Wiberg, H. A. Loeliger, and R. Kötter, Codes and Iterative Decoding on General Graphs, *European Transactions on Telecommunications*, vol. 6, pp. 513–526, 1995.
196. M. Willett, Arithmetic in a Finite Field, *Mathematics of Computation*, vol. 35, pp. 1353–1359, 1980.
197. J. K. Wolf, Adding Two Information Symbols to Certain Nonbinary BCH Codes and Some Applications, *Bell System Technical Journal*, vol. 48, pp. 2405–2424, 1969.
198. J. K. Wolf, Efficient Maximum-Likelihood Decoding of Linear Block Codes Using a Trellis, *IEEE Transactions on Information Theory*, vol. 24, pp. 76–80, 1978.
199. J. M. Wozencraft, Sequential Decoding for Reliable Communication, *1957 National IRE Convention Record*, vol. 5, Part 2, pp. 11–25, 1957.
200. J. M. Wozencraft and B. Reiffen, *Sequential Decoding*, MIT Press, Cambridge, MA, 1961.
201. W. W. Wu, New Convolutional Codes, *IEEE Transactions on Communication Theory*, Part I, vol. COM-23, pp. 942–956, 1975; Part II, vol. COM-24, pp. 19–33, 1976; Part III, vol. COM-24, pp. 946–955, 1976.
202. A. D. Wyner and R. B. Ash, Analysis of Recurrent Codes, *IEEE Transactions on Information Theory*, vol. IT-9, pp. 143–156, 1963.
203. N. Zierler, *On a Variation of the First Order Reed–Muller Codes*, Massachusetts Institute of Technology Lincoln Laboratory Group Report 34-80, Lexington, MA, 1958.
204. K. Zigangirov, Sequential Decoding Procedures, *Problemy Peredachi Informatsii*, vol. 2, pp. 13–25, 1966.

Index

Page numbers in bold refer to the most important page, usually where the index entry is defined or explained in detail.